KEYS TO THE NEMATODE PARASITES OF VERTEBRATES

Supplementary Volume

This volume is dedicated to the late Dr Sheila Willmott,
a former Director of the CAB International Institute of Parasitology
and one of the editors of the *CIH Keys to the Nematode Parasites of Vertebrates*
with Professors R.C. Anderson and A.G. Chabaud

KEYS TO THE NEMATODE PARASITES OF VERTEBRATES

Supplementary Volume

Lynda M. Gibbons

Department of Pathology and Infectious Diseases
The Royal Veterinary College
University of London
Hawkshead Lane, North Mymms
Hatfield, Herts AL9 7TA, UK

www.cabi.org

CABI is a trading name of CAB International

CABI Head Office
Nosworthy Way
Wallingford
Oxfordshire OX10 8DE
UK

Tel: +44 (0)1491 832111
Fax: +44 (0)1491 833508
E-mail: cabi@cabi.org
Website: www.cabi.org

CABI North American Office
875 Massachusetts Avenue
7th Floor
Cambridge, MA 02139
USA

Tel: +1 617 395 4056
Fax: +1 617 354 6875
E-mail: cabi-nao@cabi.org

A catalogue record for this book is available from the British Library, London, UK.

Library of Congress Cataloging-in-Publication Data

Gibbons, Lynda M.
Keys to the nematode parasites of vertebrates. Supplementary volume / Lynda M. Gibbons.
p. cm.
Supplement to: CIH keys to the nematode parasites of vertebrates / edited by Roy C. Anderson, Alain G. Chabaud, Sheila Willmott. 1974-
Includes bibliographical references and index.
ISBN 978-1-84593-571-9 (alk. paper)
1. Nematodes--Classification. 2. Nematodes--Identification. 3. Vertebrates--Parasites--Classification. I. CIH keys to the nematode parasites of vertebrates. II. Title.

QL391.N4C636 1974 Suppl.
636.089'6965--dc22

2009023393

ISBN-13: 978 1 84593 571 9

Typeset by SPi, Pondicherry, India
Printed and bound in the UK by Cambridge University Press, Cambridge

CONTENTS

INTRODUCTION

The *CIH Keys to the Nematode Parasites of Vertebrates* were published in ten parts between 1974 and 1983 to 'provide a working tool' and to 'help establish a classification which is not simply a series of pigeon-holes into which nematodes can be separated but which gives a real insight into their relationships and is, therefore, of value to the specialist'. The *Keys* have realized both of these goals and have become a standard reference work. A number of taxa were deliberately omitted, as their validity was doubtful and the type species of dubious status. The three editors, namely Roy C. Anderson, Alain G. Chabaud and Sheila Willmott, realized the difficulties of resolving the conflicts between 'lumpers and splitters' and produced a classification as 'nearly natural as possible', suggesting that early synonymies and references prior to 1960 should be accessed from the 'classical works'.

In this compilation volume of new, revised and redescribed taxa I have not attempted to rewrite the *Keys*, nor have I attempted to undertake a major revision of nematode classification. My aim is to draw attention to new taxa published by many authors after the *Keys* were completed and to attempt to identify the current position of some of the older taxa not included in the original. I am indebted to the publication by Spencer Jones and Gibson (1987), which gave me a good starting point for this update. Genera listed in this publication are noted in the text, but I should point out that, as Spencer Jones and Gibson (1987) commented (which is also valid in this supplement), 'inclusion of a name does not mean that it is a recognized taxon'.

The layout of the orders and superfamilies follows that in the Chabaud (1974) *Keys to Subclasses, Orders and Superfamilies* (in Anderson *et al.*, 1974).

The classification used is based on the original *Keys* and, where new suborders or superfamilies have been proposed since their publication, these are reviewed under the relevant taxa. These are to be found between square brackets, as are comments on the history and any changes in status of any of the taxa. Each of the taxa listed has a brief diagnosis based on the original reference (in Bibliography), unless otherwise stated within the square brackets. A few – where translations were not available – list only details of name and authority. The diagnoses of some taxa previously included in the *Keys* have been emended to include additional description and revised classification. These emended diagnoses have been included where relevant; however, as with the taxa not previously included, the original references should be consulted for a complete description.

Classifications are not static, changing as new taxa are described. Checklists such as Spratt *et al.* (1991) are important tools that have led to the description of new genera. Similarly, the checklist by Bruce *et al.* (1994) gives a comprehensive list of Ascaridoid parasites from fish indicating their taxonomic status, making it a useful tool for further studies. Indeed, several classifications have been proposed from systematic studies during, as well as since, the publication of the *Keys*, some including nematodes as a whole and not just the parasitic groups. A brief review of some of these classifications is given below.

Mozgovoi and Shakhmatova (1971) revised the ascarid group and gave an alternative classification to that of Hartwich (1975), and Sonin (1971) revised the classification of the suborder Filariata, including the Aproctoidea, Diplotriaenoidea and Filarioidea.

Ryzhikov and Sonin (1981), following the examination of three classification systems (Chitwood, 1933, 1937; Skrjabin and Schultz, 1940; Chabaud, 1974), propose a new classification based on Chabaud (1974).

Malakhov *et al.* (1982) give an integrated classification, dividing the class Nematoda into three subclasses.

Inglis (1983) outlines an integrated classification of free-living and parasitic nematodes. The author uses three classes, seven subclasses and 25 orders.

Maggenti (1982, 1983) also presents an integrated classification system, in which he regards the nematodes as a separate phylum, the Nemata.

Gibson (1992) comments that 300 genera of nematodes have been erected in recent years, while Gibson (1998) reviews the classification of the parasitic helminths and bases his discussion of the Nematoda on the *Keys*. He briefly reviews the main morphological features of the superfamilies, their life cycles and symptoms, if any, in the host species.

De Ley and Blaxter (2002) produced an integrated classification based on phylogenetic analysis. Okulewicz and Perec (2004) also review the use of molecular phylogenetics to 'examine the interrelationships between animal parasites, free-living nematodes and plant parasites versus traditional classification based on morphological–ecological characters'.

Nematodes are one of the most numerous groups of multicellular animals and have an important impact on both wild and domesticated animals and humans. Poinar (1982) estimated that there were in excess of 15,000 species of nematodes identified, and there are 500,000 species of both parasitic and free-living nematodes in existence. Since Poinar's estimate the number of new species identified and described has expanded well beyond 15,000. Parasitic species cause disease to animals and humans, resulting in economic losses as well as having medical and public health importance. Identification is the key that allows the planning of treatment and control measures.

Roundworms or nematodes were mentioned in the Egyptian *Papyrus Ebers c.*155 BC (*Ascaris* in Grove, 1990) and, since then, an amazing amount of knowledge has been accumulated on their biology, ecology and evolution. For years much of our knowledge of nematodes was based purely on comparative morphological studies at the light microscopy level. However, with the development of transmission and scanning electron microscopy, our understanding of structures and their functions has expanded and assisted our ability to identify and classify these organisms. Now classification is being challenged by new techniques, including those at the molecular level that enable us to further examine evolutionary relationships. However, a meaningful classification – which also reflects phylogenetic relationships – cannot be achieved by any one technique alone, be it morphological or molecular-based: it will require comparative studies that integrate data available to us from meticulous studies of both morphology and molecular techniques.

As the references mentioned above show, classifications are already changing as these new techniques are developed. New studies using cladistics and the examination of polymorphism are also being undertaken. All these techniques when applied will challenge current classifications and will inevitably lead to further changes and the description of new taxa.

As a result of ongoing studies, there may be changes or new taxa not included here that may have been published while this manuscript is in press, and it would be appreciated if these were drawn to the attention of the author.

Lynda M. Gibbons

ACKNOWLEDGEMENTS

The author wishes to acknowledge the assistance given by Dr D.I. Gibson and his colleagues, especially Mrs Eileen Harris, at the Zoology Department, Natural History Museum for the use of the host–parasite taxonomic database and for their assistance in alerting me to new genera that I might not otherwise have seen.

I would also like to acknowledge: Dr I. Beveridge for his helpful editing of some of the manuscript; Dr E.P. Hoberg and his colleague Ms P.A. Pilitt for their extensive help and useful comments on sections of the manuscript; Dr O. Bain and Dr M.-C. Durette-Desset for reviewing parts of the manuscript and their helpful suggestions and additions; Dr F. Moravec for his help with information and providing references covering his extensive work on parasitic nematodes, which I would otherwise not have been able to access; and Professor Heng Jian and Dr Vitaliy Kharchenko for assistance with translation of some of the descriptions.

I am also grateful to the many authors and colleagues who have kindly sent or made available copies of offprints, including: O. Bain, I. Beveridge, R.B. Blaylock, J. Cassone, M.-C. Durette-Desset, H.P. Fagerholm, M.T. Fox, S.L. Gardner, D.I. Gibson, E. Harris, H. Hasegawa, E. Hoberg, J.P. Hugot, V. Kharchenko, R.C. Ko, F. Moravec, R. Overstreet, D.L. Ribu, T. Sakamoto, J.W. Smith and L. Warner.

A publication of this type involves a lot of time accumulating literature, and I acknowledge the assistance of former colleagues at the CAB International Institute of Parasitology, especially the librarian Mrs Lynne Fox, in tracing and acquiring references. I also acknowledge the assistance of Dr L.F. Khalil, who was involved in the early stages of discussion and preparation of this supplement.

I also wish to acknowledge the support of my colleagues Professor D.E. Jacobs, Dr M.T. Fox and Dr C. Hermosilla at The Royal Veterinary College, University of London during the various stages of this project.

Finally, I wish to acknowledge the scientists who are too numerous to name here but whose painstaking description of new taxa and revisions of classification have been published since the *Keys*, making this additional volume necessary.

ADENOPHOREA

1 ENOPLIDA

DIOCTOPHYMATOIDEA

Dioctophymatidae (Railliet, 1915)

Dioctophymatinae (Castellani and Chalmers, 1910)

Kheraia Rautela and Malhotra, 1985 (Fig. 1)

DIAGNOSIS: Cuticle smooth. Mouth opening with 4 poorly developed lips. Six cephalic papillae present. Oesophagus long, not very wide. *Male*: Bursa bell-shaped without bursal rays. Spicule single. *Female*: Vulva opens anterior to midbody. Monodelphic. Oviparous. Eggs with polar plugs. Anus terminal. Tail bluntly rounded. Parasite in the small intestine of the Hill bird, *Sturnus pagodarum* in Kotdwara, Pauri-Garhwal District, U.P., India.

Type species: *Kheraia himalayii* Rautela and Malhotra, 1985

[The authors place their genus in the Dioctophymatinae and consider it is close to the genus *Dioctophyma* Collet-Meygret, 1802 and differs from *Hystrichis* Dujardin, 1845 in the subfamily Hystrichinae Skrjabin and Shikhobalova, 1954, a subfamily not accepted in Anderson and Bain (1982), by the position of the vulva.]

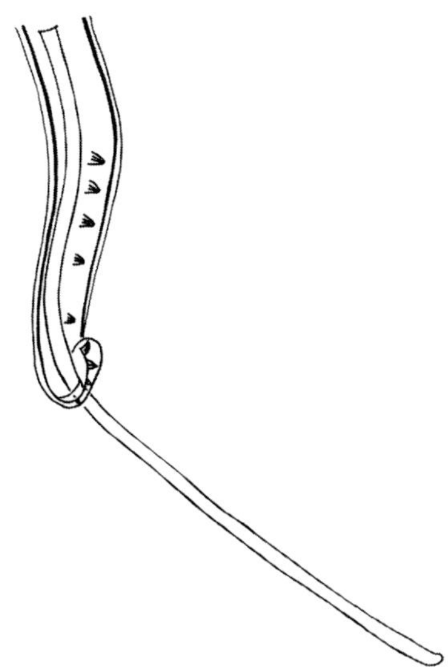

FIG. 1 *Kheraia*, posterior end of male. (Redrawn after Rautela and Malhotra, 1985).

Eustrongylinae Chitwood and Chitwood, 1937

Neostrongyloides Rathore and Nama, 1985 (Fig. 2)

DIAGNOSIS: Cuticle transversely striated. Mouth simple, opens into thick-walled buccal capsule. Inner and outer circle of cephalic papillae present. Oesophagus long. *Male*: 'Bursa bell-shaped with muscular walls not supported by rays. Spicule single and very long. Gubernaculum elongated and prominent. Pedunculated and sessile papillae present.' *Female*: Vulva opens near anus. Tail short. Parasite in crop of passeriforme birds (*Dicrurus adsimilis*, black drango) in Jodhpur, Rajasthan, India.

Type species: *Neostrongyloides dicrurus* Rathore and Nama, 1985

[The authors distinguish their genus from *Hystrichis* Dujardin, 1845 and *Eustrongylides* Jägerskiöld, 1909 in the family Dioctophymatidae and place it in the subfamily Hystrichinae Skrjabin and Shikhobalova, 1954. Following Anderson and Bain (1982), who do not include this subfamily, we have temporarily placed this genus in the subfamily Eustrongylinae. Further study may revise the placement of this genus, as the illustrated features of the type species are nearer to the spirurids.]

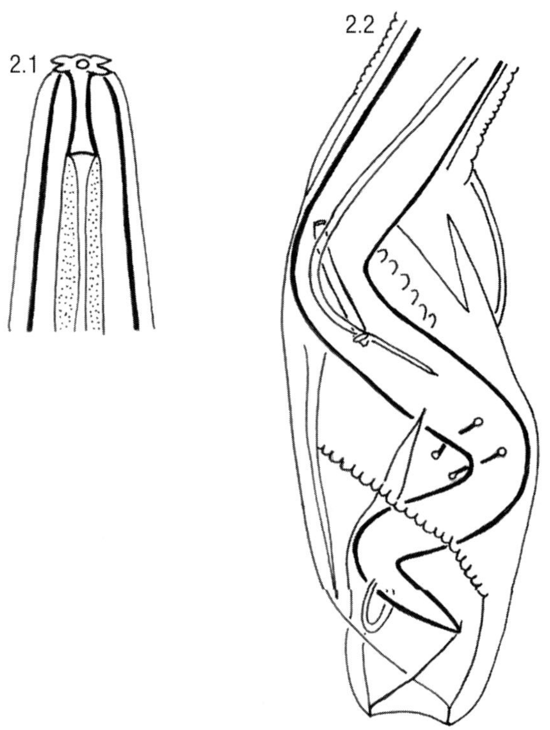

FIG. 2 *Neostrongyloides* 2.1 Anterior end of male; 2.2 Posterior end of male. (Redrawn after Rathore and Nama, 1985.)

TRICHINELLOIDEA

[Moravec (2001a) reviews the superfamily and includes the families Capillariidae Railliet, 1915 (subfamily of the Trichuridae in Anderson and Bain, 1982), Trichuridae (Railliet, 1911) Railliet, 1915 (for the single genus *Trichuris* Roederer, 1761, in the subfamily Trichurinae in Anderson and Bain, 1982), Trichinellidae Ward, 1907 (for the single genus *Trichinella* Railliet, 1895), Cystoopsidae Skrjabin, 1923 (for the single genus *Cystoopsis* Wagner, 1867, included in the subfamily Cystoopsinae in Anderson and Bain, 1982), Dioctowittidae Chabaud and Le Van Hoa, 1960 (a subfamily, Dioctowittinae, of the Cystoopsidae in Anderson and Bain, 1982 for the single genus *Dioctowittus* Chabaud and Le Van Hoa, 1960) and Trichosomoididae Hall, 1916 (a subfamily of the Trichuridae in Anderson and Bain, 1982).]

Trichuridae (Ransom, 1911) Railliet, 1915

[Anderson and Bain (1982) based their classification on Roman, 1965 (in Grassé, 1965) and accepted the family Trichuridae with three subfamilies, namely Capillariinae, Trichurinae and Trichosomoidinae. Moravec (2001a) includes only the genus *Trichuris* Roederer, 1761.]

Capillariinae Railliet, 1915

[Moravec (1982a, 2001a) maintains the use of the family name Capillariidae Railliet, 1915. Lomakin and Romashov (1987) use the family Capillariidae Railliet, 1915 and emend the diagnosis. They split the family into three subfamilies, two new, namely Capillariinae Railliet, 1915 for the genera: *Capillaria* Zeder, 1800, type genus; *Schulmanela* Ivaschkin, 1964; *Eucoleus* Dujardin, 1945; *Echinocoleus* López-Neyra, 1947; *Pterothominx* Freitas, 1959 (all listed, except *Echinocoleus*, as synonyms of *Capillaria* in Anderson and Bain (1982), Baruscapillariinae Lomakin and Romashov, 1987 for the genera: *Baruscapillaria* Moravec, 1982, type genus; *Paracapillaria* Medonça, 1963*; *Pseudocapillaria* Freitas, 1950*; *Liniscus* Dujardin, 1845*; *Pearsonema* Freitas and Medonça, 1960*; *Aonchotheca* López-Neyra, 1947*; *Calodium* Dujardin, 1845*; *Capillostrongyloides* Freitas and Lent, 1935*; *Gessyella* Freitas, 1959*; *Pseudocapillaroides* Moravec and Cosgrove, 1982 (*all genera regarded as synonyms of *Capillaria* in Anderson and Bain, 1982) and Skrjabinocapillariinae Lomakin and Romashov, 1987 for the type genus *Skrjabinocapillaria* Skarbilovich, 1946. Moravec (1982a) considers the validity of *Gessyella* Freitas, 1959 (a parasite of Australian fish distinguished by the female having eggs enclosed in special capsules, type species: *Gessyella latridopsis* (Johnston and Mawson, 1945) Freitas, 1959) as doubtful and the position of *Skrjabinocapillaria* Skarbilovich, 1946 (a parasite of bats in the former USSR distinguished by the absence of a spicule, a sheath without spines and the presence of lateral caudal alae, type species: *Skrjabinocapillaria eubursata* Skarbilovich, 1946) as unclear. He indicates that, after further study, the latter genus may be found to have similarities with *Aonchotheca*. Raina and Raina (1985) consider the genus *Skrjabinocapillaria* a synonym of *Capillaria* on the basis that a thin, poorly sclerotized spicule is present, a feature which can be easily overlooked. These two genera are listed as synonyms of *Capillaria* in Anderson and Bain (1982). Moravec (2001a) validates *Gessyella* and lists *Skrjabinocapillaria* as a synonym of *Aoncotheca*.

Gargarin and Nazarova (1966), Baruš and Segreeva (1990a) and De and Maity (1995) also use the family Capillariidae.]

Type genus: *Capillaria* Zeder, 1800 (Fig. 3)

DIAGNOSIS: *Male*: Caudal alae absent. Distal end rounded with two lateral, ventro-lateral or dorso-lateral lobes. Pseudobursa absent. Two small, sessile precloacal papillae generally present. Spicular sheath with spines. *Female*: Vulvar appendage present or absent. Parasites of the intestine of fish, amphibians, reptiles, birds and mammals.

Type species: *Capillaria anatis* (Schrank, 1790) Travassos, 1915

[Emended diagnosis based on Moravec (1982a). Moravec (2001a) lists *Trichosoma* Rudolphi, 1819, *Trichosomum* Creplin, 1829, *Thominx* Dujardin, 1845 and *Orthothominx* Freitas and Silva, 1960 as synonyms of *Capillaria*.]

Subgenus *Capillaria* Zeder, 1800

Type species: *Capillaria anatis* (Schrank, 1790) Travassos, 1915

[Moravec (2001a) lists this subgenus but does not include a diagnosis.]

Subgenus *Procapillaria* Moravec, 1987 (Fig. 4)

DIAGNOSIS: Stichosome, elongate stichocytes with large nuclei. *Male*: Posterior end with two large ventro-lateral lobes which have wide cuticular margins, large papilla at base of each lobe. Spicule well sclerotized. *Female*: Vulvar appendage present when gravid. Parasites in intestine of fish.

Type species: *Capillaria* (*Procapillaria*) *gracilis* (Bellingham, 1840) Travasoso, 1915

[Diagnosis based on Moravec (1987). Listed in Moravec (2001a).]

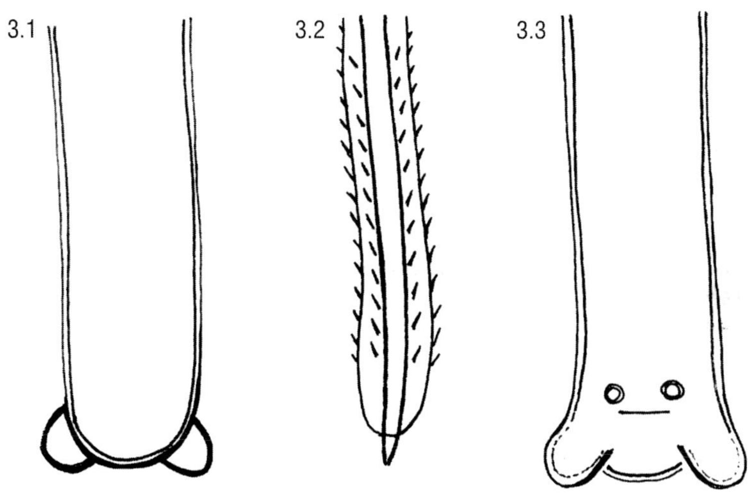

FIG. 3 *Capillaria* 3.1 Posterior end of male, dorsal view; 3.2 Posterior end of spicule and sheath; 3.3 Posterior end of male, ventral view. (Redrawn after Moravec, 1982a.)

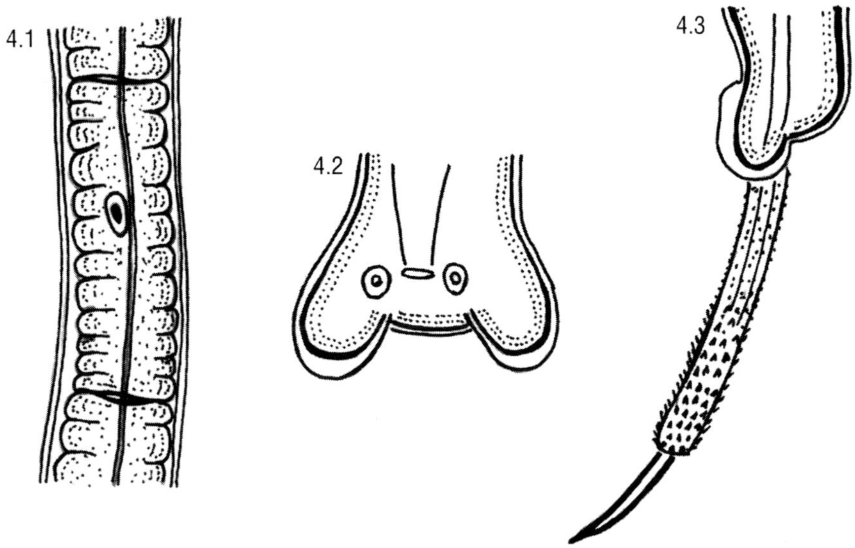

FIG. 4 *Procapillaria* 4.1 Stichosome; 4.2 Posterior end of male, ventral view; 4.3 Posterior end of male, lateral view. (Redrawn after Moravec, 1987.)

Subgenus *Neocapillaria* Moravec, 1987 (Fig. 5)

DIAGNOSIS: Stichosome, elongate stichocytes with large nuclei. *Male*: Posterior end with two subventral lobes without cuticular margins, pair of small papillae near cloaca. Spicules well sclerotized. *Female*: Vulvar appendage absent. Parasites in intestine of fish.

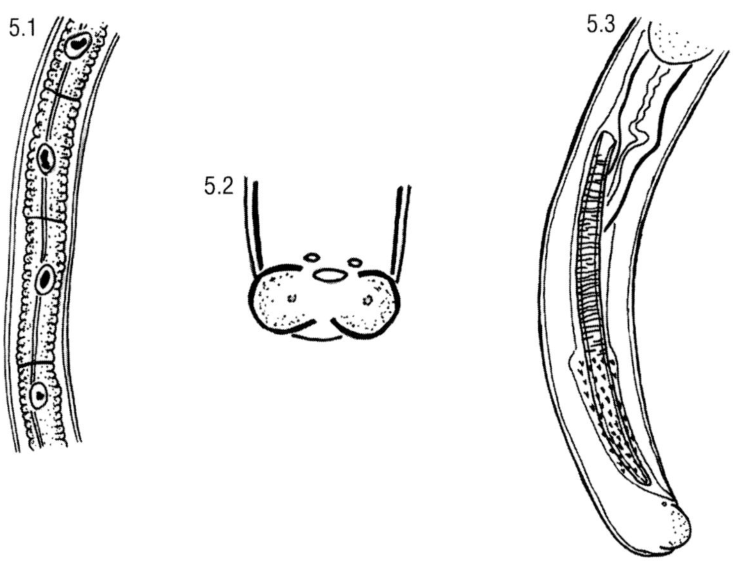

FIG. 5 *Neocapillaria* 5.1 Anterior portion of male showing stichosomes; 5.2 Posterior end of male, ventral view; 5.3 Posterior end of male, lateral view. (Redrawn after Moravec, 1987.)

Type species: *Capillaria* (*Neocapillaria*) *pterophylli* Heinze, 1933

[Diagnosis based on Moravec (1987). Listed in Moravec (2001a).]

Subgenus *Capillaroides* Moravec, 1987 (Fig. 6)

DIAGNOSIS: Stichosome, elongate stichocytes with large nuclei. *Male*: Posterior end without subventral lobes, pair of papillae-like projections present. Spicule weakly sclerotized. *Female*: Vulvar appendage absent. Parasites of fish.

Type species: *Capillaria* (*Capillaroides*) *catenata* (Van Cleave and Mueller, 1932) Moravec, 1987

[Diagnosis based on Moravec (1987). Listed in Moravec (2001a).]

Subgenus *Hepatocapillaria* Moravec, 1987 (Fig. 7)

DIAGNOSIS: Stichosome, short stichocytes with small nuclei. *Male*: Posterior end rounded, subventral lobes absent, pair of small subventral papillae present. Spicules well sclerotized. *Female*: Vulvar appendage absent. Parasites in liver of fish.

Type and only species: *Capillaria* (*Hepatocapillaria*) *cyprinodonticola* Huffman and Bullock, 1973

[Diagnosis based on Moravec (1987). Listed in Moravec (2001a).]

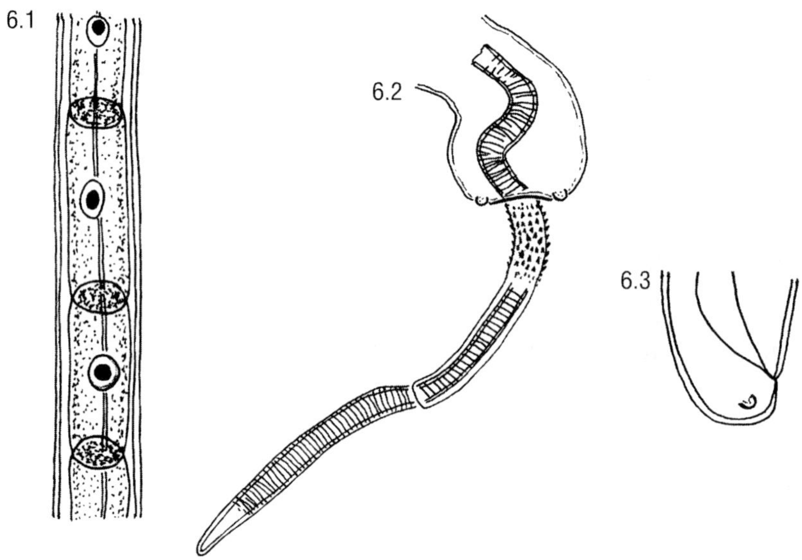

FIG. 6 *Capillaroides* 6.1 Anterior end of male showing stichosomes; 6.2 Posterior end of male, ventral view; 6.3 Posterior end of male, lateral view. (Redrawn after Moravec, 1987.)

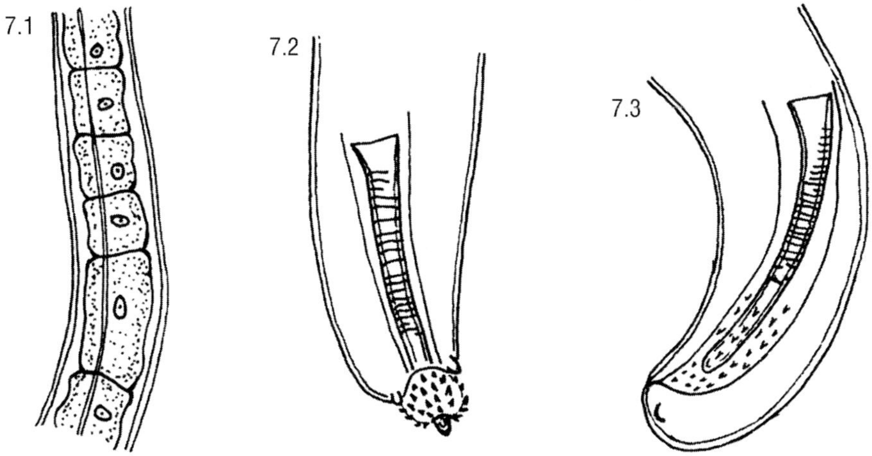

FIG. 7 *Hepatocapillaria* 7.1 Middle region of female stichosome; 7.2 Posterior end of male, ventral view; 7.3 Posterior end of male, lateral view. (Redrawn after Moravec, 1987.)

Echinocoleus López-Neyra, 1947 (Fig. 8.1)

DIAGNOSIS: *Male*: Caudal alae absent. Small pseudobursa may form two lobes, supported by one or more lobe-like dorso-lateral projections on each side. Dorsal caudal projection absent. Spicule medium length. Spicular sheath with spines. *Female*: Vulvar appendage present or absent. Parasites in gastrointestinal tract of birds and mammals.

Type species: *Echinocoleus cyanopicae* López-Neyra, 1947

[Skrjabin *et al.* (1957) list this genus as a synonym of *Thominx* Dujardin, 1845. Anderson and Bain (1982) include *Thominx* as a synonym of the genus *Capillaria* but do not include the genus *Echinocoleus*. Spencer Jones and Gibson (1987) include this genus in their list. Moravec (1982a) validates the genus and the emended diagnosis is based on his publication. Listed in Moravec (2001a).]

Gessyella Freitas, 1959 (Fig. 8.2)

DIAGNOSIS: Stichosome, 1 row of stichocytes. *Male*: Lateral caudal alae present. Membranous pseudobursa with two large lobe-like lateral papillae; two dorso-lateral projections with distal ends extending beyond margin also present. Spicule with numerous transverse grooves on surface. Spicular sheath spinous. *Female*: Eggs contained in special capsules. Parasites of fish.

Type species: *Gessyella latridopsis* (Johnston and Mawson, 1945) Freitas, 1959

[Emended diagnosis based on Moravec (2001a).]

Tridentocapillaria Baruš and Sergeeva, 1990 (Fig. 9)

DIAGNOSIS: Stichosome, 1 row of stichocytes. Bacillary bands 4. *Male*: Lateral caudal alae absent. Bilobed pseudobursa rudimentary. Distal end of tail with 2 lateral processes, each with

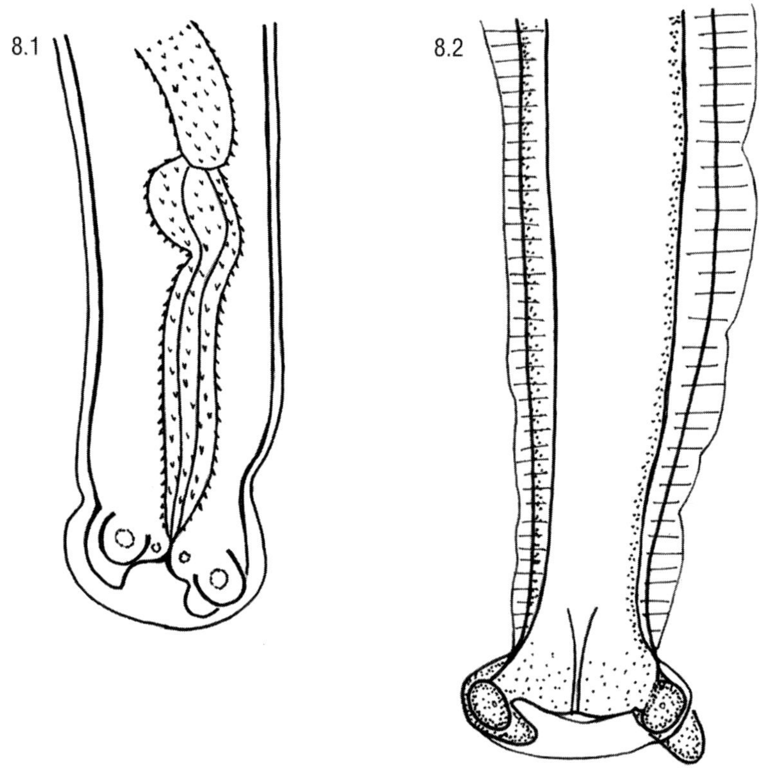

FIG. 8.1 *Echinocoleus* Posterior end of male, ventral view. (Redrawn after Freitas and Almeida, 1935 from Skrjabin *et al.*, 1957.)
FIG. 8.2 *Gessyella* Posterior end of male, ventral view. (Redrawn after Moravec, 1987 from Moravec, 2001a.)

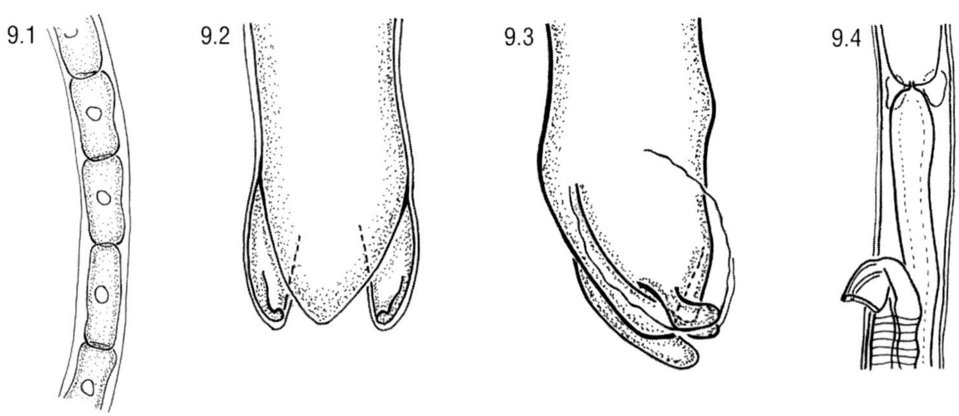

FIG. 9 *Tridentocapillaria* 9.1 Part of anterior showing stichocytes; 9.2 Posterior end of male, dorsal view; 9.3 Posterior end of male, lateral view; 9.4 Vulvar region, lateral view. (Redrawn after Baruš and Sergeeva, 1990a.)

papilla and 1 dorsal process. Spicule well sclerotized. Spicular sheath with spines on distal part. *Female*: Vulva with appendage. Parasites in the intestine of Passeriformes and Pisciformes birds.

Type species: *Tridentocapillaria tridens* (Dujardin, 1845) Baruš and Sergeeva, 1990

[Diagnosis based on Baruš and Sergeeva (1990a). Moravec (2001a) lists this genus as a subgenus of *Capillaria*.]

Pterothominx Freitas, 1959 (Figs 10.1 and 10.2)

DIAGNOSIS: Cephalic region tapered, rounded. Mouth slit-like. Bacillary bands 4. *Male*: Lateral caudal alae present. Pseudobursa membraneous, large, 1–2 processes each side. Spicule sheath with minute spines at distal end. *Female*: Vulva opens at end of stichosome, appendage present. Parasite of intestine of birds.

Type species: *Pterothominx meleagrisgallopavo* (Barile, 1912) Freitas, 1959

[Diagnosis based on Baruš and Sergeeva (1990c), who validate the genus previously considered a synonym of *Capillaria* in Anderson and Bain (1982). Moravec (1982a) also validates the genus. Moravec (2001a) lists *Armocapillaria* Gargarin and Nazarova, 1966 as a synonym of *Pterothominx*. *Armocapillaria* was listed as a synonym of *Capillaria* in Anderson and Bain, 1982.]

Aonchotheca López-Neyra, 1947

DIAGNOSIS: Cephalic region tapered, rounded. Mouth slit-like. Stichosome, 1 row of stichocytes. Bacillary bands 4. *Male*: Lateral caudal alae present. Pseudobursa with one plus pairs of lateral processes. Spicular sheath unarmed, may be transversely striated. *Female*: Vulva with or without vulvar appendage. Parasites in alimentary canal of mammals, birds and amphibians.

Type species: *Aonchotheca putorii* (Rudolphi, 1819) López-Neyra, 1947

[Diagnosis based on Baruš and Sergeeva (1990c), who validate the genus previously considered a synonym in Anderson and Bain (1982). Earlier, Moravec (1982a) had also validated the genus indicating that some of the species may be seen to represent different genera in the future. Moravec (2001a) lists *Skrjabinocapillaria* Skarbilovich, 1946 as a synonym of *Aonchotheca*.]

Subgenus *Aonchotheca* (López-Neyra, 1947 genus) (Figs 10.3 and 10.4)

DIAGNOSIS: Parasite of mammals.

Type species: *Aonchotheca* (*Aonchotheca*) *putorii* (Rudolphi, 1819) López-Neyra, 1947

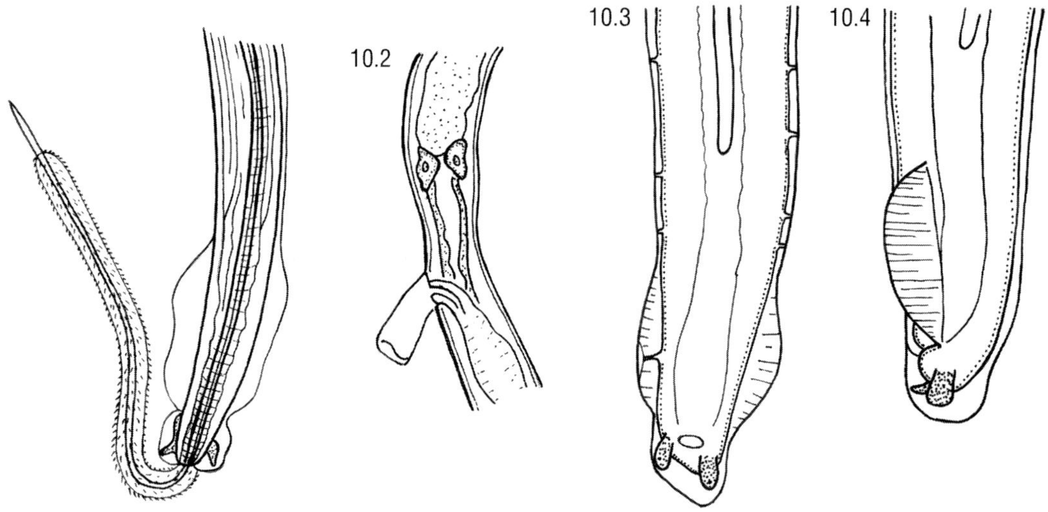

FIGS 10.1 and 10.2 *Pterothominx* 10.1 Posterior end of male. (Redrawn after López-Neyra, 1947.); 10.2 Vulvar region, lateral view. (Redrawn after Baruš and Sergeeva, 1990c.)
FIGS 10.3 and 10.4 *Aonchotheca* 10.3 Posterior end of male, ventral view; 10.4 Posterior end of male, lateral view. (Redrawn after Moravec and Sey, 1985 from Moravec, 2001a.)

[Baruš and Sergeeva (1990c) divided the genus *Aonchotheca* on the basis of host groups. The nominative subgenus *Aonchotheca* was erected to take the species from mammals subject to future revisions. Species in amphibians not included in authors' study. Listed in Moravec (2001a).]

Subgenus *Avesaonchotheca* Baruš and Sergeeva, 1990 (Fig. 11)

DIAGNOSIS: *Male*: Pseudobursa with 1–2 pairs of processes projecting caudally or laterally. Spicule lightly sclerotized. Parasites of intestine of birds.

Type species: *Aonchotheca* (*Avesaonchotheca*) *caudinflata* (Molin, 1958) Moravec, 1982

[Diagnosis based on Baruš and Sergeeva (1990c). Listed in Moravec (2001a).]

Subgenus *Armocapillaria* (Gargarin and Nazarova, 1966 genus) Moravec and Baruš, 1991 (Fig. 12)

DIAGNOSIS: *Male*: Spicule sheath armed by small tubercles. Spicule cylindrical.

Type species: *Aonchotheca* (*Armocapillaria*) *moschiferi* (Gargarin and Nazarova, 1966) Moravec and Baruš, 1991

[Anderson and Bain (1982) list *Armocapillaria* as a synonym of *Capillaria*. Moravec (1982a) considers *Armocapillaria* may be a synonym of *Pterothominx* Freitas, 1959. Later, Moravec and Baruš (1991) propose that *Armocapillaria* is distinct from *Pterothominx* and reinstate it as a subgenus of *Aonchotheca*. Moravec (2001a) lists *Armocapillaria* as a synonym of *Pterothominx* Freitas, 1959.]

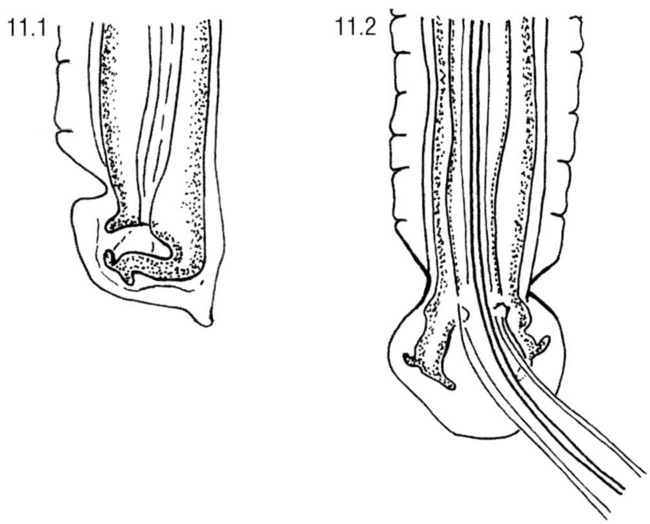

FIG. 11 *Avesaonchotheca* 11.1 Posterior end of male, lateral view; 11.2 Posterior end of male, ventral view. (Redrawn from Baruš and Sergeeva, 1990c, after Rysàvý *et al.*, 1982.)

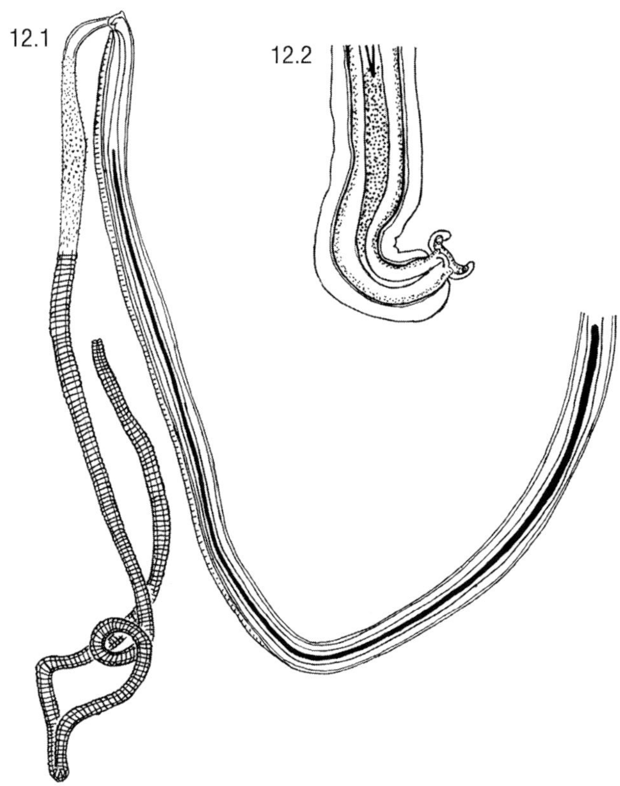

FIG. 12 *Armocapillaria* 12.1 Posterior end of male, showing spicular sheath with tubercles on proximal end. (Redrawn after Moravec and Baruš, 1991.) 12.2 Posterior end of male. (Redrawn from Gargarin and Nazarova, 1966.)

Tenoranema Mas-Coma and Esteban, 1985 (Fig. 13)

DIAGNOSIS: Stichosome, 1 row of stichocytes. *Male*: Ventrally directed caudal alae present. One lobed terminal membrane supported by 2 pairs of digitiform papillae: unbranched antero-ventral pair; branched postero-ventral pair; each with 4 processes which may divide into short projections at distal tips. Genital cone between antero-ventral papillae. Spicule slender. Spicular sheath without spines. *Female*: Pre- and post-vulvar appendages present. Parasite in intestine of *Eliomys quercinus*, Rodentia, Gliridae in Mallorca, Balearic Islands.

Type species: *Tenoranema alcoveri* Mas-Coma and Esteban, 1985

[Baruš and Sergeeva (1990c) consider the genus *Tenoranema* is valid. They consider that some species of the genus *Aonchotheca* may also belong in this genus following their redescription. Listed in Spencer Jones and Gibson (1987) and Moravec (2001a).]

Schulmanela Ivashkin, 1964 (Figs 14.1, 14.2 and 14.3)

DIAGNOSIS: Stichosome, 3 rows of stichocytes. *Male*: Caudal alae absent. Small membraneous pseudobursa, pair of dorso-lateral digitiform projections, curve parallel to pseudobursal margin. Dorsal caudal projection absent. Pair of large adcloacal papillae present. Spicule well sclerotized, superficial transverse grooves in mid-region may be present. Spicular sheath spiny. *Female*: Vulvar appendage present or absent. Parasites in intestine and liver of fish and amphibians.

Type species: *Schulmanela petruschewskii* (Shulman, 1948) Ivashkin, 1964

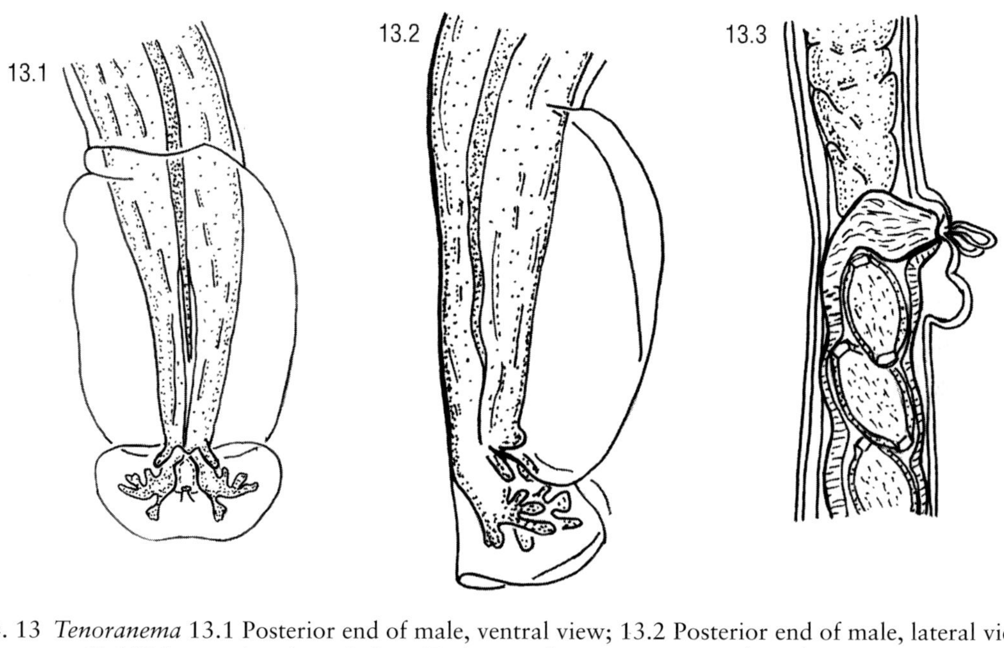

FIG. 13 *Tenoranema* 13.1 Posterior end of male, ventral view; 13.2 Posterior end of male, lateral view; 13.3 Vulvar region, lateral view. (Redrawn after Mas-Coma and Esteban, 1985.)

[Moravec (1982a) validates the genus listed as a synonym of *Capillaria* by Anderson and Bain (1982), and the emended diagnosis is based on his description. Moravec splits the genus into three subgenera but later (Moravec, 1987) upgrades one, *Piscicapillaria* to generic status. Moravec (2001a) upgrades all previous subgenera of the genus to generic rank.]

Genus *Amphibiocapillaria* Moravec, 1982 (Fig. 14.4)

DIAGNOSIS: Stichosome, 1 row of stichocytes, few may be double at posterior end. *Male*: Dorsal caudal projection absent. Parasites in intestine and liver of amphibians.

Type species: *Amphibiocapillaria tritonispunctati* (Diesing, 1851) Moravec, 1982

[Baker (1987) synonymizes this subgenus with *Capillaria*. Spencer Jones and Gibson (1987) include this as a subgenus in their list. Moravec (2001a) lists this former subgenus of *Schumanela* as a genus in the Capillariidae.]

Paracapillaria Mendonça, 1963 (Fig. 15)

DIAGNOSIS: Stichosome, 1 row of stichocytes. *Male*: Caudal alae absent. Pseudobursa with two dorso-lateral digitiform projections, curve dorsally along pseudobursal margin. One pair of large, sessile adcloacal or post-cloacal papillae present. Spicule well sclerotized without transverse grooves on surface. Spicular sheath without spines. *Female*: Vulvar appendage absent. Parasite in digestive tract of fish and reptiles.

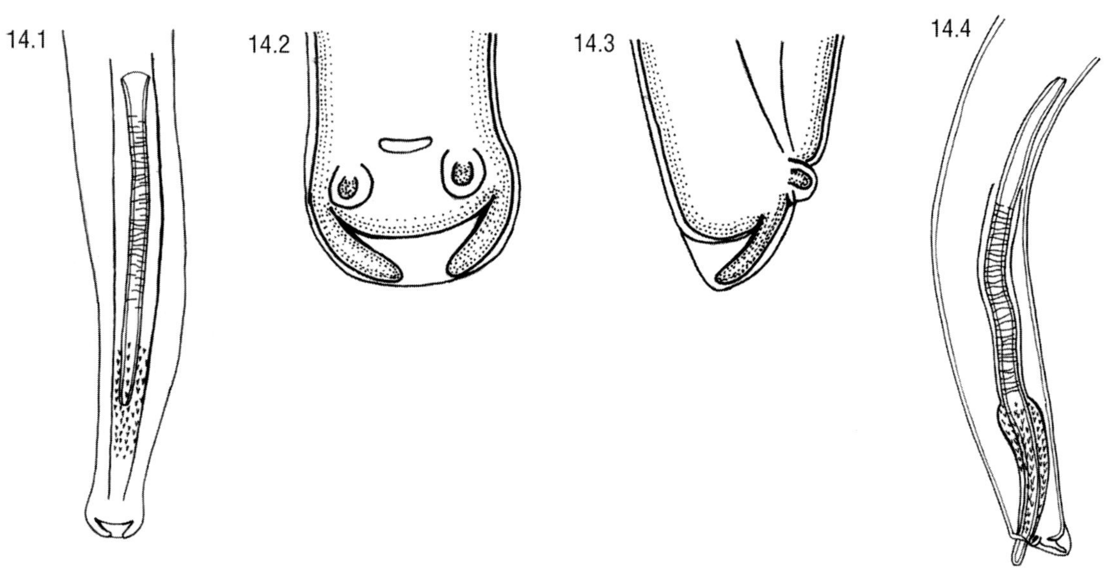

FIGS 14.1, 14.2 and 14.3 *Schulmanela* 14.1 Posterior end of male; 14.2 Posterior end of male, ventral view; 14.3 Posterior end of male, lateral view. (Redrawn after Moravec, 1980 from Moravec, 1994.)
FIG. 14.4 *Amphibiocapillaria* 14.4 Posterior end of male, lateral view. (Redrawn after Moravec, 1986 from Moravec, 2001a.)

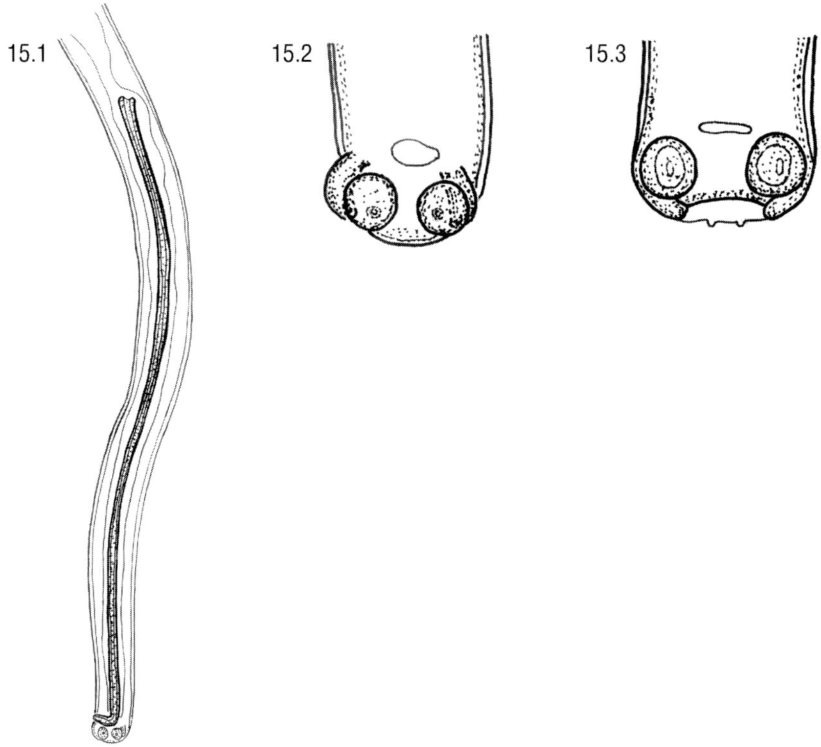

FIG. 15 *Paracapillaria* 15.1 Posterior end of male; 15.2 Posterior end of male, ventral view; 15.3 Posterior end of male, ventral view. (Redrawn after Moravec, 1987 from Moravec, 1994.)

Type species: *Paracapillaria piscicola* (Travassos, Artigas and Pereira, 1928) Mendonça, 1963

[Moravec (1982a, 1986a,b) validates this genus listed as a synonym of *Capillaria* in Anderson and Bain (1982), and the emended diagnosis is based on his publications. Moravec (2001a) includes three subgenera.]

Subgenus *Paracapillaria* Mendonça, 1963

DIAGNOSIS: Stichocytes with medium-sized nuclei, nucleoli without distinct corpuscles. *Female*: Oviparous. Parasites in digestive tract of fish and amphibians.

Type species: *Paracapillaria (Paracapillaria) piscicola* (Travassos, Artigas and Pereira, 1928) Mendonça, 1963

[Diagnosis based on Moravec (2001a).]

Subgenus *Ophidiocapillaria* Moravec, 1986

DIAGNOSIS: Stichocytes with distinctly large nuclei, nucleoli with several corpuscles. *Female*: Oviparous. Parasites of snakes.

Type species: *Paracapillaria* (*Ophidiocapillaria*) *sonsinoi* (Parona, 1897) Mendonça, 1963

[Diagnosis based on Moravec (1986b, 2001a).]

Subgenus *Crossicapillaria* Moravec, 2001

DIAGNOSIS: *Female*: Release eggs and larvae. Adults parasitic in digestive tract of birds and mammals. Intermediate host in fish, larvae occur in small intestine.

Type species: *Paracapillaria* (*Crossicapillaria*) *philippinensis* (Chitwood, Velasquez and Salazar, 1968) Moravec, 2001

[Diagnosis based on Moravec (2001a,b).]

Capillostrongyloides Freitas and Lent, 1935 (Figs 16.1 and 16.2)

DIAGNOSIS: Stichosome, 1 row of stichocytes. *Male*: Caudal alae absent. Pseudobursa extends beyond posterior end of body. Two subventral lobes present posterior to cloaca. Spicule short. Spicular sheath without spines. *Female*: Vulvar appendage absent. Parasites in gastrointestinal tract of freshwater fish.

Type species: *Capillostrongyloides sentinosa* (Travassos, 1927) Moravec, 1982

[Moravec (1982a, 2001a) validates this genus listed as a synonym of *Capillaria* in Anderson and Bain (1982), and the emended diagnosis is based on his publications.]

Pseudocapillaria Freitas, 1959

DIAGNOSIS: Stichosome, 1 row of stichocytes. *Male*: Caudal alae absent. Rounded distal end. Two subventral lobes posterior to cloaca. Dorsal pseudobursal membrane absent or very reduced. Spicular sheath without spines. *Female*: Vulvar appendage absent. Parasites in intestine and stomach of fish, reptiles, birds and mammals.

Type species: *Pseudocapillaria tomentosa* (Dujardin, 1843)

[Moravec (1982a, 2001a) validates this genus listed as a synonym of *Capillaria* in Anderson and Bain (1982), and the emended diagnosis is based on his publications. Moravec (2001a) includes four subgenera.]

Subgenus *Pseudocapillaria* (Freitas, 1959 genus) (Figs 16.3 and 16.4)

DIAGNOSIS: *Male*: Dorsal pseudobursal membrane absent.

Type species: *Pseudocapillaria* (*Pseudocapillaria*) *tomentosa* (Dujardin, 1843)

[Diagnosis based on Moravec (1982a) and listed in Moravec (2001a).]

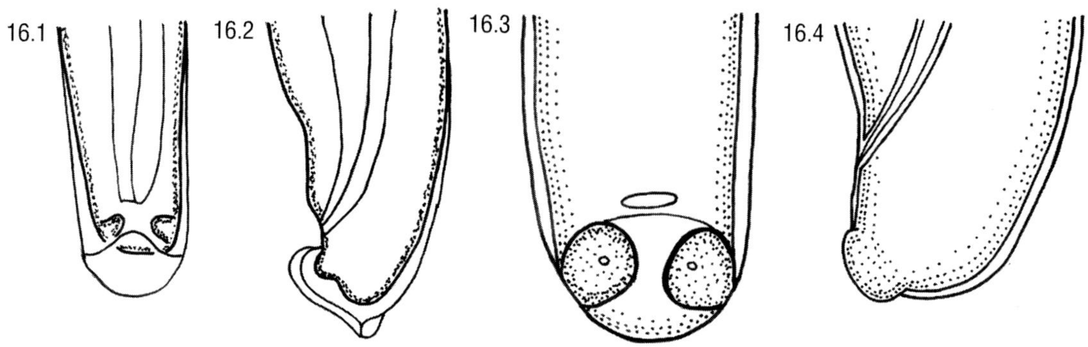

FIGS 16.1 and 16.2 *Capillostrongyloides* 16.1 Posterior end of male, ventral view; 16.2 Posterior end of male, lateral view. (Redrawn after Freitas and Lent, 1935.)
FIGS 16.3 and 16.4 *Pseudocapillaria* 16.3 Posterior end of male, ventral view; 16.4 Posterior end of male, lateral view. (Redrawn after Moravec, 1980 from Moravec, 2001a.)

Subgenus *Ichthyocapillaria* Moravec, 1982 (Fig. 17)

DIAGNOSIS: *Male*: Dorsal pseudobursal membrane present, connects subventral lobes.

Type species: *Pseudocapillaria* (*Ichthyocapillaria*) *salvelini* (Polyansky, 1952) Moravec, 1982

[Diagnosis based on Moravec (1982a). Listed in Spencer Jones and Gibson (1987) and Moravec (2001a).]

Subgenus *Indocapillaria* De and Maity, 1995 (Fig. 18)

DIAGNOSIS: Stichosome, 1 row of stichocytes. Bacillary bands 2, lateral. *Male*: Generally smaller in size than gravid females. Lateral caudal alae absent. Posterior end: wide, 2 dorso-

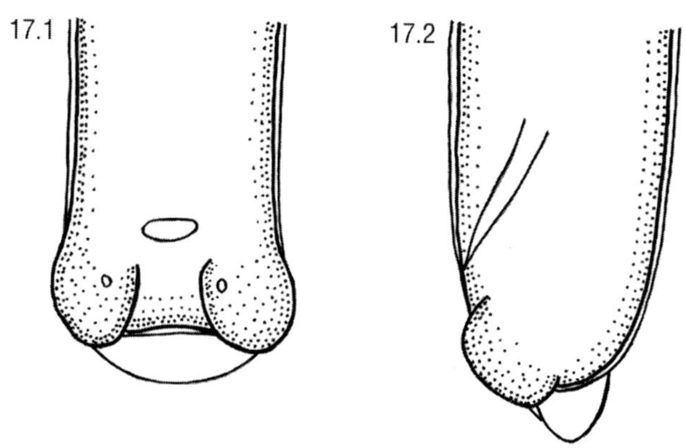

FIG. 17 *Icthyocapillaria* 17.1 Posterior end of male, ventral view; 17.2 Posterior end of male, lateral view. (Redrawn after Moravec, 1987 from Moravec, 1994.)

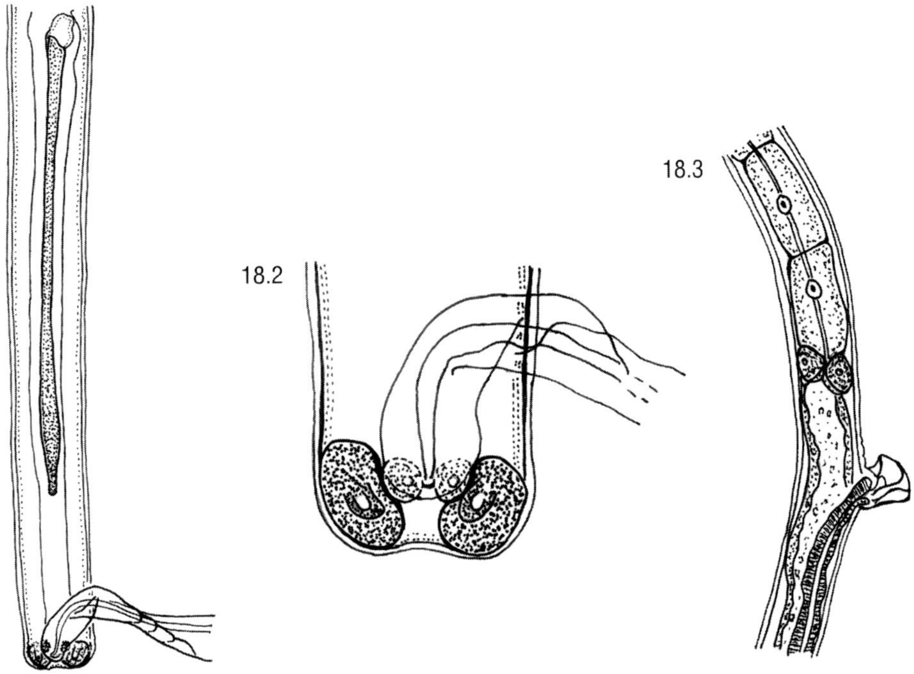

FIG. 18 *Indocapillaria* 18.1 Posterior end of male, ventral view; 18.2 Distal end of male, ventral view; 18.3 Vulvar region, lateral view. (Redrawn after De and Maity, 1995.)

lateral lobes present each with papilla; 2 sub-ventral papillae at level of cloaca, ventral precloacal 'lobe-like elevation' of body wall absent. Spicule: well sclerotized; wide; smooth surface; rounded distal tip. Dorsal cuticular pseudobursal membrane on tail absent. Spicular sheath without spines. *Female*: Vulva opens posterior to oesophagus. Vulvar appendage large, bell-shaped in gravid female. Mature egg thin-walled, unsegmented. Distal end rounded, anus subterminal. Parasite in the stomach of fish.

Type and only species: *Pseudocapillaria* (*Indocapillaria*) *gibsoni* (De and Maity, 1995) Moravec and Spratt, 1998

[Diagnosis based on De and Maity (1995) and Moravec (2001a). Moravec (2001a) lists *Indocapillaria* as a subgenus of *Pseudocapillaria*.]

Subgenus *Discocapillaria* De and Maity, 1996 (Figs 19.1 and 19.2)

DIAGNOSIS: *Male*: Ventral precloacal 'lobe-like elevation' of body wall present with additional pair of papillae. Dorsal cuticular pseudobursal membrane on tail absent.

Type species: *Pseudocapillaria* (*Discocapillaria*) *margolisi* De and Maity, 1996

[Diagnosis based on De and Maity (1996) and Moravec (2001a). Moravec (2001a) also lists this subgenus in *Pseudocapillaria*.]

Paratrichosoma Ashford and Muller, 1978 (Fig. 19.3)

DIAGNOSIS: Cuticle thick with extensive pores from hypodermal gland cells. Stichosome, 1 row of elongate stichocytes, ends some distance from oesophago-intestinal junction. Caudal alae absent. Membraneous pseudobursa, lobes on distal end, defined genital papillae absent. Cloaca terminal. Spicule distinctly sclerotized. Spicular sheath spinose. *Female*: Vulvar appendage absent. Parasites in skin of crocodiles.

Type species: *Paratrichosoma crocodylus* Ashford and Muller, 1978

[Moravec (2001b) validates this genus considered a synonym of *Capillaria* in Anderson and Bain (1982). Emended diagnosis based on Moravec (2001a).]

Freitascapillaria Moravec, 1982 (Fig. 20)

DIAGNOSIS: *Male*: Caudal alae absent. Distal end rounded, laterally expanded. Cloaca terminal. Spicule absent. Spicular sheath without spines. *Female*: Vulvar appendage present on anterior lip. Parasites in stomach and pyloric caeca of freshwater fish.

Type species: *Freitascapillaria maxillosa* (Vaz and Pereira, 1934) Moravec, 1982

[Diagnosis based on Moravec (1982a). Moravec *et al.* (2008b) consider *Neocapillaria* Yu and Wang, 1994 and *Sinocapillaria* Moravec and Spratt, 1998 synonyms of *Freitascapillaria*. Listed in Spencer Jones and Gibson (1987) and Moravec (2001a).]

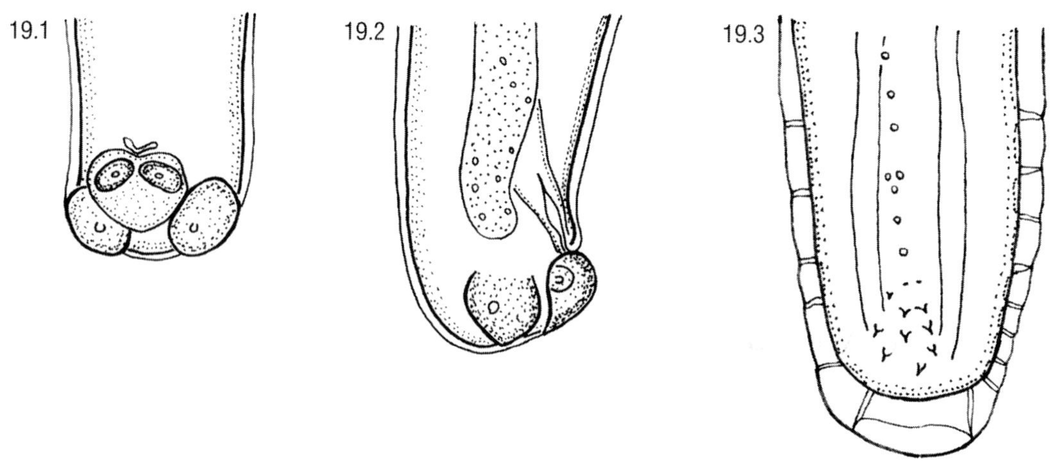

FIGS 19.1 and 19.2 *Discocapillaria* 19.1 Posterior end of male, ventral view; 19.2 Posterior end of male, lateral view. (Redrawn after De and Maity, 1996.)
FIG. 19.3 *Paratrichosoma* Posterior end of male, ventral view. (Redrawn after Moravec, 1986 from Moravec, 2001a.)

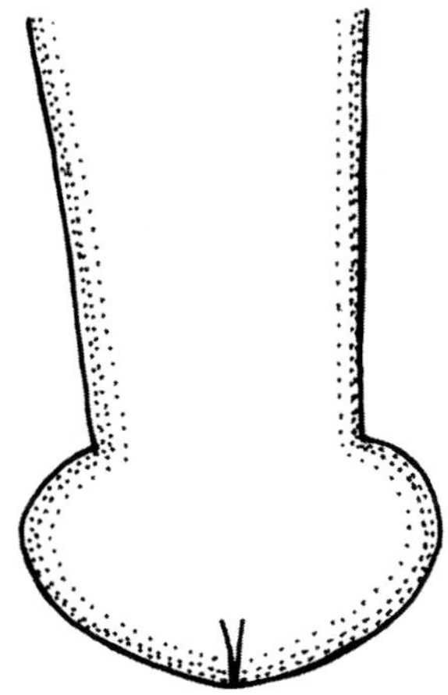

FIG. 20 *Freitascapillaria* Posterior end of male. (Redrawn after Moravec, 1982a.)

Baruscapillaria Moravec, 1982 (Fig. 21)

DIAGNOSIS: *Male*: Caudal alae absent. Pseudobursa may be lobed or unlobed, supported by 1–2 small, rounded lobes, narrower at base, each with tiny projection generally bent ventrally. Spicule long. Spicule sheath without spines. *Female*: Vulvar appendage present or absent. Parasites of intestine and stomach of birds and mammals.

Type species: *Baruscapillaria obsignata* (Madsen, 1945) Moravec, 1982

[Diagnosis based on Moravec (1982a). Listed in Spencer Jones and Gibson (1987) and Moravec (2001a).]

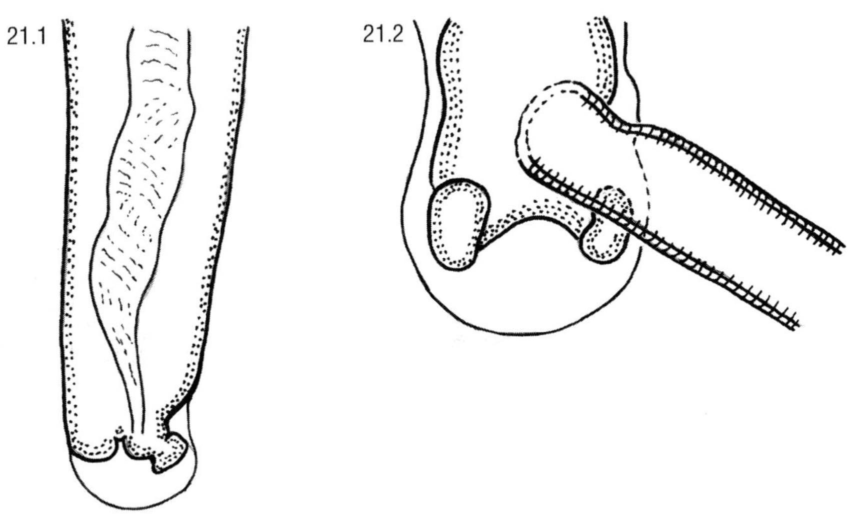

FIG. 21 *Baruscapillaria* 21.1 Posterior end of male, lateral view; 21.2 Posterior end of male, ventral view. (Redrawn after Graybill, 1924 from Madsen, 1945.)

Ornithocapillaria Baruš and Sergeeva, 1990 (Fig. 22)

DIAGNOSIS: Stichosome, 1 row of stichocytes. Bacillary bands 4. *Male*: Lateral caudal alae absent. Pseudobursa membraneous, wide. Tail bifid, each branch with dorsal and ventral process. Spicule well sclerotized. Spicule sheath unarmed. *Female*: Vulva with long tubular appendage. Parasite in intestine of Passeriformes, Falconiformes, Strigiformes and Piciformes birds.

Type species: *Ornithocapillaria ovopunctata* (Linstow, 1873) Baruš and Sergeeva, 1990.

[Diagnosis based on Baruš and Sergeeva (1990b), who placed the genus in the subfamily Baruscapillariinae Lomakin and Romashov, 1987. Moravec (2001a) lists this genus as a synonym of *Baruscapillaria* in the Capillariidae.]

Liniscus Dujardin, 1845 (Fig. 23)

DIAGNOSIS: *Male*: Caudal alae absent. Small bursa, supported by 2 short, round dorso-lateral projections. Spicule thin, long. Spicular sheath without spines. *Female*: Vulvar appendage present or absent. Parasites in urinary bladder and kidneys of insectivores and rodents.

Type species: *Liniscus incrassatus* Diesing, 1851

[Moravec (1982a, 2001a) validates this genus listed as a synonym of *Capillaria* in Anderson and Bain (1982), and the emended diagnosis is based on his publication.]

Pearsonema Freitas and Mendonça, 1960 (Fig. 24)

DIAGNOSIS: *Male*: Caudal alae absent. Small bursa present. Distal end with 2 reduced lobes. Two elongate projections arise dorso-laterally, curve ventrally towards median line along bursal

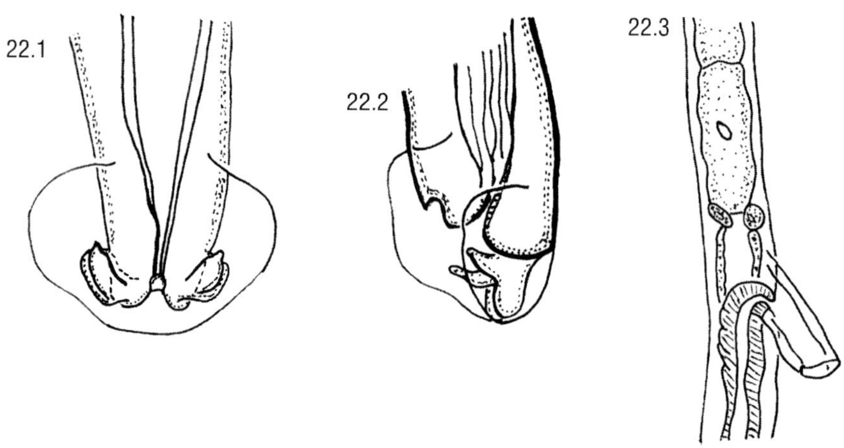

FIG. 22 *Ornithocapillaria* 22.1 Posterior end of male, ventral view; 22.2 Posterior end of male, lateral view; 22.3 Vulvar region, lateral view. (Redrawn after Baruš and Sergeeva, 1990b.)

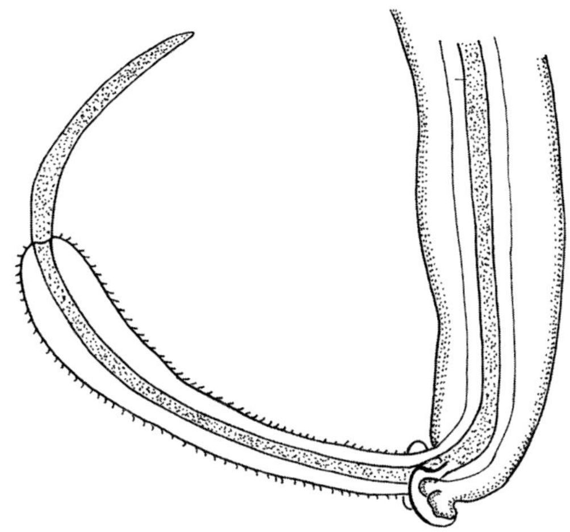

FIG. 23 *Liniscus* Posterior end of male. (Redrawn after Joyeux and Bayeux, 1937 from Skrjabin *et al*., 1957.)

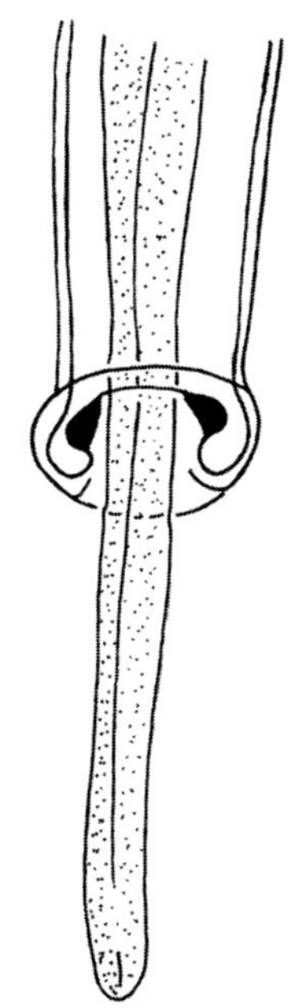

FIG. 24 *Pearsonema* Posterior end of male. (Redrawn after Freitas and Mendonça, 1960.)

margin. Spicule thin, long. Spicular sheath without spines. *Female*: Vulvar appendage present or absent. Parasites of urinary bladder of carnivores.

Type species: *Pearsonema pearsoni* Freitas and Mendonça, 1960

[Moravec (1982a, 2001a) validates this genus listed as a synonym of *Capillaria* in Anderson and Bain (1982), and the emended diagnosis is based on his publication.]

Eucoleus Dujardin, 1845 (Fig. 25)

DIAGNOSIS: *Male*: Caudal alae absent. Distal end with 2 small, posteriorly directed rounded lobes, connected by reduced membrane. Spicule lightly sclerotized, may be difficult to see. Spicular sheath long, dense spines. *Female*: Vulvar opening not elevated. Parasites in respiratory system and mucous cover of oesophagus, oral cavity and stomach of birds and mammals.

Type species: *Eucoleus aerophilus* (Creplin, 1839) Dujardin, 1845

[Moravec (1982a, 2001a) validates this genus listed as a synonym of *Capillaria* in Anderson and Bain (1982), and the emended diagnosis is based on his publication. Moravec (2001a) lists *Ritaklossia* Freitas, 1959 as a synonym of this genus. Anderson and Bain (1982) list *Ritaklossia* as a synonym of *Capillaria*.]

Calodium Dujardin, 1845 (Figs 26.1 and 26.2)

DIAGNOSIS: *Male*: Caudal alae present. Distal end with 2 small, rounded papillae. Dorsal cuticular membrane absent. Spicule spherical in transverse section. Spicular sheath without spines. *Female*: Vulvar appendage present or absent. Parasites in spleen and liver of mammals.

Type species: *Calodium splenaecum* (Dujardin, 1843) Dujardin, 1845

[Moravec (1982a, 2001a) validates this genus listed as a synonym of *Capillaria* in Anderson and Bain (1982), and the emended diagnosis is based on his publication. Moravec (1982a, 2001a) lists *Hepaticola* Hall, 1916 as a synonym of this genus. Anderson and Bain (1982) also list *Hepaticola* as a synonym of *Capillaria*.]

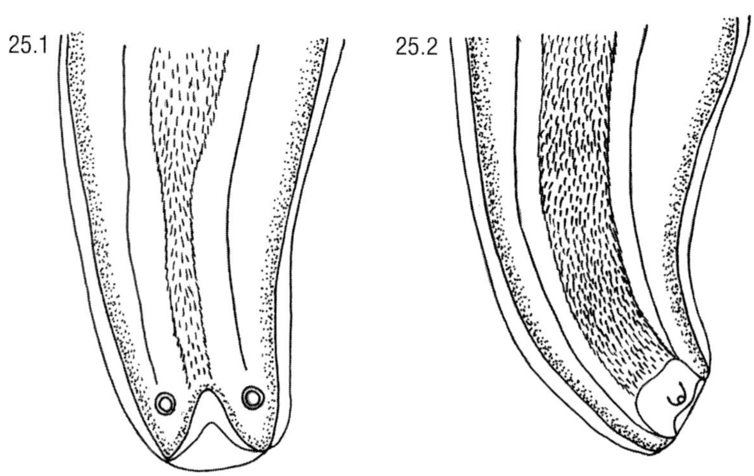

FIG. 25 *Eucoleus* 25.1 Posterior end of male, ventral view; 25.2 Posterior end of male, lateral view. (Redrawn after Moskvin, 1939 from Skrjabin *et al.*, 1957.)

Piscicapillaria (Moravec, 1982 subgenus) Moravec, 1987

DIAGNOSIS: Stichosome, 1 row stichocytes. *Male*: Lateral caudal alae absent. Small membraneous pseudobursa with pair of dorso-lateral projections, curving dorsally near margin. Single dorsal caudal projection present. Pair of large caudal papillae present. Spicule with numerous transverse grooves on surface. Spicular sheath with spines. *Female*: Vulvar appendage present or absent. Parasites in intestine of fish.

Type species: *Piscicapillaria freemani* (Moravec, Margolis and McDonald, 1981) Moravec, 1982

[Listed in Spencer Jones and Gibson (1987) and Moravec (2001a).]

Subgenus *Piscicapillaria* Moravec, 1982 (Figs 26.3 and 26.4)

DIAGNOSIS: Stichocytes, fewer than 60, large, elongate, nuclei in 1 longitudinal row, may alternate with short stichocytes towards distal end. Parasites in intestine of marine elasmobranchs (sharks and rays).

Type species: *Piscicapillaria* (*Piscicapillaria*) *freemani* (Moravec, Margolis and McDonald, 1981) Moravec, 1982

[Listed in Spencer Jones and Gibson (1987) and Moravec (2001a).]

Subgenus *Lomakinela* Moravec, 1987 (Fig. 27)

DIAGNOSIS: Stichocytes, more than 120, short, transverse nuclei arranged in 2 longitudinal rows. Parasites in freshwater bony fishes (sturgeons).

Type species: *Piscicapillaria* (*Lomakinela*) *tuberculata* (Linstow, 1914) Moravec, 1987

[Listed in Moravec (2001a).]

Pseudocapillaroides Moravec and Cosgrove, 1982 (Fig. 28)

DIAGNOSIS: Body thread-like, small. Lateral bacillary bands present. Stichosome, single row of stichocytes. Oesophago-intestinal junction well developed. *Male*: Caudal alae absent. Posterior end with two small, rounded lobes without dorsal membraneous pseudobursa. Spicules well sclerotized. Spicular sheath without spines. *Female*: Body length 2–4 times male. Vulva opens near distal end of oesophagus, appendage absent. Eggs embryonated in uterus. Parasite in the skin of the South African clawed frog, *Xenopus laevis*.

Type species: *Pseudocapillaroides xenopi* Moravec and Cosgrove, 1982

[Baker (1987) synonymizes this genus with *Capillaria*. Listed in Spencer Jones and Gibson (1987). Moravec (2001a) validates this genus.]

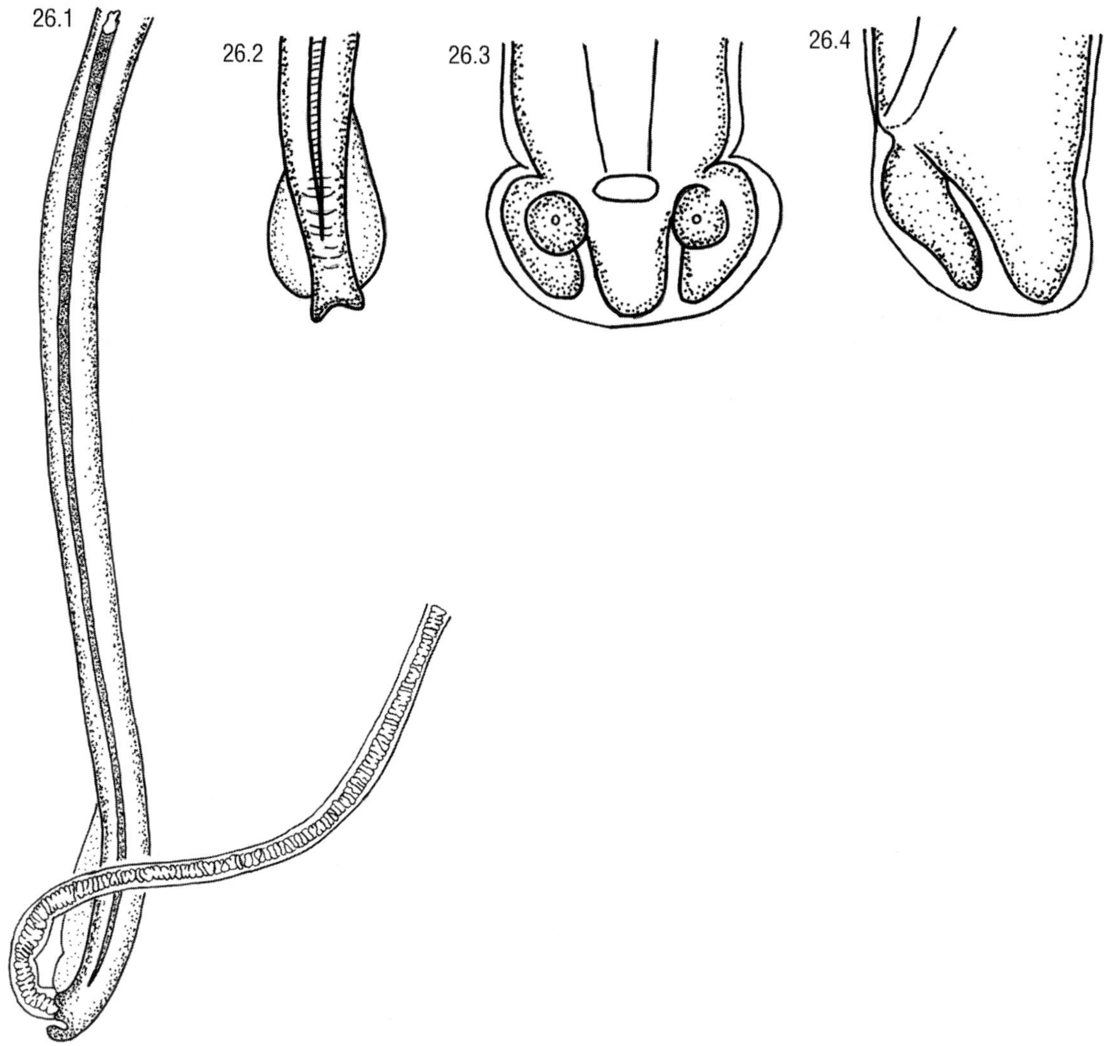

FIGS 26.1 and 26.2 *Calodium* 26.1 Posterior end of male, lateral view; 26.2 Posterior end of male, ventral view. (Redrawn after Dujardin, 1843 from Skrjabin *et al.*, 1957.)
FIGS 26.3 and 26.4 *Piscicapillaria* 26.3 Posterior end of male, ventral view; 26.4 Posterior end of male, lateral view. (Redrawn after Moravec *et al.*, 1981 from Moravec, 2001a.)

Brevithominx Travassos, Teixeira de Freitas and Machado de Mendonça, 1964 (Fig. 29)

DIAGNOSIS: Cephalic region not differentiated. Anterior body smaller than posterior. *Male*: Caudal alae absent. Small pseudobursa with two lobes, each with small ventral papilla. Spicule absent. Spicule sheath with spines. *Female*: Vulva slightly raised without evagination. Anus subterminal. Parasite of birds (*Capella paraguiae*) in Brazil.

Type and only species: *Brevithominx asperodorsus* Travassos, Teixeira de Freitas and Machado de Mendonça, 1964.

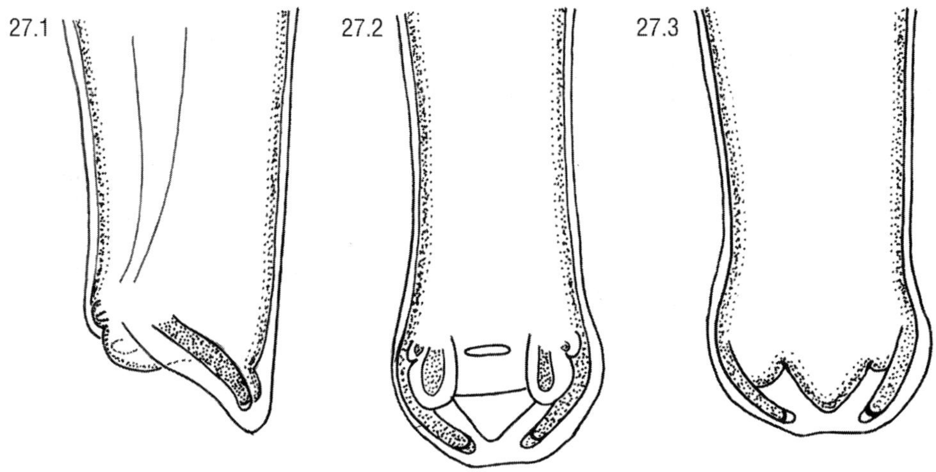

FIG. 27 *Lomakinela* 27.1 Posterior end of male, lateral view; 27.2 Posterior end of male, ventral view; 27.3 Posterior end of male, dorsal view. (Redrawn from Moravec, 1980 from Moravec, 1994.)

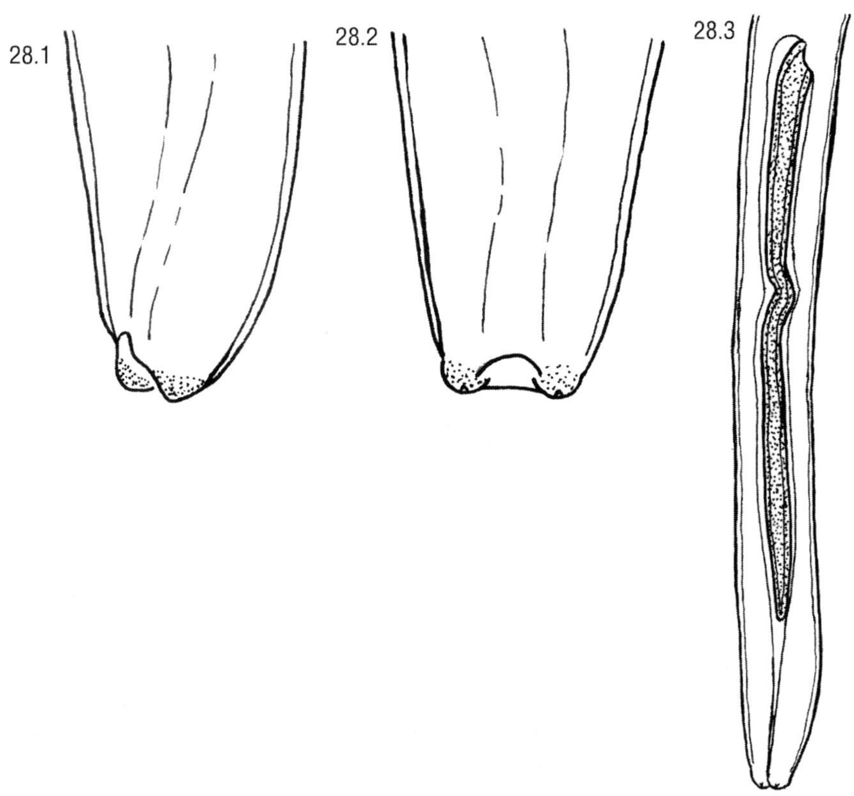

FIG. 28 *Pseudocapillaroides* 28.1 Posterior end of male, lateral view; 28.2 Posterior end of male, ventral view; 28.3 Posterior end of male, ventral view. (Redrawn after Moravec and Cosgrove, 1982.)

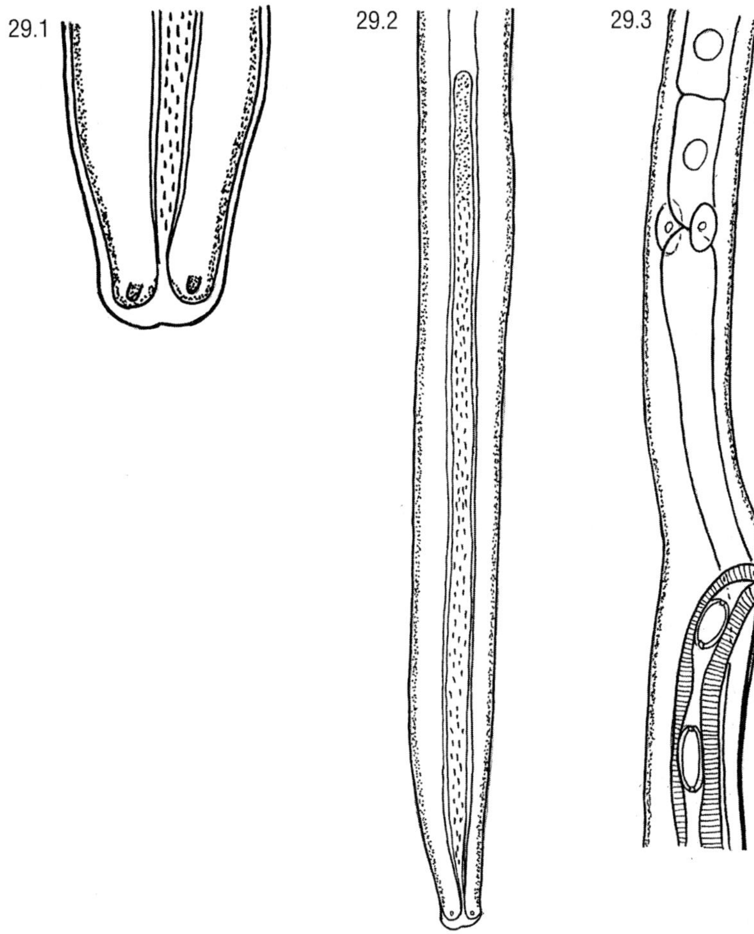

FIG. 29 *Brevithominx* 29.1 Distal end of male, ventral view; 29.2 Posterior end of male, ventral view; 29.3 Vulvar region, lateral view. (Redrawn after Travassos *et al.*, 1964.)

[Listed in Spencer Jones and Gibson (1987). Nearest genus considered to be *Ritaklossia* Freitas, 1959, which is listed as a synonym of *Capillaria* in Anderson and Bain (1982) and as a synonym of *Eucoleus* in Moravec (2001a). The genus is not listed in the classification of Moravec (2001a).]

Hepatospina Thieme, 1961

DIAGNOSIS: Not available. Parasite in *Acerina cernua*.

Types species: *Hepatospina acerinae* (Thieme, 1961) Thieme, 1961

[Listed in Spencer Jones and Gibson (1987). Moravec (1994, 2001a) lists *Capillaria acerinae* Thieme, 1961 as a synonym of *Schulmanela petruschewskii* (Schulman, 1948) Ivaschkin, 1964 the type and only species of *Schulmanela* Ivaschkin, 1964. *Hepatospina* is therefore a synonym of *Schulmanela* Ivaschkin, 1964.]

Neocapillaria Yu and Wang, 1994 (Figs 30.1 and 30.2)

DIAGNOSIS: Cephalic region tapered. Oesophagus shorter or nearly equal to posterior region of body. Stichocytes approximately 35. *Male*: Spicule slightly sclerotized, pointed distal tip. Sheath without spines, 'saccular'. Distal end expands. Two small ventral papillae present. Cloaca subterminal. *Female*: Vulva opens just posterior to oesophago-intestinal junction. Parasites in intestine of *Phoxinus legowskii variegatus* in Qingjiang River, Hubei Province, China.

Type species: *Neocapillaria phoxini* Yu and Wang, 1994

[Diagnosis based on Yu and Wang (1994). The authors are unaware that the name for this genus is preoccupied by *Neocapillaria* Moravec, 1987 subgenus. Moravec and Spratt (1998) rename the genus *Sinocapillaria* nom. nov. for *N. phoxini*, which had previously been moved to *Pseudocapillaria*. Moravec *et al.* (2008b), following examination of type specimens, now consider *Neocapillaria* and *Sinocapillaria* synonyms of *Freitascapillaria* Moravec, 1982.]

Crocodylocapillaria Moravec and Spratt, 1998 (Figs 30.3, 30.4 and 30.5)

DIAGNOSIS: Stichosome, 1 row of elongate stichocytes. *Male*: Caudal alae absent. Distal end rounded with 2 ventro-lateral lobes. Pseudobursa absent. Single pair of subventral papillae adjacent to cloaca. Spicule slightly sclerotized, long. Spicular sheath spinose 'on proximal and distal parts'. *Female*: Vulva opening with appendage. Eggs with extended polar plugs. Parasites in stomach of crocodiles.

Type species: *Crocodylocapillaria longiovata* Moravec and Spratt, 1998

[Diagnosis based on Moravec (2001a).]

Paracapillaroides Moravec, Salgado-Maldonado and Caspeta-Mandujano, 1999 (Figs 30.6 and 30.7)

DIAGNOSIS: Stichosome, 1 row of stichocytes. *Male*: Caudal alae absent. Membranous pseudobursa present with 2 projections along dorsal margins. Dorsal caudal projection absent. Pair of sessile adcloacal or post-cloacal papillae present. Spicule with smooth surface. Spicular sheath long, spinose at posterior end. *Female*: Vulva without appendage. Parasites in gastrointestinal tract of fish.

Type species: *Paracapillaroides agonostomi* Moravec, Salgado-Maldonado and Caspeta-Mandujano, 1999

[Diagnosis based on Moravec (2001a).]

Trichosomoididae Hall, 1916

DIAGNOSIS: Marked sexual dimorphism present or absent. Cuticle often with 'cephalic vesicles and bosses'. Oesophagus with posterior muscular part only with aligned stichocytes. *Male*: May live in or deeply penetrates female uterus or vagina. Cloaca short. Spicule, spicular sheath absent or vestigial. Pseudobursa absent. *Female*: Vulva opens near distal end of oesophagus.

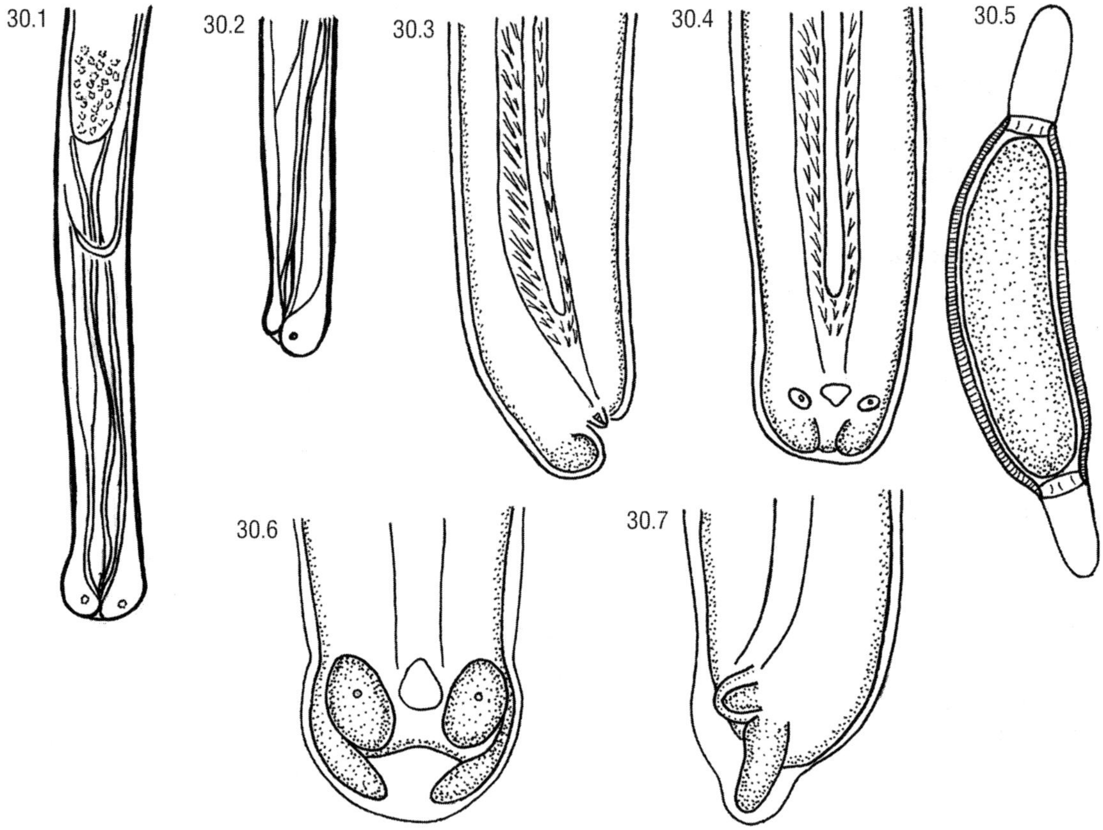

FIGS 30.1 and 30.2 *Neocapillaria* 30.1 Posterior end of male, ventral view; 30.2 Posterior end of male, lateral view. (Redrawn after Yu and Wang, 1994.)
FIGS 30.3, 30.4 and 30.5 *Crocodylocapillaria* 30.3 Posterior end of male, lateral view; 30.4 Posterior end of male, ventral view; 30.5 Egg from uterus. (Redrawn after Moravec and Spratt, 1998 from Moravec, 2001a.)
FIGS 30.6 and 30.7 *Paracapillaroides* 30.6 Posterior end of male, ventral view; 30.7 Posterior end of male, lateral view. (Redrawn after Moravec *et al.*, 1999 from Moravec, 2001a.)

Oviparous. Eggs thick-shelled with polar plugs, embryonated. Immature eggs expelled without fully formed shells. Anus terminal. Parasites in mucosa, paracloacal glands, cornea, musculature, swimbladder wall of fish and mammals.

[Moravec (2001a) validates the family considered a subfamily of the Trichuridae in Anderson and Bain (1982). Emended diagnosis based on Moravec (2001a), who includes three subfamilies including the Trichosomoidinae.]

Trichosomoidinae Hall, 1916

[Huffman and Moravec (1988), in their description of *Huffmanela huffmani* adults, refer to this subfamily as the family Trichosomoididae. Anderson and Bain (1982) consider this a subfamily of the Trichuridae. Moravec (2001a) also regards this subfamily to be of family rank.]

Trichosomoides Railliet, 1895

Type species: *Trichosomoides crassicauda* (Bellingham, 1840) Railliet, 1895

[Moravec (2001a) includes this single genus in the subfamily and lists *Trichodes* Linstow, 1874 as a synonym.]

Huffmanelinae Moravec, 2001

DIAGNOSIS: Sexual dimorphism not marked. *Male*: Spicule, spicular sheath and pseudobursa absent. *Female*: Immature eggs expelled without fully formed shell, more mature eggs in host tissue embryonated, dark, enveloped in thin layer with spines. Parasites in tissues of fish.

Type genus: *Huffmanela* Moravec, 1987

[Diagnosis based on Moravec, 2001a.]

Huffmanela Moravec, 1987 (Fig. 31)

DIAGNOSIS: Stichosome: single row of stichocytes. Coelomocytes present post-oesophagus. *Male*: Cloaca long, spicule and spicular sheath absent. Posterior end of body dorso-ventrally depressed, membraneous pseudobursa absent. One pair of adcloacal papillae present. *Female*: Vulva opens near posterior end of oesophagus. Eggs in uterus, non-embryonated; eggs in host tissue contain larva, strongly pigmented, dark, not translucent, wall often distinctly thick, walls

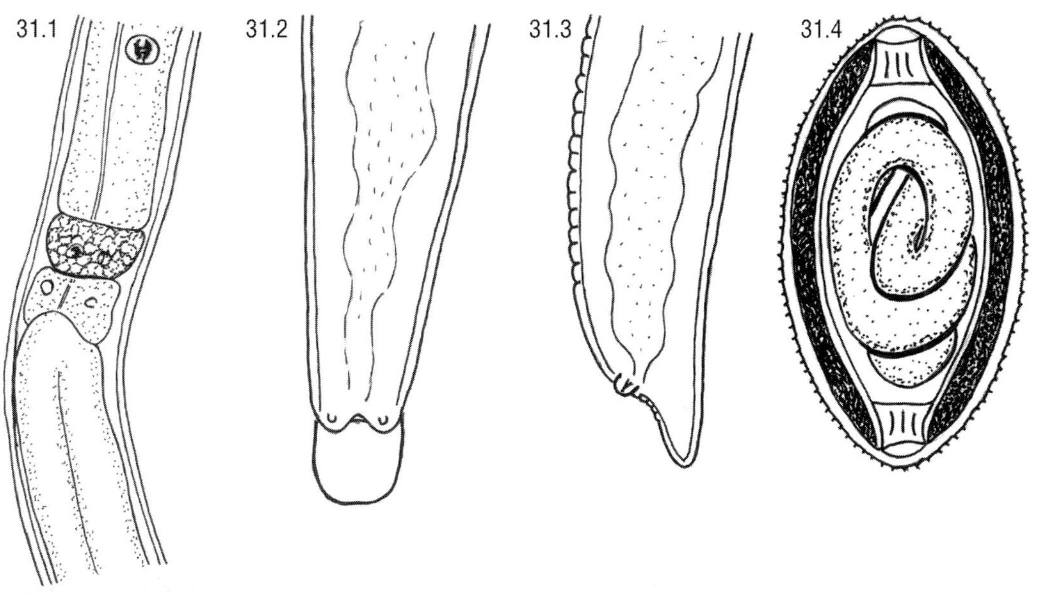

FIG. 31 *Huffmanela* 31.1 Oesophago-intestinal junction; 31.2 Posterior end of male, ventral view; 31.3 Posterior end of male, lateral view; 31.4 Mature egg. (Redrawn after Huffman and Moravec, 1988.)

completely covered, including polar plugs with clear membrane covered in minute spines or ornamentation. Anus terminal. Parasites in skin, mucosa of gill arches, muscles, swimbladder, serosal cover of intestine of elasmobranchs and bony fishes.

Type species: *Huffmanela carcharhini* (MacCallum, 1925) Moravec, 1987

[Moravec (1987) established this genus on the basis of the morphology of the eggs, as adults were unknown and initially placed it in the Capillariidae. Huffman and Moravec (1988) described the adults *Huffmanela huffmani* Moravec, 1987 from Centrarchid fish in the Upper San Marcos River, Central Texas, USA, and the generic diagnosis is based on both these publications. Moravec (2001a) places this genus in a new subfamily.]

Anatrichosomatinae Smith and Chitwood, 1954

Anatrichosoma Smith and Chitwood, 1954

Type species: *Anatrichosoma cynamolgi* Smith and Chitwood, 1954

[Moravec (2001a) validates the subfamily Anatrichosomatinae in the family Trichosomoididae for this single genus. Anderson and Bain (1982) include this genus in the subfamily Trichosomoidinae in the family Trichuridae.]

MUSPICEOIDEA

Muspiceidae

Maseria Rausch and Rausch, 1983 (Fig. 32)

DIAGNOSIS: Body small, translucent. Cuticle with transverse striations. Mouth slightly ventral. Amphids 'cyathiform'. Oesophagus long, narrow, distal end just posterior to 2 latero-ventral stichocytes. Trophosome and oesophagus linked by small duct with 'Demanian system'. Single subventral oesophageal gland just posterior to nerve ring. Posterior dorsal osmoregulatory cell near distal end of oesophagus. Alimentary canal non-functional in adults. *Male*: Unknown. *Female*: 'Proterandric hermaphrodite'. Didelphic, ovaries generally reflexed. Vagina well developed. Vulva closed to exterior. Phasmids small, simple. Anus absent. *Mature larvae*: Cephalic armature present. Parasite in subcutaneous tissues of rear feet of Neartic bats (Vespertilionidae) in Oregon and Alaska.

Type species: *Maseria vespertilionis* Rausch and Rausch, 1983

[Listed in Spencer Jones and Gibson (1987).]

Robertdollfusidae Chabaud and Campana, 1950

Lappnema Bain and Nikander, 1983 (Fig. 33)

DIAGNOSIS: *Male*: Unknown. *Female*: Body very short; epithelial – muscular sheath thick. Mouth vestigial, on tissue cone. Cephalic vesicle with 4 large pedunculate papillae, 6 small

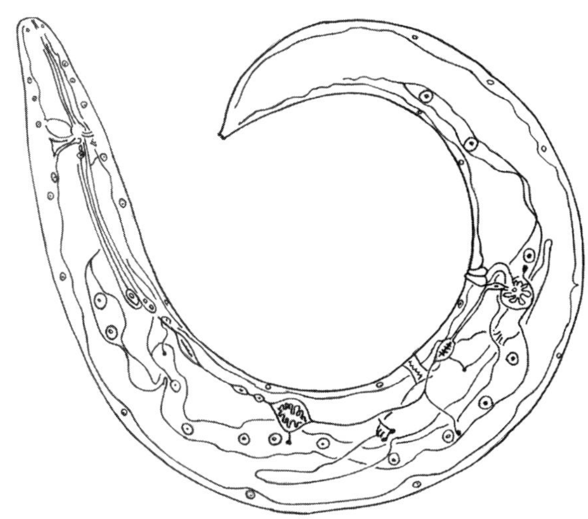

FIG. 32 *Maseria* 32.1 Whole immature nematode. (Redrawn after Rausch and Rausch, 1983.)

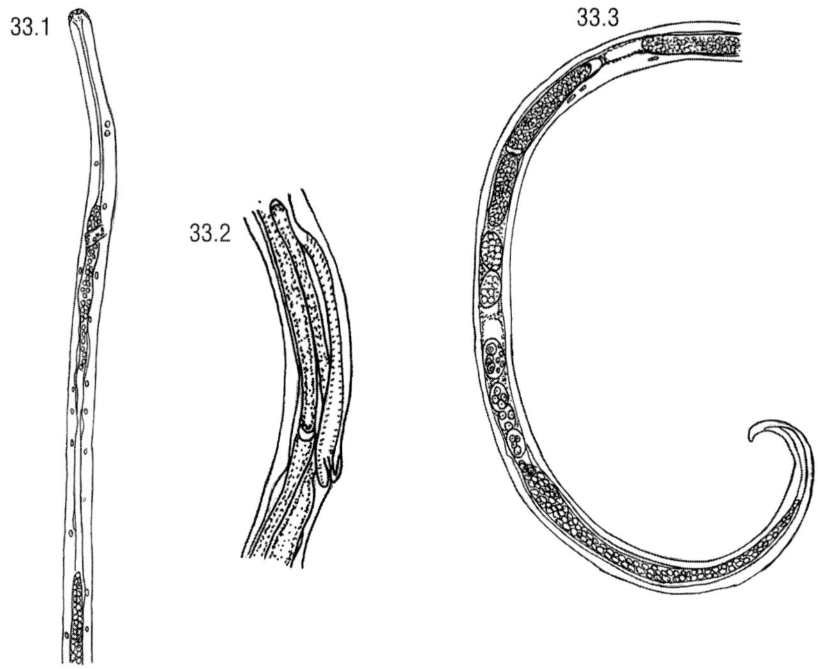

FIG. 33 *Lappnema* 33.1 Anterior part of body; 33.2 Vulvar region; 33.3 Posterior part of body. (Redrawn after Bain and Nikander, 1983.)

papillae, 2 lateral papillae, amphids at base of vesicle. Digestive tract atrophied. Didelphic. Vulva opens in posterior part of body. Anus atrophied. Parasite of subcutaneous capillaries and nodules in ears of reindeer, *Rangifer tarandus* in Finland.

Type species: *Lappnema auris* Bain and Nikander, 1983

[Listed in Spencer Jones and Gibson (1987).]

SECERNENTEA

2 RHABDITIDA

RHABDITOIDEA

Rhabdiasidae Railliet, 1916

Neoentomelas Hasegawa, 1989 (Figs 34.1 and 34.2)

DIAGNOSIS: *Female*: Parasitic females only present. Cephalic region broadly rounded with distal margin formed of 1 dorsal, 1 ventral lobe, 2 dorso-lateral and 2 ventro-lateral lobes. Body narrows posterior to cephalic region. Mouth opening elongate, dorso-ventrally flattened, with two massive muscular lips. Each lip with 2 double and 2 smaller cephalic papillae. Amphids near lateral margins of mouth. Buccal capsule globular, thick-walled, indented laterally in cross-section. Teeth absent. Oesophagus short, clavate. Vulva opens slightly posterior to middle of body. Vagina weakly developed. Amphidelphic. Eggs thin-walled. Parasites in lungs of skinks, *Ateuchosaurus pellopleurus* in Ryukyu Archipelago, Japan.

Type species: *Neoentomelas asatoi* Hasegawa, 1989

Chabirenia Lhermitte-Vallarino, Bain, Deharo, Bertani, Voza, Attout, and Gaucher, 2005 (Figs 34.3, 34.4 and 34.5)

DIAGNOSIS: *Female*: Parasitic females only present, few mm long. Body helically coiled. Longitudinal cuticular ridges present. Cephalic region flat without external ornamentation. Buccal cavity tiny without thick walls, 3 oesophageal onchia (teeth) present at its base. Excretory glands paired, long, conspicuous. Homogonic free-living cycle. Infective larvae enclosed in a sheath with chequered ornamentation. Parasites in salivary glands of teiid saurians.

Type species: *Chabirenia cayennensis* Lhermitte-Vallarino, Bain, Deharo, Bertani, Voza, Attout, and Gaucher, 2005

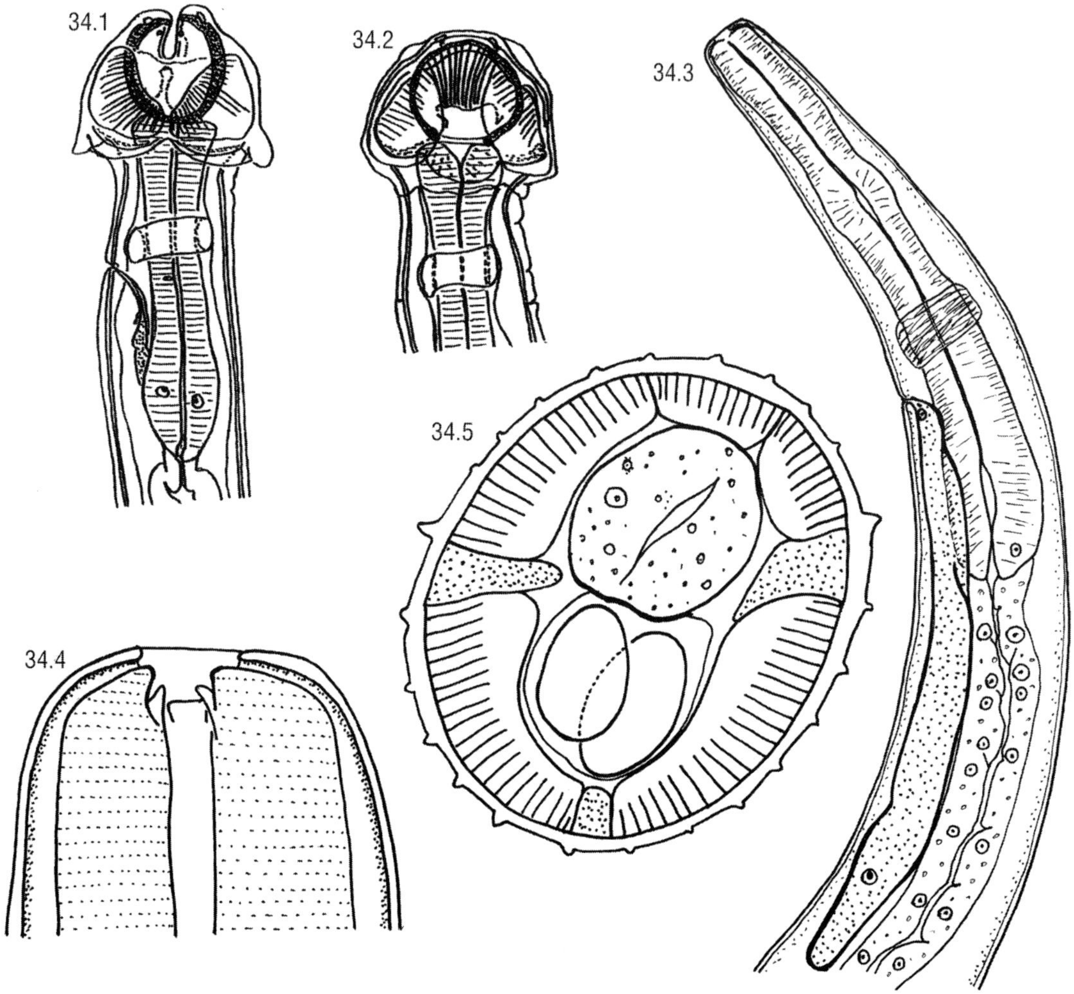

Figs 34.1 and 34.2 *Neoentomelas* 34.1 Anterior end, lateral view; 34.2 Anterior end, ventral view. (Redrawn after Hasegawa, 1989.)

Figs 34.3, 34.4 and 34.5 *Chabirenia* 34.3 Anterior end, left lateral view; 34.4 Cephalic region, dorsal view; 34.5 Transverse section of body just anterior to vulva. (Redrawn after Lhermitte-Vallarino *et al.*, 2005.)

3 STRONGYLIDA

[Durette-Desset and Chabaud (1993) propose the creation of four suborders: Ancylostomatina for Ancylostomatoidea and Diaphanocephaloidea; Strongylina for Strongyloidea; Trichostrongylina for Trichostrongyloidea and two new superfamilies (Molineoidea and Heligmosomoidea) and Metastrongylina for Metastrongyloidea. No differential diagnoses for these suborders was given.]

DIAPHANOCEPHALOIDEA

[Durette-Desset and Chabaud (1993) propose that this superfamily should be moved to a new suborder Ancylostomatina in the order Strongylida. Differential diagnosis not given.]

Diaphanocephalidae Travassos, 1920

Kalicephalus Molin, 1861

Subgenus *Schadius* Lichtenfels, 1980

[Baker (1987) synonymizes the subgenus *Schadius* Lichtenfels, 1980 with the subgenus *Kalicephalus* (Molin, 1861 genus).]

Paracylicostrongylus Srivastava and Gupta, 1976 (publ. 1978) (Fig. 35)

DIAGNOSIS: Buccal capsule cup-shaped, 3 chitinous ridges present. Oesophagus well developed, clavate. *Male*: Bursa: dorsal ray simple, accessory branches absent; externo-dorsal ray arises near base of dorsal ray; ventral rays with common stem, parallel, remain close together; lateral rays with common stem, only slightly separated distally. Spicules subequal, slender. *Female*: Unknown. Parasite in intestine of marine teleost, *Trichiurus muticus* from Pentkota, Puri, Orissa, India.

Type species: *Paracylicostrongylus trichiurusi* Srivastava and Gupta, 1976 (publ. 1978)

[The genus was placed in the family Cylicostrongylidae Yamaguti, 1961, a family erected for *Cylicostrongylus* Yamaguti, 1961, which has similarities to the hookworm *Ancylostoma* and is regarded as a genus *incertae sedis* in Spencer Jones and Gibson (1987). *Paracylicostrongylus* is based on a single male specimen. The figure of the specimen shows similarities to species of the genus *Kalicephalus*, which are generally parasites of snakes or, rarely, lizards. In view of this the genus is also considered *incertae sedis*. Spencer Jones and Gibson (1987) list this genus in the superfamily Diaphanocephaloidea.]

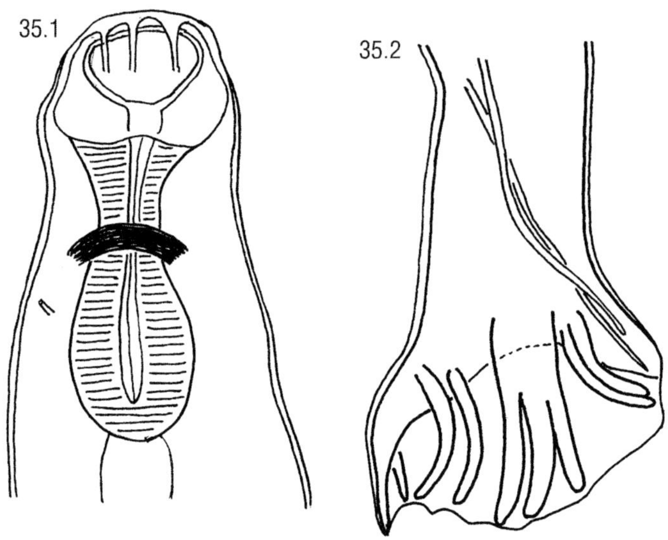

FIG. 35 *Paracylicostrongylus* 35.1 Anterior end; 35.2 Posterior end of male, lateral view. (Redrawn after Srivastava and Gupta, 1976 (publ. 1978).)

ANCYLOSTOMATOIDEA

[Durette-Desset and Chabaud (1993) propose that this superfamily should be included in a new suborder Ancylostomatina in the order Strongylida. No differential diagnosis for this suborder was given.]

Ancylostomatidae (Looss, 1905)

Bunostominae (Railliet and Henry, 1909 tribe) Looss, 1911

Bunostominea Railliet and Henry, 1909

Bunostomoides Pinto, 1935

DIAGNOSIS: Buccal capsule: dorsal cone short; subventral lancets paired. Cervical papillae at level of anterior third of oesophagus. *Male*: Spicules very long.

Type species: *Bunostomoides phlebotomum* (Railliet, 1900) Pinto, 1935

[Pinto (1935) separates this genus from *Bunostomum* on the size of the dorsal cone (long in *Bunostomum*), position of the cervical papillae (middle third in *Bunostomum*) and spicule size (short in *Bunostomum*). Lichtenfels (1980b), in his key, separates the genus *Bunostomum* from the other genera of the subfamily by the presence of a globular buccal capsule with large dorsal cone; presence of subventral lancets; filiform spicules; externo-dorsal rays arising anterior to dorsal bursal rays; dorsal ray asymmetrical; and vulva opening at junction of anterior and middle thirds of body. The characters used by Pinto are considered specific, not generic, and *Bunostomoides* is considered a synonym of *Bunostomum*. Listed in Spencer Jones and Gibson (1987).]

Rhinocerotonema Jiang, Yin and Kung, 1986 (Fig. 36)

DIAGNOSIS: Mouth opening antero-dorsal. Buccal capsule subglobular, proximal rim without cutting plates or teeth. Dorsal cone and 2 pairs of subventral lancets present. *Male*: Bursal rays: ventral rays with common stem, parallel, remain close together; lateral rays with common stem, medio- and postero-laterals remain close together, antero-lateral diverges; externo-dorsal rays asymmetrical. Genital cone long. Spicules long, filiform. Gubernaculum absent. *Female*: vulva opens in anterior third of body. Tail sharply pointed. Parasite in large intestine of *Rhinoceros unicornis* in Beijing Zoo, China.

Type species: *Rhinocerotonema rhinocerotis* Jian, Yin and Kung, 1986

[Jiang *et al.* (1986) place their genus in the subfamily Globocephalinae, which they place in the Strongylidae but which now forms the tribe Globocephalinea (Travassos and Vogelsang, 1932 subfamily) of the Ancylostomatinae, Ancylostomatidae. Lichtenfels (1980b) includes the single genus *Globocephalus* Molin, 1861 in the tribe characterized by the absence of teeth and cutting plates. Jiang *et al.* (1986) separate their genus from a number of genera that are now placed in different superfamilies. The presence of a dorsal cone and asymmetrical externo-dorsal rays,

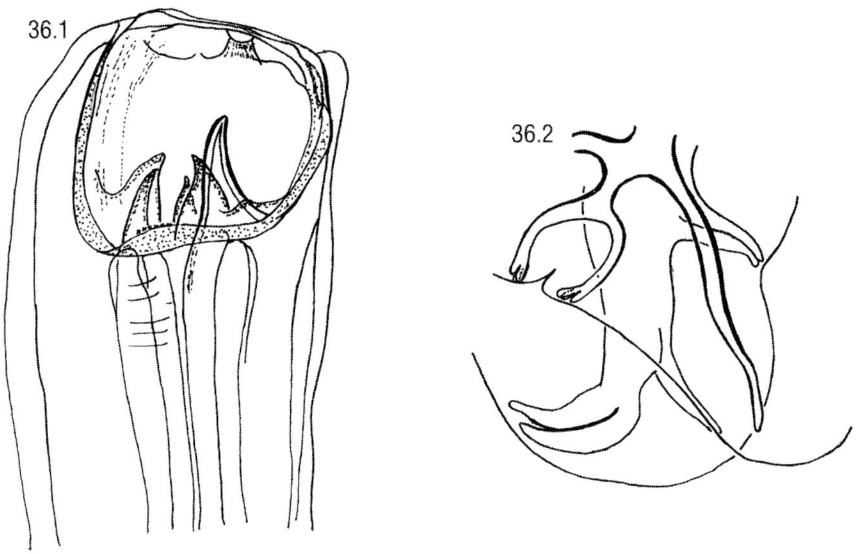

FIG. 36 *Rhinocerotonema* 36.1 Anterior end, lateral view; 36.2 Part of male bursa showing dorsal lobe and one lateral lobe, dorsal view. (Redrawn after Jiang *et al.*, 1986.)

however, shows some affinity to *Bunostomum*. The 2 pairs of subventral lancets and long spicules are also seen in *Bunostomum phlebotomum*. For these reasons, until further studies of the material are undertaken, this genus is placed in the Bunostominae rather than in the Globocephalinea.]

Cylicostrongylidae Yamaguti, 1961

DIAGNOSIS: Buccal capsule with 'denticular prominences' on internal surface. Oesophagus muscular. *Male*: Bursa: circular, membraneous. Spicules: 2 present. *Female*: Unknown. Parasites of fish.

Type genus: *Cylicostrongylus* Yamaguti, 1961

Cylicostrongylus Yamaguti, 1961 (Fig. 37)

DIAGNOSIS: Head: rounded, wider than cervical region. Mouth: oval; opens dorsally; lips absent. Buccal capsule: 'cup-shaped; spinose median stylet; horn-like thickenings, 1 each ventral and lateral, 2 dorsally'. Structures described probably relate to the dorsal gutter, dorso- and ventro-lateral teeth, respectively. *Male*: Bursa: ventral and lateral rays 'simple'; externo-dorsal rays arise from base of dorsal ray; dorsal ray with tiny branches at distal tip. Spicules equal, similar. *Female*: unknown. Parasite in stomach of salmonids in Rumania.

Type species: *Cylicostrongylus ciureai* (Dinulescu, 1942) Yamguti, 1961

[Spencer Jones and Gibson (1987) cite this genus as *incertae sedis*. The illustrations of the type genus show similarities to species of the genus *Ancylostoma*. Re-examination of the type specimens may reveal this genus to be a synonym of *Ancylostoma*.]

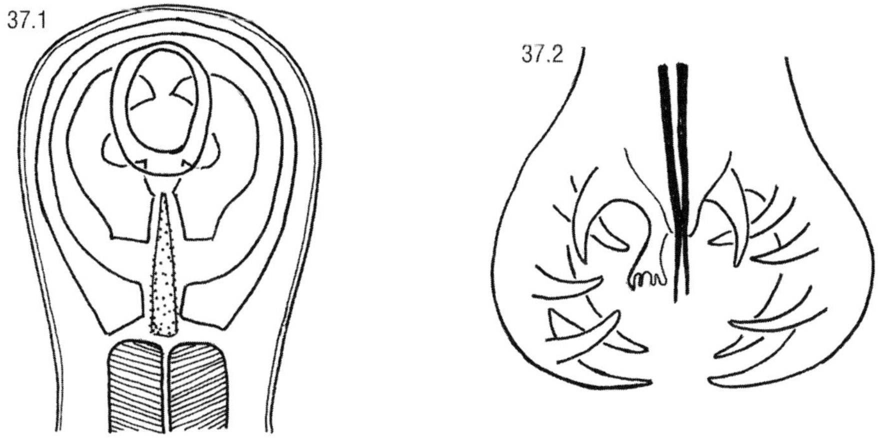

Fig. 37 *Cylicostrongylus* 37.1 Anterior end, dorsal view; 37.2 Male bursa, ventral view. (Redrawn after Yamaguti, 1961.)

STRONGYLOIDEA

[Durette-Desset and Chabaud (1993) propose that this superfamily should be moved to a new suborder Strongylina in the order Strongylida. No differential diagnosis for this suborder is given.]

Strongylidae

[Dvoïnos (1982) created an additional subfamily, the Gyalocephalinae Dvoïnos, 1982, for Popova's tribe, Gyalocephalini Popova, 1952, and the type genus *Gyalocephalus* Looss, 1900. Lichtenfels *et al.* (2008) in their key to strongylid parasites place *Gyalocephalus* in the Cyathostominae.]

Cylicostomias Cram, 1925

Type species: *Cylicostomias aegyptiaca* (Railliet, 1923)

[Opinion 1972 on Case 3075 submitted by Gibbons and Lichtenfels, 1997 to the ICZN and listed in the *Bulletin of Zoological Nomenclature* 55, 1 (March, 1998), case published in the *Bulletin of Zoological Nomenclature* 56, 4 (December, 1999), with the decision published in the *Bulletin of Zoological Nomenclature* 58, 2 (June, 2001), places the name '*aegyptiaca* Railliet, 1923' on the Official Index of Rejected and Invalid Species Names in Zoology. The generic name *Cylicostomias* Cram, 1925 is no longer accepted as valid and is a synonym of *Cyathostomum* Molin, 1861. *Cylicostomias* is listed in Spencer Jones and Gibson (1987) and as a synonym of *Cyathostomum* in Lichtenfels *et al.* (2008).]

Strongylinae Railliet, 1885

[Dvoïnos (1982) divided the subfamily into three tribes, two new and Strongylini Railliet and Henry, 1912 for *Strongylus* Muller, 1780, *Alfortia* (Railliet, 1923) Skrjabin, 1933 and *Delafondia* (Railliet, 1923) Skrjabin, 1933. *Alfortia* and *Delafondia* are widely regarded as synonyms of *Strongylus* (see Lichtenfels *et al.*, 2008). Dvoïnos used the ending '-ini' rather than '-ea' favoured by other workers. The two new tribes are as written by Dvoinos and the genera included listed: Triodontophorini Dvoïnos, 1982, type genus *Triodontophorus* Looss, 1902 (Dvoïnos (1982) also includes *Craterostomum* Boulenger, 1920); Oesophagodontini Dvoïnos, 1982, type genus *Oesophagodontus* Railliet and Henry, 1902. The genera *Decruzia* Lane, 1914 and *Equinurbia* Lane, 1914 are removed from the subfamily Strongylinae by Dvoïnos (1982), with no details being given as to their position. Only parasites of horses are retained in the subfamily. Dvoïnos (1982) reinstates a third tribe, the Strongylini Railliet and Henry, 1912. Lichtenfels (1980a) did not use tribes in his key for this subfamily.]

Parastrongylus Yin, Jiang and Kung, 1986 (Fig. 38)

DIAGNOSIS: Mouth terminal. Buccal capsule: subglobular; dorsal gutter extends to proximal margin of buccal capsule; duct of dorsal oesophageal gland opens into buccal capsule through numerous pores in dorsal gutter. Teeth absent. External leaf crown: over 200 elements; long, slender. Internal leaf crown: minute elements; numerous. *Male*: Bursa: membrane margin angular; ventral

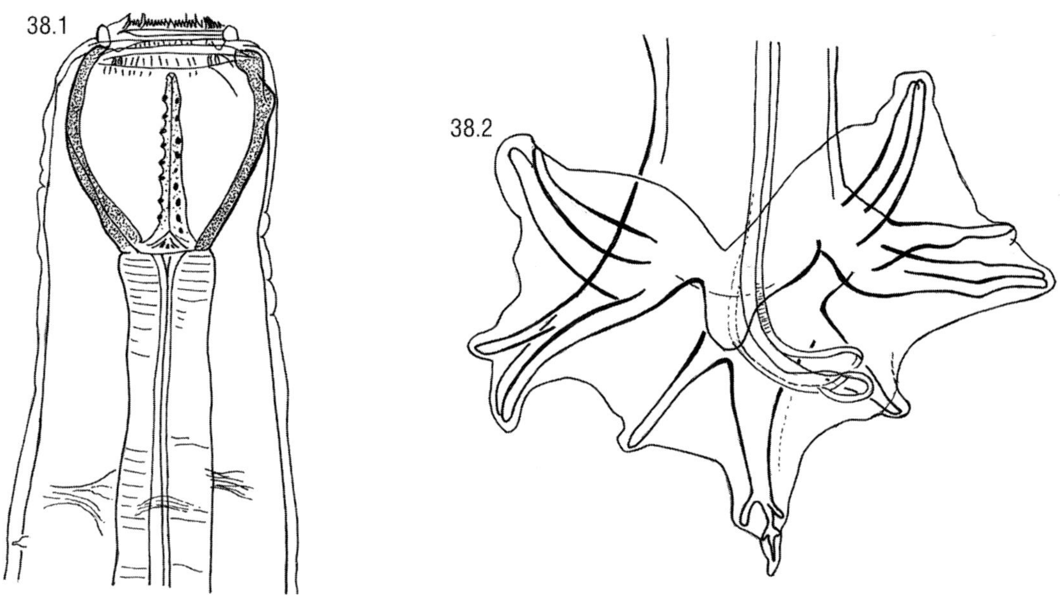

FIG. 38 *Parastrongylus* 38.1 Anterior end, ventral view; 38.2 Male bursa, ventral view. (Redrawn after Yin *et al*., 1986.)

and lateral rays typical of family; dorsal ray atypical with 2 pairs of short lateral branches near distal end of single median point, median fissure absent. Spicules: long; slender; alate. Gubernaculum absent. *Female*: Vulva opens at proximal end of posterior third of body. Amphidelphic. Tail short, conical. Parasite in large intestine of *Rhinoceros unicornis* in Beijing Zoo, China.

Type species: *Parastrongylus paradoxus* Yin, Jiang and Kung, 1986

[The generic name is preoccupied by *Parastrongylus* Baylis, 1928, a metastrongyloid genus recovered from the arterial circulatory system of rodents. A re-examination of the type specimens of *Parastrongylus paradoxus* and *Strongylus tremletti*, Round, 1962 (from *Diceros bicornis*) may reveal similarities. Round (1962) placed his species in a new subgenus *Dicerocola* in the genus *Strongylus*. A further species, *Strongylus additictus* Railliet, Henry and Bauche, 1914 (from *Elephas indicus*) has a similar dorsal ray, but the presence of 2 subventral teeth at the base of the buccal capsule separates it from the other two species. The formation of the dorsal ray of these three species differs from other members of the Strongylidae and indicates that further studies are required to identify their taxonomic position. Listed in Spencer Jones and Gibson (1987).]

Cyathostominae Nicoll, 1927

Trichonema

[Spencer Jones and Gibson (1987) list four subgenera within this genus. Following the classification of Lichtenfels (1980a), the validity of the genus *Trichonema* is no longer accepted. The subgenera which have not gained acceptance are listed below.]

Subgenus *Zebrinicola* Ricci, 1939

DIAGNOSIS: Cephalic region with 2 long lateral papillae. Elements of internal leaf crown triangular. Buccal capsule not symmetrical. Anterior margin of oesophagus with 7 irregularly sized teeth.

Type species: *Trichonema* (*Zebrinicola*) *zavattarii* Ricci, 1939

[Lichtenfels *et al.* (1998) consider the type species is *species inquirenda*, the validity of the subgenus is therefore in doubt. Listed in Spencer Jones and Gibson (1987).]

Subgenus *Skladnikia* Ricci, 1939

DIAGNOSIS: Anterior margin of oesophagus with 9 teeth, 3 long teeth, 1 on each lobe with 6 shorter teeth, 2 per lobe. Buccal capsule cylindrical, symmetrical. Elements of internal leaf crown triangular.

Type species: *Trichonema* (*Skladnikia*) *symmetrum* Ricci, 1939

[Lichtenfels *et al.* (1998) consider the type species is *species inquirenda*, and the validity of the subgenus is therefore in doubt. Listed in Spencer Jones and Gibson (1987).]

Subgenus *Borania* Ricci, 1939

DIAGNOSIS: Internal leaf crown difficult to see or absent. *Male*: unknown. *Female*: Vulva opens on deep transverse groove.

Type species: *Trichonema* (*Borania*) *maestrii* Ricci, 1939

[Lichtenfels *et al.* (1998) consider the type species is *species inquirenda*, and the validity of the subgenus is therefore in doubt. Listed in Spencer Jones and Gibson (1987).]

Subgenus *Javella Ricci, 1939*

DIAGNOSIS: Internal leaf crown small and numerous. *Male*: Unknown. *Female*: Tail short, narrow cone.

Type species: *Trichonema* (*Javella*) *aequatorialis* Ricci, 1939

[Lichtenfels *et al.* (1998) consider the type species is *species inquirenda*, and the validity of the subgenus is therefore in doubt. Listed in Spencer Jones and Gibson (1987).]

Skrjabinodentus Tshoijo, in Popova, 1958 (Fig. 39)

DIAGNOSIS: Body 7.5–11.0 mm long. Oral collar depressed. Cephalic sensory organs: amphids do not project above oral collar; submedian cephalic papillae prominent, well developed. Leaf crowns: external leaf crown with 8 broad, triangular elements, extra chitinous supports absent; internal leaf crown with an equal or larger number of elements than external leaf crown. Buccal capsule: slightly narrower anteriorly but with thicker, swollen anterior walls; asymmetrical with ventral wall longer than dorsal wall; dorsal gutter present. Oesophageal funnel with well-

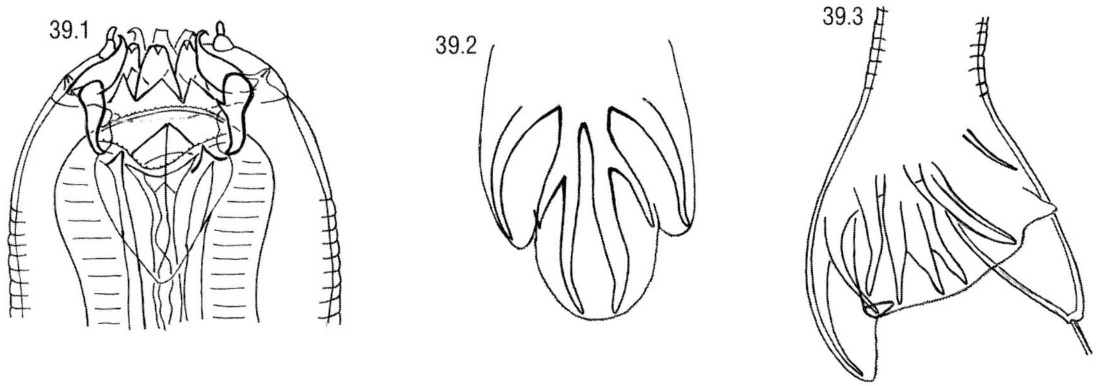

FIG. 39 *Skrjabinodentus* 39.1 Anterior end; 39.2 Dorsal lobe of bursa; 39.3 Male tail, lateral view. (Redrawn after Dvoĭnos and Kharchenko, 1986.)

developed triangular teeth. *Male*: Bursa: dorsal ray split to origin of externo-dorsal ray, each branch divides into two additional branches. Genital cone extends beyond bursa. Gubernaculum with 2 lateral wings on distal margin. Spicules: equal; filiform; hook-shaped distal tips. *Female*: Vulva opens close to anus. Ovejectors type I. Tail straight. Parasite of horses.

Type species: *Skrjabinodentus caragandicus* (Funikova, 1939) Tshoijo, in Popova, 1958

[Emended diagnosis based on Dvoĭnos and Kharchenko (1986, 1994).]

Hsiungia K'ung and Yang, 1964 (Fig. 40)

DIAGNOSIS: Body length 15–20 mm. Oral collar medium height. Cephalic sensory organs: amphids do not project above oral collar; submedian cephalic papillae short, well developed. Leaf crowns: external leaf crown with numerous narrow, long elements, extra chitinous supports absent; internal leaf crown with a similar number of elements to external leaf crown, inserted near anterior margin of buccal capsule. Buccal capsule: short, cylindrical, width several times larger than length; walls thin with hoop-shaped margin around posterior margin; ventral wall more or less equal in length to dorsal wall; dorsal gutter rudimentary. Oesophageal funnel without teeth. *Male*: Bursa: edge with fine denticulation; dorsal ray split to origin of its first secondary branch. Spicules: equal; filiform. *Female*: Vulva opens some distance from anus. Ovejectors type I. Tail straight. Parasite of horses.

Type species: *Hsiungia pekingensis* K'ung and Yang, 1964

[Dvoĭnos and Kharchenko (1988) raised the rank of the subgenus *Hsiungia* to genus. The emended diagnosis is based on Dvoĭnos and Kharchenko (1988, 1994).]

Cyathostominea

[Dvoĭnos (1982) used the subfamily Trichonematinae Railliet, 1916 and the tribe Trichonematini Popova, 1952. He also used two other tribes: Cylicocyclini Dvoĭnos, 1982

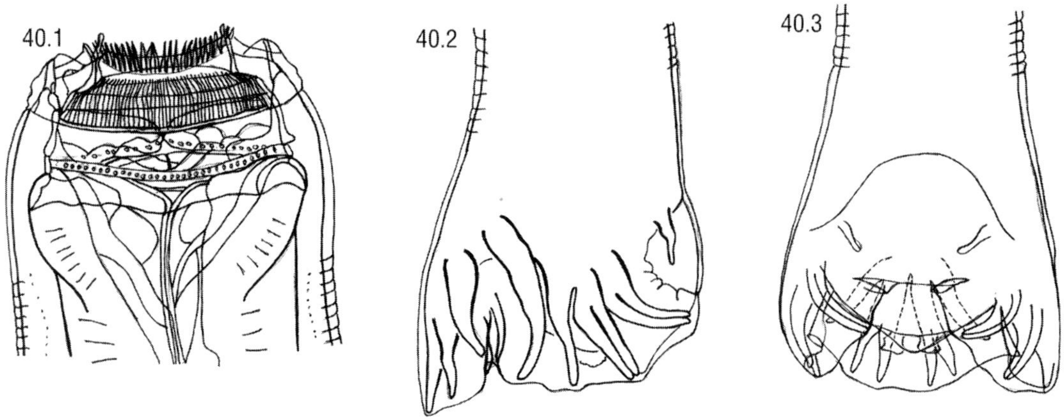

FIG. 40 *Hsiungia* 40.1 Anterior end; 40.2 Male tail, lateral view; 40.3 Male tail, ventral view. (Redrawn after Dvoĭnos and Kharchenko, 1988.)

including the type genus *Cylicocyclus* (Ihle, 1922) Erschow, 1943, *Petrovinema* Erschow, 1943 and *Poteriostomum* Quiel, 1919 (Lichtenfels *et al.*, 1998 included these genera in the tribe Cyathostominea); Schulzitrichonematini Dvoĭnos, 1982, including the type genus *Schulzitrichonema* Erschow, 1943 (in part a synonym of *Cylicostephanus* (Ihle, 1922) Lichtenfels, 1975 in Lichtenfels, 1980a), *Bidentostomum* Tshoijo, 1957 (in the Strongylinae in Lichtenfels, 1980a and Lichtenfels *et al.*, 2008). Dvoĭnos (1982) also reinstated a third tribe, Trichonematini Popova, 1952. Dvoĭnos (1982) placed his three tribes in the subfamily Trichonematinae, which is no longer accepted and is replaced by the Cyathostominae. The genera *Khalilia* Neveu-Lemaire, 1924 (= *Amira* Lane, 1914), *Bourgelatia* Railliet, Henry and Bauche, 1919 (placed in Oesophagostominae in Lichtenfels, 1980a), *Phacochoerostrongylus* Schwartz, 1928 (placed in Oesophagostominae in Lichtenfels, 1980a) and *Trachypharynx* Leiper, 1911 (placed in Oesophagostominae in Lichtenfels, 1980a) from the Proboscidea, pigs and rodents are removed by Dvoĭnos (1982), and only parasites of horses are retained by him in the subfamily Trichonematinae.

The subfamily Cyathostominae and tribe Cyathostominea are now universally accepted. Lichtenfels *et al.* (1998) published an annotated checklist recognizing 13 genera in the tribe Cyathostominea. The details of this checklist were published following consensus recommendations of a workshop on the *Systematics of the Cyathostominea of Horses*, convened 11 August 1997 at the 16th International Conference of the World Association for the Advancement of Veterinary Parasitology, held in Sun City, South Africa. The genera accepted are *Cyathostomum* Molin, 1861, *Coronocyclus* Hartwich, 1986, *Cylicodontophorus* Ihle, 1922, *Cylicocyclus* Ihle, 1922, *Cylicostephanus* Ihle, 1922, *Skrjabinodentus* Tshoijo, in Popova, 1958, *Tridentoinfundibulum* Tshoijo, in Popova, 1958, *Petrovinema* Erschow, 1943, *Poteriostomum* Quiel, 1919, *Parapoteriostomum* Hartwich, 1986, *Hsiungia* K'ung and Yang, 1964, *Cylindropharynx* Leiper, 1911 and *Caballonema* Abuladze, 1937. *Schulzitrichonema* Erschow, 1943 is considered a *nomen nudum*. For full details of synonyms and species assigned to the genera, Lichtenfels *et al.* (1998, 2008) should be consulted.

Lichtenfels *et al.* (2002) recommend that the common name 'cyathostomin(s)' be used for the subfamily Cyathostominae and the tribe Cyathostominea rather than 'cyathostomes' and 'small strongyles'.]

Cyathostomum (Molin, 1861) Hartwich, 1986

DIAGNOSIS: Oral collar: moderately high; distinct demarcation from body, near proximal rim of buccal capsule. Cephalic sensory organs: submedian papillae short, distal end separated by a notch; amphids not prominent. Buccal capsule: slightly dorso-ventrally oval in cross-section; wider than deep; walls thickening posteriorly; extra-chitinous supports originate from proximal rim of buccal capsule to distal rim of internal wall of mouth collar. Leaf crowns: external leaf crown (ELC) 18–25 elements; 2–3 times longer than internal leaf crown elements (ILC); ILC elements inserted in anterior half of buccal capsule wall, posterior to proximal rim; ILC 1.5–3.5 times more elements than ELC, may be inserted in an uneven line. Dorsal gutter absent or dome-like. Oesophageal funnel: 3 teeth present; do not protrude into buccal capsule. Cervical papillae and excretory pore at same level. *Male*: Body length up to 11 mm. Bursa: dorsal ray with 3 branches either side of median fissure, may have shorter additional branches; externo-dorsal ray arises from base of dorsal; lateral rays with common stem; ventral rays with common stem, remain parallel and close together. Prebursal papillae present. Spicules with barbed tips. Gubernaculum present. *Female*: Body length up to 13 mm long. Vulva opens close to anus. Posterior end may be markedly bent dorsally. Parasite of caecum and colon of equids.

Type species: *Cyathostomum tetracanthum* (Mehlis, 1831) Molin, 1861 in part Looss, 1900 nec *Cyathostomum tetracanthum sensu* Hartwich, 1986

[The emended diagnosis is based on Hartwich (1986). The use of the type species as shown is used following an application to the ICZN to conserve the usage of this name and the designation of a neotype for *C. tetracanthum* in the interest of stabilizing the nomenclature (see Gibbons and Lichtenfels, 1999).]

Coronocyclus Hartwich, 1986 (Fig. 41)

DIAGNOSIS: Oral collar: round; inflated, moderately high; distinct demarcation from body. Cephalic sensory organs: submedian papillae short, distal end separated by notch, not longer than external leaf crown elements; amphids not prominent. Leaf crowns: external leaf crown 16–24 elements, long, broad, protrude beyond mouth; internal leaf crown 44–80 elements inserted in straight line in anterior 1/5–1/2 of buccal capsule, shorter, thinner, more numerous than external leaf crown; extra chitinous supports extend from base of oral collar to junction of external and internal leaf crowns, join leaf crowns and buccal capsule by connective tissue bands; buccal capsule: cylindrical; as wide as is long or wider than long; dorsal oesophageal gland opens in dorsal floor of buccal capsule or extends into cavity in slightly dome-shaped dorsal gutter. Oesophageal funnel: 3 teeth present; do not protrude into buccal capsule. Cervical papillae and excretory pore at level of mid-oesophagus. *Male*: Body up to 11 mm. Bursa: distinct dorsal lobe; dorsal ray with 3 main branches either side of median fissure, may have smaller additional branches; externo-dorsal ray arises from main stem of dorsal ray, markedly curved posteriorly; lateral rays with common stem, distal tips separated; ventral rays with common stem, remain parallel. Prebursal papillae present. Spicules with barbed tips. *Female*: Body up to 12.8 mm long, may widen anterior to vulva. Vulva opens close to anus. Tail narrows to fine point. Parasite of equids.

Type species: *Coronocyclus coronatus* (Looss, 1900) Hartwich, 1986

[The diagnosis is based on Hartwich (1986) and Dvoĭnos *et al.* (1994). Listed in Spencer Jones and Gibson (1987).]

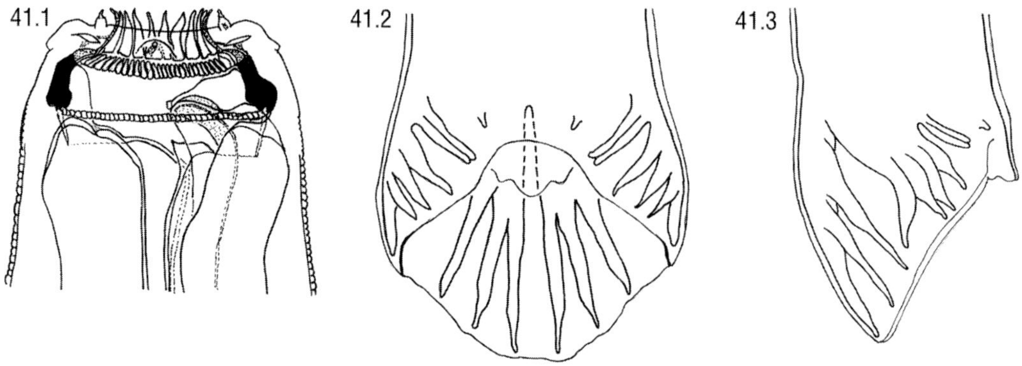

FIG. 41 *Coronocyclus* 41.1 Anterior end, lateral view; 41.2 Male tail, ventral view; 41.3 Male tail, lateral view. (Redrawn after Dvoĭnos *et al.*, 1994.)

Parapoteriostomum Hartwich, 1986 (Fig. 42)

DIAGNOSIS: Oral collar: large; distinct demarcation from body; distal margin at level of proximal rim of buccal capsule. Cephalic sensory organs: submedian papillae large, conical with narrow tip separated by notch; amphids not prominent. Buccal capsule: circular to dorso-ventrally oval in cross-section; wider than deep; narrows anteriorly; wall of even thickness, curves medially, distal edge with indentation; dorsal gutter absent or slightly dome-shaped; extra-chitinous support of external leaf crown extends from external margin of proximal rim of buccal capsule to posterior edge of mouth collar. Leaf crowns: external leaf crown (ELC) 38–60 elements, sharply pointed, formed directly from oral collar; internal leaf crown elements (ILC) 30–46, longer than ELC elements, formed from proximal rim of buccal capsule. Oesophagus short; oesophageal funnel with 3 teeth which do not protrude into buccal capsule. Excretory pore opens, in region of distal half of oesophagus. Cervical papillae near excretory pore. *Male*: Body length up to 10.9 mm. Bursa: large, rounded dorsal lobe present; dorsal ray divided into 3 branches either side of a median fissure, extends to level of anterior most pair of branches; externo-dorsal ray arises from main stem of dorsal ray; lateral rays with common stem, postero-lateral separate from antero- and medio-laterals rays; ventral rays with common stem, parallel, remain close together. Pre-bursal papillae present. Spicules with barbed tips. Gubernaculum simple, groove-like. Genital cone occasionally prominent, long. *Female*: Body length up to 16.2 mm. Vulva opens near anus. Tail length longer than distance from vulva to anus, conical. Parasites of caecum and colon of equids.

Type species: *Parapoteriostomum mettami* (Leiper, 1913) Hartwich, 1986

[Listed in Spencer Jones and Gibson (1987).]

Syngamidae Leiper, 1912

Syngaminae Baylis and Daubney, 1926

Cyathostoma Blanchard, 1849

DIAGNOSIS: Oral collar absent. Buccal capsule: cup-shaped; teeth at base. Males and females separate, rarely in permanent copulation. *Male*: Bursa with dorsal lobe longer than lateral lobes;

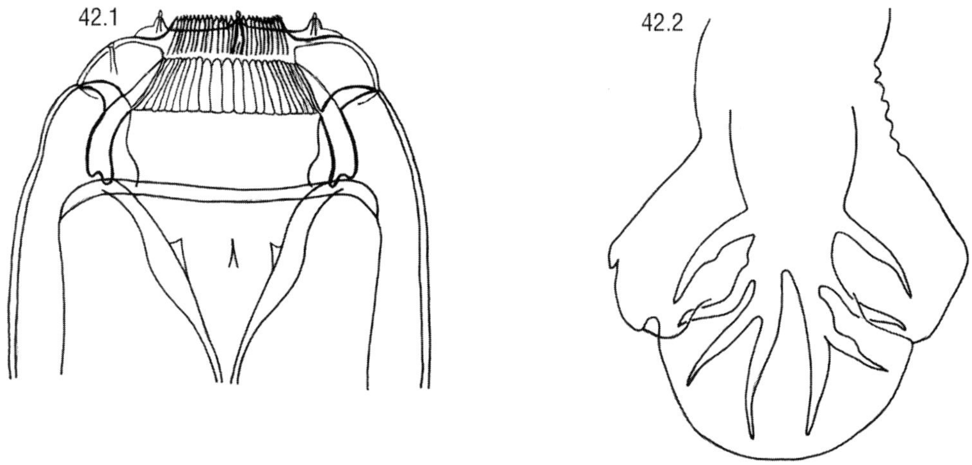

FIG. 42 *Parapoteriostomum* 42.1 Anterior end, latero-dorsal view; 42.2 Dorsal ray and lobe of bursa. (Redrawn after Hartwich, 1994.)

externo-dorsal rays shorter than dorsal ray; dorsal ray may project beyond bursal margin in thorn-like projection. Spicules: 0.08–0.88 mm long. *Female*: Eggs with 1–2 opercula. Parasites of the respiratory tract of birds.

Type species: *Cyathostoma lari* Blanchard, 1849

[The emended diagnosis to include species with spicule lengths greater than 0.08 mm is based on Bowie (1985). In his key, Lichtenfels (1980a), divides the species into two subgenera on the basis of spicule length and whether the dorsal ray projects beyond the bursal margin.]

Chabertiidae (Popova, 1952 subfamily) Lichtenfels, 1980

Chabertiinae Popova, 1952

Ancistronema Smales, 1992 (Fig. 43)

DIAGNOSIS: Oral collar present with lateral amphids and 4 submedian cephalic papillae. Mouth: circular; terminal; external leaf crown with 16 slender 'triangular' elements; internal leaf crown with 16 hook-like elements. Buccal capsule wider than deep. Oesophageal funnel with 3 teeth. Oesophagus: swollen proximal end; broadly expanded distal end. Excretory pore opens at level of nerve ring. Nerve ring encircles mid-oesophagus. *Male*: Bursa: lobes joined; ventral rays with common stem, parallel, close together; lateral rays with common stem, antero-lateral separate from other laterals; externo-dorsal rays arise from main stem of dorsal; dorsal ray divides in distal half, each branch dividing to form a short external branch. Genital cone: single ventral papilla (0); 2 rounded dorsal 'raylets' (7). *Female*: Vulva opens near anus. Ovejectors type II, parallel to longitudinal axis of body, vestibule and sphincters equal in length, infundibulum shorter, *vagina vera* enters ovejector ventrally. Tail tapers to point. Parasite of colon and caecum of *Rattus sordidus*, Muridae, in Australia.

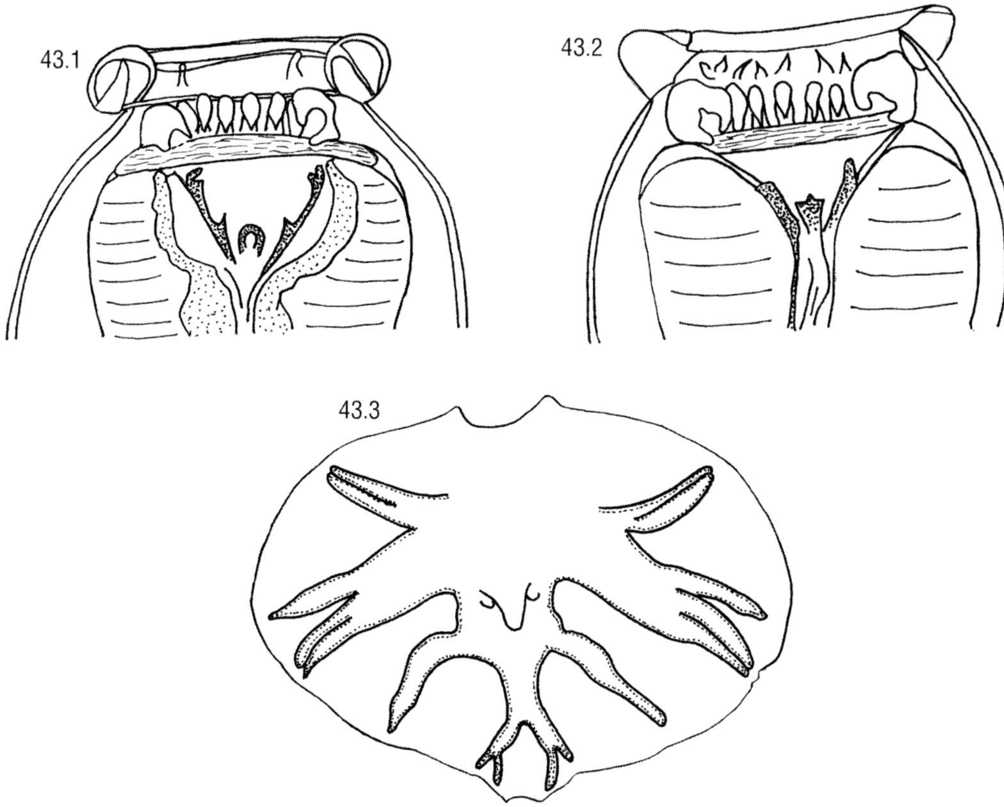

FIG. 43 *Ancistronema* 43.1 Anterior end, ventral view; 43.2 Anterior end, lateral view; 43.3 Bursa. (Redrawn after Smales, 1992.)

Type species: *Ancistronema coronatum* Smales, 1992

[Smales (1992) established the genus for *Ancistronema coronatum* from the Cane field rat, *Rattus sordidus* in Australia. The genus was distinguished from *Cyclodontostomum* Adams, 1933 by the presence of an external leaf crown. Hasegawa and Syafruddin (1994b) established that *Cyclodontostomum* has an external leaf crown, that *Ancistronema* has a slightly inclined cephalic collar and confirmed that both genera have different-sized hook-like elements in the internal leaf crown. Consequently, Hasegawa and Syafruddin (1994b) consider *Ancistronema coronatum*, the type species of the genus, a synonym of *Cyclodontostomum purvisi*, and the genus *Ancistronema* therefore becomes a synonym of *Cyclodontostomum* Adams, 1933.]

Chabertiella Tadros, 1964 (Fig. 44)

DIAGNOSIS: Mouth opens antero-ventrally. Internal and external leaf crowns 'armed with spines'. Internal leaf crown with twice the number or elements of the external leaf crown. Internal surface of buccal capsule with 'spine-like markings'. Teeth absent in oesophageal funnel. *Male*: Bursa: ventral rays parallel, remain close together, reach bursal margin; antero-lateral thicker and separate from other laterals, medio- and postero-laterals parallel, remain close together;

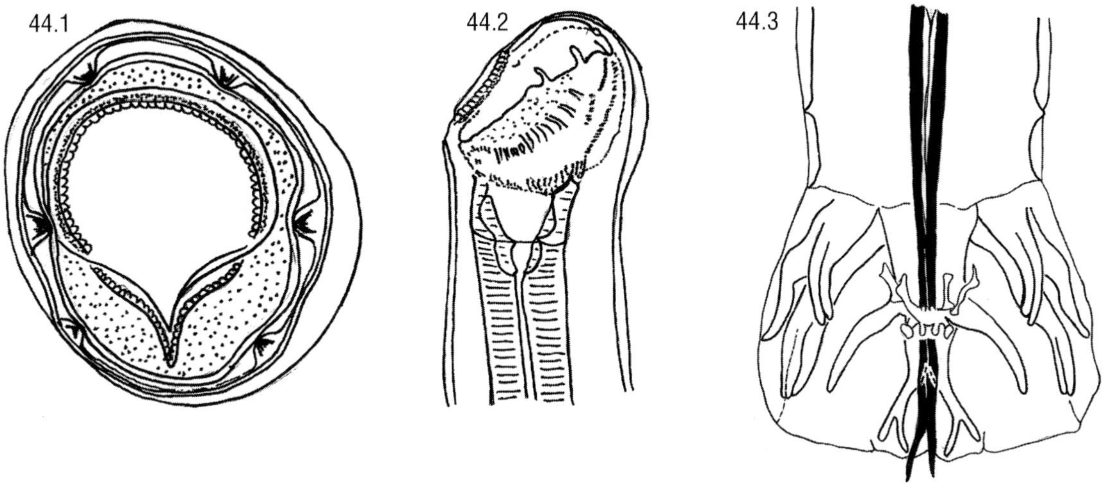

FIG. 44 *Chabertiella* 44.1 Head, *en face* view; 44.2 Anterior end, lateral view; 44.3 Bursa, ventral view. (Redrawn after Tadros, 1964.)

externo-dorsal ray arises from main stem of dorsal ray; dorsal ray divides in distal half, each branch with short external branch. Spicules equal, alate, simple distal tips. Gubernaculum and telamon present. Female: Vulva opens near anus. *Vagina vera* very long. Parasite in intestine of the giraffe, *Giraffa camelopardalis*, in Thika, Kenya.

Type species: *Chabertiella pesteri* Tadros, 1964

[Tadros (1964) distinguishes this genus from *Chabertia* on the basis of spines on the leaf crowns, the spine-like markings on the buccal capsule, both the ventral rays reaching the bursal margin, the shape of the gubernaculum, presence of a telamon, size of gubernaculum, spicules and eggs. The author placed the genus in the subfamily Oesophagostominae but, in view of its similarities to *Chabertia*, it is moved to the subfamily Chabertiinae. Listed in Spencer Jones and Gibson (1987).]

Oesophagostominae Railliet, 1916

Oesophagostomum Molin, 1861

[Le Roux (1940) proposed two additional subgenera, *Hudsonia* and *Pukuia*. Popova (1958) mentioned these subgenera in the 'historical review of the genus', and they are listed in Spencer Jones and Gibson (1987). Popova (1958) did not include them in the classification of *Oesophagostomum*.]

Subgenus *Lerouxiella* Chabaud and Durette-Desset, 1973 (Fig. 45)

DIAGNOSIS: Ten to 12 elements in external leaf crown. Buccal capsule with internal projections, narrows anteriorly. Oesophageal funnel with 3 poorly armed concave valves. Cervical papillae median. Parasite of African rodents.

Type species: *Oesophagostomum* (*Lerouxiella*) *xeri* Ortlepp, 1922

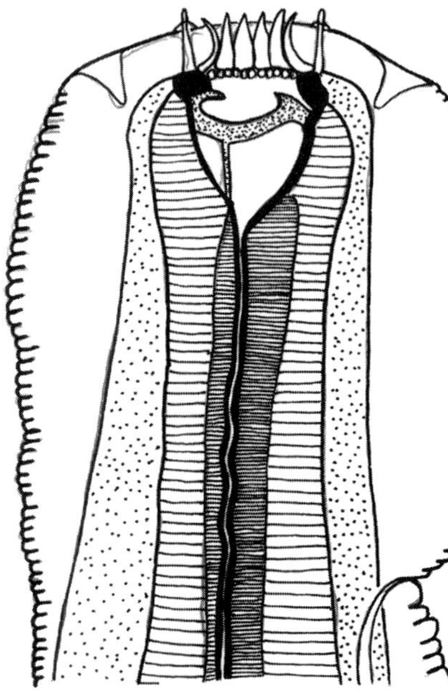

FIG. 45 *Lerouxiella* Anterior end. (Redrawn after Ortlepp, 1922.)

Wuia K'ung, 1959 (Fig. 46)

DIAGNOSIS: Mouth terminal. Oral collar: distinctly separated from body, incorporates cephalic papillae. External and internal leaf crowns present. Buccal capsule: short; cylindrical. Oesophageal funnel: small; teeth absent. Cephalic vesicle and cervical groove absent. *Male*: Bursa: ventral rays parallel, close together; laterals arise from common stem; externo-dorsal and dorsal rays arise from common stem; dorsal ray divides for approximately half its length, each branch dividing again to form short, lateral branch. Spicules equal. Gubernaculum present. *Female*: Vulva opens near anus. Parasite of pigs, *Sus scrofa*.

Type species: *Wuia hsiungi* (Ling, 1959) K'ung, 1959

[K'ung (1959) separates the genus from *Oesophagostomum* on the basis of the absence of a cephalic vesicle and cervical groove. This genus was not referred to in the *CIH Keys* but listed in Spencer Jones and Gibson (1987).]

Neorhabditostomum Gibbons and Kan, 1987 (Fig. 47)

DIAGNOSIS: Leaf crown: single; 10 elements surrounding mouth opening. Transverse cervical groove entirely circles body. Oesophageal funnel non-segmented, elongate. Cervical papillae close to distal end of oesophagus. *Male*: Spicules: slender; alate at distal end. *Female*: Vagina short. Vulva opens near anus, not evaginated. Parasites of barking deer, *Muntiacus muntjak*.

Type species: *Neorhabditostomum yeni* Gibbons and Kan, 1987

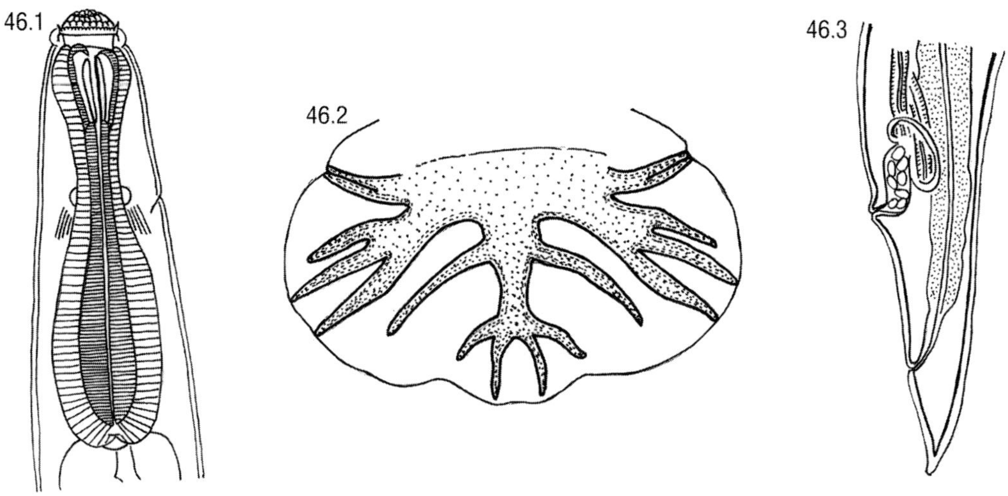

FIG. 46 *Wuia* 46.1 Anterior end; 46.2 Bursa; 46.3 Posterior end of female, lateral view. (Redrawn after Ling, 1959.)

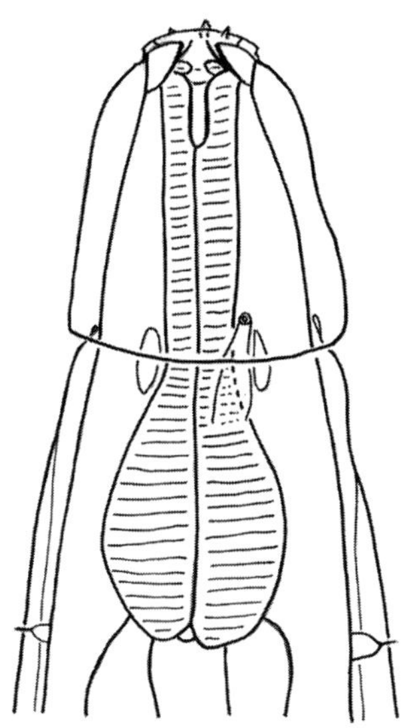

FIG. 47 *Neorhabditostomum* Anterior end of female, ventral view. (Redrawn after Gibbons and Kan, 1987.)

Cloacinidae (Stossich, 1899 subfamily) Travassos, 1919

DIAGNOSIS: Two leaf crowns absent or, if present, associated with type I ovejectors. Buccal capsule: cylindrical or globular. *Male*: Bursa: dorsal ray with 2 pairs of branches each side of median fissure. Paired elongate lateral thickening with central cordate thickening at the junction of the 2 spicular sheaths present. *Female*: Ovejectors of type I or II, transitional forms may also be present.

[Beveridge (1987) reinstated the family Cloacinidae following phylogenetic studies on Australian Strongyloidea occurring in the stomach, oesophagus and large intestine of kangaroos and wallabies and in the large intestine of wombats. The classification is reassessed on the hypothesis that the Australian Strongyloidea are monophyletic.]

Cloacininae Stossich, 1899

DIAGNOSIS: Buccal capsule: cylindrical or ring-like; dorsal gutter absent. *Male*: Bursa: externodorsal ray which arises near base of lateral rays. *Female*: Ovejectors type II.

[The emended diagnosis is based on Beveridge (1987). He includes the tribes Cloacininea (Stossich, 1899 subfamily); Macropostrongylinea Lichtenfels, 1980; Zoniolaiminea (Popova, 1952 subfamily); Labiostrongylinea Beveridge, 1983; Pharyngostrongylinea Popova, 1952; and Coronostrongylinea Beveridge, 1986.]

Genera not assigned to a tribe

Maplestonema Johnston and Mawson, 1939b (Fig. 48)

DIAGNOSIS: Oral collar absent. Cephalic sensory organs: 6 equal, elongate papillae. Buccal capsule deeper than wide, narrow chitinous ring near base. Oseophagus uniform width along length. *Male*: Unknown. *Female*: Vulva opens near anus. Tail tapers to a point. Parasite in stomach of marsupials, *Macropus* spp.

Type species: *Maplestonema typicum* Johnston and Mawson, 1939b

[The authors place the genus in the subfamily Trichonematinae and distinguish their genus from *Buccostrongylus* Johnston and Mawson, 1939a and *Spirostrongylus* Yorke and Maplestone, 1926 (not Mönnig, 1926). *Buccostrongylus* was considered by Kung (1948) to be a synonym of *Zoniolaimus* (see Lichtenfels, 1980a). *Zoniolaimus* and *Spirostrongylus* are placed in the subfamily Cloacininae in Lichtenfels (1980a), and so we have placed this genus in the same subfamily. This genus is listed in Spencer Jones and Gibson (1987). However, Smales (1983) recorded that, following examination of additional specimens by Mawson (Mawson personal communication in Smales, 1983), the type species represented larval

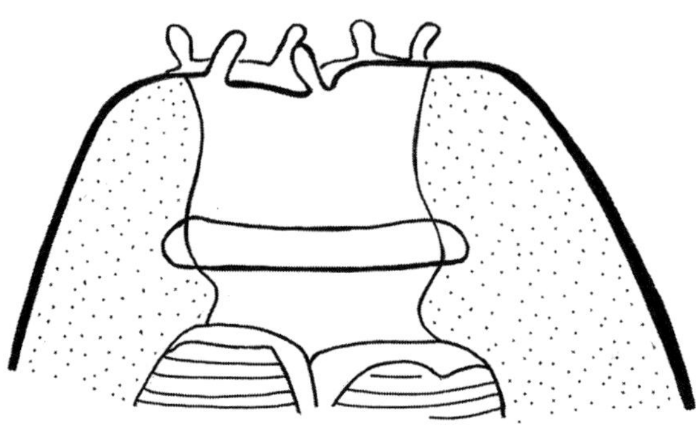

FIG. 48 *Maplestonema* Cephalic region. (Redrawn after Johnston and Mawson, 1939.)

stages of an unknown *Cloacina* species and, as such, is regarded as a *species inquirenda* by Beveridge (1998b).]

Alocostoma Mawson, 1979 (Fig. 49)

DIAGNOSIS: Oral collar well developed. Cephalic sensory papillae: submedian papillae cylindrical, truncated. Circumoral membrane and proximal part of buccal capsule finely striated. Buccal capsule lightly sclerotized. Oesophagus: long, slender with distal bulb. *Male*: Bursa: not joined ventrally; ventral rays with common stem, parallel, close together; lateral rays with common stem, antero-lateral shorter and divergent from other laterals, which remain close together; externo-dorsal arises from base of laterals; dorsal ray divides for half its length, each division with short lateral branch. Spicules: equal; alate. Gubernaculum present. *Female*: Vulva opens close to anus. Ovejectors opposed, parallel to longitudinal axis. Tail tapers to a blunt point. Parasites of macropodid marsupials.

Type species: *Alocostoma clelandi* (Johnston and Mawson, 1939) Mawson, 1979

[Lichtenfels (1980a), in his discussion of the subfamily Cloacininae, draws attention to Mawson (1977), in which she indicates that a new genus is to be described for *Cyclostrongylus clelandi* Johnston and Mawson, 1939. Mawson (1979) places the genus in the family Trichonematidae; however, the classification of Lichtenfels (1980a) is now widely accepted and the genus is included here in the Cloacininae following his classification. Listed in Spencer Jones and Gibson (1987).]

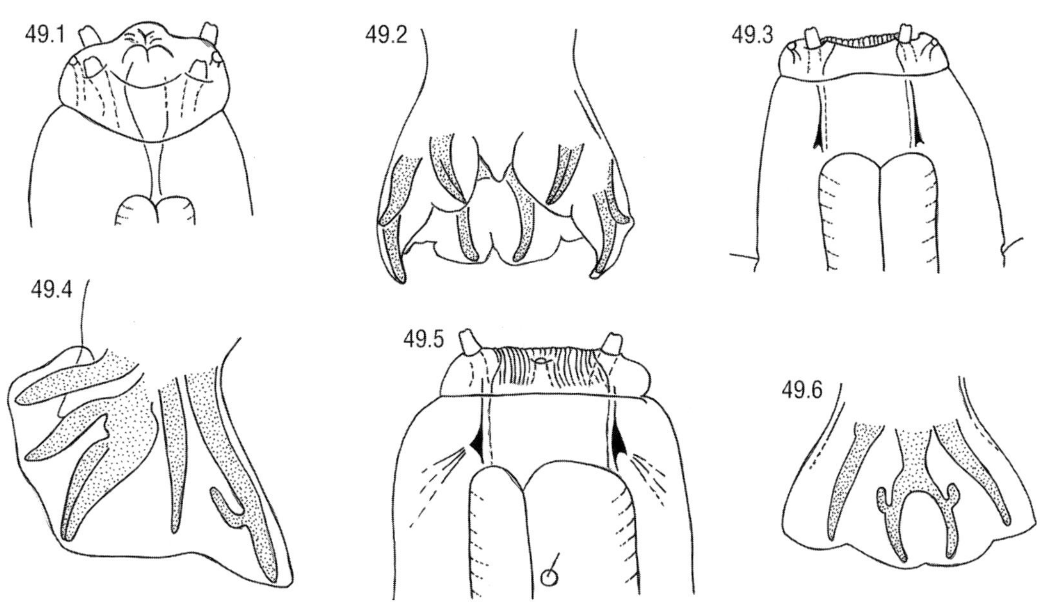

FIG. 49 *Alocostoma* 49.1 Cephalic region, mouth closed; 49.2 Bursa, ventral view; 49.3 Cephalic region, ventral view; 49.4 Bursa, lateral view; 49.5 Cephalic region, lateral view; 49.6 Dorsal lobe of bursa. (Redrawn after Mawson, 1979.)

Beveridgea Mawson, 1980 (Fig. 50)

DIAGNOSIS: Oral collar well developed. Cephalic sensory organs: 4 submedian papillae project above mouth collar, constriction separates distal third of papillae. Buccal capsule: cylindrical, deeper than wide; anterior margin notched. Eight leaf crown elements arise near anterior rim of buccal capsule. Cervical papillae needle-like, at level of anterior half of oesophagus. Dorsal gutter just projects into base of buccal capsule. *Male*: Bursa: not deeply lobed; ventral rays with common stem, parallel, remain close together; lateral rays with common stem, anterolateral separate from other laterals; externo-dorsal arises separately between lateral and dorsal rays; dorsal ray divides for half its length, each branch dividing at distal tip. Spicules alate. Gubernaculum present. *Female*: Vulva opens near anus. Ovejectors type II, uteri parallel with longitudinal axis of body. Tail short, tapers to blunt point. Parasite in the stomach of macropodid marsupials.

Type species: *Beveridgea corneri* Mawson, 1980

[Listed in Spencer Jones and Gibson (1987).]

Cloacininea (Stossich, 1899 subfamily) Beveridge, 1987

Cloacina Linstow, 1898

[Beveridge (1998b) revised the genus *Cloacina*, evaluated all the morphological characters, recognizing 90 species. A key to species and host/parasite checklist are included.]

Cervonemella Beveridge, 2001 (Fig. 51)

DIAGNOSIS: Cuticle with fine annulations. Oral collar present. Cephalic sensory organs: 4 submedian papillae with proximal and distal segments, incurved; 2 lateral amphids.

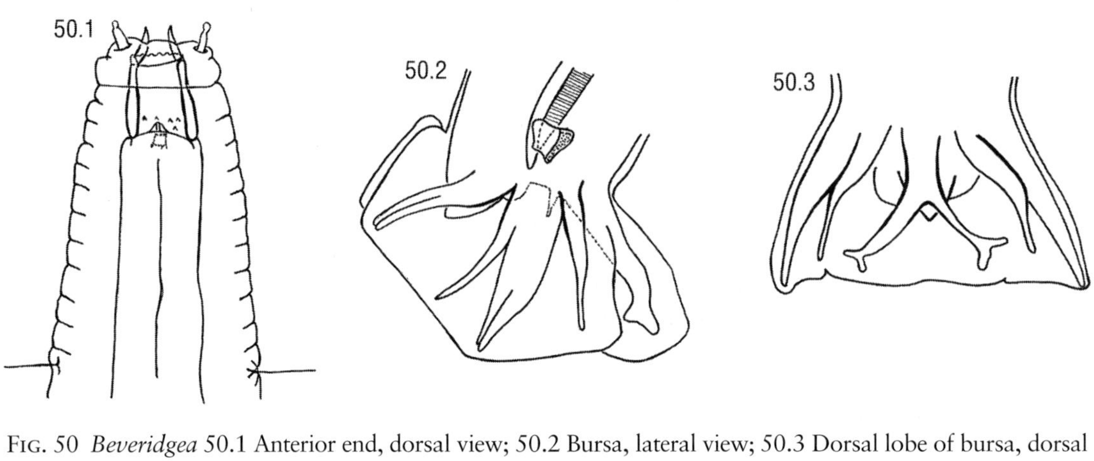

FIG. 50 *Beveridgea* 50.1 Anterior end, dorsal view; 50.2 Bursa, lateral view; 50.3 Dorsal lobe of bursa, dorsal view. (Redrawn after Mawson, 1980.)

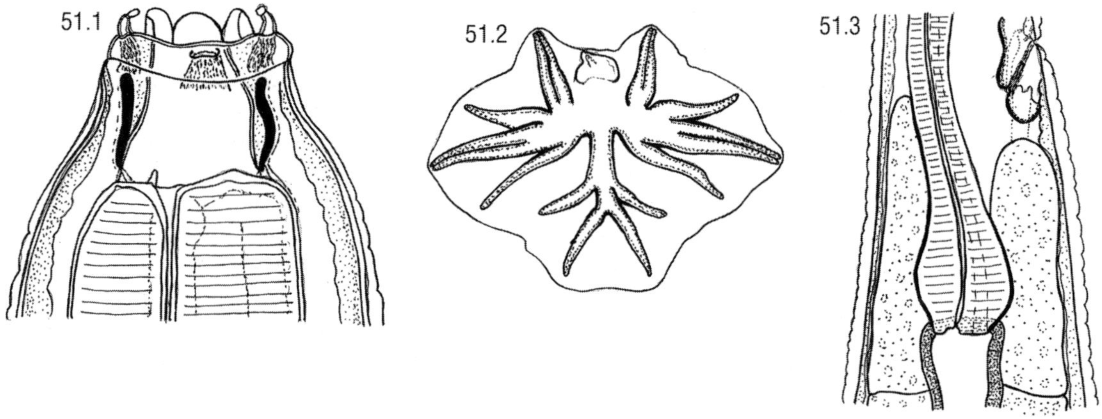

FIG. 51 *Cervonemella* 51.1 Anterior end; 51.2 Bursa; 51.3 Distal end of oesophagus surrounded by elongation of intestinal cells. (Redrawn and modified after Beveridge, 2001.)

Mouth: oval; internal leaf crown with 8 elements. Buccal capsule: sclerotized; as wide as deep. Oesophagus: clavate; surrounded by extension of intestinal cells at oesophago-intestinal junction. *Male*: Bursa: short; distinct lobes; ventral rays with common stem, parallel, remain close together; lateral rays with common stem, antero-lateral separate and shorter, medio- and postero-lateral rays parallel, remain close together; externo-dorsal ray arises from base of laterals; dorsal ray with 4 branches. Genital cone: single ventral papilla on cone, dorsal raylets form papillae. Spicules: alate; lateral and median sheaths thickened. Gubernaculum present. *Female*: Vulva opens close to anus. Ovejectors: type II; sphincters, infundibula short. Tail conical. Parasite in stomach of macropodid marsupials in Papua New Guinea.

Type species: *Cervonemella reardoni* Beveridge, 2001

Zoniolaiminea (Popova, 1952 subfamily)

DIAGNOSIS: Small body. Alae absent. Mouth: with 4 lips; teeth arranged in multiples of 4 at proximal margin of buccal capsule; or external leaf crown with many tiny elements. Buccal capsule: cylindrical; walls maybe lightly striated. Oesophagus: elongate corpus and ovoid distal bulb or broad corpus and short, narrow cylindrical distal bulb. *Male*: Bursa: lobes distinct. Bursal rays: ventrals with common stem, parallel, remain close together; laterals with common stem, antero-lateral separate from other laterals, medio- and postero-laterals parallel, remain close together; externo-dorsal arises close to base of laterals; dorsal ray divides in mid-region, each branch with short lateral branch at level of division. Spicules: alate. *Female*: Type II ovejectors, *vagina vera* opening on ventral side. Parasites in the stomach of macropodid marsupials.

Type genus: *Zoniolaimus* Cobb, 1898

[Emended diagnosis after Beveridge (1983).]

Zoniolaimus Cobb, 1898 (= *Buccostrongylus* Johnston and Mawson, 1939a)

DIAGNOSIS: Cuticle: may be thickened in cervical region; fine transverse striations present. Oral collar: present or absent. Cephalic sensory organs: four setigerous submedian papillae level with or raised above body surface; 2 lateral amphids. Mouth: circular or quadrangular with 4 submedian lips; 2 lateral or 4 interlabia (2 lateral, 1 dorsal, 1 ventral) present or absent; submedian lips generally associated with tooth-like projections from internal wall of buccal capsule. Buccal capsule: cylindrical or bell-shaped; moderately sclerotized. Oesophagus: long, narrow; fusiform median bulb posterior to nerve ring; distal ovoid bulb with 3 pairs of sclerotized plates associated with valve. Excretory pore and setiform cervical papillae at level of nerve ring. Intestine: proximal end thick-walled. *Male*: Bursa: lobes equal in length; bosses absent on ventral surface. Bursal rays: ventrals with common stem, parallel, remain close together; laterals with common stem, antero-lateral separate and shorter than other laterals, medio- and postero-laterals parallel, remain close together; externo-dorsal arises close to base of laterals; dorsal ray divides in mid-region, each branch with a shorter lateral branch at level of division. Genital cone: ventral cone with papilla (0) at tip, pair of bilobed dorsal raylets (7), additional genital appendages may be present. Spicules: main stem with wide, transversely striated alae, narrow towards distal tip or finish abruptly; distal tips bluntly pointed. Gubernaculum: present or absent. *Female*: Vulva opens near anus. Ovejectors type II, vestibules short, *vagina vera* thick-walled, enters ovejector on ventral side. Tail conical. Parasites of stomach of macropodid marsupials.

Type species: *Zoniolaimus setifera* Cobb, 1898

[Emended diagnosis based on Beveridge (1983). Beveridge (1983) upholds the synonymy of *Buccostrongylus* Johnston and Mawson, 1939 with *Zoniolaimus* proposed by Kung (1948) on the grounds that the type species, *Buccostrongylus buccalis* Johnston and Mawson, 1939 and *Zoniolaimus setifera*, are similar. *Buccostrongylus australis* Johnston and Mawson, 1939 and *B. labiatus* Johnston and Mawson, 1939 are transferred by Beveridge (1983) to his new genus *Wallabinema*, and *Buccostrongylus setifer* Johnston and Mawson, 1939 to his new genus *Thallostonema* (see below). Lichtenfels (1980a) did not include *Buccostrongylus* in his key as the systematics of the group were far from settled at the time of its preparation.]

Wallabinema Beveridge, 1983 (Fig. 52)

DIAGNOSIS: Cuticle: sometimes thickened in cephalic region; fine annular striations present. Oral collar present. Cephalic sensory organs: 4 submedian cephalic sensory papillae, slightly raised above body surface with 1–2 small setae; 2 lateral amphids. Mouth: circular or dorso-ventrally elongate; 4 lips (2 lateral, 1 dorsal, 1 ventral) without tooth-like supports; interlabia absent. Buccal capsule: cylindrical, moderately sclerotized, deeper than wide. Oesophagus: long or short; median fusiform bulb absent; distal bulb with 3 pairs of sclerotized plates. Nerve ring located in anterior oesophageal region or isthmus. Excretory pore opens in mid-oesophageal region or isthmus. Cervical papillae located in mid-oesophageal region. Proximal end of intestine thick-walled. *Male*: Bursa: lobes similar in length, bosses absent from ventral surface. Bursal rays: ventrals with common stem, parallel, remain together; laterals with common stem, antero-lateral shorter and separate from other laterals, medio- and postero-laterals remain close together, externo-dorsal arises close to base of laterals, dorsal ray divides in mid-region, each branch with a shorter lateral branch at level of division. Genital cone: ventral cone with papilla (0) at tip; dorsal raylets (7) associated with occasional additional genital appendages. Spicules:

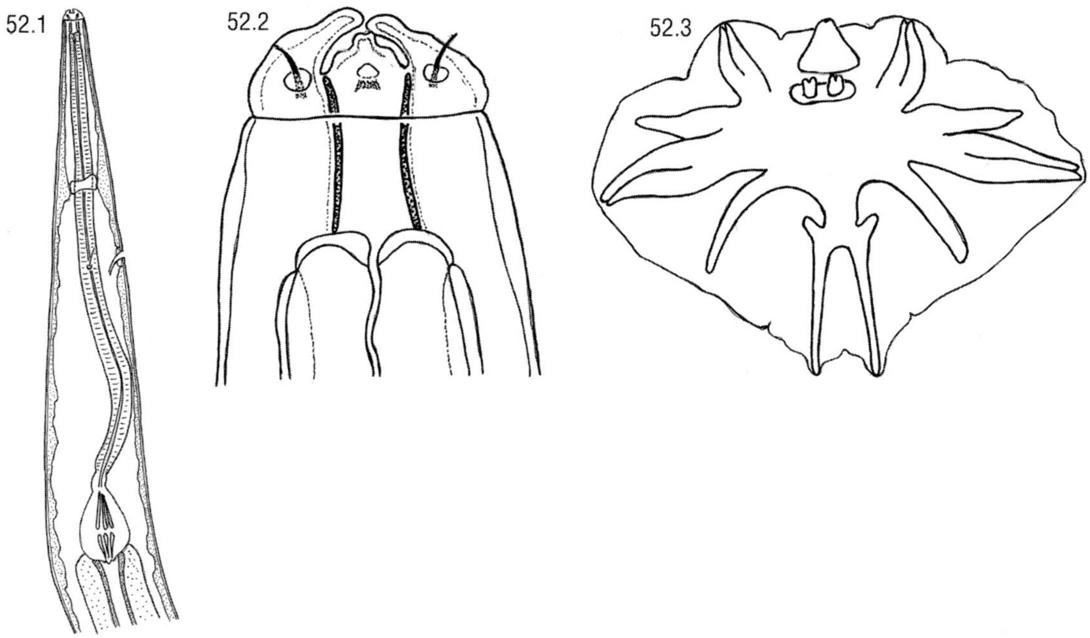

FIG. 52 *Wallabinema* 52.1 Anterior end, lateral view; 52.2 Cephalic region, lateral view; 52.3 Bursa. (Redrawn after Beveridge, 1983.)

main stem with wide, transversely striated alae, narrowing towards distal tip or finishing abruptly; distal tips blunt. Gubernaculum absent. *Female*: Vulva opens near anus. Ovejectors type II, vestibules short. Tail conical. Parasites in stomach of macropodid marsupials.

Type species: *Wallabinema australis* (Johnston and Mawson, 1939) Beveridge, 1983

[Listed in Spencer Jones and Gibson (1987).]

Thallostonema Beveridge, 1983 (Fig. 53)

DIAGNOSIS: Cuticle: fine annular striations present. Oral collar poorly developed. Cephalic sensory organs: 4 submedian cephalic papillae, slightly raised above body surface each with single or bifid seta(e); 2 lateral amphids. Mouth: circular, trapezoidal or dorso-ventrally elongate; lips rudimentary or absent. 'Tooth-like projections' present on proximal rim of buccal capsule. Buccal capsule: cylindrical or dorso-ventrally elongate in cross-section; deeper than wide; thick-walled, fine, irregular transverse striations present or absent. Oesophagus: long, wide corpus; short, narrow isthmus; slender, cylindrical bulb. Nerve ring and excretory pore at level of isthmus. Cervical papillae near distal margin of buccal capsule. *Male*: Bursa: lobes similar in length; lateral lobes divided between ventral and lateral rays; ventral surface with sclerotized bosses. Bursal rays: ventral rays with common stem, parallel, remain close together; laterals with common stem, antero-lateral shorter than other laterals, medio- and postero-laterals parallel, remain close together; externo-dorsal ray arises close to base of laterals; dorsal ray thick-stemmed, divides in mid-region, each branch with a shorter lateral branch. Genital cone: ventral cone, papilla (0) at tip; dorsal bilobed raylets (7) with occasional additional genital appendages. Spicules: transversely striated

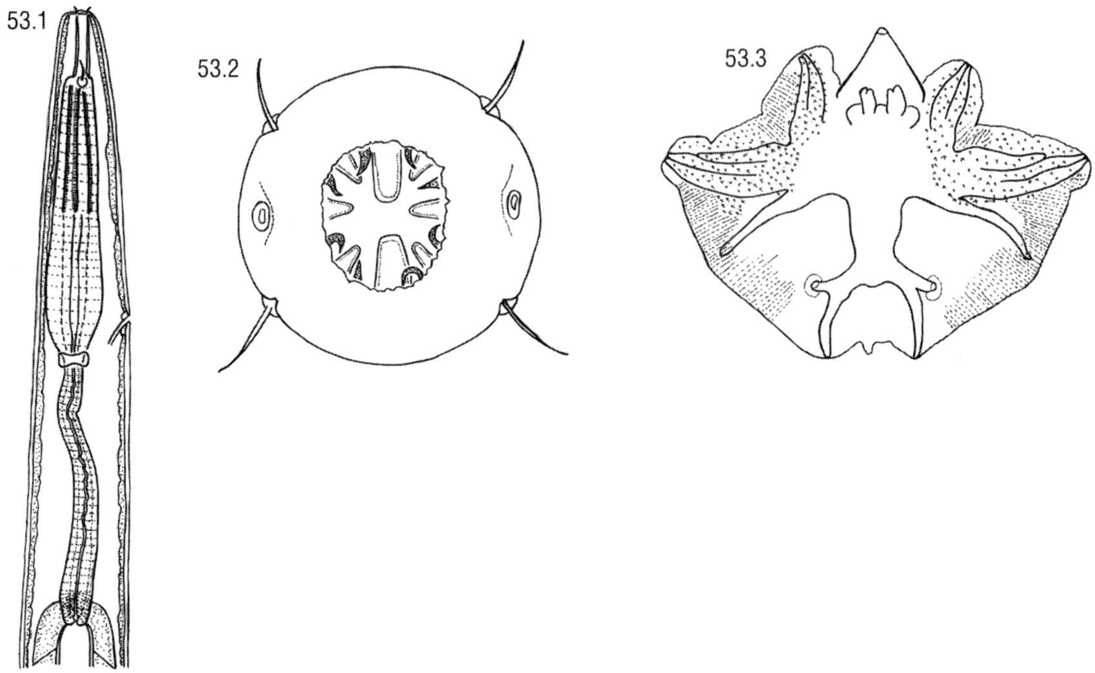

FIG. 53 *Thallostonema* 53.1 Anterior end, lateral view; 53.2 Cephalic region, *en face* view; 53.3 Bursa. (Redrawn after Beveridge, 1983.)

alae present; distal tips blunt. Gubernaculum present. *Female*: Vulva opens near anus. Ovejectors type II, vestibules short. Tail conical. Parasites in stomach of macropodid marsupials.

Type species: *Thallostonema lichtenfelsi* Beveridge, 1983

[Listed in Spencer Jones and Gibson (1987).]

Tethystrongylus Beveridge, 1983 (Fig. 54)

DIAGNOSIS: Cuticle: fine annular striations present. Oral collar absent. Cephalic sensory organs: 4 rounded, setiferous submedian papillae; 2 lateral amphids. Mouth: circular; lips and interlabia absent. Approximately 45 small elements in single leaf crown. Teeth absent. Buccal capsule: cylindrical, deeper than wide; wall well sclerotized, 3 triangular processes at base. Oesophagus: long, slender corpus; distal bulb with 3 pairs of sclerotized plates associated with valve. Nerve ring, excretory pore, cervical papillae located in anterior oesophageal region. *Male*: Bursa: lobes similar in length; ventral surface without bosses. Bursal rays: ventrals with common stem, parallel, remain close together; laterals with common stem, antero-lateral short and separate from other laterals, medio- and postero-laterals parallel, remain close together; externo-dorsal arises near base of laterals; dorsal ray bifurcates in mid-region, each branch with short lateral branch. Genital cone: well-developed ventral cone with papilla (0) at tip; dorsal raylets (7) bilobed, additional genital appendages present. Spicules: wide transversely striated alae present, narrow towards distal tips; distal tips blunt. Gubernaculum absent. *Female*: Vulva

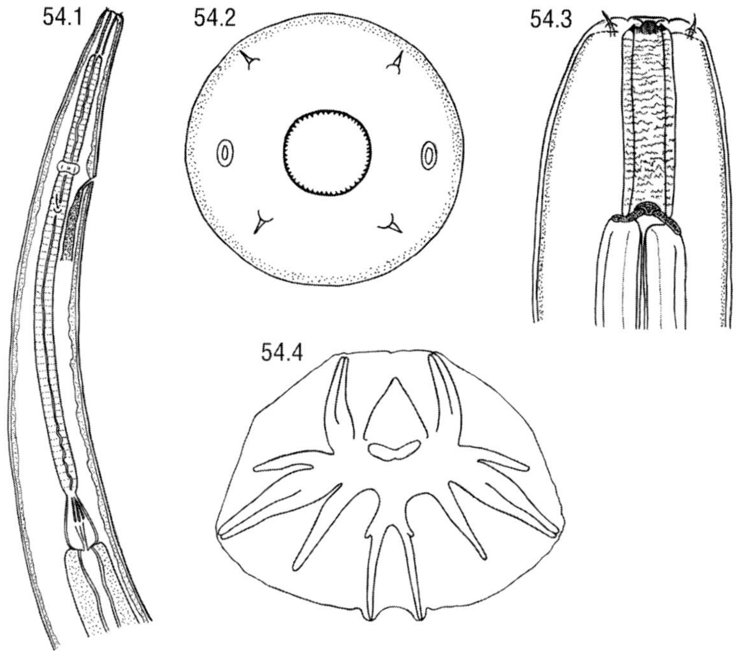

FIG. 54 *Tethystrongylus* 54.1 Anterior end, lateral view; 54.2 Cephalic region, *en face* view; 54.3 Cephalic region, lateral view; 54.4 Bursa. (Redrawn after Beveridge, 1983.)

opens close to anus. Ovejectors type II, vestibules short. Tail conical. Parasites in stomach of macropodid marsupials.

Type species: *Tethystrongylus coronatus* Beveridge, 1983

[Listed in Spencer Jones and Gibson (1987).]

Dorcopsinema Mawson, 1977

DIAGNOSIS: Oral collar wide. Cephalic sensory organs: 4 submedian papillae; 2 lateral amphids. Mouth with 6–8 'lip-like processes'. Buccal capsule cylindrical. Oesophagus not divided into corpus, isthmus and bulb. Oesophago-intestinal junction with paired 'glandular structures arising from intestine'. *Male*: Bursa: not divided; ventral rays with common stem, parallel, remain close together; laterals with common stem, antero-lateral separate, medio- and postero-laterals parallel, remain close together; externo-dorsal arises from base of lateral rays; dorsal ray divides in mid-region, each branch with short external branch. Spicules alate. *Female*: Vulva opens near anus. Ovejectors type II. Tail conical. Parasites in stomach of macropodid marsupials from Papua New Guinea.

Type species: *Dorcopsinema dorcopsis* (Baylis, 1940) Mawson, 1977

[Smales (1982a) emended the diagnosis of this genus but retained it in the same tribe and subfamily.]

Pharyngostrongylinea Popova, 1952

Parapharyngostrongylus Beveridge, 1982 (Fig. 55)

DIAGNOSIS: Body up to 10mm long. Cuticle: numerous transverse striations, alae absent. Oral collar well developed. Cephalic sensory organs: 4 setigerous submedian papillae; 2 lateral amphids; associated with cephalic collar. Labial collar: '4 lateral elements apposed in pairs, 4 submedian elements overlying lateral elements' internal to cephalic collar. Mouth opening laterally flattened. Buccal capsule: dorso-ventrally elongate in cross-section; walls cylindrical, transversely striated; dorsal and ventral teeth on proximal margin. Oesophagus: narrow; corpus long; bulb elongate; walls of intestine form anteriorly directed projections at oesophago-intestinal junction. *Male*: Bursa with distinct lobes, bosses absent on ventral surface. Bursal rays: ventral rays with common stem, close together; laterals with common stem, antero-lateral shorter and slightly separated from other laterals, medio- and postero-laterals closed together; externo-dorsal arises near base of laterals; dorsal ray bifurcates in mid-region, 2 shorter external lateral branches at level of bifurcation. Genital cone: large ventral cone with single papilla (0); 2 paired dorsal raylets (7). Spicules: alate, distal tips bluntly pointed. Gubernaculum present. *Female*: Vulva opens near anus. Type II ovejectors. Tail conical. Parasites in stomach of macropodid marsupials (*Macropus dorsalis*).

Type species: *Parapharyngostrongylus dentatus* Beveridge, 1982

[Listed in Spencer Jones and Gibson (1987).]

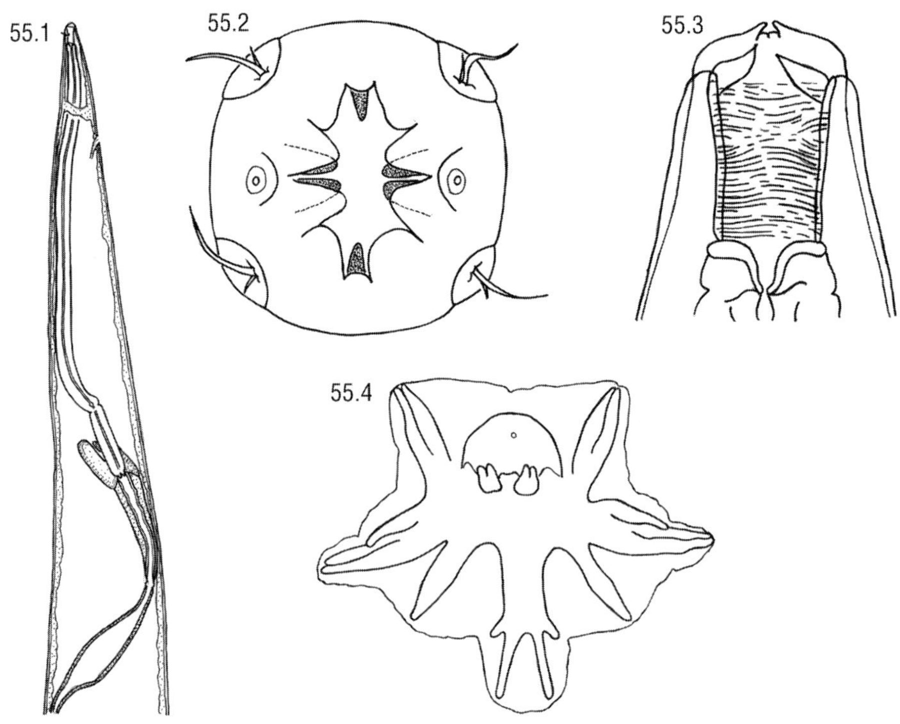

FIG. 55 *Parapharyngostrongylus* 55.1 Anterior end, lateral view; 55.2 Cephalic region, *en face* view; 55.3 Cephalic region, lateral view; 55.4 Bursa. (Redrawn after Beveridge, 1982.)

Rugostrongylus Beveridge, 1982 (Fig. 56)

DIAGNOSIS: Body up to 10 mm long. Cuticle: numerous transverse striations, alae absent. Oral collar well developed. Cephalic sensory organs: 4 setigerous submedian papillae; 2 lateral amphids. Labial collar: not lobed. Mouth opening and buccal capsule (in cross-section) dorso-ventrally elongate. Buccal capsule walls cylindrical, prominently striated, cavity deeper than wide. Oesophagus: narrow; corpus long; bulb elongate; walls of intestine project anteriorly at oesophago-intestinal junction. *Male*: Bursa short, lobes distinct; ventral surface with conical, sclerotized bosses. Bursal rays: ventrals with common stem, parallel, remain close together; laterals with common stem, antero-lateral short, medio- and postero-laterals parallel, remain close together; externo-dorsal ray arises near base of laterals; dorsal ray divides for two-thirds of its length, 2 short branches at level of bifurcation. Genital cone: large ventral cone with single papilla (0), 2 paired dorsal raylets (7). Spicules: alate, distal tips bluntly pointed. Gubernaculum absent. *Female*: Vulva opens near anus. Type II ovejectors. Tail conical. Parasites in stomach of macropodid marsupials.

Type species: *Rugostrongylus labiatus* (Davey and Wood, 1938) Beveridge, 1982

[Listed in Spencer Jones and Gibson (1987).]

Thylostrongylus Beveridge, 1982 (Fig. 57)

DIAGNOSIS: Body up to 5 cm long. Cuticle: fine transverse striations; alae absent. Cephalic collar present. Cephalic sensory organs: 4 setigerous submedian papillae; 2 lateral amphids.

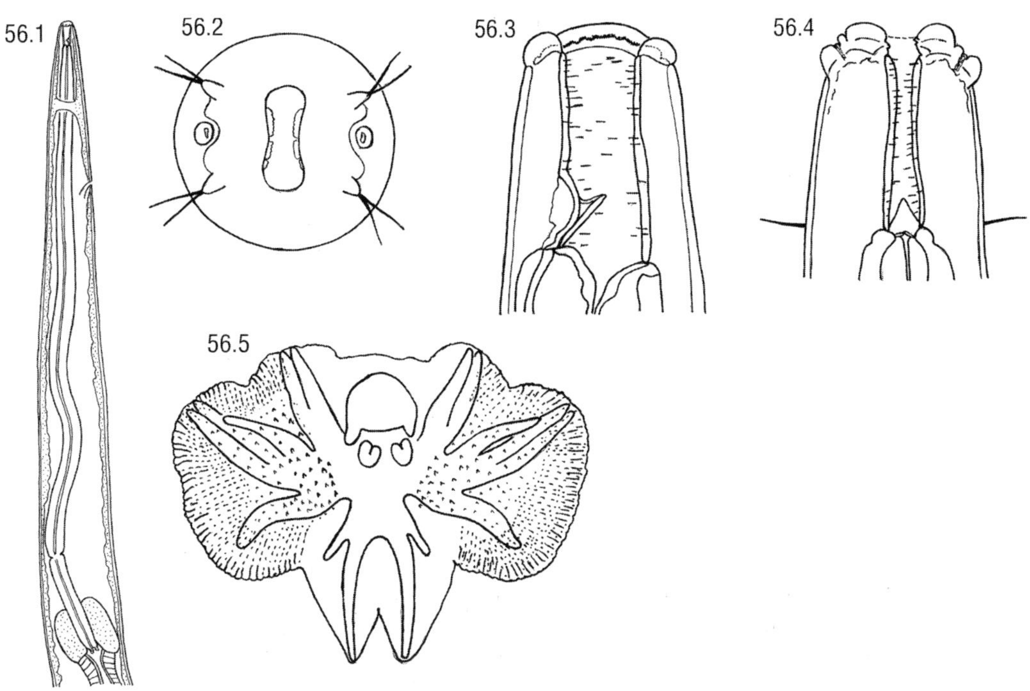

FIG. 56 *Rugostrongylus* 56.1 Anterior end; 56.2 Cephalic region, *en face* view; 56.3 Optical section of cephalic region, lateral view; 56.4 Cephalic region, ventral view; 56.5 Bursa. (Redrawn after Beveridge, 1982.)

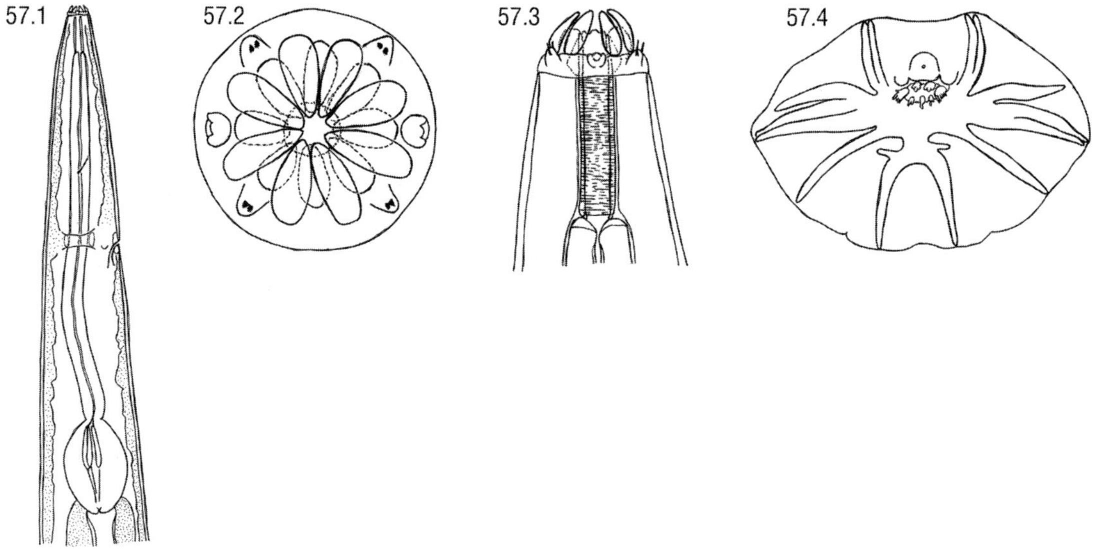

FIG. 57 *Thylostrongylus* 57.1 Anterior end, lateral view; 57.2 Cephalic region, *en face* view; 57.3 Cephalic region, lateral view; 57.4 Bursa. (Redrawn after Beveridge, 1982.)

External leaf crown: 8 elements; internally 8 small lips present or absent. Mouth and buccal capsule (in cross-section) circular. Buccal capsule much deeper than wide, walls cylindrical, heavily sclerotized. Oesophagus: narrow corpus; oval bulb with elongate, sclerotized valve; no anterior projection of intestinal walls at oesophago-intestinal junction. *Male*: Bursa short, lobes indistinct; ventral surface without refractile bosses. Bursal rays: ventral rays with common stem, parallel, remain close together; laterals with common stem, antero-lateral short, separate from other laterals, medio- and postero-laterals parallel, remain close together; externo-dorsal ray arises near base of laterals; dorsal ray divided forming 2 long branches, 2 smaller lateral branches at level of bifurcation. Genital cone: large ventral cone, single ventral papilla (0), dorsal lip with 2 paired dorsal raylets (7) and several branched genital appendages. Spicules: alate, distal tips bluntly pointed. Gubernaculum absent. *Female*: Vulva opens near anus. Type II ovejectors. Tail conical. Parasites in stomach of macropodid marsupials.

Type species: *Thylostrongylus parvus* Beveridge 1982

[Listed in Spencer Jones and Gibson (1987).]

Woodwardostrongylus Wahid, 1964

DIAGNOSIS: Oesophagus: corpus; isthmus; elongate bulb. *Male*: Gubernaculum absent. *Female*: ‘Modified type II ovejectors present’, consistent with modification required by site in host. Parasite in tunnels of superficial squamous epithelium of stomach and oesophagus of macropodid marsupials.

Type species: *Woodwardostrongylus woodwardi* (Wood, 1931)

[Beveridge (1998a) proposes the emendments to the generic diagnosis outlined above. He considers the genus a highly modified member of the tribe.]

Rugonema Beveridge, 1999 (Fig. 58)

DIAGNOSIS: Oral collar present. Mouth with 1 dorsal, 1 ventral, 2 lateral 'flap-like' lips. Cephalic sensory organs: 4 submedian cephalic papillae, dome-like with setae. Buccal capsule transversely striated, proximal region circular in cross-section, appears constricted in mid-region in dorso-ventral view. Oesophagus: corpus sub-cylindrical; isthmus short; bulb elongate without sclerotized plates. *Male*: Bursa: ventral surface with bosses; ventral rays equal, parallel, remain close together; antero-lateral short, distal tip curves away from other laterals, medio- and postero-laterals equal, parallel, remain close together; externo-dorsal arises near base of laterals; dorsal ray divides in distal half, each branch with outer lateral branch. Genital cone: single ventral papilla (0), dorsal raylets (7) bifid. Spicules equal, slender, alate. *Female*: vulva opens near anus. Ovejectors type II. Tail conical. Parasites in stomach of macropodid marsupials.

Type species: *Rugonema labiatum* Beveridge, 1999

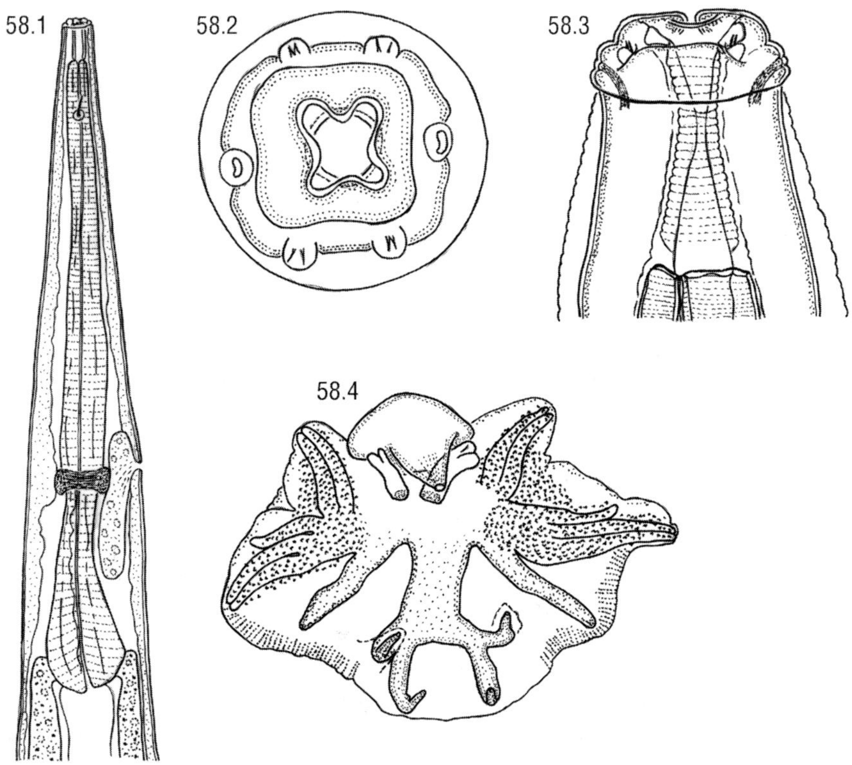

FIG. 58 *Rugonema* 58.1 Anterior end, lateral view; 58.2 Cephalic region, *en face* view; 58.3 Cephalic region, dorsal view; 58.4 Bursa. (Redrawn after Beveridge, 1999.)

Dorcopsistrongylus Smales, 1982 (Fig. 59)

DIAGNOSIS: Oral collar little developed with 2 amphids and 4 submedian cephalic papillae. Mouth with 4 'lip-like elements'. Labial collar present 'arising from internal lining of buccal capsule'. Buccal capsule: cylindrical; deeper than wide; transverse striations present. Oesophagus: corpus elongate; bulb clavate. Oesophago-intestinal junction surrounded by 'extensions of intestine'. *Male*: Bursa: lobes joined; dorsal lobe with medium notch; ventral rays with common stem, parallel, close together; lateral rays with common stem, antero-lateral separate, medio- and postero-laterals parallel, close together; externo-dorsal arises from base of laterals; dorsal ray divides for most of its length, each branch dividing again to form short external branch. Spicules slender, alate. *Female*: Vulva opens near anus. Vestibules orientated at 90° to longitudinal axis of body, uteri directed anteriorly, thick-walled *vagina vera* enters ovejectors on posterior margin. Parasite of macropodid marsupials in Papua New Guinea.

Type species: *Dorcopsistrongylus labiacarinatus* Smales, 1982

[Listed in Spencer Jones and Gibson (1987).]

Macropostrongylinea Lichtenfels, 1980

DIAGNOSIS: Small body. Alae absent. Mouth: round or triangular. Labial collar, labial crown, lobed labial region or external leaf crown present or absent. Buccal capsule: slightly sclerotized with transparent lining; circular, triangular or dorso-ventrally flattened in cross-section;

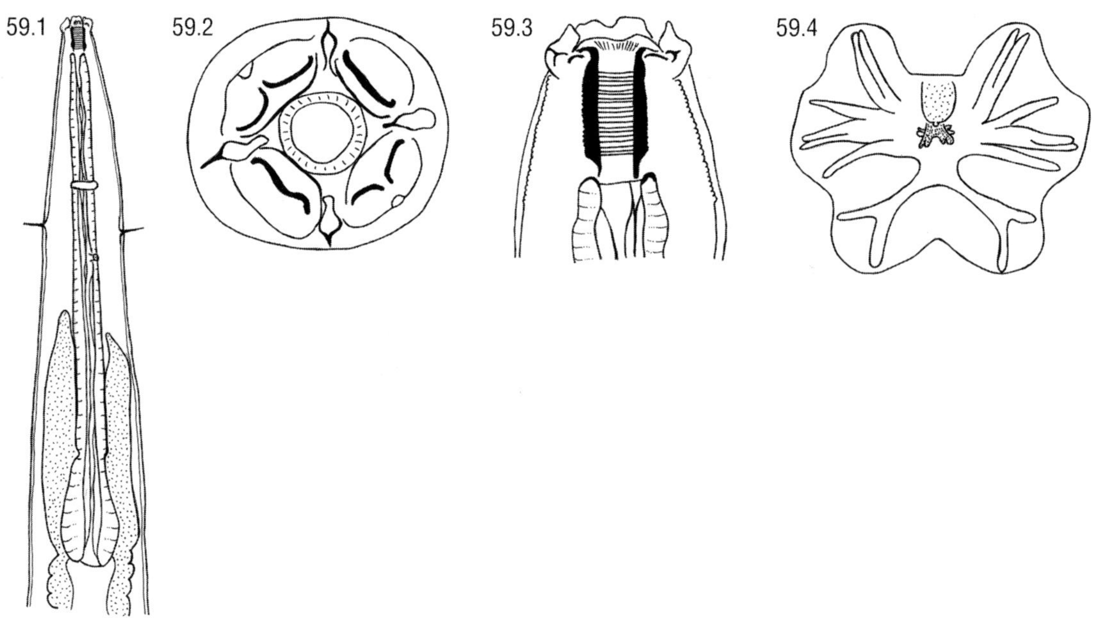

FIG. 59 *Dorcopsistrongylus* 59.1 Anterior end, ventral view; 59.2 Cephalic region, *en face* view; 59.3 Cephalic region, lateral view; 59.4 Bursa. (Redrawn after Smales, 1982b.)

supporting musculature present, generally with muscular bands in eights extending from external surface of buccal capsule to somatic musculature. Oesophagus: elongate corpus; bulb elongate, claviform without sclerotized plates. Proximal region of intestine enlarged. *Male*: Bursa: dorsal ray with two pairs of branches. Spicules alate. *Female*: Vulva opens near anus. Vestibule arranged longitudinally or transversely, prodelphic. Eggs ellipsoidal. Parasites in stomach of macropodid marsupials.

Type genus: *Macropostrongylus* Yorke and Maplestone, 1926 emend. Mawson, 1977

[Emended diagnosis based on Beveridge (1986), who divided the tribe Macropostrongylinea on the basis of the morphology of the buccal capsule. He also includes *Alocostoma* Mawson, 1979, *Foliostoma* Beveridge and Johnson, 1981, *Macroponema* Mawson, 1978, *Monilonema* Beveridge and Johnson, 1981 and *Trigonostonema* Beveridge, 1981 on the basis of the presence of a non-sclerotized buccal capsule supported externally by sets of muscles.]

Trigonostonema Beveridge, 1981 (Fig. 60)

DIAGNOSIS: Mouth opening triangular. Oral collar present with 4 conical submedian cephalic papillae, 2 lateral amphids. Leaf crown with numerous minute elements. Buccal capsule: poorly sclerotized; triangular in cross-section; wall with annular thickening; lining longitudinally striated. Oesophagus: long corpus; isthmus poorly defined; bulb clavate. *Male*: Bursal lobes not

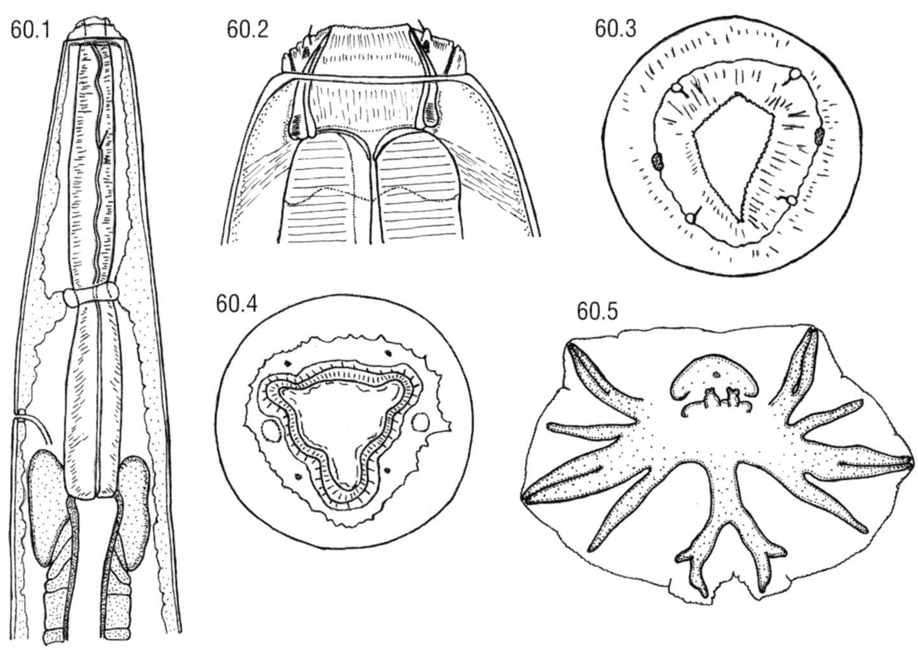

FIG. 60 *Trigonostonema* 60.1 Anterior end, lateral view; 60.2 Cephalic region, dorsal view; 60.3 Cephalic region, *en face* view; 60.4 Buccal capsule, optical transverse section; 60.5 Bursa. (Redrawn after Beveridge, 1981a.)

divided. Bursal rays: ventral rays with common stem, close together; antero-lateral shorter and separate from other laterals, medio- and postero-laterals parallel with common stem; externo-dorsal arises near base of laterals; dorsal ray bifurcates in distal half, each branch with shorter lateral branch. Spicules equal, similar. Gubernaculum absent. *Female*: Vulva opens near anus. Ovejectors parallel to longitudinal axis, vagina enters on ventral side. Tail conical. Parasites in stomach of macropodid marsupials.

Type species: *Trigonostonema trigonostoma* Beveridge, 1981

[Listed in Spencer Jones and Gibson (1987).]

Cassunema Beveridge and Johnson, 1981 (Fig. 61)

DIAGNOSIS: Mouth: oval; dorso-ventrally elongate. oral collar present. Cephalic sensory organs: 4 submedian papillae; 2 lateral amphids. Leaf crowns absent. Buccal capsule: not sclerotized; cylindrical; as deep as wide; internal surface with longitudinal ridges in proximal half; teeth absent. Oesophagus: narrow corpus; small distal bulb. *Male*: Bursa: small; lobes separate at distal margin; ventral rays with common stem, parallel, remain close together; laterals with common stem; antero-lateral ray separate from other laterals; dorsal ray divides in mid-region, each branch with shorter lateral branch at level of division. Spicules: equal; transversely striated

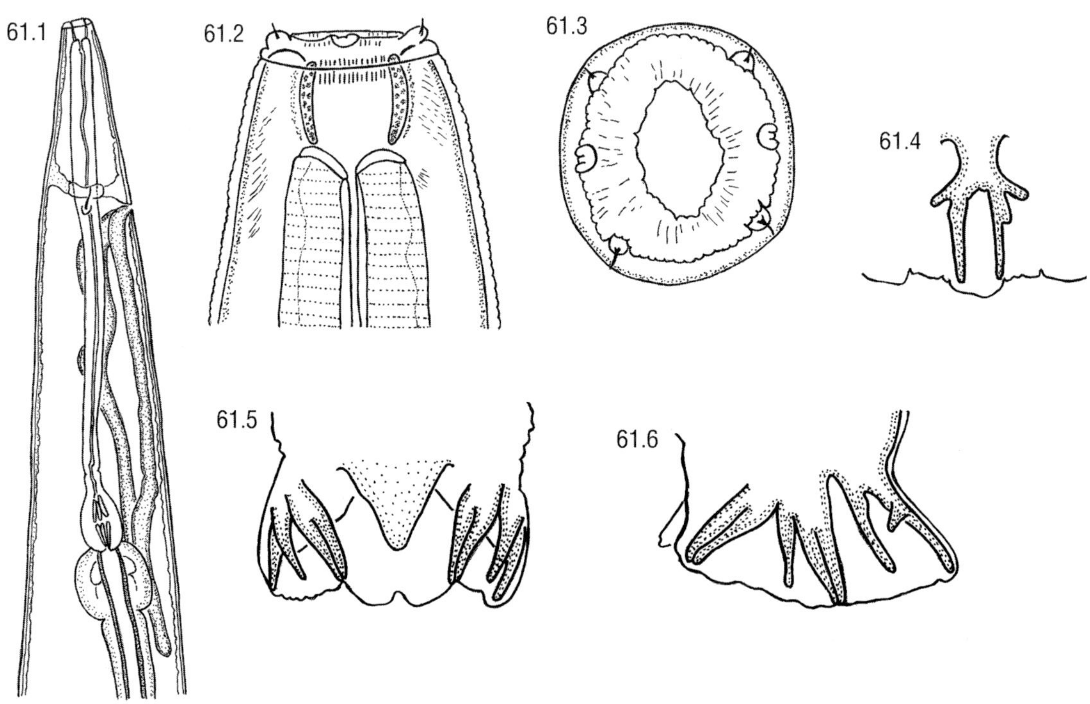

FIG. 61 *Cassunema* 61.1 Anterior end, lateral view; 61.2 Cephalic region, lateral view; 61.3 Cephalic region, *en face* view; 61.4 Dorsal lobe of bursa; 61.5 Bursa, ventral view; 61.6 Bursa, lateral view. (Redrawn after Beveridge and Johnson, 1981.)

alae present, narrow towards distal tips. Gubernaculum absent. *Female*: Vulva opens near to anus. Tail: short; conical. Parasites in stomach of macropodid marsupials.

Type species: *Cassunema exigua* Beveridge and Johnson, 1981

[Listed in Spencer Jones and Gibson (1987).]

Monilonema Beveridge and Johnson, 1981 (Fig. 62)

DIAGNOSIS: Mouth: oval; dorso-ventrally elongate. Oral collar absent. Cephalic sensory organs: 4 conical submedian papillae, each with 2 setae; 2 lateral amphids. Cervical collar: swollen; extends posterior to buccal capsule. Buccal capsule: cylindrical; wider than deep; walls slightly sclerotized; membraneous flange protrudes from proximal margin with leaf crown-like edge; teeth absent. Paired, blindly ending sac-like tubes extend posteriorly from level of cervical collar nearly to nerve ring on dorsal and ventral sides of oesophagus. Oesophagus: short; clavate. Cervical papillae: needle-like; posterior to indentation of cervical collar. Excretory pore posterior to oesophago-intestinal junction. *Male*: Bursa: separate lobes; ventral refractile bosses present; ventral rays with common stem, parallel, remain close together; laterals with common stem, antero-lateral shorter and separate from other laterals, medio- and postero-laterals parallel, remain close together; externo-dorsal arises near base of laterals; dorsal ray divides in mid-region, each branch with a smaller lateral branch at level of division. Spicules: equal; alae narrow towards distal tips.

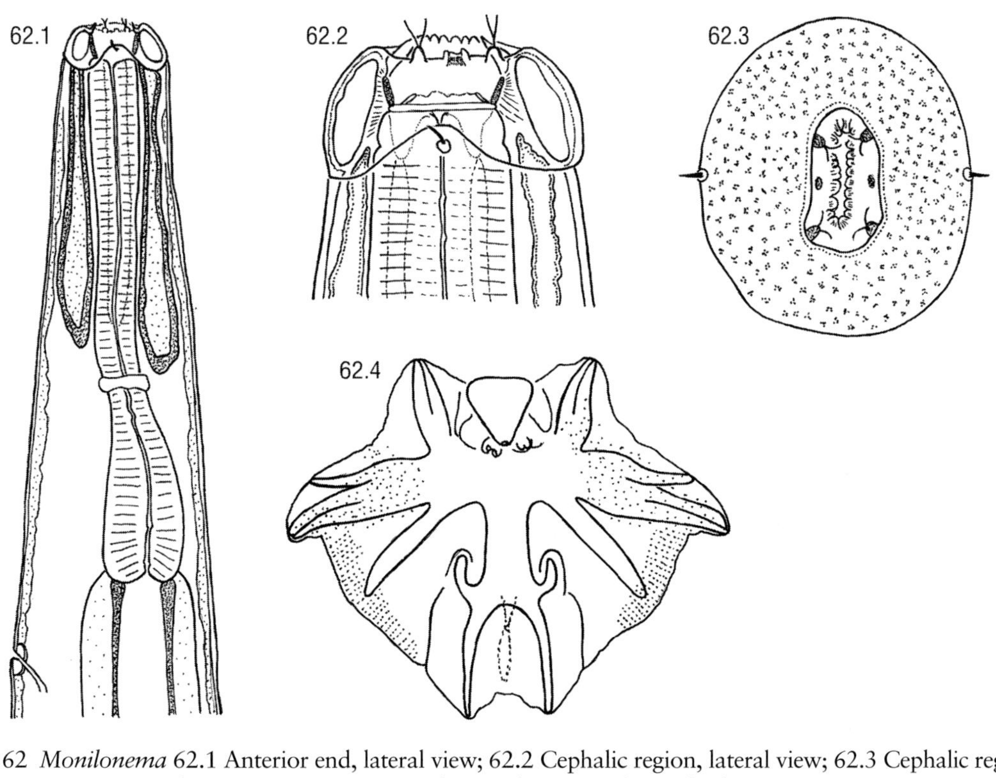

FIG. 62 *Monilonema* 62.1 Anterior end, lateral view; 62.2 Cephalic region, lateral view; 62.3 Cephalic region, *en face* view; 62.4 Bursa. (Redrawn after Beveridge and Johnson, 1981.)

Gubernaculum present. Genital cone: ventral cone with apical papilla (0); bidigitate dorsal raylets (7). *Female*: Vulva opens near to anus. Parasites in stomach of macropodid marsupials.

Type species: *Monilonema lacunosa* Beveridge and Johnson, 1981

[Listed in Spencer Jones and Gibson (1987).]

Foliostoma Beveridge and Johnson, 1981 (Fig. 63)

DIAGNOSIS: Mouth: oval; dorso-ventrally elongate. Oral collar absent. Cephalic sensory organs: 4 small, conical submedian papillae with single seta; 2 lateral amphids. Leaf crown: present; numerous tiny elements. Buccal capsule: cylindrical; walls lightly sclerotized; annular thickening in mid-region. Oesophagus: long corpus, widens distally; narrow, elongate distal bulb. *Male*: Bursa: bell-shaped lobes only slightly separated; ventral refractile bosses present; ventral rays with common stem, parallel, remain close together; laterals with common stem, antero-lateral short, separate from other laterals, medio- and postero-laterals parallel remain close together; externo-dorsal ray arises close to laterals; dorsal ray divides in mid-region, each branch with shorter lateral branch at level of division. Spicules: equal; transversely striated alae present, stop abruptly near distal tips. Gubernaculum present, ovoid in ventral view. Genital cone: prominent ventral cone with papilla (0) at distal tip; two bilobed dorsal raylets (7). *Female*: Vulva opens close to anus. Tail short, conical. Parasites in stomach of macropodid marsupials.

Type species: *Foliostoma macropodis* Beveridge and Johnson, 1981

[Listed in Spencer Jones and Gibson (1987).]

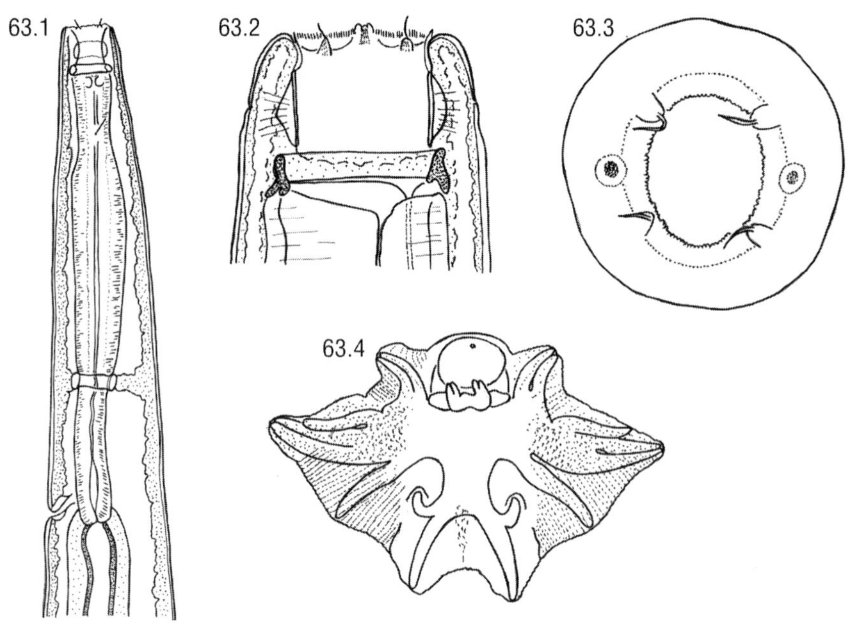

FIG. 63 *Foliostoma* 63.1 Anterior end, lateral view; 63.2 Cephalic region, lateral view; 63.3 Cephalic region, *en face* view; 63.4 Bursa. (Redrawn after Beveridge and Johnson, 1981.)

Labiostrongylinea Beveridge, 1983

DIAGNOSIS: Stout body. Oral collar present or absent. Mouth opening with 2 or 6 prominent, fleshy lips, each with distinct pulp cavity. Interlabia present or with '2–8 lip-like processes'. Cephalic sensory organs: 4 papillae on submedian lips or oral collar; amphids on lateral lips or oral collar. Buccal capsule wider than deep, prominent extrachitinous supports present. Oesophagus: clavate or with distal bulb; paired intestinal diverticula surround oesophago-intestinal junction. *Male*: Bursal rays: ventrals with common stem, parallel, remain close together; laterals with common stem, antero-lateral ray separate from other laterals; externo-dorsal arises close to base of laterals; dorsal bifurcates, each branch with an external branch. *Female*: Ovejectors type II. *Vagina vera* enters ovejectors on ventral side. Eggs thin-shelled, ellipsoidal. Parasites in stomach of macropodid marsupials.

Type genus: *Labiostrongylus* Yorke and Maplestone, 1926

[Beveridge (1983) also includes the genera *Parazoniolaimus* Johnston and Mawson, 1939, *Potorostrongylus* Johnston and Mawson, 1940 and *Dorcopsinema* Mawson, 1977. These genera were previously placed in the tribe Zoniolaiminea in the key by Lichtenfels (1980a). Beveridge follows Lichtenfels in using the ending '-ea' rather than the '-ini' as proposed by the International Code of Zoological Nomenclature. For further details on the changes in this tribe, Beveridge (1983) should be consulted. Smales (2002) also includes *Paralabiostrongylus* Smales, 1982, *Labiosimplex* Smales, 1994 and *Labiomultiplex* Smales, 1994.]

Labiostrongylus Yorke and Maplestone, 1926 (Fig. 64)

DIAGNOSIS: Oral collar absent. Mouth with 6 fleshy lips each with pulp cavity. Cephalic sensory organs: amphids 1 on each of 2 lateral lips; 1 submedian papilla on each of 4 submedian lips. Single dorsal and ventral interlabia present. Buccal capsule cylindrical, wider than deep. Oesophagus clavate. Oesophago-intestinal junction encircled by paired intestinal diverticula. Excretory pore, nerve ring and cervical papillae up to one-quarter of oesophageal length from anterior end. *Male*: Bursa: dorsal lobe with 'lappets'; ventral rays with common stem, parallel, close together; antero-lateral short, separates from other laterals, medio- and postero-laterals remain close together; externo-dorsal arises near base of laterals; dorsal ray divided in middle third, each branch extends into lappets, each branch with shorter external branch. Gubernaculum present. Spicules alate. *Female*: Vulva opens near anus. Type II ovejectors, posterior branch directed anteriorly, *vagina vera* 'slightly sinuous' enters ventrally, vestibule parallel to longitudinal axis of body, infundibula smaller and thinner than sphincters. Tail conical, pointed tip. Parasite in stomach of macropodid marsupials.

Type species: *Labiostrongylus labiostrongylus* Yorke and Maplestone, 1926

[Emended diagnosis based on Smales (1982a, 1994, 2002). Smales (1994, 2002) revised the genus *Labiostrongylus*.]

Paralabiostrongylus Smales, 1982 (Fig. 65)

DIAGNOSIS: Oral collar: little developed; 4 submedian papillae present. Mouth with 2 fleshy lips lobed, incorporating amphids, 4 'lip-like lobed flaps'. Buccal capsule cylindrical, wider than

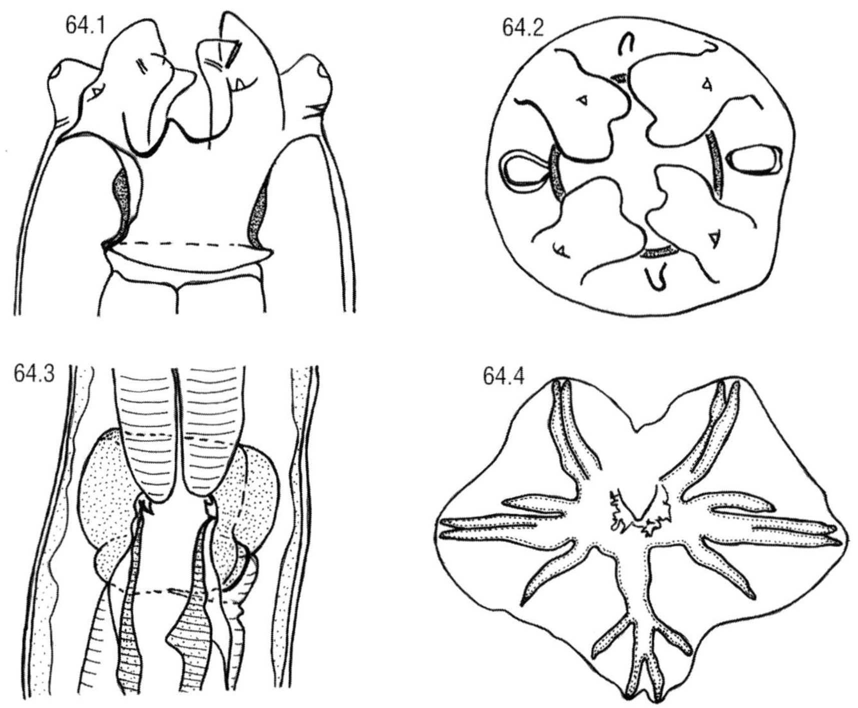

FIG. 64 *Labiostrongylus* 64.1 Cephalic region, dorsal view; 64.2 Cephalic region, *en face* view; 64.3 Oesophago-intestinal junction, lateral view; 64.4 Bursa. (Redrawn after Smales, 1994.)

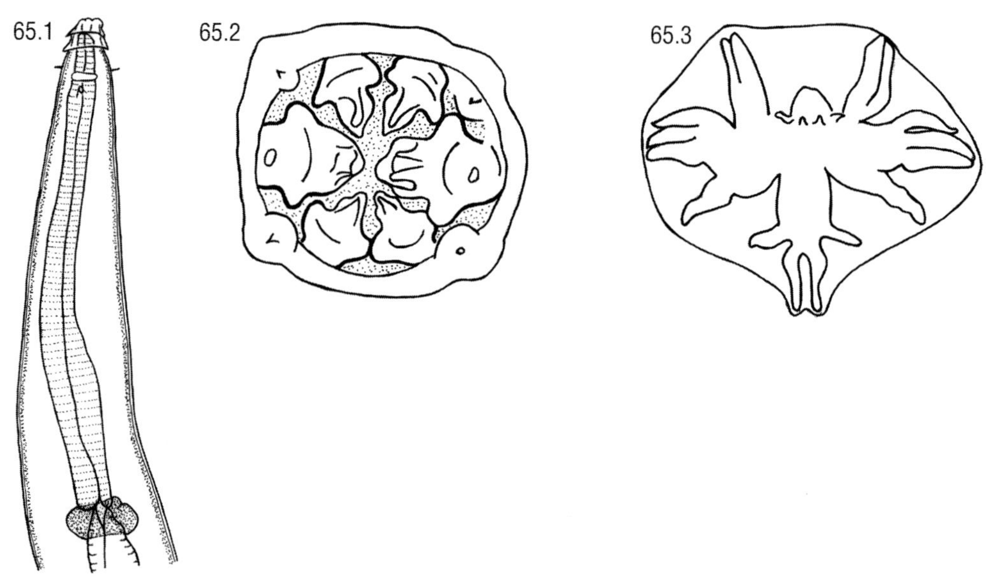

FIG. 65 *Paralabiostrongylus* 65.1 Anterior end, lateral view; 65.2 Cephalic region, *en face* view; 65.3 Bursa. (Redrawn after Smales, 1982.)

deep. Oesophagus clavate. Oesophago-intestinal junction surrounded by intestinal diverticula. Excretory pore, nerve ring and cervical papillae up to one-fifth from anterior end of oesophagus. *Male*: Bursa: lobes joined; ventral rays with common stem, parallel, close together; lateral rays with common stem, medio- and postero-laterals parallel, remain close together; externo-dorsal rays arise from base of laterals; dorsal ray divides in middle third, 2 lateral branches arise prior to division. Spicules slender, alate. Gubernaculum present. *Female*: Vulva opens near anus. Ovejectors, type II, parallel to longitudinal axis of body, uteri parallel. *Vagina vera* thick-walled, enters ovejectors ventrally. Parasites in stomach of macropodid marsupials in Papua New Guinea.

Type species: *Paralabiostrongylus bicollaris* Smales, 1982

[Diagnosis based on Smales, 1982a, 2002. Smales (2002) moves *Paralabiostrongylus* from the Zoniolaiminea to the Labiostrongylinea. Listed in Spencer Jones and Gibson (1987).]

Genus *Labiosimplex* Smales, 1994 (Fig. 66)

DIAGNOSIS: Oral collar absent. Mouth with 6 prominent fleshy lips, each with pulp cavities, submedian lips with ridges on outer surface. Interlabia absent. Buccal capsule wider than deep. Cephalic sensory organs: amphids on 2 lateral lips; submedian papillae on 4 submedian lips. Oesophagus clavate. Oesophago-intestinal junction encircled by intestinal diverticula. Cervical papillae, nerve ring, excretory pore close together up to one-third of oesophageal length from anterior end. *Male*: Bursa: ventral rays with common stem, parallel, close together; antero-lateral separate from other laterals, medio- and postero-laterals remain close together; externo-dorsal ray arises near base of laterals, dorsal lobe and dorsal ray variable. Gubernaculum present. Spicules slender, alate. *Female*: Vulva opens close to anus. *Vagina vera* generally sinuous, enters ovejectors ventrally. Ovejectors, type II, parallel to longitudinal axis of body. Parasites of stomach of kangaroos and wallabies in Australia and Papua New Guinea.

Type species: *Labiosimplex australis* (Kung, 1948)

[Diagnosis based on Smales (1994, 1995a, 2002).]

Genus *Labiomultiplex* Smales, 1994 (Fig. 67)

DIAGNOSIS: Oral collar absent. Mouth with 6 prominent fleshy lips, each with pulp cavities. Interlabia absent in adults. Buccal capsule wider than deep. Cephalic sensory organs: amphids on 2 lateral lips; submedian papillae on 4 submedian lips. Oesophagus: widens abruptly at approximately one-third of its length, forming terminal bulb. Cervical papillae position variable, nerve ring and excretory pore near distal end of oesophagus. Diverticula present or absent at oesophago-intestinal junction. *Male*: Bursa: ventral rays with common stem, parallel, close together; antero-lateral ray separate from other laterals, medio- and postero-laterals remain close together; externo-dorsal ray arises near base of laterals; dorsal ray stem thick, divides in posterior half, each with shorter lateral branches. Gubernaculum present. Spicules slender, alate. *Female*: Vulva opens near to anus. *Vagina vera* straight enters ovejectors ventrally. Ovejectors, type II, parallel to longitudinal axis of body. Parasites of stomach of wallabies, hare wallabies and potoroids in Australia and Papua New Guinea.

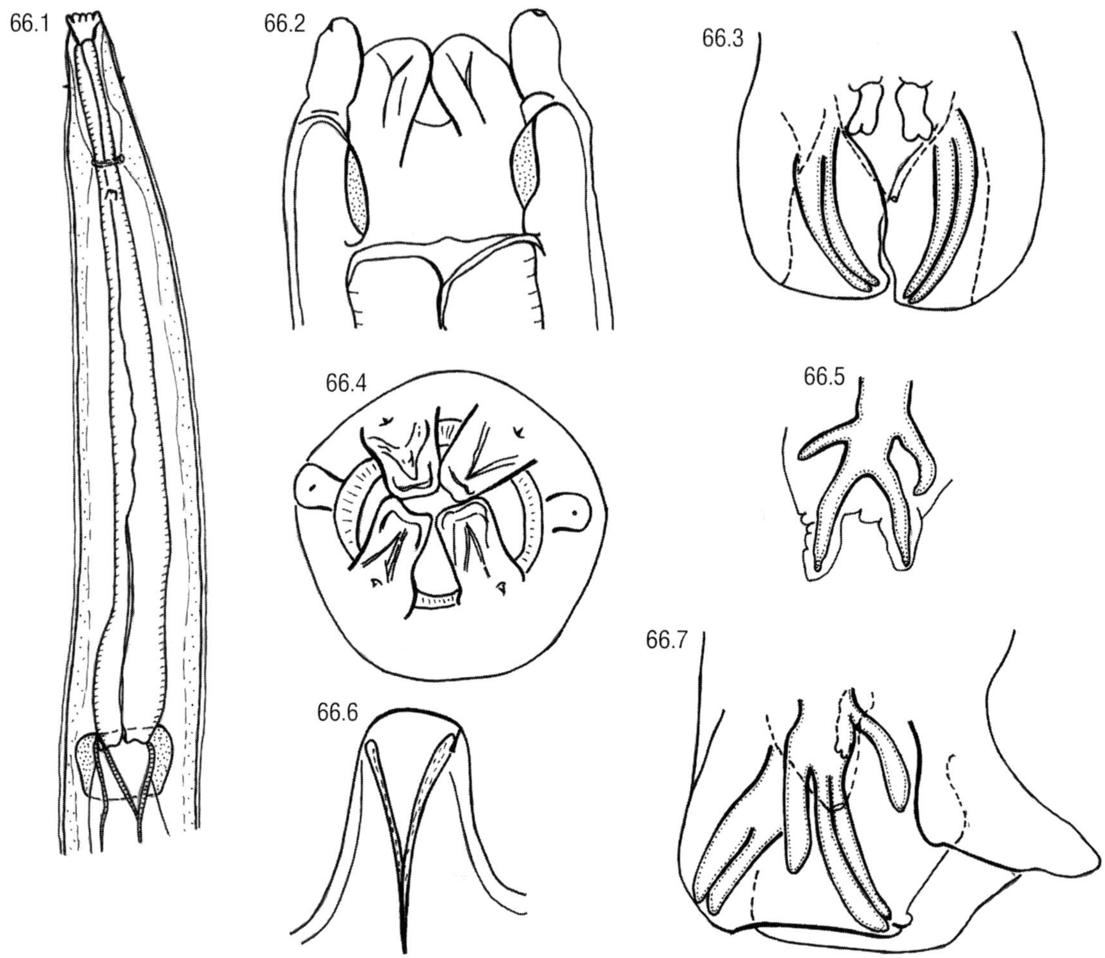

FIG. 66 *Labiosimplex* 66.1 Anterior end, ventral view; 66.2 Cephalic region, ventral view; 66.3 Bursa, ventral view; 66.4 Cephalic region, *en face* view; 66.5 Dorsal lobe and ray, dorsal view; 66.6 Submedian lip, external view; 66.7 Bursa, lateral view. (Redrawn after Smales, 1995.)

Type species: *Labiomultiplex eugenii* (Johnston and Mawson, 1940)

[Diagnosis based on Smales (1994, 2002).]

Potorostrongylus Johnston and Mawson, 1939 (Fig. 68)

DIAGNOSIS: Cervical cuticle inflated. Mouth with 6 prominent fleshy lips, each with distinct pulp cavities. Cephalic sensory organs: amphids on 2 lateral lips; submedian papillae on each submedian lip. Dorsal and ventral interlabia present. Buccal capsule: short; cylindrical. Oesophagus: long; cylindrical; posterior elongate bulb with interrupted internal sclerotized plates. Cervical papillae posterior to excretory pore. *Male*: Bursa: slightly lobed; ventral rays with common stem, parallel, remain close together; antero-lateral ray separate from other laterals, medio- and postero-lateral rays with common stem, parallel, remain close together; externo-

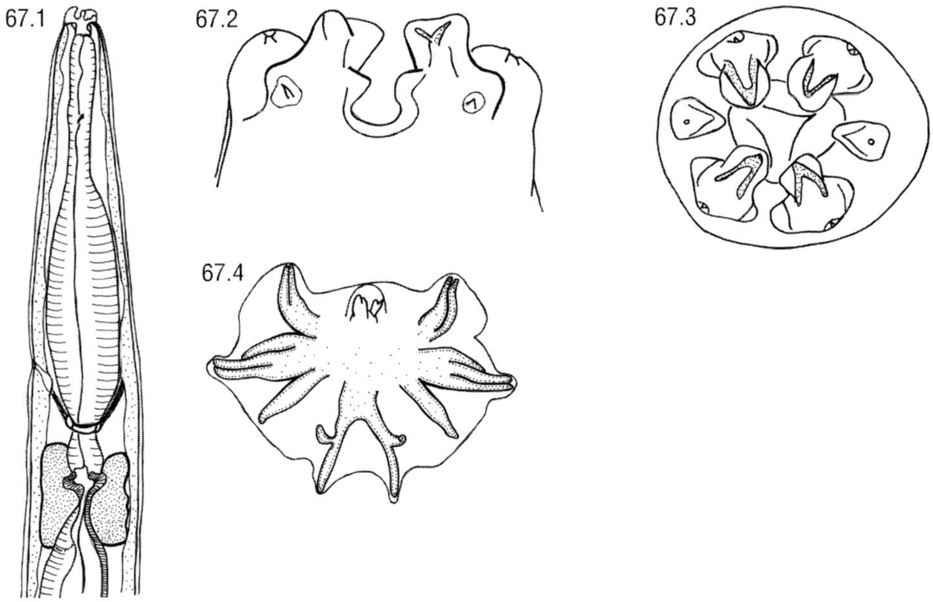

FIG. 67 *Labiomultiplex* 67.1 Anterior end, lateral view; 67.2 Cephalic region, dorsal view; 67.3 Cephalic region, *en face* view; 67.4 Bursa. (Redrawn after Smales, 1994.)

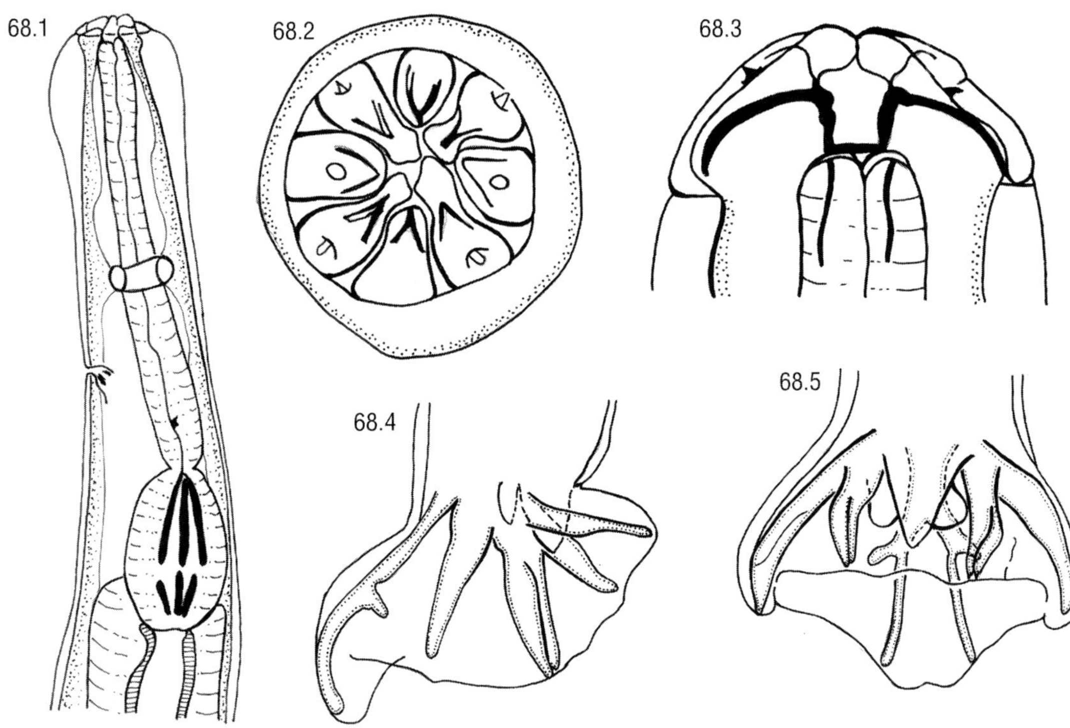

FIG. 68 *Potorostrongylus* 68.1 Anterior end, lateral view; 68.2 Cephalic region, *en face* view; 68.3 Cephalic region, dorsal view; 68.4 Bursa, lateral view; 68.5 Bursa, ventral view. (Redrawn after Smales, 1997a.)

dorsal ray arises at base of laterals; dorsal divides in mid-region, each branch with a shorter external branch. Spicules: equal; long; alate. Gubernaculum absent. Genital cone: single ventral papilla (0) at apex of cone; dorsal raylets (7) composed of several appendages; rounded lateral genital appendages present. *Female*: Vulva opens close to anus. Ovejectors type II, posterior branch directed anteriorly. Tail conical. Parasites in stomach of potoroid marsupials.

Type species: *Potorostrongylus finlaysoni* Johnston and Mawson, 1939

[Originally placed in the tribe Zoniolaiminea by Lichtenfels (1980a). The emended diagnosis is based on Smales (1997a).]

Parazoniolaimus Johnston and Mawson, 1939

DIAGNOSIS: Cephalic region constricted at level of base of buccal capsule. Mouth with 6 fleshy lips, each with distinct pulp cavity, 2 larger lateral lips with amphids, 4 submedian lips each with cephalic papilla bearing bifid setae. Buccal capsule wider than deep. 'Cuticular frill' situated between distal end of lips and cephalic constriction. Oesophagus with proximal cylindrical region and distal clavate region. Oesophago-intestinal junction encircled by intestinal diverticula. Excretory pore, nerve ring, cervical papillae near distal end of oesophagus. *Male*: Bursa with separate lobes. Ventral rays with common stem, parallel, close together; antero-lateral ray separated from other laterals, medio- and postero-laterals remain close together; externo-dorsal ray arises near base of laterals; dorsal ray with lateral branches prior to division into 2 main branches. Gubernaculum present. Spicules slender and alate. *Female*: Vulva opens near anus. *Vagina vera* sinuous, enters ovejectors ventrally. Ovejectors, type II, parallel to longitudinal axis of body. Parasites of stomach of macropodid marsupials (*Wallabia bicolour*).

Type species: *Parazoniolaimus collaris* Johnston and Mawson, 1939

[Emended diagnosis based on Smales (2002). The genus is moved from the Zoniolaiminea, where it was placed in Lichtenfels (1980a) to the Labiostrongylinea by Smales (2002).]

Coronostrongylinea Beveridge, 1986

DIAGNOSIS: Body small. Alae absent. Mouth: round, triangular or oval; lips and teeth absent. Labial collar present or absent. Buccal capsule: wall with 2 distinct layers; inner layer thick, transparent, may be folded; outer layer sclerotized, cylindrical, may be reduced to an annulus; supporting musculature absent. Oesophagus: undivided corpus; bulb spherical, may be claviform without sclerotized plates. Proximal region of intestine not modified. *Male*: Bursa: dorsal ray with 2 pairs of branches. Spicules alate. *Female*: Vulva opens near anus. Vestibule arranged longitudinally, rarely obliquely, prodelphic. Eggs ellipsoidal. Parasites in stomach of macropodid marsupials.

Type genus: *Coronostrongylus* Johnston and Mawson, 1939

[Beveridge (1986) splits the tribe Macropostrongylinea created by Lichtenfels (1980a) on the basis of the morphology of the buccal capsule. He includes in his new tribe *Papillostrongylus* Johnston and Mawson, 1939, *Popovastrongylus* Mawson, 1977 and *Thylonema* Beveridge, 1981 on the basis of the presence of a sclerotized outer layer in the buccal capsule. Beveridge (pers. comm.) designated *Coronostrongylus* as type genus.]

Thylonema Beveridge, 1981 (Fig. 69)

DIAGNOSIS: Mouth opening dorso-ventrally, flattened with 2 'lip-like projections', each with two cephalic papillae (bilobed with 'medially directed setae'). Amphids between lips, may open on conical projections. Buccal capsule: poorly sclerotized; dorsal and ventral walls thickened anteriorly. Oesophagus: corpus long, cylindrical; isthmus narrower or wider than corpus; bulb clavate. *Male*: Bursal lobes not divided. Bursal rays: ventral rays with common stem, close together; antero-lateral shorter and separate from other laterals, medio- and postero-laterals with common stem, close together; externo-dorsal arises close to lateral stem; dorsal ray bifurcates in distal half, each branch with smaller external lateral branch at level of bifurcation. Spicules equal, similar. Gubernaculum absent. *Female*: Vulva opens near to anus. Ovejectors parallel to longitudinal axis, *vagina vera* enters on ventral side. Tail conical. Parasites in stomach of macropodid marsupials.

Type species: *Thylonema thylonema* Beveridge, 1981

[Listed in Spencer Jones and Gibson (1987).]

Mbaisonema Beveridge, 1997 (Fig. 70)

DIAGNOSIS: Buccal capsule: reduced; forms sclerotized ring; lining forms 4 lobes extending to mouth. Mouth with leaf crown of numerous small elements. Oesophagus: elongate corpus; short isthmus and bulb. Excretory pore opens at level of corpus. Nerve ring surrounds mid-region of

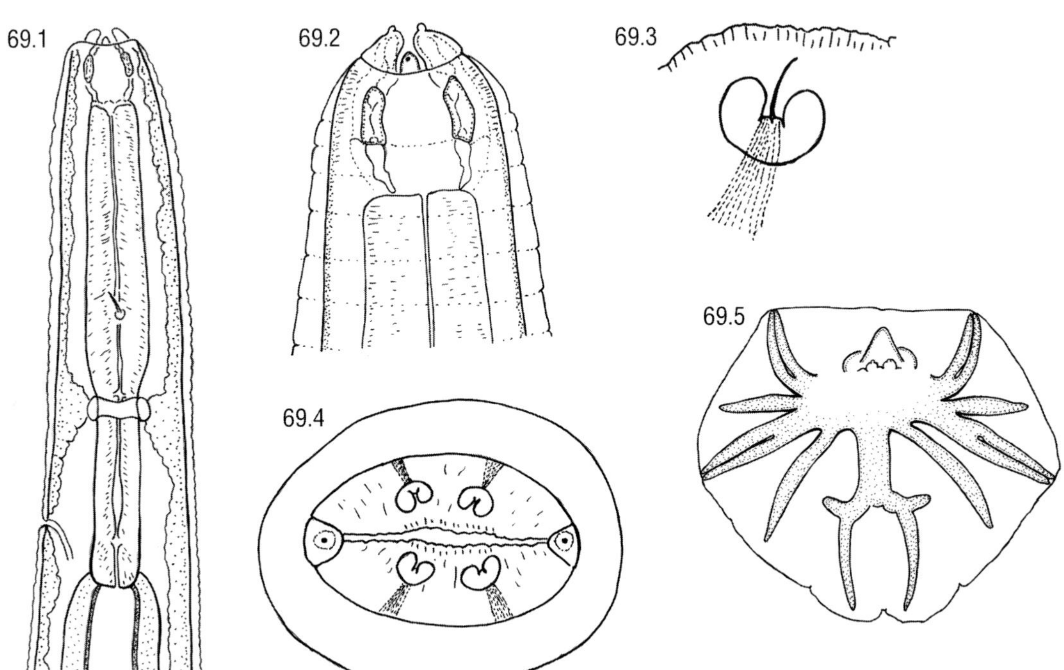

FIG. 69 *Thylonema* 69.1 Anterior end, lateral view; 69.2 Cephalic region, lateral view; 69.3 One of submedian cephalic papillae; 69.4 Cephalic region *en face* view; 69.5 Bursa. (Redrawn after Beveridge, 1981b.)

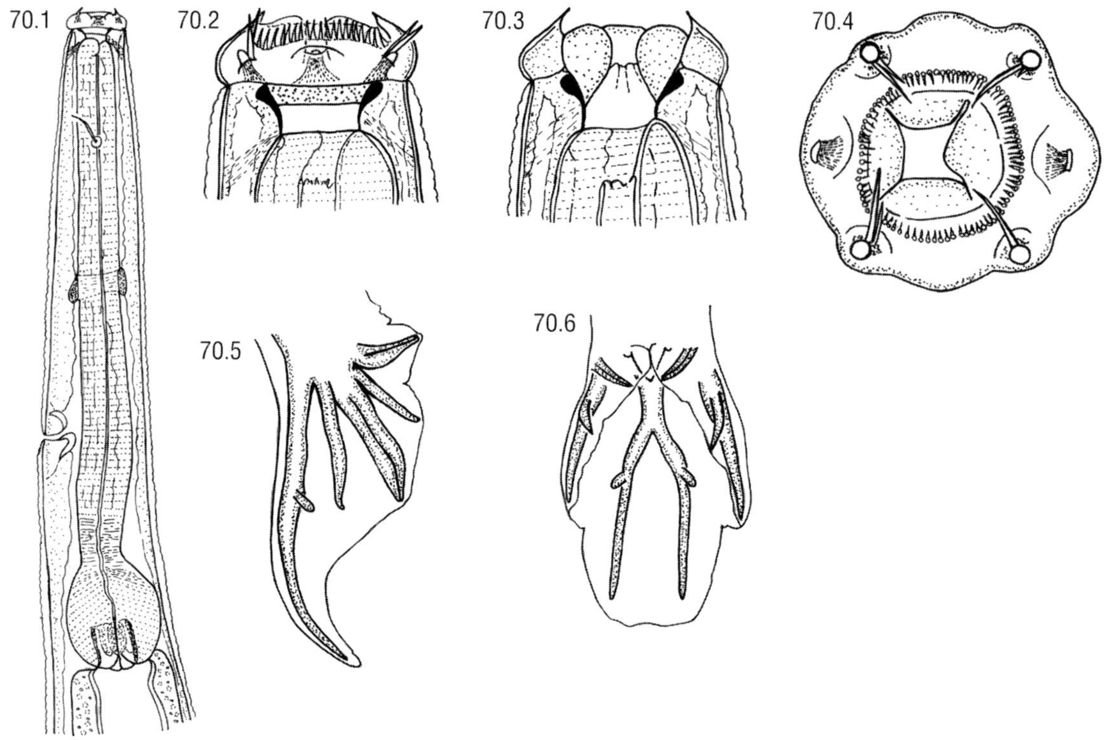

FIG. 70 *Mbaisonema* 70.1 Anterior end, lateral view; 70.2 Cephalic region, lateral view; 70.3 Cephalic region, lateral view, optical section through buccal capsule; 70.4 Cephalic region, *en face* view; 70.5 Bursa, lateral view; 70.6 Bursa, ventral view. (Redrawn after Beveridge, 1997.)

corpus. Cervical papillae near anterior end of corpus. *Male*: Bursa with dorsal ray constructed of 4 branches. Gubernaculum absent. *Female*: Vulva opens just anterior to anus. Vagina short, leads to type I ovejectors. Tail sharply tapers posterior to anus. Parasites in stomach of macropodid marsupials in Papua New Guinea.

Type species: *Mbaisonema coronatum* Beveridge, 1987

Corollonema Beveridge and Chilton, 1998 (Fig. 71)

DIAGNOSIS: Cuticle: numerous annulations present. Oral collar present. Cephalic sensory organs: 4 dome-like submedian papillae with setae; 2 lateral amphids. Mouth with 8 elements in leaf crown attached at base of buccal capsule. Buccal capsule: proximal region thin-walled; posterior region encircled by thick-walled 'annulus'. Oesophagus: corpus widens slightly posteriorly; isthmus distinct; bulb subglobular without sclerotized lining. Cervical papillae at level of corpus. *Male*: Bursa: short; ventral rays with common stem, parallel, remain close together; laterals with common stem, antero-lateral shorter and separate, medio- and postero-laterals remain close together; externo-dorsal ray arises separately from dorsal; dorsal ray divides in distal half, each division with short lateral branch. Spicules equal, slender alate. Gubernaculum present. *Female*: Vulva opens near anus. Ovejectors type II. Tail conical. Parasites in stomach of macropodid marsupials.

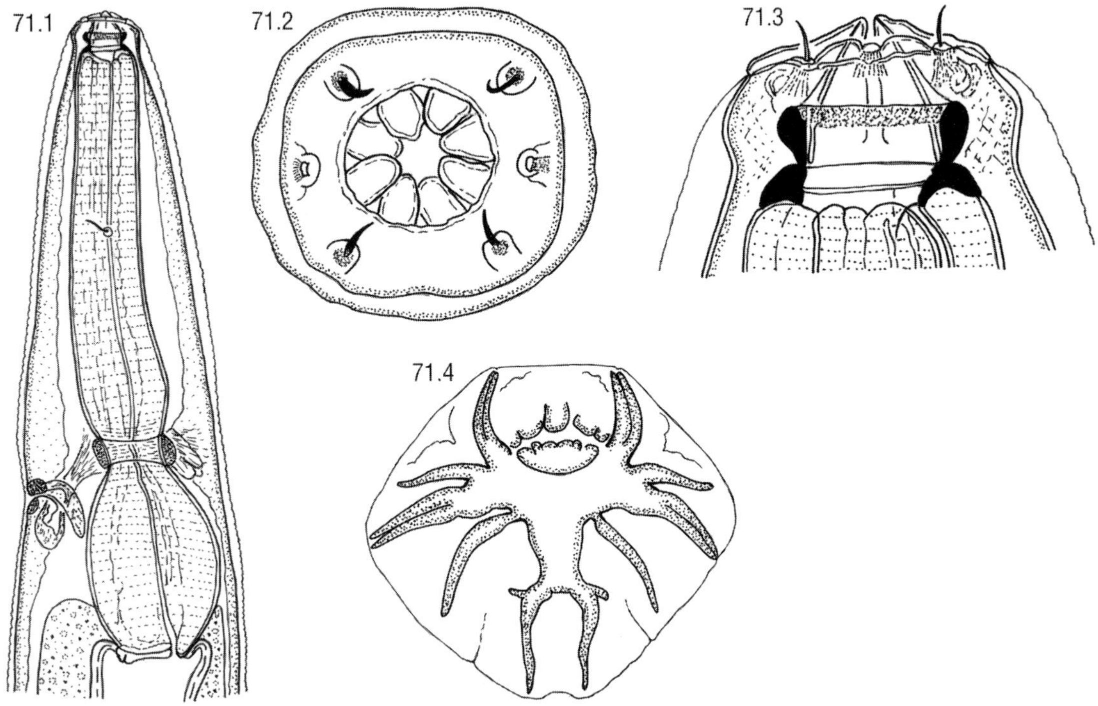

FIG. 71 *Corollonema* 71.1 Anterior end, lateral view; 71.2 Cephalic region, *en face* view; 71.3 Cephalic region, lateral view; 71.4 Bursa. (Redrawn after Beveridge and Chilton, 1998.)

Type species: *Corollonema thylogale* Beveridge and Chilton, 1998

Phascolostrongylinae Lichtenfels, 1980

DIAGNOSIS: Buccal capsule: cylindrical or globular; prominent dorsal gutter present. *Male*: Bursa: externo-dorsal ray arises from main stem of dorsal ray. *Female*: Ovejectors type I or modified type I.

[Emended diagnosis based on Beveridge (1987). This subfamily was erected by Lichtenfels (1980a) for genera that generally occur in the intestine of vombatid and macropodid Australian marsupials. It was originally placed by Lichtenfels (1980a) in the family Strongylidae Baird, 1853.]

Phascolostrongylinea (Lichtenfels, 1980 subfamily) Beveridge, 1987

DIAGNOSIS: External and internal leaf crowns present. Buccal capsule: large; cylindrical; straight sided with dorsal gutter extending to proximal rim. *Female*: Ovejectors type I. Parasites in the large intestine of Vombatidae.

Type genus: *Phascolostrongylus* Canavan, 1931 emended Beveridge, 1978

[Apart from *Macropostrongyloides* Yamaguti, 1961 and *Paramacropostrongylus* Johnston and Mawson, 1940, which are moved to the Macropostrongyloidinea, Beveridge (1987) bases the generic content of this tribe on the Lichtenfels (1980a) subfamily. The type genus is designate by Beveridge (pers. comm.).]

Hypodontinea Beveridge, 1987

DIAGNOSIS: Mouth: directed anteriorly; antero-dorsally or antero-ventrally. Oesophageal teeth or leaf crowns present. Buccal capsule: large; subglobular; dorsal gutter Y-shaped. *Female*: Ovejectors type I or modified. Parasites of caecum or large intestine of Macropodidae.

Type genus: *Hypodontus* Mönnig, 1929

[Beveridge (1987) also includes *Macropicola* Mawson, 1978 (formerly placed in the subfamily Strongylinae by Lichtenfels (1980a) and *Corollostrongylus* Beveridge, 1978 (formerly placed in the Chabertiinae by Lichtenfels (1980a). The type genus is designated by Beveridge (pers. comm.).]

Macropostrongyloidinea Beveridge, 1987

DIAGNOSIS: Mouth opening terminal or slightly ventral. Buccal capsule: small; cylindrical; dorsal gutter Y-shaped; teeth or denticles present. *Female*: Ovejectors type I or modified. Parasites of stomach and caecum of Macropodidae and colon of Vombatidae.

Type genus: *Macropostrongyloides* Yamaguti, 1961 emended Mawson and Beveridge, 1978

[Beveridge (1987) also includes *Paramacropostrongylus* Johnston and Mawson, 1940 emended Beveridge and Mawson, 1978. These genera are retained in the same subfamily as in Lichtenfels (1980a) but are now placed in a new tribe. The type genus is designated by Beveridge (pers. comm.).]

TRICHOSTRONGYLOIDEA

[The constituents of this superfamily have undergone many revisions using different criteria, resulting in variations in the classification. Durette-Desset and Chabaud (1993) proposed that this superfamily should be in a new suborder, the Trichostrongylina. The authors included three superfamilies in the suborder, namely Trichostrongyloidea and two new superfamilies that they proposed should be erected – Molineoidea (see under Molineidae for the differential diagnosis of the superfamily proposed by Durette-Desset and Chabaud, 1993) and Heligmosomoidea (see under Heligmosomidae for the differential diagnosis of the superfamily proposed by Durette-Desset and Chabaud, 1993). No differential diagnosis for this suborder is given. Durette-Desset and Chabaud (1993) redefine the diagnosis of the superfamily Trichostrongyloidea as: Body uncoiled; lips and buccal capsule absent; cephalic vesicle absent in the primitive forms; synlophe absent, bilaterally symmetrical or sometimes at the most evolved an axis of orientation appears; *Male* with bursa basically 1-3-1 type; *Female*, didelphic, parasites mainly of ratite and neotropical birds, Australian marsupials, lagomorphs and ruminants. Two families are included – Dromaeostrongylidae (Skrjabin and Schulz, 1937) and Trichostrongylidae (Leiper, 1908). Later, Durette-Desset *et al.* (1999) applied cladistic analysis to the superfamily and proposed three families and eight subfamilies. For full details of the cladistic analysis and features assigned to the various taxa, Durette-Desset *et al.* (1999) should be consulted. The proposed changes of Durette-Desset *et al.* (1999) are indicated under the relevant taxa.

The synlophe is an important character in this group, and involves detailed analysis of the longitudinal cuticular ridge patterns both in gross morphology and in cross-section. Durette-Desset (1971) pioneered the study of this character in her examination of the heligmosomes. Later, Durette-Desset (1985) used this character in conjunction with the bursa to examine the morphological evolution of this superfamily and identified three distinct branches, namely: (i) the Trichostrongylid type, including the Dromaeostrongylidae and Trichostrongylidae; (ii) the Molineid type, including the Amphibiophilidae, Dictyocaulidae, Molineidae and Mackerrastrongylidae; and (iii) the Heligmosomid type, including the Nicollinidae, Herpetostrongylidae, Viannaiidae, Ornithostrongylidae, Heligmosomidae and Heligmonellidae. Hoberg *et al.* (2004) underline the value of this character at both the generic and specific level for determining relationships and identification, citing other studies where this characteristic has been applied, namely Lichtenfels *et al.* (1994), Lichtenfels and Pilitt (2000) and Hoberg *et al.* (2002). Assigning females to genera and species using general morphological features has also created difficulties in this group. Studies of the synlophe in gross morphology in the cervical region have proved a valuable character in the assignment of females to species in studies where this has been applied. It is essential to consult the original descriptions for the complete analysis of this character, and the figures illustrating the arrangement of the ridges.

Please note that all figures of the synlophe redrawn after Durette-Desset and colleagues are orientated with the dorsal field to the top of the page and the left lateral field to the left-hand margin of the page.]

Trichostrongylidae (Leiper, 1908 subfamily) Leiper, 1912

[Hoberg and Lichtenfels (1994) completed the first phylogenetic analysis of the Trichostrongylidae and recognized six subfamilies, namely Trichostrongylinae, Cooperiinae, Libyostrongylinae, Graphidiinae, Ostertagiinae and Haemonchinae.

Durette-Desset *et al.* (1999) recognized only three subfamilies in their cladistic analysis, namely Trichostrongylinae, Amidostomatinae and Filarinematinae, elevating the Haemonchinae and

Cooperiinae to family status. The Haemonchidae includes the Haemonchinae and Ostertagiinae, and the Cooperiidae includes the Cooperiinae, Libyostrongylinae and Obeliscoidinae. These differences relate to the sets of characters that each group of authors used for their studies, and the original references should be consulted.]

Trichostrongylinae Leiper, 1912

[*Graphidioides* Cameron, 1923 was included in the subfamily Graphidiinae by Gibbons and Khalil (1982a), but transferred to the Trichostrongylinae by Durette-Desset (1983) and Durette-Desset *et al.* (1999). *Graphinema* Guerrero and Chavez, 1984, although not part of the cladistic study, was included by Durette-Desset *et al.* (1999) in the Trichostrongylinae. Durette-Desset *et al.* (1999) also transferred *Parostertagia* Schwartz and Alicata, 1933 to this subfamily from the Graphidiinae. Earlier, Hoberg *et al.* (1993a) considered this genus should be in the Cooperiinae (see under *Parostertagia*, Graphidiinae).]

Trichostrongyella Dikov, 1961 (Fig. 72)

DIAGNOSIS: Buccal capsule, cephalic vesicle absent. Cervical papillae small. *Male*: Large symmetrical bursa, dorsal lobe not separated from lateral lobes; ventral rays parallel, unequal; dorsal ray long, dividing into 2 distal branches, each bifurcating halfway along length. Genital cone: telamon well developed, 2 dorsal papillae, large ventral papilla. Spicules short, dorsal alae strongly developed, fused, ventral alae weakly developed. *Female*: Unknown. Parasite of abomasum of sheep in Kazakhstan.

Type species: *Trichostrongyella schulzi* Dikov, 1961

[Jansen (1989 publ. 1990) considers the taxonomic position of this genus unclear, but considered it may have some affinities with *Mazamastrongylus*. Listed in Spencer Jones and Gibson (1987).]

Graphidioides Cameron, 1923

DIAGNOSIS: Synlophe: longitudinal cuticular ridges perpendicular to body surface; absent in lateral fields. Cervical papillae hook-shaped. *Male*: Bursa with ventral surface of lateral lobes ornamented. Bursal rays arranged with ventrals separated from laterals, i.e. 2–3 type; postero-lateral and externo-dorsal rays parallel; dorsal ray long. Spicules long, thin, joined at tip (except *G. berlai*). *Female*: Vulva opening transverse. Ovejectors asymmetrical (except *G. affinis*). Tail conical. Parasites in stomach and intestine of Caviomorpha, primates, Cricetidae in South America.

Type species: *Graphidioides affinis* (Mégnin, 1895)

[Emended diagnosis based on Sutton and Durette-Desset (1995).]

Libyostrongylinae Durette-Desset and Chabaud, 1977

[Durette-Desset *et al.* (1999) transfer this subfamily to their new family Cooperiidae. The genus *Pseudostertagia* (Orloff, 1933) Travassos, 1937 is transferred by them to the Cooperiinae; *Obeliscoides* Graybill, 1924 is transferred to a new subfamily Obeliscoidinae as type genus,

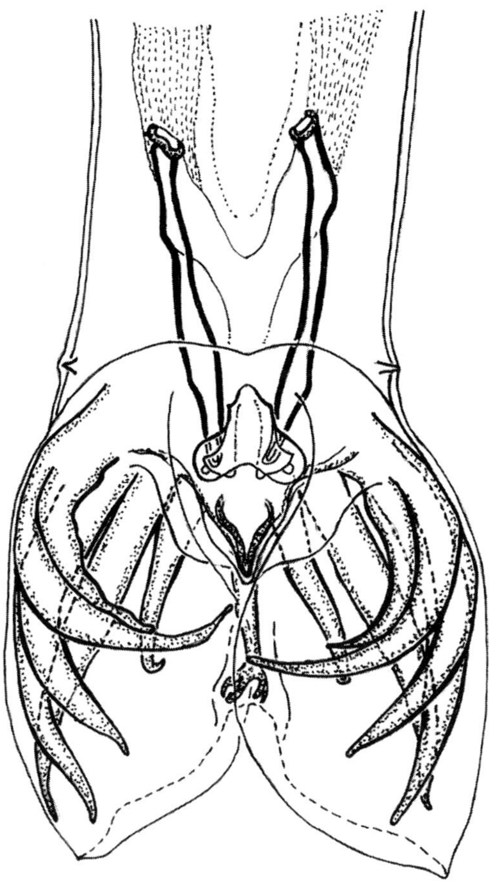

FIG. 72 *Trichostrongyella* Male tail. (Redrawn after Dikov, 1961.)

and *Laurostrongylus* Durette-Desset and Chabaud, 1992 is transferred to this family from the Graphidiinae. The remaining genera are unchanged from Durette-Desset (1983).]

Teporingonema Harris, 1985 (Fig. 73) (= *Lamothiella* González-Ortega, 1984)

DIAGNOSIS: Cephalic region broad, flat. Mouth opening hexagonal. Cephalic sensory organs: inner circle 4 submedian papillae, 2 lateral amphids, outer circle of 4 submedian cephalic papillae 'indistinct'. Buccal capsule wider than deep, curved wall decreases in thickness anteriorly, proximal rim with numerous hooked teeth. Proximal lobes of oesophagus protrude into buccal cavity, covered in numerous 'granulations'. Synlophe: 48–50 longitudinal cuticular ridges, equal in height, may be discontinuous. Cervical papillae project above body surface. *Male*: Bursa elongate, dorsal lobe reduced, ventral surface 'papillate'. Prebursal papillae distinct. Ventral rays diverge, close together at distal tips, antero-ventral shorter than postero-ventral. Antero-lateral ray longer and separate from medio- and postero-lateral rays, which are equal, parallel and close together. Externo-dorsal shorter than laterals. Dorsal ray slender, divides in mid-region to form 2 branches, each dividing at distal tips, an additional pair of tiny branches arising from main stem between base and main branches. Genital cone simple. Spicules equal, slender, alate,

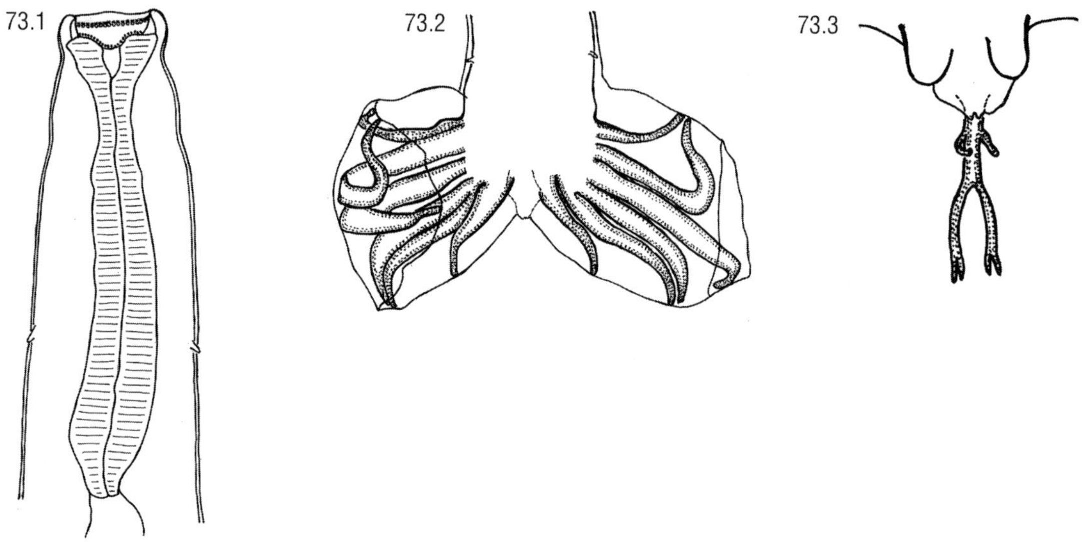

FIG. 73 *Teporingonema* 73.1 Anterior end; 73.2 Bursa; 73.3 Dorsal ray. (Redrawn after Harris, 1985.)

form 2 branches distally, each with terminal barb, main branch with largest barb. *Female*: Vulva opens in posterior half of body. *Vagina vera* leads to opposed ovejectors. Didelphic, amphidelphic. Parasite in stomach of *Romerolagus diazi* in Mexico.

Type species: *Teporingonema cerropeladoensis* Harris, 1985

[Listed in Spencer Jones and Gibson (1987).]

Lamothiella González-Ortega, 1984

DIAGNOSIS: Mouth opening with *corona radiata*, 51–53 elements. Buccal capsule small. *Male*: Bursa with large lateral lobes, small dorsal lobe. *Female*: Vulva opens 1–2 mm from anus. Tail with small spine. Parasite in stomach of *Romerolagus diazi*.

Type species: *Lamothiella romerolagi* González-Ortega, 1984

[This genus, listed in *Helminthological Abstracts* 57, 2901 (published in 1988), was described in a thesis and, although it indicates in the text that the description of the species is 'in press', no primary publication of the description of this genus has been traced. Under Article 9 of the Rules of Zoological Nomenclature, 1999, a thesis does not constitute a publication and the genus is therefore considered invalid. The genus is listed here as a synonym of *Teporingonema* Harris, 1985.]

Hoazinstrongylus Pinto and Gomes, 1985 (Fig. 74)

DIAGNOSIS: Cuticle with fine transverse striations. Cephalic vesicle and synlophe absent. Buccal capsule poorly developed. *Male*: Bursa with large lateral lobes, reduced dorsal lobe; ventral rays with common stem, diverge and come together at distal tips; lateral rays with common stem, antero-lateral curves towards ventrals in distal half, medio- and postero-laterals slightly

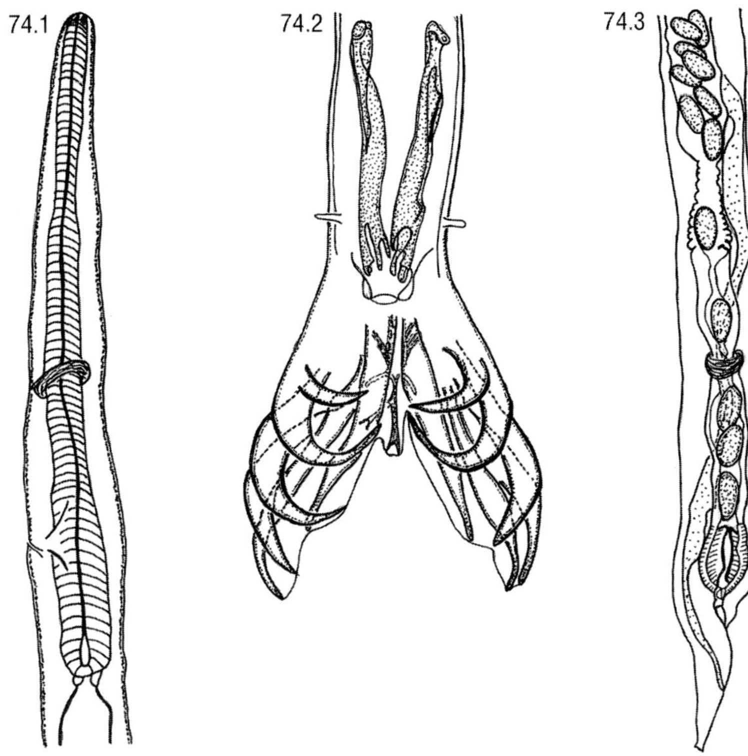

FIG. 74 *Hoazinostrongylus* 74.1 Anterior end, lateral view; 74.2 Posterior end of male, ventral view; 74.3 Posterior end of female, lateral view. (Redrawn after Pinto and Gomez, 1985.)

diverge, close at distal tips; externo-dorsal rays arise near base of dorsal ray, almost twice as long as dorsal ray; dorsal ray with pair of lateral branches in mid-region, bifurcates distally, each branch dividing again at tips. Spicules short, stout, complex, equal. Gubernaculum present. Prebursal papillae prominent. *Female*: Monodelphic. Vulva opens near anus. Eggs thin-walled, oval, morulated before laid. Tail bluntly pointed. Parasites in proventriculus and gizzard of birds, Opisthocomidae in Amazonian region (Ilha do Careiro, Paraná de Cambixa, Brazil).

Type species: *Hoazinstrongylus amazonensis* Pinto and Gomes, 1985

[Listed in Spencer Jones and Gibson (1987).]

Cooperiinae (Skrjabin and Schulz, 1937 tribe) Skrjabin and Schikhobalova, 1952

[Durette-Desset *et al.* (1999) elevate this subfamily to family status following cladistic analysis of the superfamily and include the subfamilies Cooperiinae, Libyostrongylinae Durette-Desset and Chabaud, 1977 and their new subfamily Obeliscoidinae Durette-Desset, Hugot, Darlu and Chabaud, 1999. Following this classification the Cooperiinae now also includes *Cooperioides* Daubney, 1933 (considered a synonym of *Cooperia* in Durette-Desset (1983), *Paracooperioides* Boomker, Horak and de Vos, 1981 and *Pseudostertagia* (Orloff, 1933) Travassos, 1937 (considered a member of the Libyostrongylinae in Durette-Desset (1983). The remaining genera are as in Durette-Desset (1983), i.e. *Gazellostrongylus*, *Chabaudstrongylus*, *Impalaia*, *Megacooperia*, *Minutostrongylus* and *Paracooperia*. *Ortleppstrongylus* Durette-Desset, 1970 was not included in the cladistics study, but it is indicated that it should be included in the Molineinae in one of the new superfamilies Molineoidea proposed in Durette-Desset and Chabaud (1993).

The new subfamily Obeliscoidinae Durette-Desset, Hugot, Darlu and Chabaud, 1999 comprises *Obeliscoides*, transferred from the Libyostrongylinae as type genus, *Biogastronema* Rohrbacher and Ehrenford, 1954 (from the Haemonchinae in Durette-Desset, 1983), *Hoazinstrongylus* Pinto and Gomes, 1985 (from Libyostrongylinae), *Tapironema* Durette-Desset, Sutton and Chabaud, 1997 and *Teporingonema* Harris, 1985 (from the Libyostrongylinae). *Hoazinstrongylus* and *Teporingonema* were placed in this subfamily but not included in the cladistics study of Durette-Desset *et al.* (1999).]

Paracooperioides Boomker, Horak and De Vos, 1981 (Fig. 75)

DIAGNOSIS: Body occasionally coiled anteriorly. *Cooperia*-type cephalic vesicle present. Mouth with 3 indistinct lips. Four submedian cephalic papillae. Synlophe: 10 longitudinal cuticular ridges in mid-body; arête-type; dorso-lateral ridges largest. Cervical papillae small. *Male*: Bursa symmetrical; ventral surface of lateral lobes with tubercles; ventral rays common stem, separate at distal tips, curve anteriorly, antero-ventral ray smaller than postero-ventral ray; lateral rays common stem, separate at distal tips, curve posteriorly; externo-dorsal rays arise near base of dorsal ray, extend to bursal margin; dorsal ray long, divides in distal quarter each branch dividing again, forming a thinner outer and thicker inner branch. Spicules equal, similar, main stem with curved barbed foot at distal tip; branch shorter, sharply pointed. Gubernaculum absent. Genital cone complex. *Female*: Vulva, simple slit with slightly thickened margins, opens in posterior third of body. Didelphic. Tail long, pointed. Parasite in small intestine of grey rhebok, *Pelea capreolus* in South Africa.

Type species: *Paracooperioides peleae* Boomker, Horak and De Vos, 1981

[Boomker *et al.* (1981) follow the classification of Skrjabin, Schikhobalova and Schulz (1954b) and place their genus in the tribe Cooperiea. Gibbons and Khalil (1982a), and Durette-Desset (1983) did not retain the tribes in their keys. Listed in Spencer Jones and Gibson (1987).]

Neocooperia Fotedar and Bambroo, 1971

DIAGNOSIS: Cephalic vesicle poorly developed. Lips rudimentary. Teeth absent. Cervical papillae well developed. *Male*: Bursa symmetrical with 2 large lateral lobes, longer than wide, reduced dorsal lobe. Ventral rays with common stem, postero-ventral longer than antero-ventral. Two lateral rays with common stem, longer than ventrals. Two externo-dorsal rays arise from base of dorsal ray. Dorsal ray bifurcates in distal quarter, each branch bifurcates at tip. Spicules short, 2–3 branches of varying lengths and thickness. Gubernaculum and telamon absent. Genital cone complex. *Female*: Vulva opens in posterior quarter of body. Vulvar flaps posterior and dorsal to vulva. Parasites of duodenum of sheep in Kashmir.

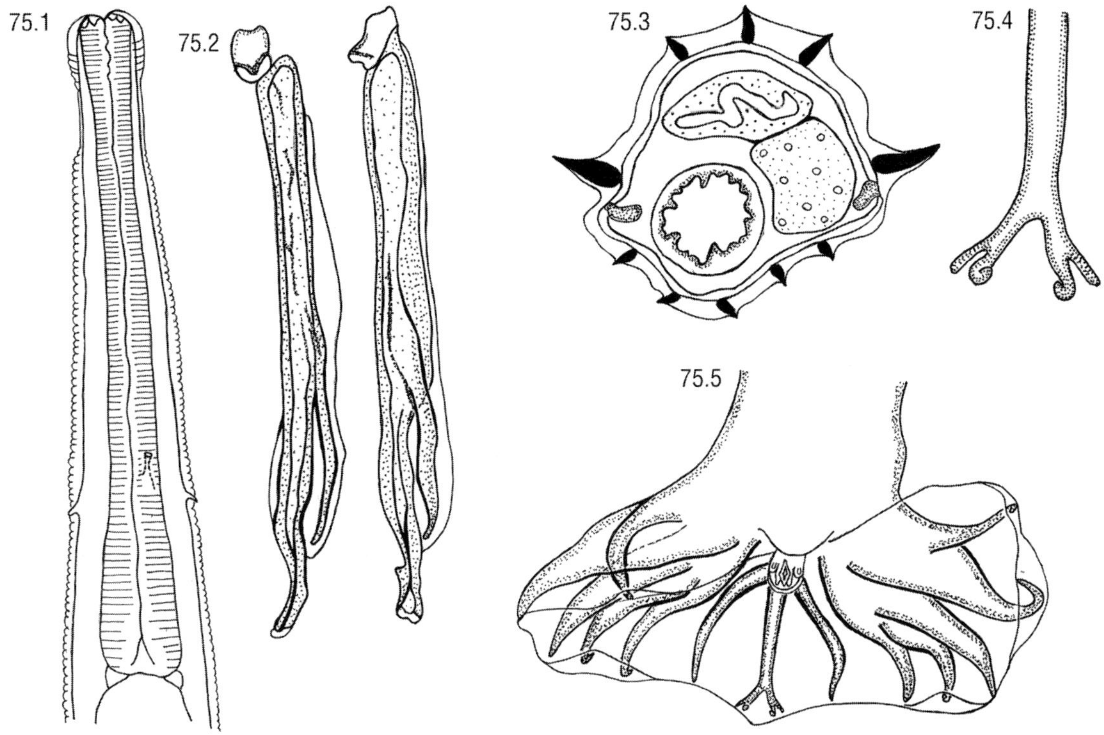

FIG. 75 *Paracooperioides* 75.1 Anterior end, ventral view; 75.2 Spicules; 75.3 Synlophe; 75.4 Distal end of dorsal ray; 75.5 Bursa, ventral view. (Redrawn after Boomker *et al.*, 1981.)

Type species: *Neocooperia magniovata* Fotedar and Bambroo, 1971

[This genus is described without illustration in the Proceedings of the 58th Session of the Indian Science Congress, Bangalore, January 1971, part III (Abstracts) pp. 554–555. Under the Rules of Zoological Nomenclature, 1999, Article 9.9, in the absence of a complete description in a traceable acceptable publication this genus is considered invalid. Listed in Spencer Jones and Gibson (1987).]

Tapironema Durette-Desset, Chabaud and Sutton, 1997 (Fig. 76)

DIAGNOSIS: Mouth opening with *corona radiata*. Buccal ring reduced, dorsal oesophageal tooth present. Synlophe: numerous, simple longitudinal cuticular ridges perpendicular to body surface, closer in lateral fields at mid-body. *Male*: Bursa symmetrical, 'transversely elongated'; antero-lateral ray shorter than other laterals; medio- and postero-lateral rays parallel, long, remain close together. Dorsal ray 'medium-sized'. *Female*: Didelphic. Tail pointed, spine absent. Parasite in the stomach of the water rat, *Holochilus brasiliensis* in Provincia de Buenos Aires, Argentina and *Tapirus terrestris* in Aratray River, French Guyana.

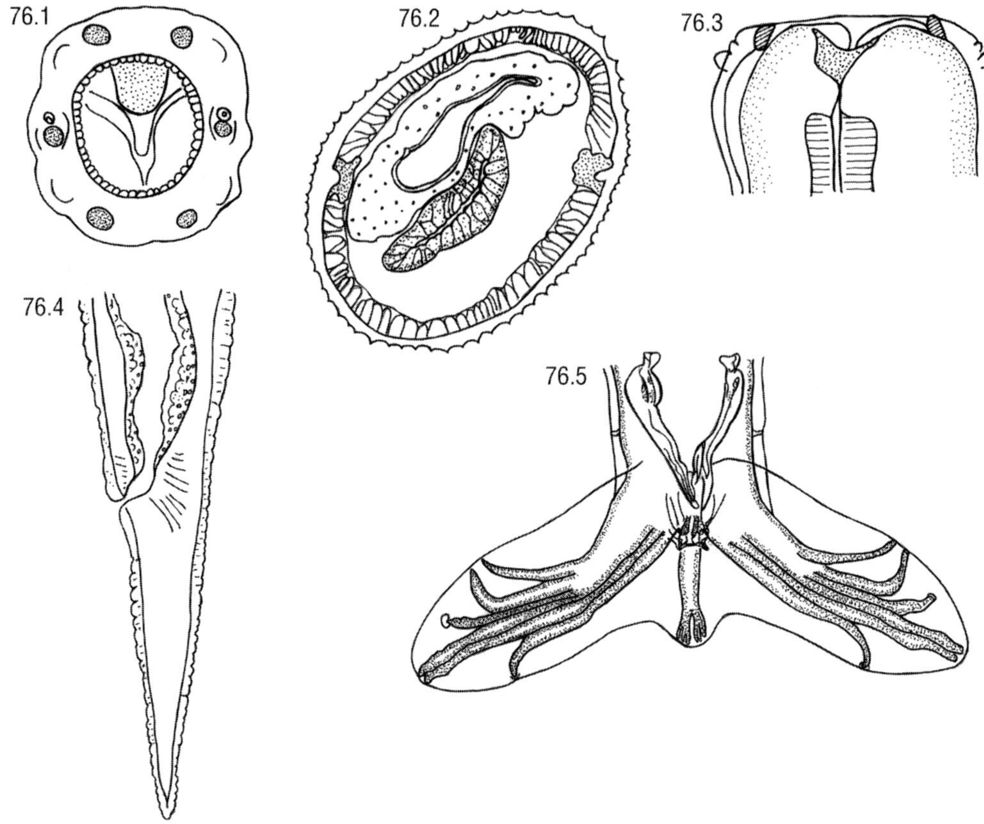

FIG. 76 *Tapironema* 76.1 Cephalic region, *en face* view; 76.2 Synlophe, mid-body region of male; 76.3 Cephalic region, lateral view; 76.4 Female tail, lateral view; 76.5 Bursa, ventral view. (Redrawn after Durette-Desset *et al.*, 1997.)

Type species: *Tapironema coronatum* Durette-Desset, Chabaud and Sutton, 1997

[Durette-Desset *et al.* (1997) place their genus in the family Cooperiidae, subfamily Obeliscoidinae; higher taxa to be established on the basis of studies on the classification of this group in preparation, and later published in 1999. See Durette-Desset *et al.* (1999).]

Graphidiinae Travassos, 1937

[Hoberg and Lichtenfels (1994), following phylogenetic studies, consider this subfamily valid. Durette-Desset *et al.* (1999), in their cladistic study, suppress this subfamily.]

Parostertagia Schwartz and Alicata, 1933

[*Parostertagia* was placed in the subfamily Graphidiinae in Durette-Desset (1983). Gibbons and Khalil (1982a) had previously considered that the genus shared some characteristics with the subfamily Ostertagiinae, but thought the taxonomic position of this genus was uncertain. Hoberg and Lichtenfels (1992) studied a range of characters and placed the genus in the subfamily Cooperiinae. This was further supported by Hoberg *et al.* (1993b) in their studies of vulval inflations and hypertrophy of cuticular ridges within the Cooperiinae and Ostertagiinae. However, Durette-Desset *et al.* (1999) placed this genus in the Trichostrongylinae following a cladistic study of the superfamily. Hoberg and Lichtenfels (1992) characterized this genus as having minute cervical papillae, rudimentary cephalic and cervical expansion and a genital cone with some homologous features with the members of the Cooperiinae. These features led them to their conclusion that this genus should be moved to the Cooperiinae.]

Laurostrongylus Durette-Desset and Chabaud, 1992 (Fig. 77)

DIAGNOSIS: Cephalic vesicle and synlophe absent. Cephalic sensory organs: 4 inner labial papillae; 6 outer labial papillae; 4 submedian cephalic papillae; 2 lateral amphids. Buccal capsule represented by a chitinoid ring. Dorsal oesophageal tooth well developed. Oesophagus claviform. Cervical papillae rounded. *Male*: Bursa small, rays equidistant. Spicules main stem and 2 pointed branches. Gubernaculum racquet-shaped. *Female*: Didelphic, uterus short and thick. Tail short, conical, bluntly pointed. Parasites in stomach of rodents (Cricetidae) in Madagascar.

Type species: *Laurostrongylus hoineffae* Durette-Desset and Chabaud, 1992

[Durette-Desset *et al.* (1999) transfer this genus to the Libyostrongylinae.]

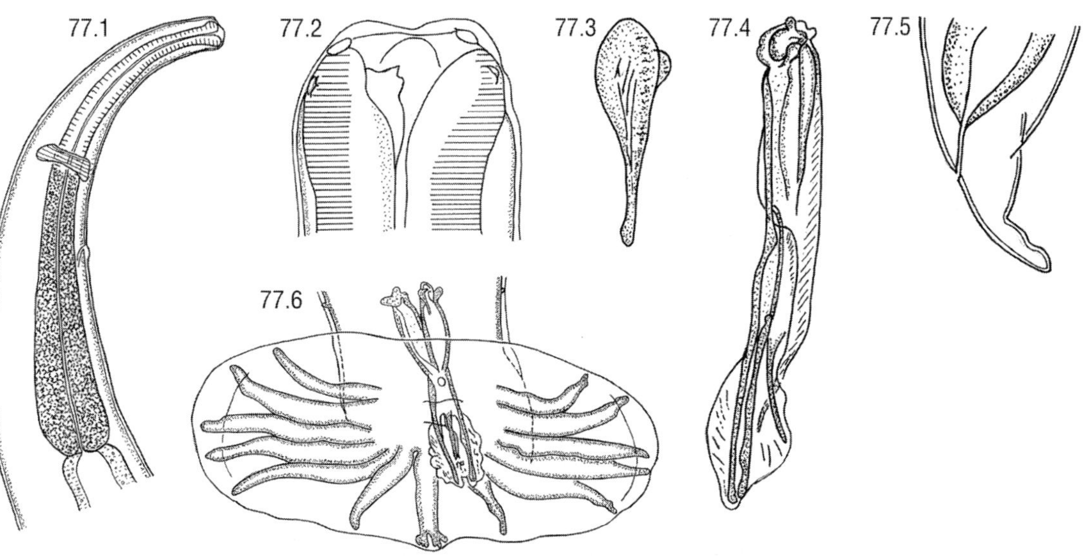

FIG. 77 *Laurostrongylus* 77.1 Anterior end, lateral view; 77.2 Cephalic region, lateral view; 77.3 Gubernaculum, ventral view; 77.4 One of two spicules, ventral view; 77.5 Female tail, lateral view; 77.6 Bursa, ventral view. (Redrawn after Durette-Desset and Chabaud, 1992.)

Graphinema Guerrero and Rojas, 1969 (Fig. 78)

DIAGNOSIS: Buccal capsule small. Oesophagus claviform. Cervical papillae present. *Male*: Antero-ventral ray very small, widely divergent from postero-ventral ray. Lateral rays long, subequal. Externo-dorsal ray arises from main stem of dorsal ray, does not reach bursal margin. Dorsal ray bifurcates near distal tip, each branch dividing distally. Spicules long, end in point. Gubernaculum absent. *Female*: Vulva opens in posterior third. Parasite of *Llama* and *Vicugna* from South America.

Type species: *Graphinema aucheniae* Guerrero and Rojas, 1969

[Guerrero and Rojas (1969) distinguish this genus from *Graphidioides* (Mégnin, 1895) Cameron, 1935 by the presence of cervical papillae and the absence of a gubernaculum. Re-examination of the type species of the two genera may show *Graphinema* to be a synonym of *Graphidoides* placed in the Trichostrongylinae in Durette-Desset (1983) and Durette-Desset *et al.* (1999). Listed in Spencer Jones and Gibson (1987).]

Ostertagiinae (Skrjabin and Schulz, 1937 tribe) Lopez-Neyra, 1947

[Polymorphism among a range of genera within the subfamily has led to many difficulties in maintaining a stable classification. Dróżdż (1995) identified 19 polymorphic species, i.e. species with both a major and minor morph often previously assigned to different genera. The morphological characteristics demonstrating polymorphism are seen in the features of the males. Acceptance of this concept has enabled the re-evaluation of genera, leading to a more correct taxonomy. Accepted genera appear in bold type.

Durette-Desset (1983) originally included five genera in the subfamily, namely *Ostertagia*, *Marshallagia*, *Longistrongylus*, *Spiculopteragia* and *Teladorsagia*. Durette-Desset (1989) proposes that the classification of the subfamily Ostertagiinae should be reorganized to include six genera, namely: (i) ***Ostertagia*** **Ransom, 1907** (= *Capreolagia*

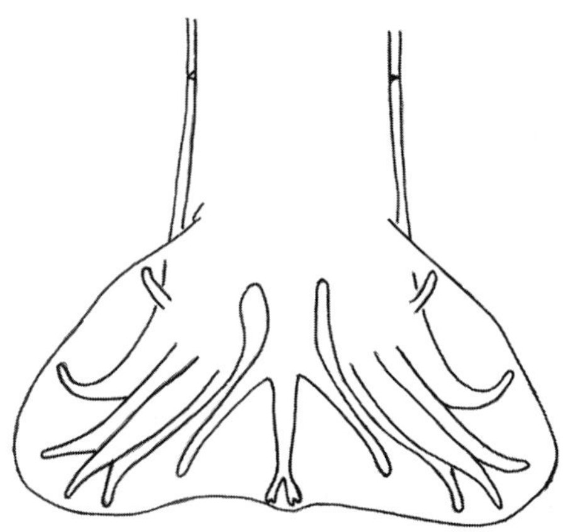

FIG. 78 *Graphinema* Posterior end of male. (Redrawn after Guerrero and Rojas, 1969.)

Schulz, Andreeva and Kadenatsii in Skrjabin, Shikhobalova and Schulz, 1954; *Gruhneria* Sarwar, 1956; *Muflonagia* Schulz, Andreeva and Kadenatsii in Skrjabin, Shikhobalova and Schulz, 1954; *Ostertagiana* Dikov, 1963; *Skrjabinagia* (Kassimov, 1942) Altaev, 1952; *Tunisostertagia* Bernard, 1987); (ii) ***Marshallagia* (Orloff, 1933)** (= *Grosspiculagia* Orloff, 1933; *Ostertagiella* Andreeva, 1957); (iii) ***Camelostrongylus* Orloff, 1933** (= *Orloffia* Dróżdż, 1965; *Ostertamia* Dikov, 1963; *Sjobergia* Sarwar, 1956; *Spiculopteroides* Jansen, 1958); (iv) ***Longistrongylus* Le Roux, 1931** (= *Bigalkea* Mönnig, 1931; *Bigalkenema* Ortlepp, 1963; *Costarcuata* Schulz and Kadenatsii, 1950; *Kobusinema* Ortlepp, 1963; *Pseudomarshallagia* Roetti, 1941); (v) ***Spiculopteragia* (Orloff, 1933)** (= *Altaevia* Sarwar, 1957; *Apteragia* Jansen, 1958; *Mazamastrongylus* Cameron, 1935; *Petrowagia* Ruchliadev, 1961; *Rinadia* Grigoryan, 1951; *Sarwaria*, Dróżdż, 1965); and (vi) ***Teladorsagia* Andreeva and Satubaldin, 1954** (= *Dorsagia* Guerrero and Chavez, 1967; *Paramecistocirrus* Roetti, 1941; *Stadelmannia* Sarwar, 1956). The proposals for these changes are based on their bursal formula of type 2-2-1 or 2-1-2. The suggested changes do not take into account polymorphic species. Suggested changes proposed following studies on polymorphic species are indicated under individual genera. Later, Jansen (1989 publ. 1990) reviewed the subfamily and proposed that it should contain seven genera, namely ***Ostertagia* Ransom, 1907; *Hyostrongylus* Hall, 1921; *Longistrongylus* Le Roux, 1931; *Marshallagia* (Orloff, 1933) Sarwar, 1956; *Mazamastrongylus* Cameron, 1935; *Spiculopteragia* (Orloff, 1933) Travassos, 1937**; and ***Teladorsagia* Andreeva and Satubaldin, 1954.** Jansen (1989 publ.1990) regards the position of *Ostertamia* Dikov, 1963 and *Trichostrongyella* Dikov, 1961 as unclear but considers they have close affinities with *Teladorsagia* and *Mazamastrongylus*, respectively. Durette-Desset *et al.* (1999) transfer this subfamily to their new family Haemonchidae, following cladistic analysis of the superfamily. They include ***Ostertagia* Ransom, 1907** (type genus), ***Camelostrongylus* Orloff, 1933, *Graphidium* Railliet and Henry, 1909** (previously in the Graphidiinae), ***Hyostrongylus* Hall, 1921** (previously considered by Durette-Desset, 1983 to be a member of the Graphidiinae) ***Longistrongylus* Le Roux, 1931, *Marshallagia* Orloff, 1933, *Spiculoptergia* Orloff, 1933** and ***Teladorsagia* Andreeva and Satubaldin, 1954** in the subfamily.

Hoberg and Abrams (2007) support *Sarwaria* Dróżdż, 1965 as an independent genus, as previously indicated by Gibbons and Khalil (1982a) and Lichtenfels, Pilitt and Craig (1996). Hoberg and Abrams (2007) also propose the subfamily should include 12 valid genera based on their bursal formula of 2-2-1, namely ***Cervicaprastrongylus***, ***Hyostrongylus***, ***Mazamastrongylus***, ***Spiculopteragia***, ***Teladorsagia***, ***Sarwaria***; and on 2-1-2, namely ***Camelostrongylus***, ***Longistrongylus***, ***Marshallagia***, ***Orloffia***, ***Ostertagia*** and ***Pseudomarshallagia***, on the understanding that these placings need 'to be confirmed in the context of phylogenetic analysis designed specifically to resolve generic-level relationships'. Hoberg and Abrams (2008), Hoberg *et al.* (2008) and Hoberg *et al.* (2009) described three more genera for the subfamily, namely ***Africanastrongylus***, ***Hamulonema*** and ***Robustostrongylus***. Currently 15 genera are recognized within the subfamily.]

Ostertagia Ransom, 1907

[Dróżdż (1995), in his study of polymorphism in the subfamily Ostertagiinae Lopez-Neyra, 1947, considers the species of *Skrjabinagia* (Kassimov, 1942) Altaev, 1952 as minor morphs of the species of the genus *Ostertagia* and that the genus should be recognized as a synonym, which confirms the view of Durette-Desset (1983), who included *Skrjabinagia* as a synonym of *Ostertagia*.]

Marshallagia Orloff, 1933

[Dróżdż (1995), in his study of polymorphism in the subfamily Ostertagiinae Lopez-Neyra, 1947, considers the species of *Grosspiculagia* Orloff, 1933 as minor morphs of the species of *Marshallagia* and that the genus should be considered a synonym, which confirms the view of Durette-Desset (1983), who included *Grosspiculagia* as a synonym of *Marshallagia*.]

Spiculopteragia Orloff, 1933

[Dróżdż (1995), in his study of polymorphism in the subfamily Ostertagiinae Lopez-Neyra, 1947, considers the species of *Rinadia* Grigorian, 1951 and *Apteragia* Jansen, 1958 as minor morphs of the species of *Spiculoptergia* and that the genera should be recognized as synonyms, which confirms the view of Durette-Desset (1983), who included *Rinadia* and *Apteragia* as synonyms of *Spiculoptergia*.]

Hyostrongylus Hall, 1921

DIAGNOSIS: *Male*: Bursa with well-developed lateral lobes, slightly reduced dorsal lobe. Ventral rays with common stem, diverge, come together at distal tips. Antero- and medio-lateral rays diverge at tips, antero-lateral curving towards ventrals. Dorsal ray with 1 pair of branches half-way along its length which curve, a second pair of thin branches arising at level of bifurcation near distal end which form third pair of branches. Genital cone with a pair of short dorsal raylets associated with a simple accessory bursal membrane. Spicules equal, branched. Parasites of suids, giraffes and pongids.

Type species: *Hyostrongylus rubidus* (Hassall and Stiles, 1892) Hall, 1921

[Emended diagnosis based on Gibbons and Khalil, 1982a, who placed the genus in the Ostertagiinae. *Hyostrongylus* was placed in the subfamily Graphidiinae by Durette-Desset (1983) in her key. Jansen (1989 publ.1990) considers the genus may have an intermediary position between the Ostertagiinae and the Graphidiinae. Hoberg *et al.* (1993b) studied the genital cone, synlophe and bursa of the type species *H. rubidus* (Hassall and Stiles, 1892) and concluded that the genus should be placed in the Ostertagiinae rather than in the Graphidiinae, and this is further supported by Hoberg *et al.* (2008).]

Mazamastrongylus Cameron, 1935 (Fig. 79)

DIAGNOSIS: Synlophe in cervical region with hood ridges on ventral side surrounding prominent margins of excretory pore, strongly tapering pattern of ridges laterally and dorsally, absence of clear subdorsal and subventral ridge patterns. *Male*: Bursa with ventral rays only slightly divergent, distal tips close together. Lateral rays with common stem, antero-lateral and medio-lateral rays close together in proximal half, diverge in distal half, postero-lateral separate for most of their length from other laterals. Genital cone with 2 dorsal raylets straight to slightly curved, surrounded by an accessory bursal membrane. Spicules with spoon-shaped dorsal process. Parasite of ruminants from South-east Asia or North and South America.

Type species: *Mazamastrongylus trinitatus* Cameron, 1935

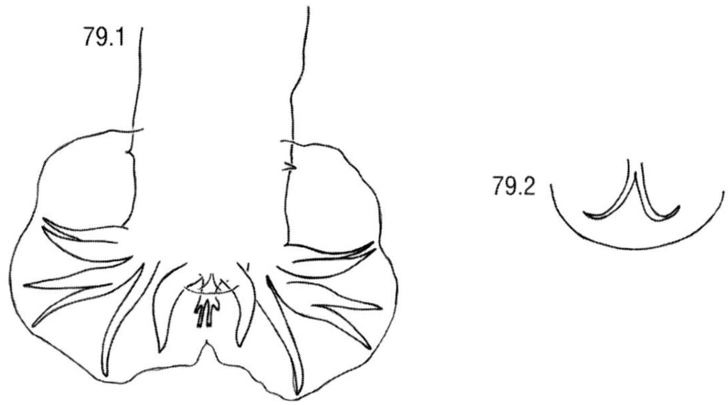

FIG. 79 *Mazamastrongylus* 79.1 Male tail (Redrawn after Cameron, 1935); 79.2 Accessory bursal membrane of genital cone (Redrawn after Jansen, 1986.)

[Jansen (1986) validates the genus, listed as a synonym of *Spiculopteragia* (Orloff, 1933) Travassos, 1937 by Durette-Desset (1983) in her key, and places it in the subfamily Ostertagiinae. The emended diagnosis is based on Jansen (1986), Hoberg (1996) and Hoberg and Khrustalev (1996).]

Cervicaprastrongylus Gibbons and Khalil, 1982

DIAGNOSIS: Cephalic vesicle of 'ostertagid-type'. Cervical papillae small, project above body surface. Synlophe: numerous longitudinal cuticular ridges, more or less even in height, symmetrically arranged. *Male*: Prebursal papillae large. Bursa symmetrical, dorsal lobe slightly reduced. Ventral rays with common stem, diverge and come together at distal tips. Antero- and medio-lateral rays close, parallel. Dorsal ray divides near distal tip forming 2 pairs of branches. Genital cone with 2 ventral papillae, complex structure dorsal to cloaca incorporating a pair of modified rays. Spicules equal with 2 branches, equal or unequal, and a main stem which curves at distal tip. Gubernaculum present. Parasites of lagomorphs and ruminants.

Type species: *Cervicaprastrongylus malviyai* (Chaturvedi and Kansal, 1977) Gibbons and Khalil, 1982.

[*Cervicaprastrongylus* was considered a synonym of *Hyostrongylus* by both Jansen (1989 publ. 1990) and Durette-Desset *et al.* (1992). Hoberg *et al.* (1993a) compared the spicules, bursa and genital cone of the type species *Hyostrongylus rubidus* with the descriptions of species assigned to *Cervicaprastrongylus* and considered that the genus was not a synonym of *Hyostrongylus*. Listed in Spencer Jones and Gibson (1987).]

Tunisostertagia Bernard, 1987 (Figs 80.1, 80.2 and 80.3)

DIAGNOSIS: Reduced buccal capsule present. Cervical papillae small. *Male*: Bursa symmetrical: ventral rays with common stem, separate from lateral rays, parallel along length, antero-ventral ray narrower at base than postero-ventral ray; lateral rays parallel, medio-lateral ray widest at base, postero-lateral ray shorter than other laterals, diverge at distal tips; externo-dorsal ray short, thick, arises from base of dorsal ray; dorsal ray slender, long, 2 branches each dividing at distal tip, shorter than externo-dorsal ray. Genital cone: 2 pairs of rays and accessory bursal membrane, outer pair of rays longer than inner pair. Spicules alate, unequal, main stem and 2 branches. *Female*: Unknown. Parasite of cattle, *Bos taurus*, in Tunis.

Type species: *Tunisostertagia heterospiculum* Bernard, 1987

[The author regards this genus as close to *Ostertagia* and, for this reason, I have moved it from the Trichostrongylinae to the Ostertagiinae. Durette-Desset (1989) and Jansen (1989 publ.1990) considers this genus a synonym of *Ostertagia*.]

Africanastrongylus Hoberg, Abrams and Ezenwa, 2008 (Figs 80.4, 80.5, 80.6, 80.7, 80.8 and 80.9)

DIAGNOSIS: Synlophe tapering bilateral symmetry. Cervical papillae minute, project above body surface. Well-defined oesophago-intestinal valve. *Male*: Monomorphic. Bursal rays symmetrical, 2-2-1 arrangement, i.e. ventral rays with common stem, diverge in mid-region, close together at distal tips; antero- and medio-lateral rays parallel, only divergent at distal tips; postero-lateral ray separate. Genital cone: accessory bursal membrane bilobed with dorsal raylets or ray 7; proconus 'weakly developed'; ventral raylets or papillae 0 paired. Telamon present at cloaca. Spicules subequal with main stem and 2 distal branches, 1 barbed. Gubernaculum present. *Female*: Amphidelphic. Vulva opens in posterior quarter. Vulva flaps or 'inflations' absent. Parasite in abomasum of African or Cape buffalo, *Syncerus caffer caffer*, in Uganda, Kenya and South Africa.

Type species: *Africanastrongylus buceros* Hoberg, Abrams and Ezenwa, 2008

[This brief diagnosis is based on Hoberg *et al.* (2008).]

Hamulonema Hoberg and Abrams, 2008

DIAGNOSIS: Synlophe bilateral parallel symmetry. Cervical papillae project above body surface. Prominent oesophago-intestinal valve. *Male*: Monomorphic. Bursal rays symmetrical, 2-2-1 arrangement, i.e. ventral rays separate, close together at distal tips, antero- and mediolateral rays parallel, postero-lateral ray separate. Genital cone: accessory bursal membrane and dorsal raylet or ray 7 reduced. Externo-dorsal ray curves towards median line. Dorsal lobe reduced. Dorsal ray with thick base divides in distal third. Spicules with main stem and 2 branches (1 pointed, 1 'hook-like'). *Female*: Amphidelphic. Vulva opens in posterior quarter. Expansions of cuticle may be present.

Type species: *Hamulonema hamata* (Mönnig, 1932) Hoberg and Abrams, 2008

[This brief diagnosis is based on Hoberg and Abrams (2008).]

Robustostrongylus Hoberg, Abrams and Pilitt, 2009 (Figs 81.1 and 81.2)

DIAGNOSIS: Synlophe bilaterally symmetrical with well-developed rounded ridges. Cephalic vesicle present. Cervical papillae minute adjacent to posterior region of oesophagus. Oesophago-intestinal valve well defined, bilobed, with 'constricted ventriculus-like' distal end. *Male*: Bursa: symmetrical without cuticular bosses; 2-1-2 ray formation; ventral rays parallel, postero-ventral

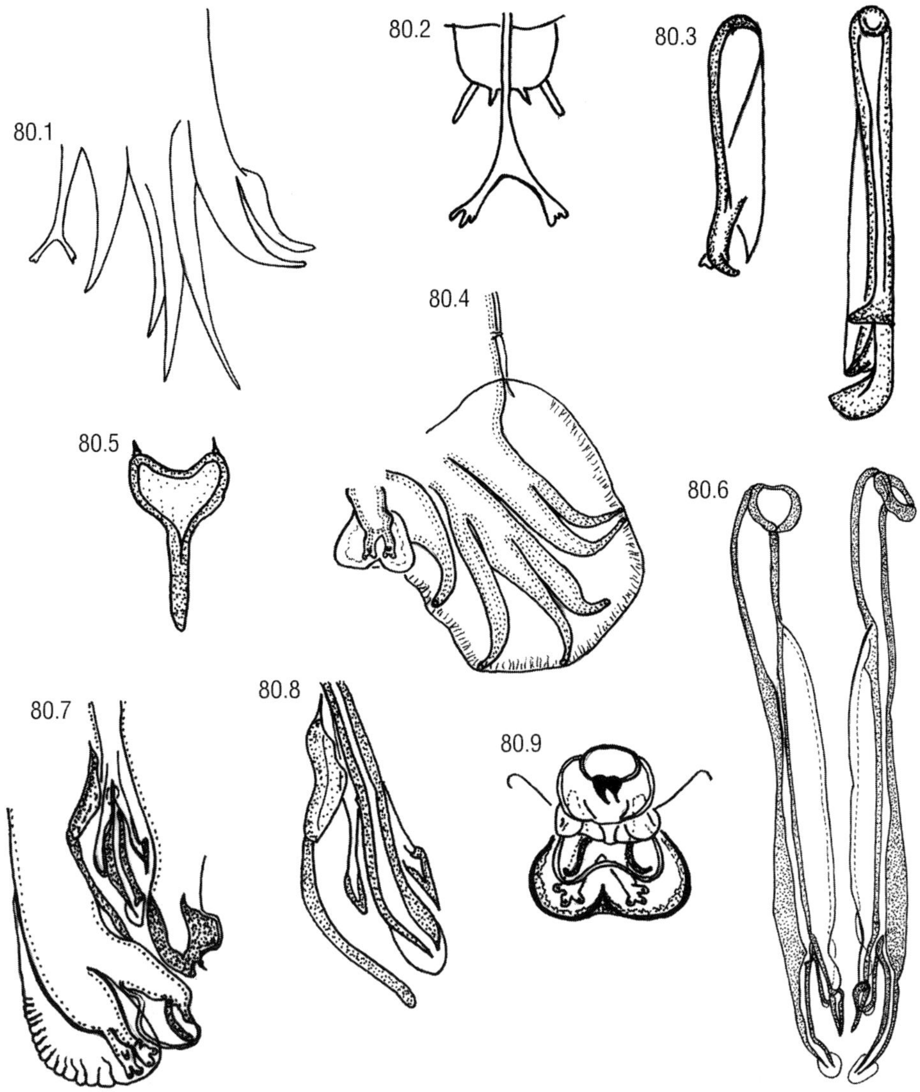

FIGS 80.1, 80.2 and 80.3 *Tunisostertagia* 80.1 Arrangement of bursal rays in one lateral lobe and dorsal lobe; 80.2 Dorsal ray and genital cone; 80.3 Spicules. (Redrawn after Bernard, 1987.)
FIGS 80.4, 80.5, 80.6, 80.7, 80.8 and 80.9 *Africanastrongylus* 80.4 Left lobe of bursa, ventral view; 80.5 Gubernaculum, ventral view; 80.6 Spicules, ventral view; 80.7 Genital cone, right lateral view; 80.8 Gubernaculum and distal tip of spicules, right lateral view; 80.9 Genital cone, ventral view. (Redrawn after Hoberg *et al.*, 2008.)

wide, similar in width or wider than antero-lateral; antero-lateral wide, separate from other laterals; medio- and postero-laterals narrow, divergent distally; externo-dorsal rays with broad bases; dorsal ray short, divides distally. Genital cone with reduced accessory bursal membrane and widely spaced dorsal raylets (papillae 7). *Female*: Amphidelphic. Vulva opens approximately at junction of middle and posterior thirds of body. Parasites in abomasum of artiodactylids in Uganda.

Type species: *Robustostrongylus aferensis* Hoberg, Abrams and Pilitt, 2009

[Brief descripton based on Hoberg *et al.* (2009).]

Haemonchinae (Skrjabin and Schulz, 1937 tribe) Skrjabin and Schulz, 1952

[Durette-Desset *et al.* (1999) raise this subfamily to family status following cladistic analysis of the superfamily. The subfamilies Haemonchinae and Ostertagiinae are included in the family. The genera *Haemonchus* Cobb, 1898 (type genus), *Ashworthius* Le Roux, 1930, *Boehmiella* Gebauer, 1932, *Leiperiatus* Sandground, 1930 and *Mecistocirrus* Railliet and Henry, 1912 are included in the subfamily Haemonchinae. Lichtenfels and Pilitt (2000) and Hoberg *et al.* (2002, 2004) describe the synlophe and its application in this subfamily for the study of relationships and identification of genera and species.]

Filarinematinae (Skrjabin and Schikhobalova, 1937 tribe)

[This subfamily is erected by Durette-Desset *et al.* (1999) with *Filarinema* Mönnig, 1929 as type genus following cladistic analysis of the superfamily. *Peramelistrongylus* Mawson, 1960 and *Profilarinema* Durette-Desset and Beveridge, 1981 are also included in the subfamily. The three genera were formerly placed in the family Dromaeostrongylidae in Durette-Desset (1983).]

***Filarinema* Mönnig, 1929 (Figs 81.3 and 81.4)**

DIAGNOSIS: Lips, buccal capsule and cephalic vesicle absent. Dorsal intra-oesophageal tooth present. Oesophagus: very long; approximately one-tenth of total body length; proximal part of oesophageal cuticle thick. Synlophe absent. *Male*: Bursa with antero-ventral very divergent from postero-ventral; antero-lateral generally short; medio- and postero-laterals diverge; dorsal ray divides in posterior half. Ventral transverse shaft of cuticle between opposing postero-ventral and antero-lateral rays. Spicules: generally divided into 3 branches, sometimes 2. *Female*: Didelphic. Vulva opens in the posterior part of the body. Branches of vestibule of equal length. Tail rounded, caudal point absent. Parasites of the pyloric region of the stomach of Macropodidae.

Type species: *Filarinema flagrifer* Mönnig, 1929

[Emended diagnosis based on Cassone and Baccam (1985).]

Amidostomatidae (Travassos, 1919 subfamily) Baylis and Daubney, 1926

[Lomakin (1991) revised this family and included two subfamilies, Amidostomatinae and Epomidiostomatinae. Later, Lomakin (1993) revised the genera of the subfamily Amidostomatinae. Durette-Desset *et al.* (1999) applied cladistic analysis to the superfamily Trichostrongyloidea and produced a classification that did not retain this family, which had previously been included in Durette-Desset (1983).]

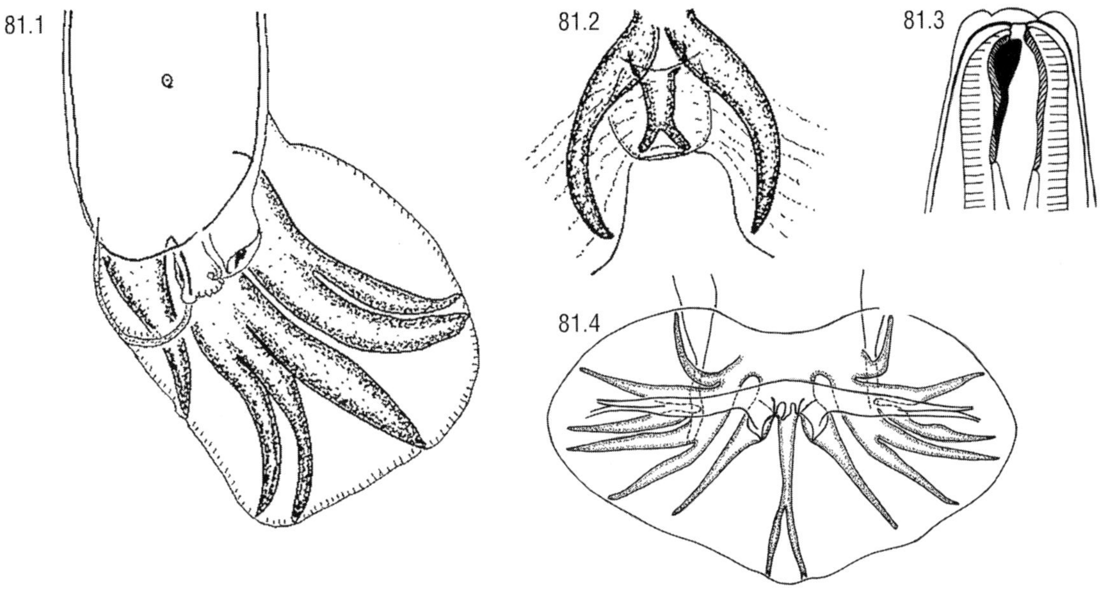

FIGS 81.1 and 81.2 *Robustostrongylus* 81.1 Bursa, right lateral lobe; 81.2 Dorsal lobe and ray with externo-dorsal rays. (Redrawn after Hoberg, Abrams and Pilitt, 2009.)
FIGS 81.3 and 81.4 *Filarinema* 81.3 Cephalic region, lateral view; 81.4 Bursa. (Redrawn after Cassone and Baccam, 1985.)

Amidostomatinae Travassos, 1919

[Durette-Desset *et al.* (1999), following cladistic analysis of the superfamily Trichostrongyloidea, transferred this subfamily to the family Trichostrongylidae. *Amidostomum* Railliet and Henry, 1909 is retained as type genus and the following genera are also included: *Epomidiostomum* Skrjabin, 1915 and *Pseudamidostomum* Boulenger, 1926 (formerly in the Epomidiostomatinae Skrjabin and Schulz, 1937), *Paramidostomum* Freitas and Mendonça, 1949 (formerly in the Dromaeostrongylidae in Durette-Desset (1983)). The subfamily Epomidiostomatinae Skjrabin and Schulz, 1937 is not retained.]

Amidostomum Railliet and Henry, 1909

[Durette-Desset *et al.* (1999), following cladistic analysis of the superfamily, transferred this genus to the Trichostrongylidae.]

Subgenus *Amidostomum* (Railliet and Henry, 1909 genus) Petrova, 1987

DIAGNOSIS: Buccal capsule: internal dimensions 0.025–0.030 mm; external dimensions 0.034–0.056 mm with 3 teeth. Synlophe absent. *Male*: Spicules with 2–4 branches at distal end; 0.186–0.376 mm long. Gubernaculum 0.089–0.160 mm long. Parasites of Anseriformes and Ciconiiformes (Aves).

Type species: *Amidostomum* (*Amidostomum*) *anseris* (Zeder, 1800)

Subgenus *Amidostomoides* Petrova, 1987 (Fig. 82)

DIAGNOSIS: Buccal capsule: internal dimensions 0.011–0.025 mm; external dimensions 0.021–0.031 mm with single tooth. Synlophe absent. *Male*: Spicules with 2–3 branches at distal end; 0.112–0.276 mm long. Gubernaculum 0.053–0.091 mm. Parasites of Anseriformes, Charadriiformes, Ralliformes (Aves).

Type species: *Amidostomum* (*Amidostomoides*) *acutum* (Lundahl, 1848)

[Lomakin (1991) upgrades this subgenus to generic rank.]

Hexapapillostomum Lomakin, 1991

DIAGNOSIS: A translation of diagnosis (Lomakin, 1991) is not available – see original description for details.

Type species: *Hexapapillostomum biziurae* (Johnston and Mawson, 1947) Lomakin, 1991

Mesamidostomum Lomakin, 1991 (Fig. 83)

DIAGNOSIS: A translation of diagnosis (Lomakin, 1991) is not available – see original description for details.

Type species: *Mesamidostomum skrjabini* (Ryzhikov and Romanova, 1958) Lomakin, 1991

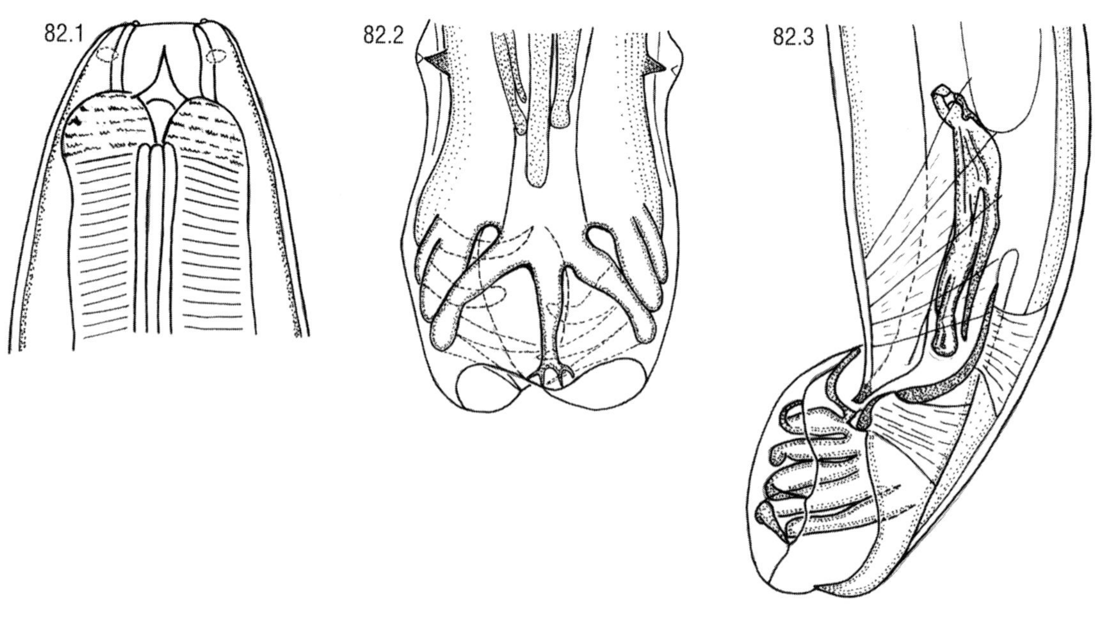

FIG. 82 *Amidostomoides* 82.1 Cephalic region; 82.2 Male tail, dorsal view; 82.3 Male tail, lateral view. (Redrawn after Lomakin, 1993.)

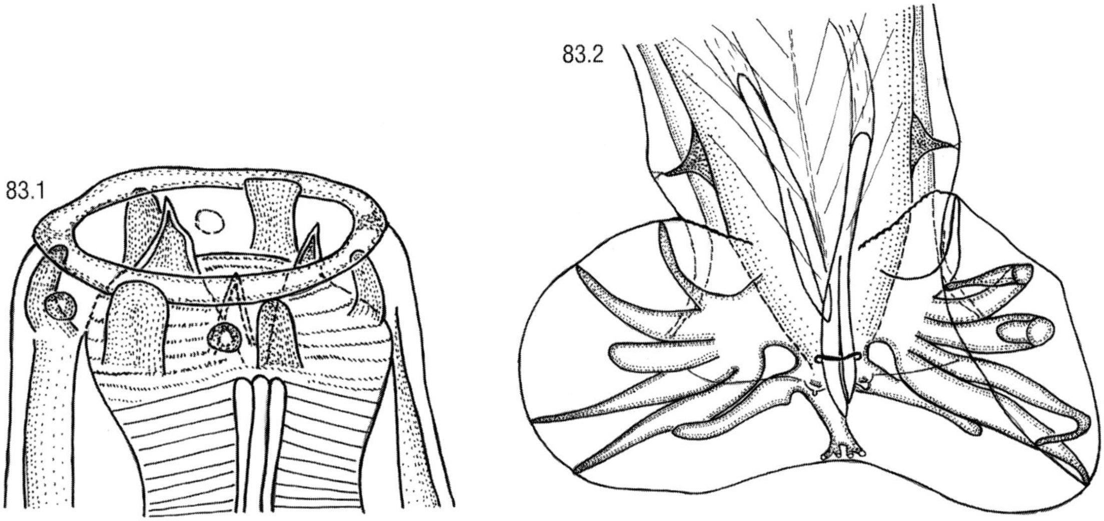

FIG. 83 *Mesamidostomum* 83.1 Cephalic region; 83.2 Male tail. (Redrawn after Lomakin, 1993.)

Quasiamidostomum Lomakin, 1991 (Fig. 84)

DIAGNOSIS: A translation of diagnosis (Lomakin, 1991) is not available – see original description for details.

Type species: *Quasiamidostomum fulicae* (Rudolphi, 1819) Lomakin, 1991

Dromaeostrongylidae (Skrjabin and Schulz, 1937 tribe) Durette-Desset and Chabaud, 1981

[Durette-Desset *et al.* (1999), following cladistic analysis for the superfamily, transfer this family to the Molineoidea Durette-Desset and Chabaud, 1993 as a subfamily containing the single genus *Dromaeostrongylus* Lubimov, 1933.]

Paramidostomum Freitas and Medonça, 1949

[Durette-Desset (1983) placed the genus in the Dromaeostrongylidae. Lomakin (1991) transfers this genus to the Amidostomatidae and the subfamily Amidostomatinae. Durette-Desset *et al.* (1999), following cladistic analysis of the superfamily Trichostrongyloidea, also transferred this genus to the Amidostomatinae, which they regard as a subfamily of the Trichostrongylidae.]

Amphibiophilidae Durette-Desset and Chabaud, 1981

[This family is placed in the new superfamily Molineoidea (see the definition of this superfamily under the family Molineidae) in the suborder Trichostrongylina in the revised classification proposed by Durette-Desset and Chabaud, 1993.]

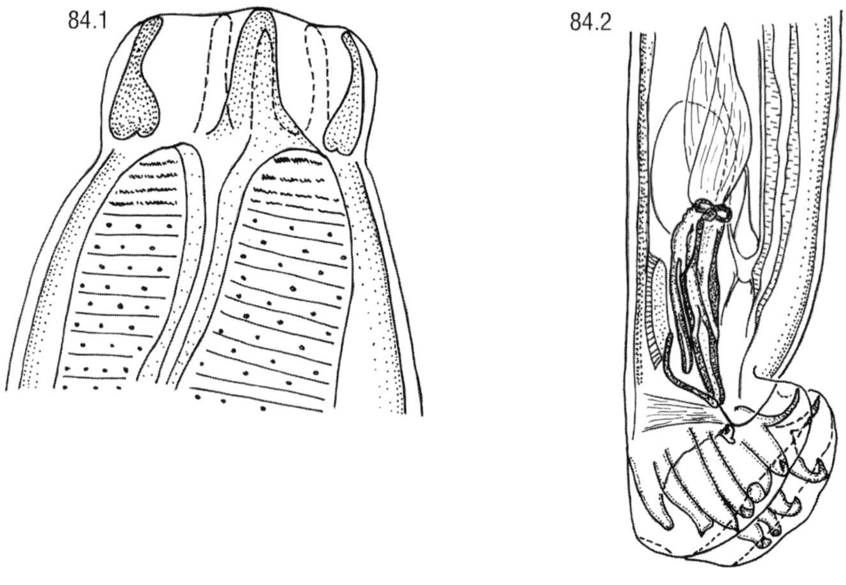

FIG. 84 *Quasiamidostomum* 84.1 Cephalic region; 84.2 Male tail. (Redrawn after Lomakin, 1993.)

Batrachostrongylus Yuen, 1963

[Baker (1983) suggest that this genus has 'trichostrongyle-like cephalic structures and an ovejector resembling a type 1 ovejector (after Lichtenfels, 1980a)'. He indicates that it could easily be separated from *Amphibiophilus* Skrjabin, 1916 in its own subfamily as no other trichostrongyle has this type of ovejector. However, until other species from the same geographical region are studied or become known, he suggests the link with this family remains on the basis of bursal morphology.]

Johnpearsoniinae Durette-Desset, Ben Slimane, Cassone, Barton and Chabaud, 1994

DIAGNOSIS: Larval synlophe bilaterally symmetrical, adult synlophe not always bilaterally symmetrical. *Male*: Bursal rays 1-3-1 arrangement, i.e. antero-ventral ray separate; postero-ventral, antero-lateral, medio-lateral rays close together; postero-lateral ray separate. *Female*: Didelphic or monodelphic. Tail with terminal spine. Parasites in Gondwanian amphibians.

Type genus: *Johnpearsonia* Durette-Desset, Ben Slimane, Cassone, Barton and Chabaud, 1994.

[Durette-Desset *et al.* (1994) also include *Batrachonema* Yuen, 1965 in their new subfamily, formerly in the family Nicollinidae in Durette-Desset (1983).]

Johnpearsonia Durette-Desset, Ben Slimane, Cassone, Barton and Chabaud, 1994 (Fig. 85)

DIAGNOSIS: Six lips surround mouth opening. Well-developed buccal capsule present. Dorsal oesophageal tooth present. Ring at base of buccal capsule absent. Larval synlophe bilaterally

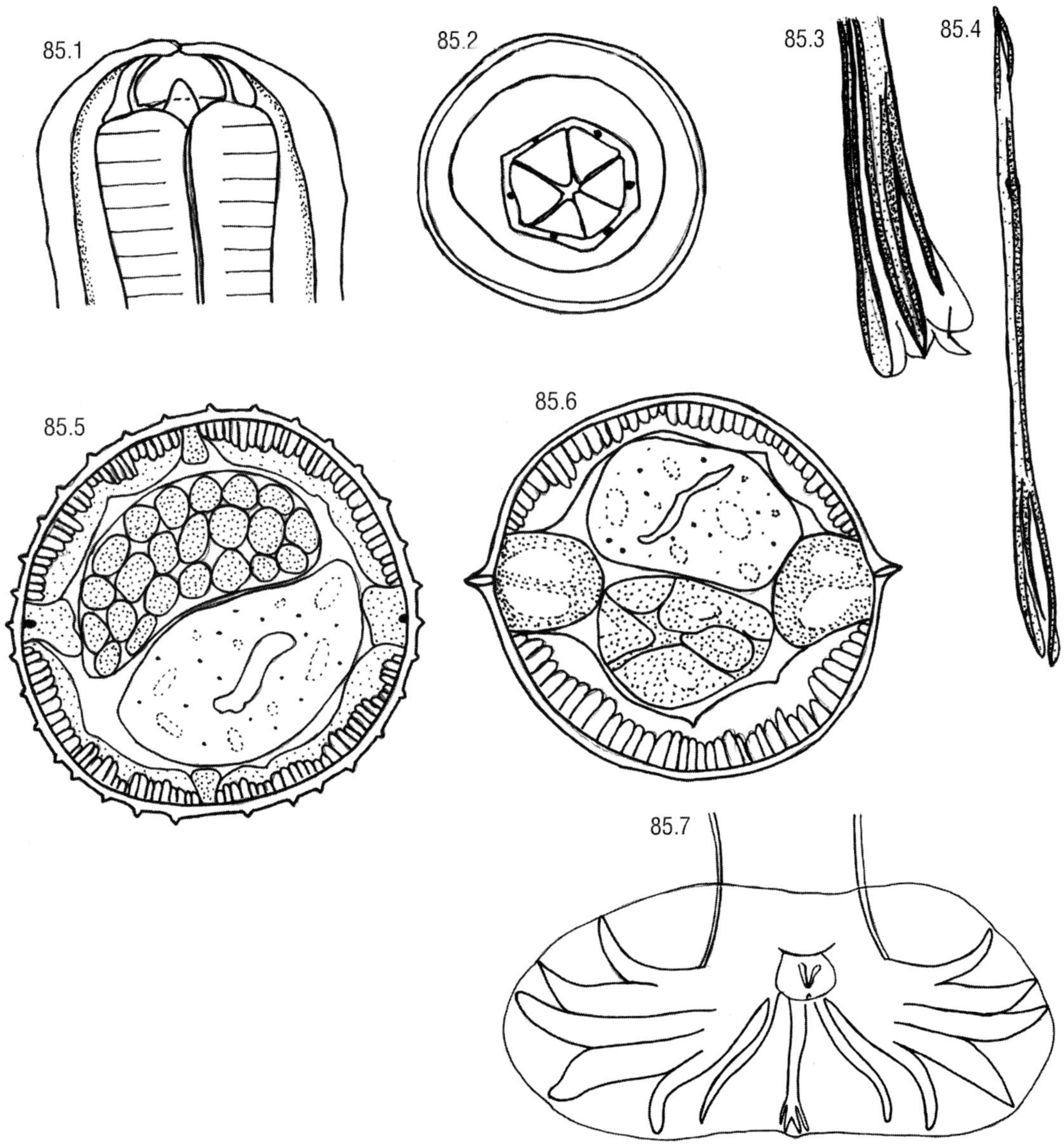

FIG. 85 *Johnpearsonia* 85.1 Cephalic region, dorsal view; 85.2 Cephalic region, *en face* view; 85.3 Distal end of right spicule; 85.4 Right spicule; 85.5 Synlophe of male, mid-body region; 85.6 Synlophe of male fourth stage larva, mid-body region; 85.7 Bursa, ventral view. (Redrawn after Durette-Desset *et al*., 1994.)

symmetrical, adult synlophe bilaterally symmetrical or with an oblique axis. *Male*: Bursal rays 1-3-1 arrangement (see subfamily diagnosis). Dorsal lobe same length as lateral lobes. Externo-dorsal ray arises at base of dorsal ray. Spicules with 2–3 points. Gubernaculum present. *Female*: Monodelphic. Tail with terminal spine. Parasites of Australian amphibians.

Type species: *Johnpearsonia pearsoni* Durette-Desset, Ben Slimane, Cassone, Barton and Chabaud, 1994.

Dictyocaulidae (Skrjabin, 1933, subfamily) Skrjabin, 1941

[Durette-Desset and Chabaud (1993) move this family to their new superfamily Molineoidea in the suborder Trichostrongylina.]

Dictyocaulus Railliet and Henry, 1907

[Chilton *et al.* (2006), using molecular techniques, discussed the evolutionary origin of nematodes within the Strongylida. Their studies show that *Dictyocaulus* was not positioned with other representatives of the Trichostrongylina and therefore does not have a close relationship with this group, but is aligned with the Metastrongylina instead.]

Arnfieldia Sarwar, 1957

DIAGNOSIS: *Male*: Dorsal ray with short proximal stem divided for most of its length, each branch divided at tip; medio- and postero-lateral rays joined for approximately one-third to one-half of their length, separate in distal half. *First stage larva*: Transparent vesicle at anterior and posterior ends. Parasite of trachea and bronchi of equids.

Type species: *Arnfieldia arnfieldi* (Cobbold, 1884) Sarwar, 1957

[Sarwar (1957) divides *Dictyocaulus* into two subgenera, namely *Dictyocaulus* and *Micrurocaulus* Skrjabin, 1934, and creates a separate genus for *D. arnfieldi* (Cobbold, 1884). Durette-Desset (1983) in her key regards *Micrurocaulus* as a synonym of *Dictyocaulus* Railliet and Henry, 1907. The splitting of the genus on the basis of dorsal and lateral ray morphology is not accepted and, consequently, *Arnfieldia* is considered a synonym of *Dictyocaulus*. Listed in Spencer Jones and Gibson (1987).]

Cardiostrongylus Sakamoto and Malgor, 1995 (Figs 86.1, 86.2 and 86.3)

DIAGNOSIS: Cephalic vesicle present. Synlophe: numerous longitudinal cuticular ridges present, even height in cross-section. Mouth circular to oval without lips. Buccal capsule wall slightly thickened. Oesophageal funnel reduced, tooth absent. *Male*: Bursa bell-shaped, lobes not separate. Dorsal ray divided to base; externo-dorsal rays separate, approximately half length of dorsals. Spicules thick, porous texture. Gubernaculum present. *Female*: Vulva opens near middle of body. Ovejectors opposed. Tail tapers, terminal spine absent. Parasite in the heart muscle of Sika deer in Japan.

Type species: *Cardiostrongylus sikae* Sakamoto and Malgor, 1995.

Molineidae (Skrjabin and Schulz, 1937 subfamily) Durette-Desset and Chabaud, 1977

[Durette-Desset and Chabaud (1993) create a new superfamily Molineoidea (Skrjabin and Schulz, 1937 subfamily) in the suborder Trichostrongylina for this family and Amphibiophilidae Durette-Desset and Chabaud, 1981, Dictyocaulidae (Skrjabin, 1933) and Mackerrostrongylidae (Inglis, 1968). The superfamily is defined by Durette-Desset and Chabaud (1993) as: primitive forms with 6 lips and well-developed buccal capsule; cephalic vesicle present; synlophe present (except Dictyocaulidae), bilaterally symmetrical with an early appearance of an axis of orientation. *Male*: bursa of 2-3 type with a tendency to the 2-1-2 type. *Female*: generally didelphic. Parasites of amphibians, reptiles, monotremes, Australian marsupials, Eutherian mammals that appeared at the beginning of the Tertiary, lagomorphs and ruminants.

Gibbons and Khalil (1988) demonstrated the presence of a synlophe in *Dictyocaulus*, using light and scanning electron microscopy.]

Molineinae Skjrabin and Schulz, 1937

[Rickard and Hoberg (2000) include *Lamanema* Becklund, 1963 in this subfamily, moving it from the Nematodirinae following the re-evaluation of morphological features including the spicules, bursa and female tail. The genus is referred to this subfamily pending phylogenetic analysis.]

Schulzia Travassos, 1937 (Figs 86.4 and 86.5)

DIAGNOSIS: Synlophe with many unsupported longitudinal cuticular ridges. *Male*: Bursa with dorsal ray longer than other rays, ventral and lateral rays evenly spaced. Gubernaculum absent.

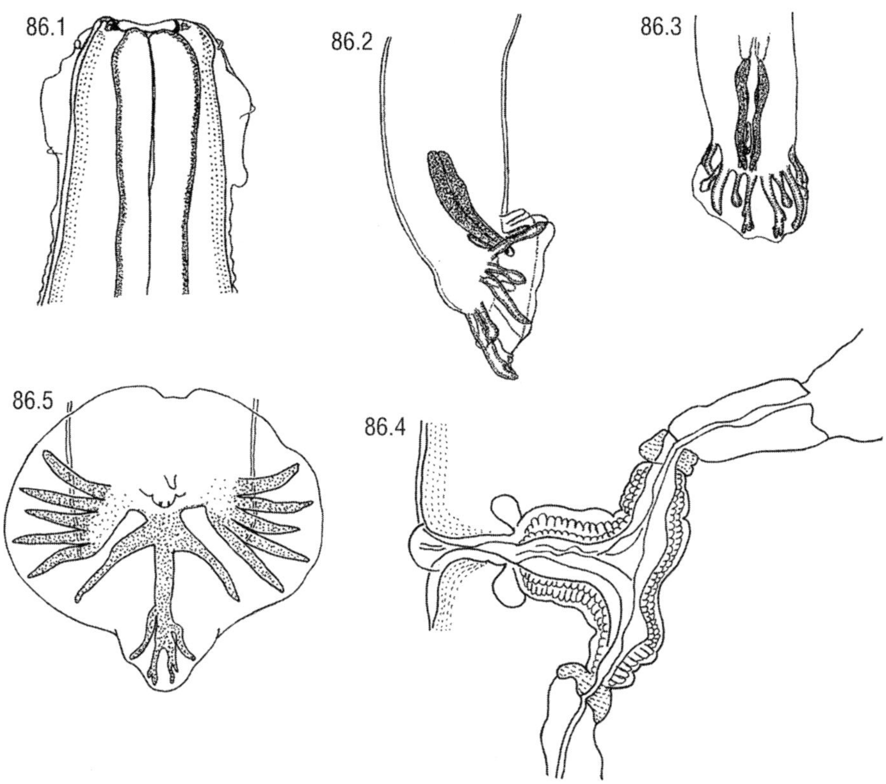

Figs 86.1, 86.2 and 86.3 *Cardiostrongylus* 86.1 Female, anterior end, lateral view; 86.2 Male, posterior end, lateral view; 86.3 Male, posterior end, ventral view. (Redrawn after Sakamoto and Malgor, 1995.) Figs 86.4 and 86.5 *Schulzia* 86.4 Female ovejector, left lateral view; 86.5 Bursa, ventral view. (Redrawn after Durette-Desset *et al.*, 1985.)

Female: Vestibule of ovejector with 3 very muscular branches, middle branch aligned with the elongate *vagina vera*. Caudal spine absent. Parasites of neotropical amphibians.

Type species: *Schulzia travassosi* Durette-Desset, Baker and Vaucher, 1985 (= *Oswaldocruzia subventricosa* sensu Travassos, 1925 nec Travassos, 1917, 1921; = *Schulzia subventricosa* Travassos, 1937 pro parte)

[Durette-Desset *et al.* (1985) redefine the genus.]

Kentropyxia Baker, 1982 (Fig. 87)

DIAGNOSIS: *Corona radiata* present, 9 unequal elements. Synlophe with numerous ridges. *Male*: Bursal rays of 2-1-2 type, i.e. ventral rays equal and parallel; antero-lateral ray separate, curves towards ventrals; medio- and postero-laterals equal and parallel; externo-dorsal ray shorter than laterals, arises from dorsal ray, parallel to postero-lateral ray; dorsal ray thick, and thin, short branches in distal third. Spicules equal: main stem ending in single point; dorsal and ventral branches divided distally into several fine points. Gubernaculum absent. Genital cone: single ventral raylet (papillae 0); conspicuous pair of dorsal raylets (papillae 7). *Female*: Vulva opens in mid-region of body. Didelphic. Tail thick with sharp distal point. Parasite in intestine of *Kentropyx calcaratus* in Brazil.

Type species: *Kentropyxia sauria* Baker, 1982

[Listed in Spencer Jones and Gibson (1987).]

Lagostonema Sutton and Durette-Desset, 1987 (Fig. 88)

DIAGNOSIS: Excretory pore opens posterior to or near distal end of oesophagus; excretory groove absent. Synlophe: 10 longitudinal cuticular ridges in mid-body; orientated ventral to dorsal; lateral ridges most developed. *Male*: Ventral bursal rays divergent; antero-ventral longer than other rays; antero-lateral ray short, closer to postero-ventral than medio-lateral ray.

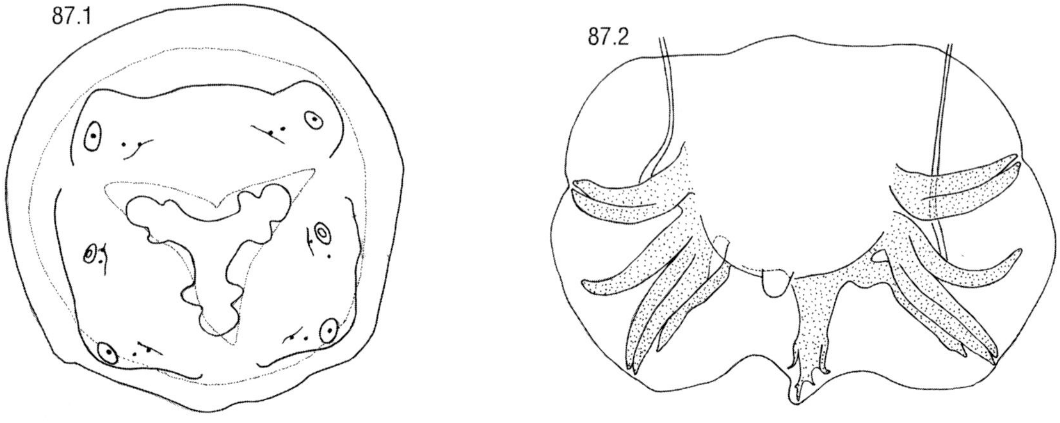

FIG. 87 *Kentropyxia* 87.1 Cephalic region, *en face* view; 87.2 Bursa, ventral view. (Redrawn after Baker, 1982a.)

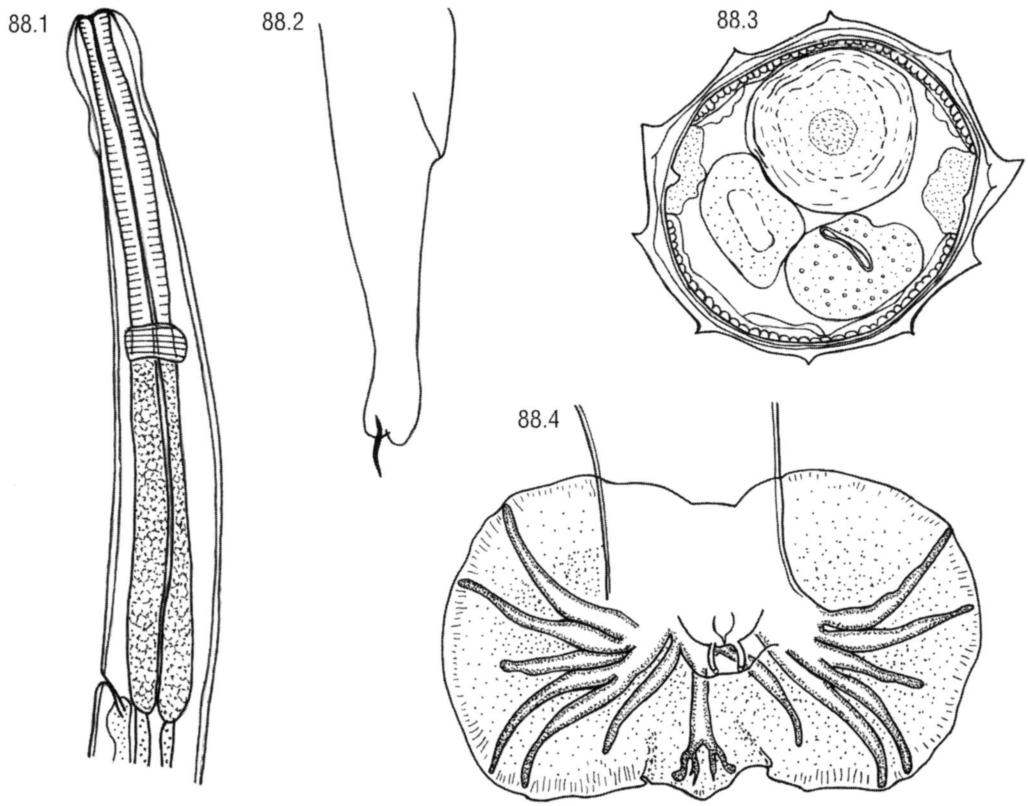

FIG. 88 *Lagostonema* 88.1 Anterior end; 88.2 Female tail, lateral view; 88.3 Cross-section in mid-body of male; 88.4 Bursa, ventral view. (Redrawn after Sutton and Durette-Desset, 1987.)

Female: Didelphic. Tail with distal spine. Parasite in stomach and small intestine of Chinchillidae, Caviomorpha (*Lagostomus maximus*) in Argentina.

Type species: *Lagostoma ecasiense* Sutton and Durette-Desset, 1987

Bakeria (Moravec and Sey, 1986 subgenus) Ben Slimane, Chabaud and Durette-Desset, 1996 (Fig. 89)

DIAGNOSIS: Synlophe present. *Male*: Bursa elongate; dorsal lobe long. Gubernaculum present, well sclerotized. Parasites of amphibians (*Phrynomantis* spp.) in Papua New Guinea.

Type species: *Bakeria bakeri* (Moravec and Sey, 1986) Ben Slimane, Chabaud and Durette-Desset, 1996

Ragenema Ben Slimane, Chabaud and Durette-Desset, 1996 (Fig. 90)

DIAGNOSIS: Cephalic vesicle present. Cephalic region without chitinous ornamentation. Buccal capsule absent. Cephalic sensory organs: internal circle of papillae well developed. Oesophageal

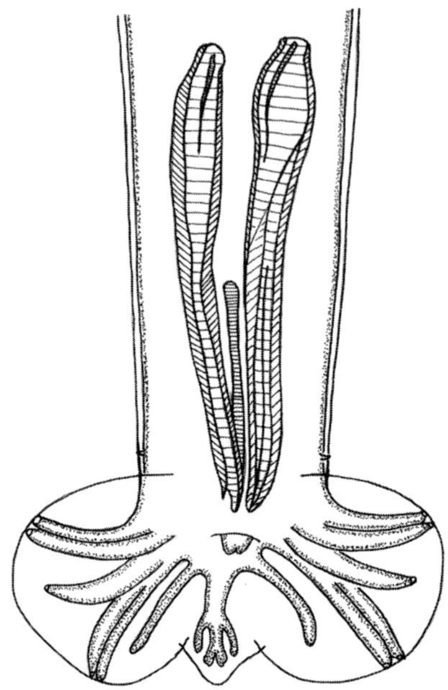

FIG. 89 *Bakeria* Posterior end of male, ventral view. (Redrawn after Moravec and Sey, 1986.)

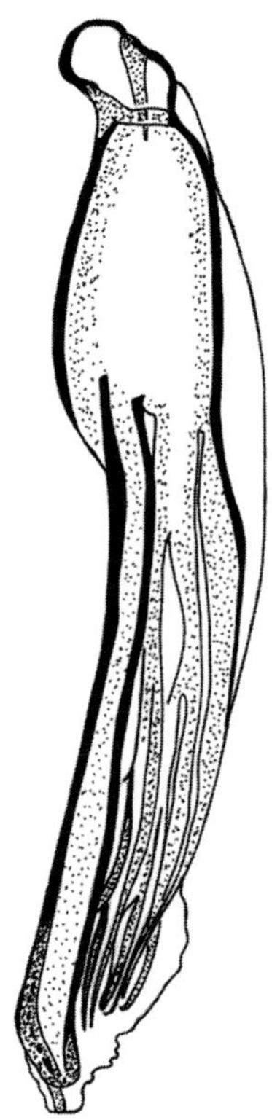

FIG. 90 *Ragenema* One of two spicules. (Redrawn after Baker, 1982 from Ben Slimane *et al*., 1996.)

tooth large. Synlophe: numerous longitudinal cuticular ridges perpendicular to body surface. *Male*: Bursa with thick dorsal ray. Gubernaculum absent. Spicules with bluntly rounded main stem and branch, very divided at distal tip. *Female*: Didelphic. Tail narrows rapidly posterior to anus, distal spine absent. Parasites of amphibians.

Type species: *Ragenema robustum* (Baker, 1982) Ben Slimane, Chabaud and Durette-Desset, 1996.

Mainspinostrongylus Kalyankar and Palladwar, 1989 (Fig. 91)

DIAGNOSIS: Body filiform, fairly stout. Cuticle with minute spines. Mouth simple, 3–4 tooth-like structures. Cephalic vesicle present. *Male*: Bursa: symmetrical, lateral lobes longer than dorsal lobe, spines and tubercles on cuticle; ventral rays with common stem; lateral rays with common stem; externo-dorsals unequal; dorsal ray divides at distal tip. Spicules: equal; alate. Gubernaculum present. *Female*: Vulva opens posterior to mid-body. Egg thin-shelled. Tail with terminal spine. Parasite in intestine of *Manis crassifiala* in India.

Type species: *Mainspinostrongylus crassifiala* Kalyankar and Palladwar, 1989

[Kalyankar and Palladwar (1989) place their genus in the subfamily Trichostrongylinae; however, the currently accepted features for this subfamily suggest that it is unlikely to belong here. In view of the host of this genus and the similarities of some of the morphological features to the genus, *Trichochenia* Kou, 1958 *Mainspinostrongylus* is placed provisionally in the subfamily Molineinae of the family Molineidae.]

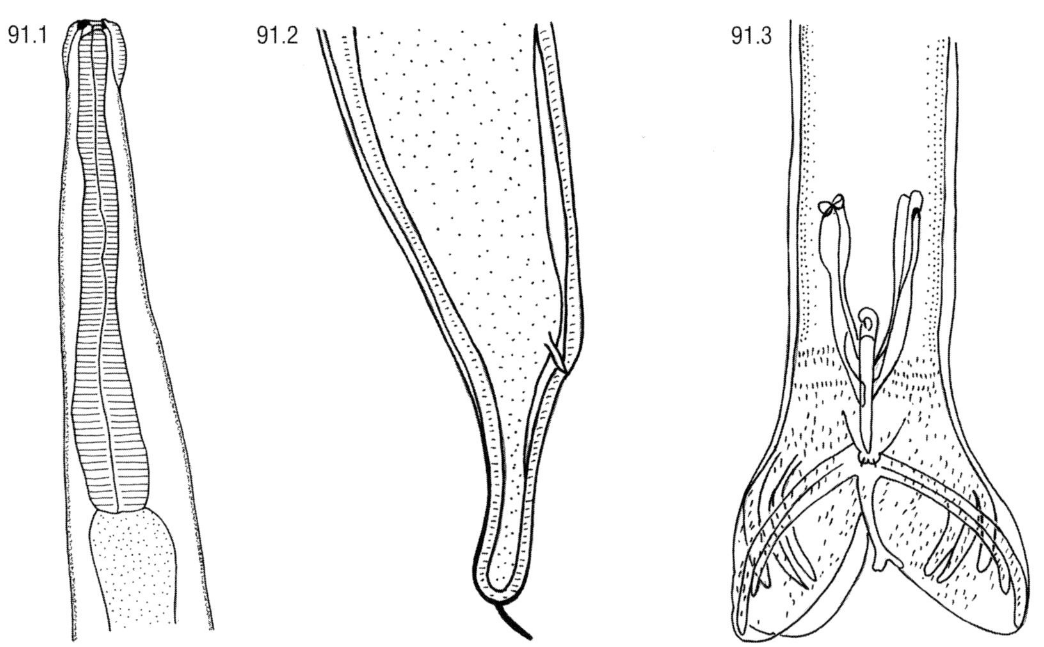

FIG. 91 *Mainspinostrongylus* 91.1 Anterior end, ventral view; 91.2 Female tail, lateral view; 91.3 Male tail. (Redrawn after Kalyankar and Palladwar, 1989.)

Anoplostrongylinae Chandler, 1938

Cheiropteronema Sandground, 1929 (Fig. 92)

DIAGNOSIS: Body length over 2 cm. Cephalic vesicle present, wider than long. Oesophagus widest in anterior part. *Male*: Bursal lobes and rays reduced. Dorsal bursal lobe separate. Genital cone well developed. Spicules with well-developed alae for two-thirds of length, 3 branches at distal tips. Gubernaculum present. *Female*: Didelphic, generally amphidelphic, rarely with ovejector opisthodelphic or prodelphic. Tail wider than long, with 4 small tubercles at distal end. Parasites of fruit-eating neotropical Chiroptera.

Type species: *Cheiropteronema globocephala* Sandground, 1929

[The emended diagnosis is based on Durette-Desset and Vaucher (1988).]

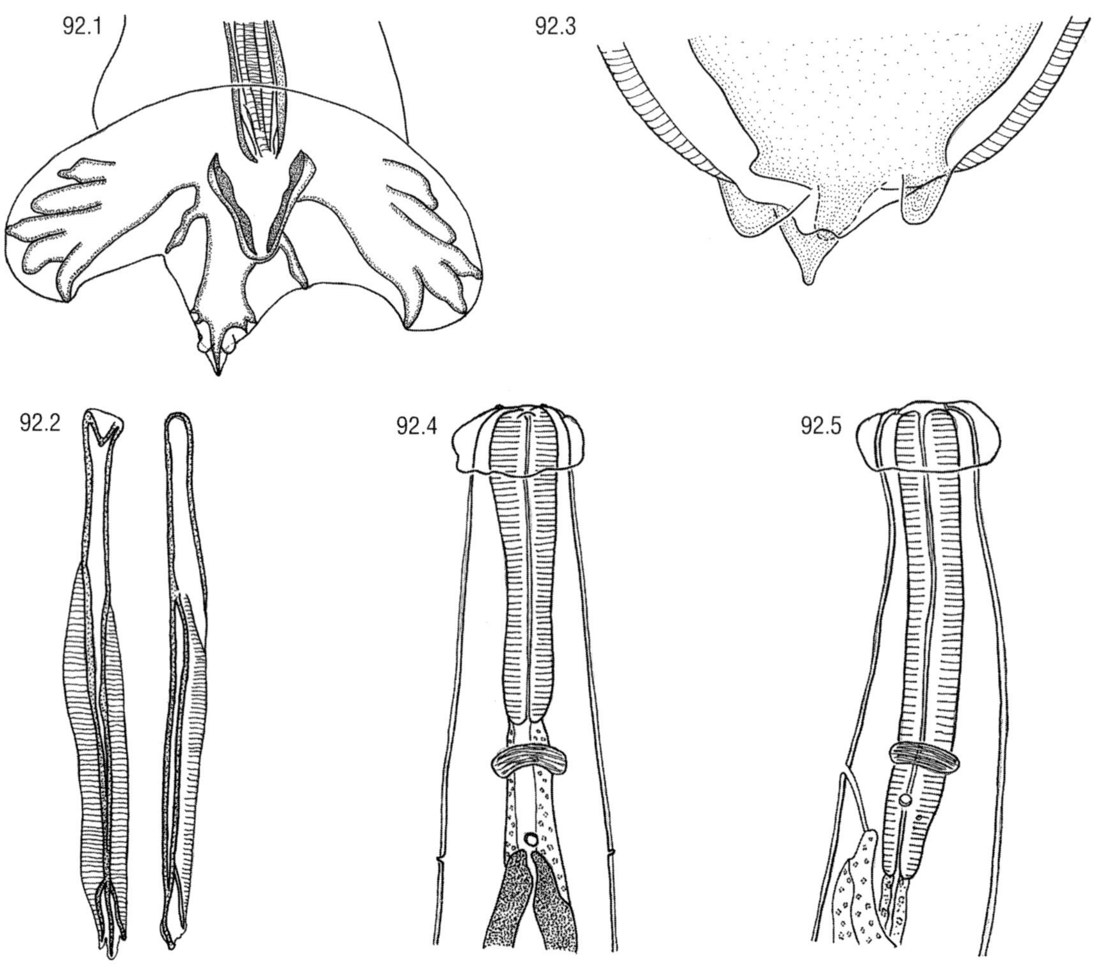

FIG. 92 *Cheiropteronema* 92.1 Bursa, ventral view; 92.2 Spicules; 92.3 Female tail, dorsal view; 92.4 Anterior end, ventral view; 92.5 Anterior end, lateral view. (Redrawn after Durette-Desset and Vaucher, 1988.)

Websternema Vaucher and Durette-Desset, 1986 (Fig. 93)

DIAGNOSIS: Cephalic vesicle forms cap, with 2 strong opposing spines arranged dorso-ventrally. Synlophe: approximately 40 longitudinal cuticular ridges of the crête-type, perpendicular to the body surface. *Male*: Bursa symmetrical; ventral rays with common stem, separate; lateral rays with common stem, shorter antero-lateral and longer medio-lateral rays stay close together, postero-lateral ray curves towards dorsal ray; dorsal ray long with 3 main distal branches. Spicules long, pointed, alate. *Female*: Didelphic. Tail with 4 points on distal tip. Parasites in neotropical Chiroptera.

Type species: *Websternema parnelli* (Webster, 1971) Vaucher and Durette-Desset, 1986

[Listed in Spencer Jones and Gibson (1987).]

Linustrongylus Vaucher and Durette-Desset, 1986 (Fig. 94)

DIAGNOSIS: Body stocky, thick, not coiled. Synlophe: absent. Head with 8 spines, divided at free distal tips, separated dorso-ventrally into 2 groups. Buccal capsule walls strongly chitinized. *Male*: Bursa small, rays short and thick; dorsal ray divides distally into 3 main branches. *Female*: Didelphic. Prodelphic. Tail with fine spines at distal tip. Parasites in neotropical Chiroptera.

Type species: *Linustrongylus pteronoti* Vaucher and Durette-Desset, 1986

[Listed in Spencer Jones and Gibson (1987).]

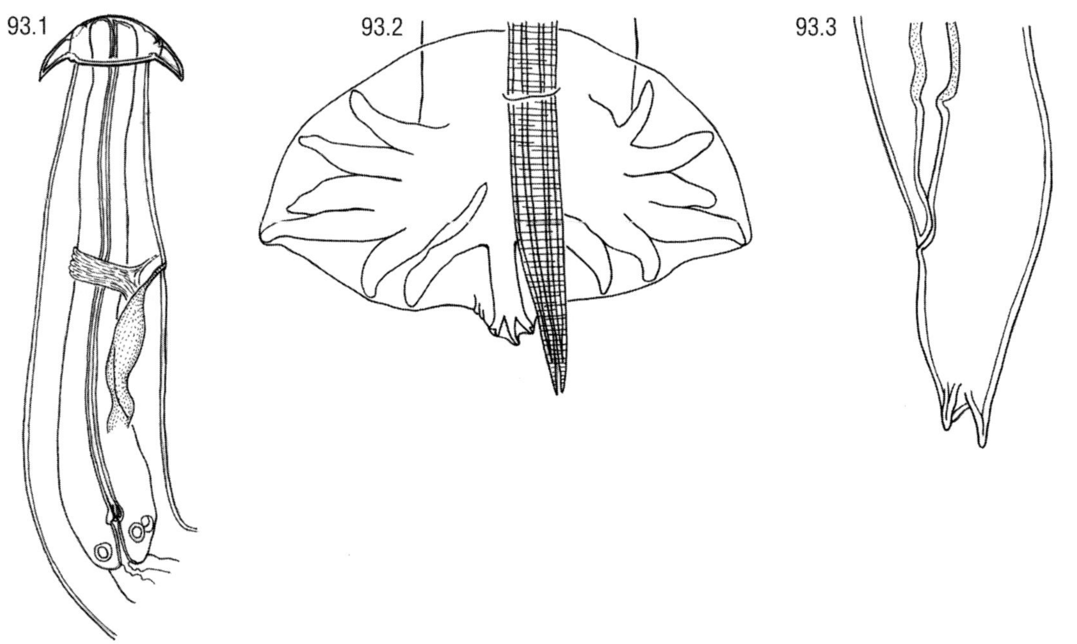

FIG. 93 *Websternema* 93.1 Anterior end, lateral view; 93.2 Bursa, ventral view; 93.3 Female tail, lateral view. (Redrawn after Vaucher and Durette-Desset, 1986.)

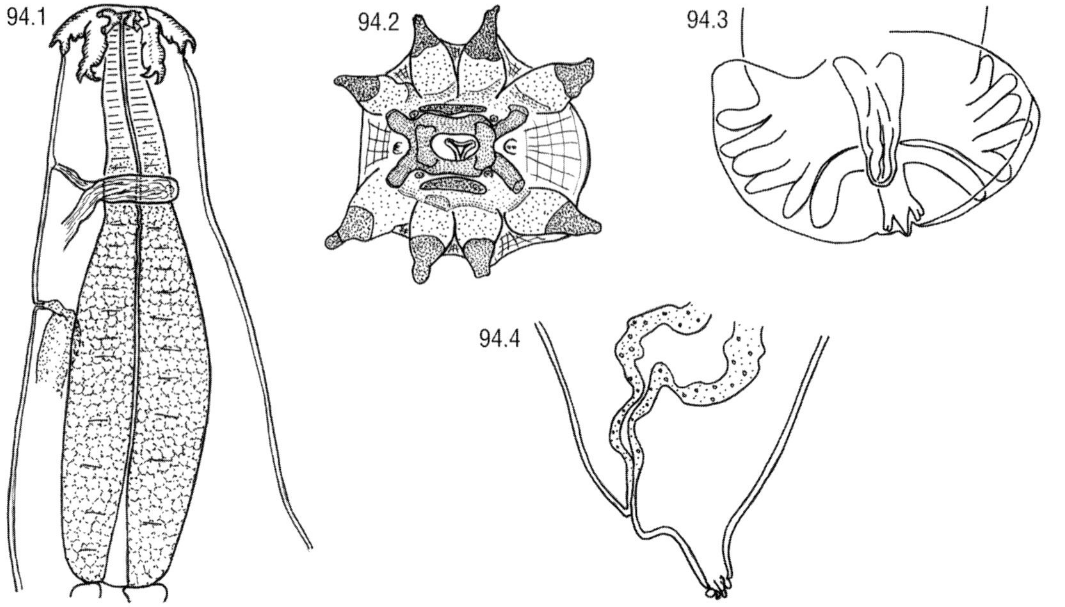

FIG. 94 *Linustrongylus* 94.1 Anterior end, lateral view; 94.2 Cephalic region, *en face* view; 94.3 Bursa, ventral view; 94.4 Female tail, lateral view. (Redrawn after Vaucher and Durette-Desset, 1986.)

Carostrongylus Durette-Desset and Vaucher, 1989 (Fig. 95)

DIAGNOSIS: Cephalic vesicle wider than long, with 4 indentations. Synlophe little developed, bilateral symmetry. *Male*: Body length over 9 mm. Lateral bursal lobes and rays reduced. Dorsal bursal lobe separate. Cuticle of bursa extends, forming membrane between dorsal lobe and genital cone. Distal tip of dorsal ray forms a spine. Genital cone well developed. Spicules with well-developed alae, 3 points at distal tips. *Female*: Body length over 18 mm. Amphidelphic, rarely opisthodelphic. Tail longer than wide, with 5–7 tubercles and median spine on its distal tip. Parasites of fruit-eating neotropical Chiroptera.

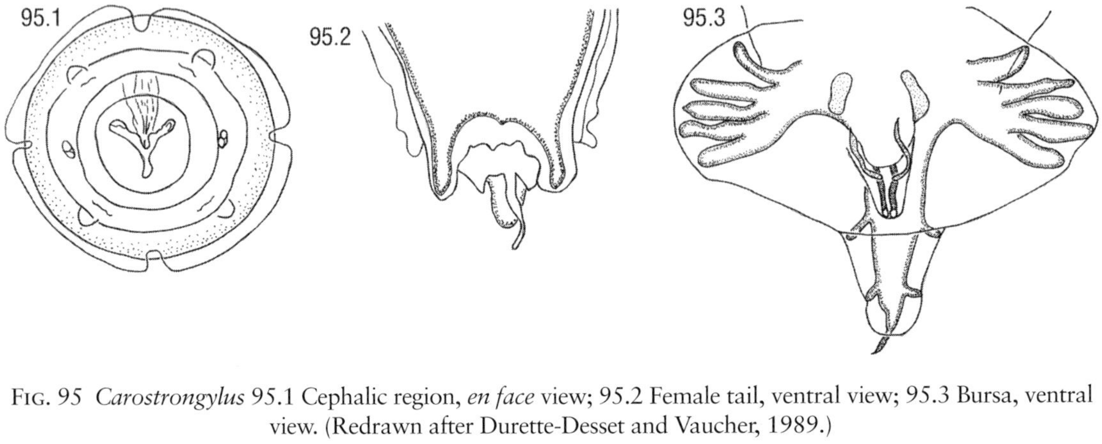

FIG. 95 *Carostrongylus* 95.1 Cephalic region, *en face* view; 95.2 Female tail, ventral view; 95.3 Bursa, ventral view. (Redrawn after Durette-Desset and Vaucher, 1989.)

Type species: *Carostrongylus touzeti* Durette-Desset and Vaucher, 1989.

[*Cheiropteronema* is considered the closest genus to *Carostrongylus* and, consequently, it is placed in the subfamily Anoplostrongylinae.]

Molostrongylus Durette-Desset and Vaucher, 1996 (Fig. 96)

DIAGNOSIS: Body length short, not coiled. Cephalic vesicle with anterior and posterior region. Synlophe: axis of orientation ventro-dorsal arranged with sagittal axis, lateral alae well developed. Excretory pore opens in region of posterior half of oesophagus, well behind anterior margin of excretory glands, which are well developed. *Male*: Bursa elongated transversely; ventral rays parallel, curve away from lateral rays; lateral rays with common stem, separate at distal tips; dorsal ray long, divided distally, each branch with shorter lateral branch; externo-dorsal ray shorter than dorsal ray. Spicules without alae, single point. Gubernaculum present. *Female*: Spines in region of vulva. Tail longer than wide, with 3 large points and a median appendix. Parasites in intestine of the bat, *Molossops temmincki*, in Paraguay.

Type species: *Molostrongylus acanthocolpos* Durette-Desset and Vaucher, 1996

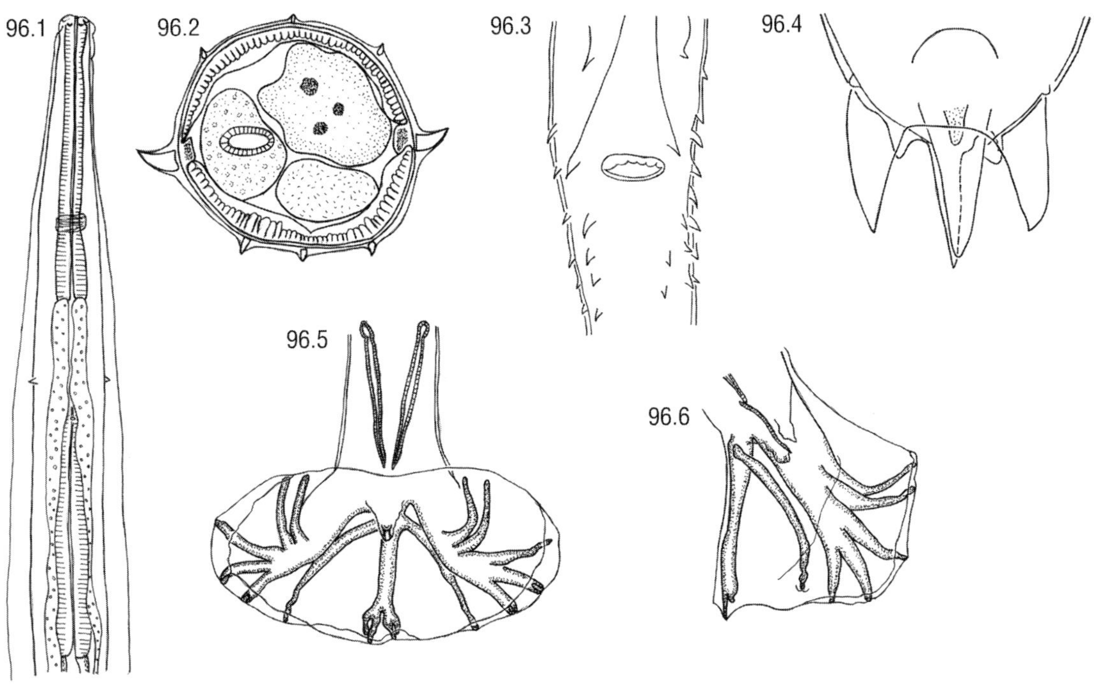

FIG. 96 *Molostrongylus* 96.1 Anterior end, ventral view; 96.2 Synlophe 1450 μm from anterior end; 96.3 Vulvar region, ventral view; 96.4 Female tail; 96.5 Bursa, ventral view; 96.6 Bursa, lateral view. (Redrawn after Durette-Desset and Vaucher, 1996.)

Tadaridanema Falcón-Ordaz, Guzmán-Cornejo, García-Prieto and Gardener, 2006 (Figs 97.1 and 97.2)

DIAGNOSIS: Synlophe with 2 well-developed dorsal longitudinal cuticular ridges in mid-body region. Cephalic vesicle simple, not distinctly divided. *Male*: Bursa symmetrical with antero-ventral ray, ray 2, longer than postero-ventral ray, ray 3. Lateral rays separate, cuticle striated. *Female*: Branches of ovejector unequal in length. Tip of tail encircled by three conical processes. Parasites in intestine of Mollossidae bats in USA and Mexico.

Type species: *Tadaridanema delicatus* (Schwartz, 1927) Falcón-Ordaz, Guzmán-Cornejo, García-Prieto and Gardener, 2006

[Falcón-Ordaz *et al.* (2006) place the genus in the Molineoidea, Trichostrongylina.]

Nematodirinae Skrjabin and Orloff, 1934

[Hoberg *et al.* (2005), following a phylogenetic study of the subfamily, proposed the inclusion of the genera *Nematodirus* Ransom, 1907, *Nematodirella* Yorke and Maplestone, 1926, *Nematodiroides* Bernard, 1967, *Murielus* Dikmans, 1939 and *Rauschia* Durette-Desset, 1979. Rickard and Hoberg (2000) had previously assigned *Lamanema* Becklund, 1963 to the subfamily Molineinae.]

Nematodirella Yorke and Maplestone, 1926

[*Nematodirella* is the new name given for *Microcephalus* on the grounds that is preoccupied. *Microcephalella* was also proposed for *Microcephalus*, but the name proposed by Yorke and Maplestone, 1926 takes precedence. See *Microcephalus* and *Microcephalella*, below.]

Microcephalus Romanovitch, 1915

DIAGNOSIS: Cephalic vesicle present. Cephalic sensory organs: 2 large lateral, 4 small submedian papillae. Synlophe: approximately 40 longitudinal cuticular ridges. Buccal capsule absent. *Male*: Spicules long, nearly half length of body. *Female*: Body widens posteriorly. Vulva opens at distal margin of anterior fourth of body. Ovejector with one branch small, the other large. [Anterior branch of reproductive system non-functional.] Parasite in small intestine of reindeer.

Type species: *Microcephalus longissime spiculatus* Romanovitch, 1915

[Yorke and Maplestone (1926) rename this genus *Nematodirella* on the grounds that *Microcephalus* is preoccupied. The name, *Microcephalus*, is unlisted in the key (Durette-Desset, 1983) and listed in Spencer Jones and Gibson (1987).]

Microcephalella Strand, 1942

[Strand (1942) proposes the generic name *Microcephallela* for *Microcephalus* Romanovitch, 1915 *nec Microcephalus* Van der Wulp, 1873, which is preoccupied. Yorke and Maplestone (1926), however, had already proposed the name *Nematodirella* (see above), with *Nematodirella longispiculata* (Romanovitch, 1915) as type species. *Microcephalella* is therefore a synonym of *Nematodirella*. The name *Microcephallela* is unlisted in the key (Durette-Desset, 1983) and listed in Spencer Jones and Gibson (1987).]

Mackerrastrongylidae (Inglis, 1968 subfamily) Durette-Desset and Chabaud, 1981

[Durette-Desset and Chabaud (1993) moved this family to their new superfamily Molineoidea in the suborder Trichostrongylina.]

Nicollinidae (Skrjabin and Schulz, 1937 tribe) Durette-Desset and Chabaud, 1981

[Durette-Desset and Chabaud (1993) move this family to their new superfamily Heligmosomoidea in the suborder Trichostrongylina.]

Batrachonema Yuen, 1965

[Baker (1983) draws attention to the mouth opening of this genus, which has cuticular structures that may represent a 'vestigial *corona radiata* or neoformations'. He points out that these do not occur in other trichostrongyles; however, 'the synlophe, bursa, spicules and ovejector... indicate a close relationship to *Nicollina*'. Durette-Desset *et al.* (1994) move this genus to their subfamily Johnpearsoniinae in the family Amphibiophilidae.]

Herpetostrongylidae (Skrjabin and Schulz, 1937 subfamily) Durette-Desset and Chabaud, 1981

[Durette-Desset and Chabaud (1993) move this family to their new superfamily Heligmosomoidea in the suborder Trichostrongylina.]

Herpetostrongylinae Skrjabin and Schulz, 1937

Austrostrongylus Chandler, 1924 (Figs 97.3, 97.4 and 97.5)

DIAGNOSIS: Body generally coiled in ventral spiral. Mouth with 6 lips. Buccal capsule subglobular, single dorsal tooth, 2 subventral teeth. Cephalic vesicle present. Synlophe: 3 or 4 ventral ridges orientated left, decreasing in size to right; 2 or more dorsal ridges present; 1–2 lateral body floats present. *Male*: Bursa with small, thickened dorsal lobe; ventral and lateral rays separate at distal tips; dorsal ray with 3 pairs of branches. Genital cone with single ventral papilla, 2 dorsal papillae. Spicules simple. *Female*: Vulva opens in posterior part of body. Didelphic, rarely monodelphic. Tail sharply pointed. Parasites in small intestine of macropodine marsupials and the marsupial mole, *Notoryctes typhlops*.

Type species: *Austrostrongylus macropodis* Chandler, 1924

[Emended diagnosis based on Beveridge and Durette-Desset (1986).]

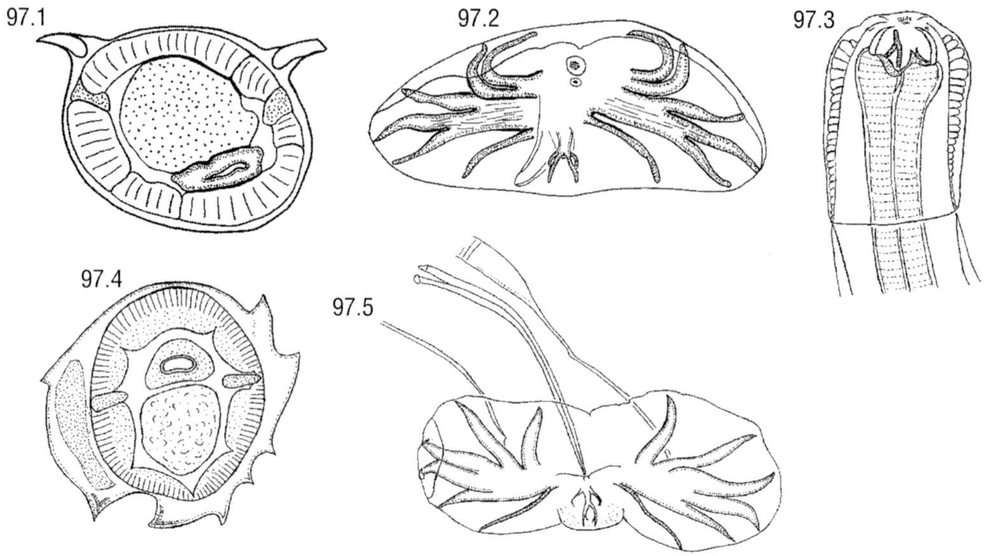

FIGS 97.1 and 97.2 *Tadaridanema* 97.1 Synlophe, mid-body; 97.2 Bursa. (Redrawn after Falcón-Ordaz, Guzmán-Cornejo, García-Prieto and Gardener, 2006.)
FIGS 97.3, 97.4 and 97.5 *Austrostrongylus* 97.3 Cephalic region, lateral view; 97.4 Synlophe, mid-body; 97.5 Bursa, ventral view. (Redrawn after Beveridge and Durette-Desset, 1986.)

Sutarostrongylus Beveridge and Durette-Desset, 1986 (Fig. 98)

DIAGNOSIS: Body with coiled ventral spiral. Mouth with 6 small lips. Buccal capsule subglobular, single dorsal tooth, 2 smaller subventral teeth. Cephalic vesicle present. Synlophe: 2 dorsal, 4 ventral ridges orientated left; lateral body floats absent. *Male*: Bursa with small, thickened dorsal lobe; ventral and lateral rays separate at distal tips; dorsal ray with 3 pairs of branches. Genital cone with single ventral papilla, 2 dorsal papillae. *Female*: Vulva opens in posterior part of body. Didelphic. Tail pointed. Parasites in small intestine of macropodine marsupials.

Type species: *Sutarostrongylus kirkpatricki* Beveridge and Durette-Desset, 1986

[Listed in Spencer Jones and Gibson (1987).]

Patricialina Inglis, 1968 (Fig. 99)

DIAGNOSIS: Cephalic vesicle present. Mouth with 6 lips. Buccal capsule well developed, dorsal tooth present. Synlophe: 10 longitudinal cuticular ridges of arête-type, orientation right to left, frontal axis. *Male*: Bursa with postero-lateral and externo-dorsal rays parallel. Spicules with 2 branches. *Female*: Didelphic. Vulva generally opens towards posterior end. Tail with distal spine. Parasites of Dasyuridae.

Type species: *Patricialina phascogale* (Mawson, 1960) Inglis, 1968

[The emended diagnosis is based on Humphrey-Smith and Durette-Desset (1981).]

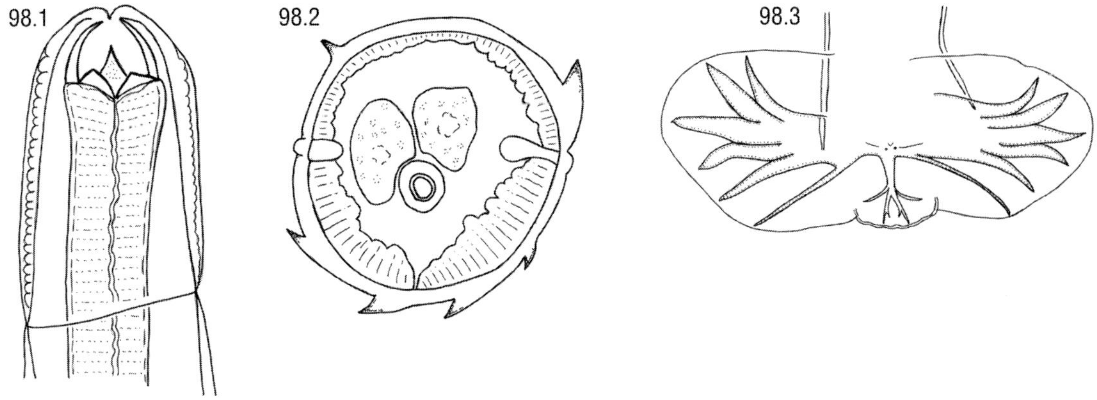

FIG. 98 *Sutarostrongylus* 98.1 Cephalic region, ventral view; 98.2 Synlophe, anterior end; 98.3 Bursa, ventral view. (Redrawn after Beveridge and Durette-Desset, 1986.)

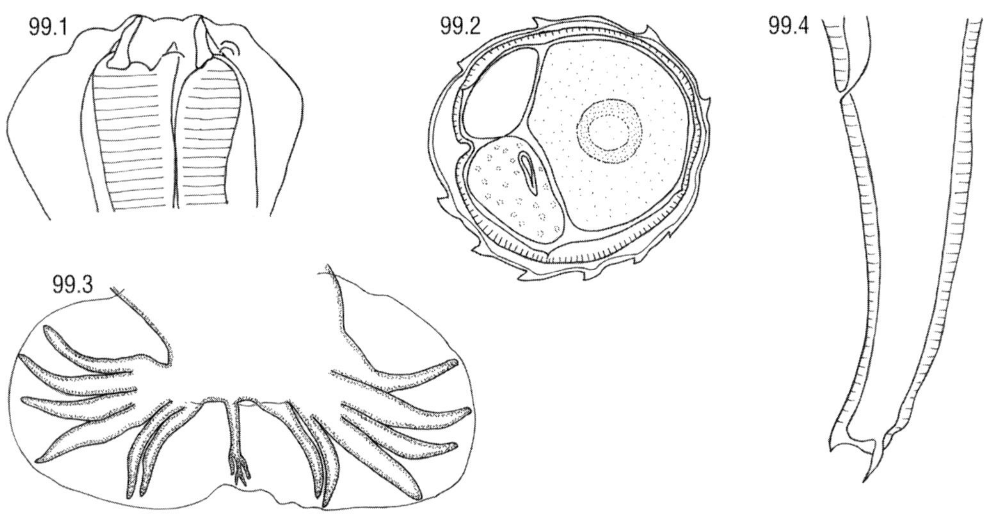

FIG. 99 *Patricialina* 99.1 Cephalic region; 99.2 Synlophe, mid-body region; 99.3 Bursa, ventral view; 99.4 Female tail, lateral view. (Redrawn after Humphrey-Smith and Durette-Desset, 1981.)

Wanaristrongylus Jones, 1987 (Fig. 100)

DIAGNOSIS: Cephalic vesicle present. Buccal capsule hexagonal in cross-section. Dorsal oesophageal tooth large. Synlophe: 10 or more small longitudinal cuticular ridges, perpendicular to body surface. *Male*: Bursa symmetrical, dorsal lobe only slightly longer than lateral lobes; ventral rays with common stem, separate at distal tips; lateral rays with common stem, separate at distal tips; externo-dorsal ray thin, arises from base of dorsal ray. Spicules variable. Gubernaculum present. *Female*: Monodelphic. Vulva opens near anus. Ventral to ventro-lateral alae anterior and lateral to vulva. Tail with 2 terminal knobs and fine subterminal spine. Tail sharply curved ventrally. Parasites of Australian lizards, Agamidae and Gekkonidae.

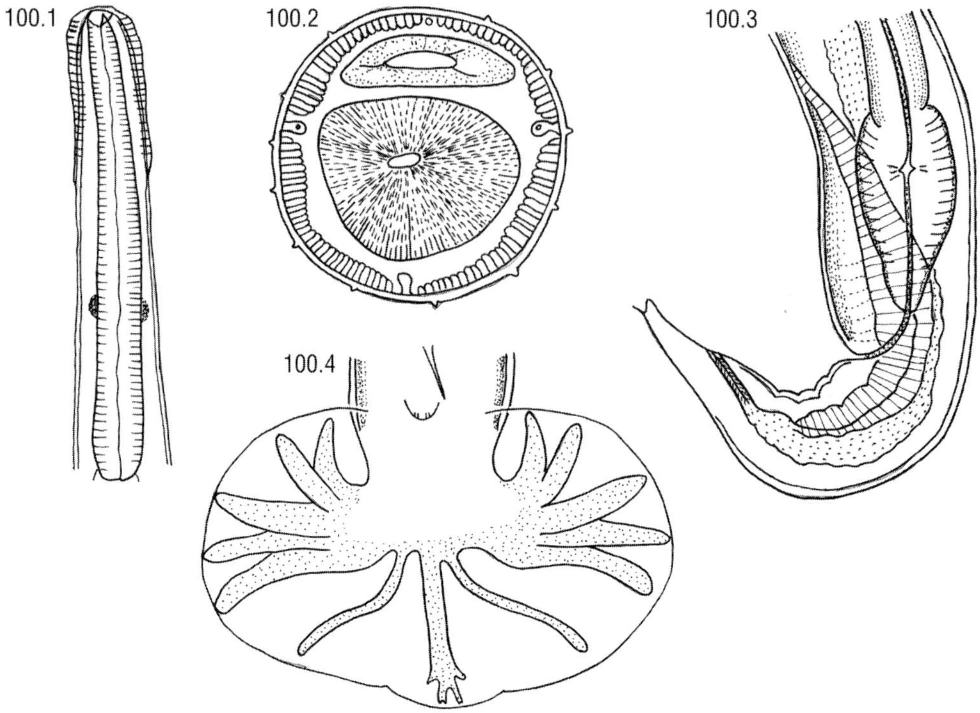

FIG. 100 *Wanaristrongylus* 100.1 Anterior end, lateral view; 100.2 Synlophe, mid-body region; 100.3 Posterior end of female; 100.4 Bursa, ventral view. (Redrawn after Jones, 1987.)

Type species: ? *Waneristrongylus pogonae* Jones, 1987

[A type species was not designated, and *W. pogonae* has been provisionally designated type species as it is the first species described by Jones (1987).]

Viannaiidae (Neveu-Lemaire, 1934 subfamily) Durette-Desset and Chabaud, 1981

[Durette-Desset and Chabaud (1993) move this family to their new superfamily Heligmosomoidea in the suborder Trichstrongylina (for definition of the superfamily see under the family Heligmosomidae).]

Viannaiinae Neveau-Lemaire, 1934

Avellaria Freitas and Lent, 1934

DIAGNOSIS: Synlophe: ventral longitudinal cuticular ridges well developed in anterior part of body. *Male*: Bursa of type 1-3-1, i.e. antero-ventral ray separate from postero-ventral, antero-lateral and medio-lateral rays, postero-lateral ray separate from other lateral rays. Antero-lateral ray, ray 4, large. Spicules straight with curved proximal end. *Female*: Didelphic. Parasites in neotropical Agoutidae and Dasyproctidae.

Type species: *Avellaria avellari* Freitas and Lent, 1934

[Durette-Desset, 1971 synonymized this genus with *Viannella* Travassos, 1918, and this was upheld in Durette-Desset (1983). Durette-Desset *et al.* (2006) revalidate this genus, and the emended diagnosis is based on their publication.]

Hydrochoerisnematinae Arantes and Artigas, 1983

DIAGNOSIS: Body coiled. Cephalic vesicle present. Synlophe: present; dorsal ridges little developed, almost perpendicular to body surface; ventral ridges well developed, not perpendicular to body surface. *Male*: Bursa with distinct lobes, dorsal lobe and ray asymmetrical. Spicule short and complex. Gubernaculum absent. Genital cone complex. *Female*: Monodelphic. Oviparous. Spine on tail absent. Parasites in *Hydrochoerus hydrochaeris hydrochaeris*.

Type genus: *Hydrochoerisnema* Arantes and Artigas, 1983

Hydrochoerisnema Arantes and Artigas, 1983 (Fig. 101)

DIAGNOSIS: Synlophe: orientated right to left, ventral ridges higher than dorsal ridges. *Male*: Bursa asymmetrical: ventral rays fused for part of length, diverge distally; lateral rays with common stem, antero-lateral separates first, rays only separate near distal tips; externo-dorsal

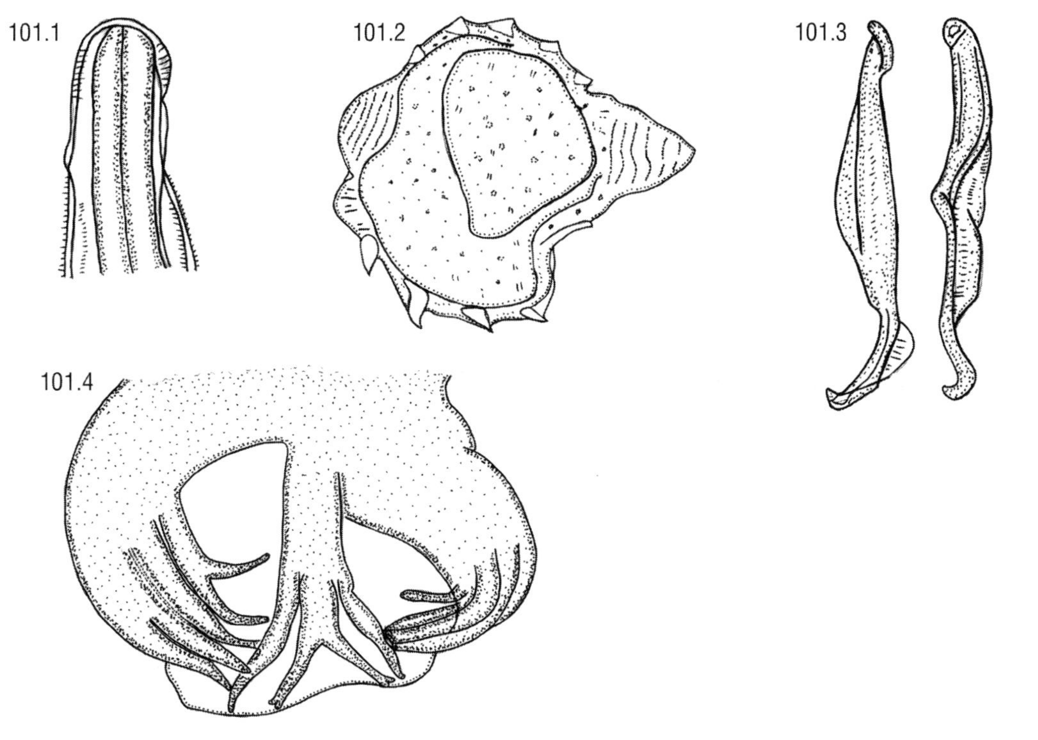

FIG. 101 *Hydrochoerisnema* 101.1 Anterior end; 101.2 Synlophe; 101.3 Spicules; 101.4 Bursa, ventral view. (Redrawn after Arantes and Artigas, 1983.)

ray arises from middle of main stem of dorsal ray; dorsal ray bifurcates in distal quarter, each tip dividing again. Spicules curved distally, forming hook-like tips. *Female*: Vulva opens near anus. Monodelphic. Tail tapers, bluntly pointed. Parasite in alimentary canal of capybara, *Hydrochoerus hydrochaeris hydrochaeris*.

Type species: *Hydrochoerisnema anomolobursata* Arantes and Artigas, 1983

[The genus was first named by Arantes and Artigas (1980) in *Congresso Brasileiro de Parasitologia*, p. 98. Listed in Spencer Jones and Gibson (1987).]

Ornithostrongylidae (Travassos, 1937 subfamily) Durette-Desset and Chabaud, 1981

[Durette-Desset and Chabaud (1993) move this family to their new superfamily Heligmosomoidea in the suborder Trichostrongylina.]

Passeristrongylus Jansen, 1985 (Figs 102.1, 102.2, 102.3, 102.4 and 102.5)

DIAGNOSIS: Cephalic vesicle present. Synlophe: poorly developed with frontal axis of orientation; longitudinal cuticular ridges arranged obliquely on right side of body. *Male*: Bursa symmetrical. Bursal rays arranged almost equidistant to each other, slightly arranged as a 2-1-2 type, i.e. ventrals parallel with a common stem, laterals with a common stem, antero-lateral slightly more separated from other laterals than medio-lateral from postero-lateral. Externo-dorsals arise from main stem of dorsal ray. Gubernaculum absent. *Female*: Didelphic. Tail without distal spine. Parasite of small intestine of the passeriform bird *Amandava subflava*.

Type species: *Passeristrongylus amandavae* Jansen, 1985.

[Listed in Spencer Jones and Gibson (1987).]

Heligmosomidae (Travassos, 1914 subfamily) Cram, 1927

[Durette-Desset and Chabaud (1993) create a new superfamily Heligmosomoidea for this family and Nicollinidae (Skrjabin and Schulz, 1937 tribe), Herpetostrongylidae (Skrjabin and Schulz, 1937 subfamily), Viannaiidae (Neveu-Lemaire, 1944 subfamily), Ornithostrongylidae (Travassos, 1937 subfamily), Heligmonellidae (Skrjabin and Schikhobalova, 1952 tribe). The superfamily is defined by Durette-Desset and Chabaud (1993) as: body generally coiled in a spiral along ventral line; primitive forms with 6 lips and buccal capsule well developed; cephalic vesicle present; synlophe always present but never with a bilateral symmetry. *Male*: with bursa type variable, seldom with 2-1-2 type. *Female*: generally monodelphic. Parasites mainly of reptiles, monotremes, marsupials, neotropical birds, modern insectivores, lagomorphs and rodents.]

Heligmosomum Railliet and Henry, 1909

Subgenus *Heligmosomum* (Railliet and Henry, 1909 genus) Asakawa and Satoh, 1987

DIAGNOSIS: Longitudinal cuticular ridges oblique along the whole body. Parasites mainly of the genus *Microtus*.

Type species: *Heligmosomum* (*Heligmosomum*) *costellatum* Dujardin, 1845

Subgenus *Paraheligmosomum* Asakawa and Satoh, 1987

DIAGNOSIS: Longitudinal cuticular ridges present in the left ventral field. Parasite of the genera *Clethrionomys* and *Eothenomys*.

Type species: *Heligmosomum* (*Paraheligmosomum*) *mixtum* Schulz, 1954

Heligmoptera Nadtochiĭ, 1977 (Figs 102.6 and 102.7)

DIAGNOSIS: Synlophe: ventral and dorsal longitudinal cuticular ridges with orientation from ventral right to dorsal left; left lateral ala complex, sclerotized structure. *Male*: Prebursal papillae very long, arise near base of ventral rays. Bursa asymmetrical; rays of left and right lateral lobes unequal, with different arrangement within each lateral lobe; externo-dorsal ray separate from dorsal ray; dorsal ray small. Spicules long, narrow. *Female*: Monodelphic. Tail with narrow, spine-like process. Parasites of Myospalacinae (Muridae) from Palaearctic region.

Type species: *Heligmoptera myospalaxi* (Nadtochiĭ, 1970) Nadtochiĭ, 1977

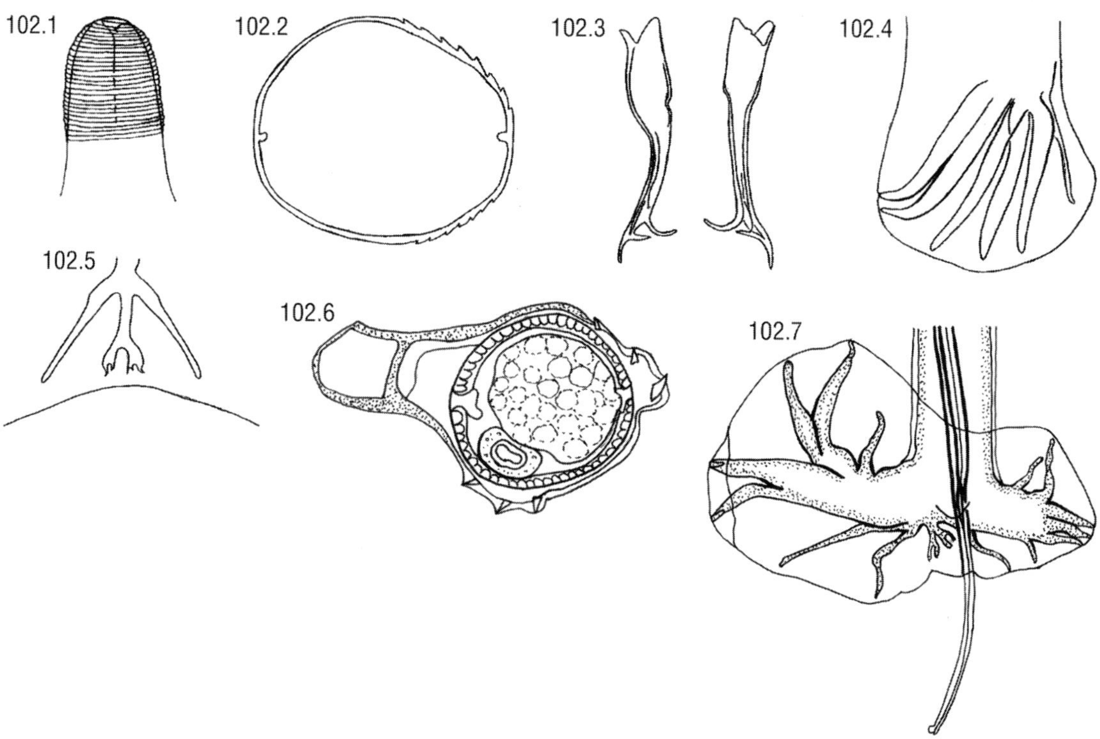

FIGS 102.1, 102.2, 102.3, 102.4 and 102.5 *Passeristrongylus* 102.1 Cephalic region; 102.2 Synlophe; 102.3 Spicules; 102.4 Lateral lobe of bursa; 102.5 Dorsal lobe of bursa. (Redrawn after Jansen, 1985.)
FIGS 102.6 and 102.7 *Heligmoptera* 102.6 Synlophe at mid-body of female; 102.7 Bursa, ventral view. (Redrawn after Elias *et al.*, 2002.)

[This genus was described in the *Summaries of the Third National Conference of Parasitology*, Albena, Bulgaria 12–14 October 1977. Elias *et al.* (2002) validate and redefine this genus, adding two new species. The diagnosis here is based on Elias *et al.* (2002).]

Heligmonellidae (Skrjabin and Schikhobalova, 1952 tribe) Durette-Desset and Chabaud, 1977

[Durette-Desset and Chabaud (1993) move this family to their new superfamily Heligmosomoidea in the suborder Trichostrongylina.]

Heligmonellinae Durette-Desset, 1983

Yatinema Asakawa and Ohbayashi, 1986 (Fig. 103)

DIAGNOSIS: Synlophe generally with fewer than 20 longitudinal cuticular ridges; orientation from sagittal axis 37–64°. *Male*: Bursa asymmetrical, right lobe larger than left lobe, rays of 2-2-1 type, i.e. ventral rays together, antero- and medio-laterals together, postero-lateral separate and short; dorsal lobe not separate. Spicules with distal tips fused, curved. Gubernaculum boat-shaped. Parasites in intestine of Oriental Cricetidae.

Type species: *Yatinema japonicum* Asakawa and Ohbayashi, 1986

[Listed in Spencer Jones and Gibson (1987).]

Pudicinae (Skrjabin and Schikhobalova, 1952 tribe) Durette-Desset, 1971

[Durette-Desset and Justine (1991) published a cladistic analysis of the genera of the subfamily Pudicinae. The authors used 13 characters based on the various features of the synlophe and bursa. Two suprageneric groups were identified, namely group 1, including *Heligmostrongylus*,

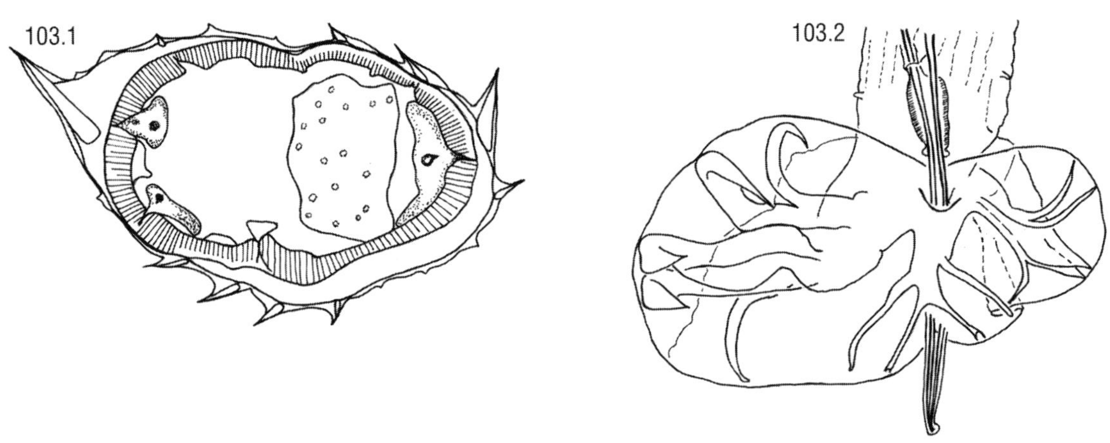

FIG. 103 *Yatinema* 103.1 Synlophe, mid-body region of male; 103.2 Bursa, ventral view. (Redrawn after Asakawa and Ohbayashi, 1986.)

Fuellebornema, *Sciurodendrium* and *Pseudoheligmosomum*, and group 2, including *Pudica*, *Acanthostrongylus*, *Justinema* and *Durettestrongylus*. For full details of the analysis, Durette-Desset and Justine (1991) should be consulted.]

Fuellebornema Travassos and Darriba, 1929 (Fig. 104)

DIAGNOSIS: Synlophe: well-developed carène present; 5 dorsal ridges; 5 or 6 ventral ridges of arête type; ventral ridges scalloped or not. *Male*: Bursa of 1-3-1 type, i.e. antero-ventral separate from postero-ventral; postero-ventral, antero-lateral and medio-lateral closer together, postero-lateral separate from other laterals; dorsal ray divided in its distal half, external branch (ray 9) shorter than internal branch (ray 10), externo-dorsal ray arises in proximal part of dorsal ray. Parasites of Caviomorphs (Dasyproctidae).

Type species: *Fuellebornema agoutii* (Neiva, Cuhna and Travassos, 1915) Travassos and Darriba, 1929

[The genus *Fuellebornema* was synonymized by Durette-Desset, 1971 with the genus *Heligmostrongylus* Travassos, 1917. Cassone and Durette-Desset (1991) reinstate the genus.]

Durettestrongylus Guerrero, 1982 (Figs 105.1 and 105.2)

DIAGNOSIS: Cephalic vesicle present. Buccal capsule reduced. Synlophe: single, hypertrophied arête forming a large lateral 'ala', small arête between 'ala' and a single large dorsal arête, two additional arêtes lateral and ventral. *Male*: Bursa symmetrical, antero-lateral ray shorter than other laterals, externo-dorsal ray arises at base of dorsal ray, which divides distally. Genital cone well developed. Spicules filiform, alate. Gubernaculum absent. *Female*: Body larger than male. Monodelphic. Vulva opens near posterior end. Parasite in the intestine of rodents (caviomorphs).

Type species: *Durettestrongylus ojastii* Guerrero, 1982

[Guerrero (1982) does not place the genus in a family but considers it in an evolutionary line with *Heligmostrongylus* and *Pudica*. Durette-Desset and Justine (1991), in their cladistic analysis of

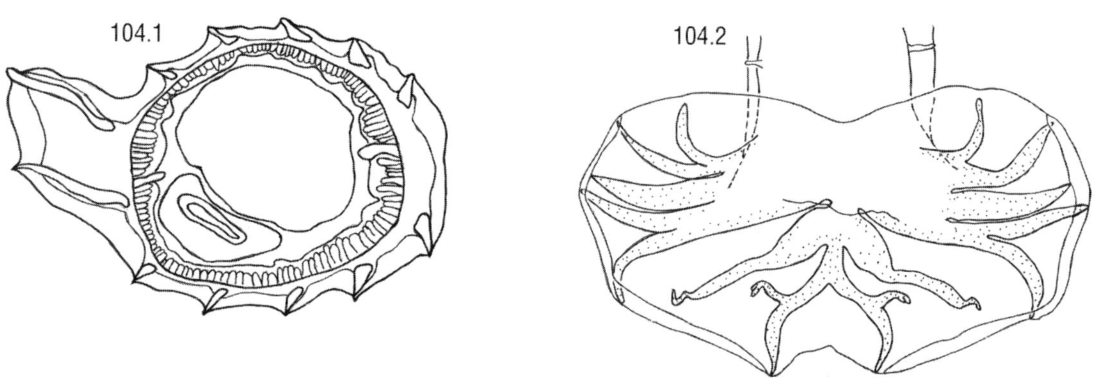

FIG. 104 *Fuellebornema* 104.1 Synlophe, mid-body region of female; 104.2 Bursa, ventral view. (Redrawn after Cassone and Durette-Desset, 1991.)

the subfamily, place this genus in the Pudicinae and, recongnizing two synapomorphies, place it in suprageneric group two (see Pudicinae). Listed in Spencer Jones and Gibson (1987).]

Justinema R'Kha and Durette-Desset, 1991 (Figs 105.3 and 105.4)

DIAGNOSIS: Synlophe: carène absent; ventral comarête present, well developed in anterior part of body; longitudinal cuticular ridges continuous. *Male*: Bursa of type 2-2-1, i.e. ventral rays separate from lateral rays, antero-lateral and medio-laterals separate from postero-lateral ray. Dorsal ray divided into its middle third, distal tips of branches divided so that ray 9 equals length of ray 10 or is slightly longer. Parasites in intestine of caviomorph rodents (Echimyidae).

Type species: *Justinema petterae* (Durette-Desset, 1970) R'Kha and Durette-Desset, 1991

[Emended diagnosis based on Durette-Desset and Justine (1991).]

Freitastrongylus Gonçalves, Pinto and Durette-Desset, 2007 (Figs 106.1 and 106.2)

DIAGNOSIS: Synlophe: carène poorly developed compared with other longitudinal ridges; 4 dorsal, 6 ventral additional ridges at mid-body, continuous, size decreasing from lateral to median sides of body. *Male*: Bursa type 1-4, i.e. antero-ventral ray separate from other rays.

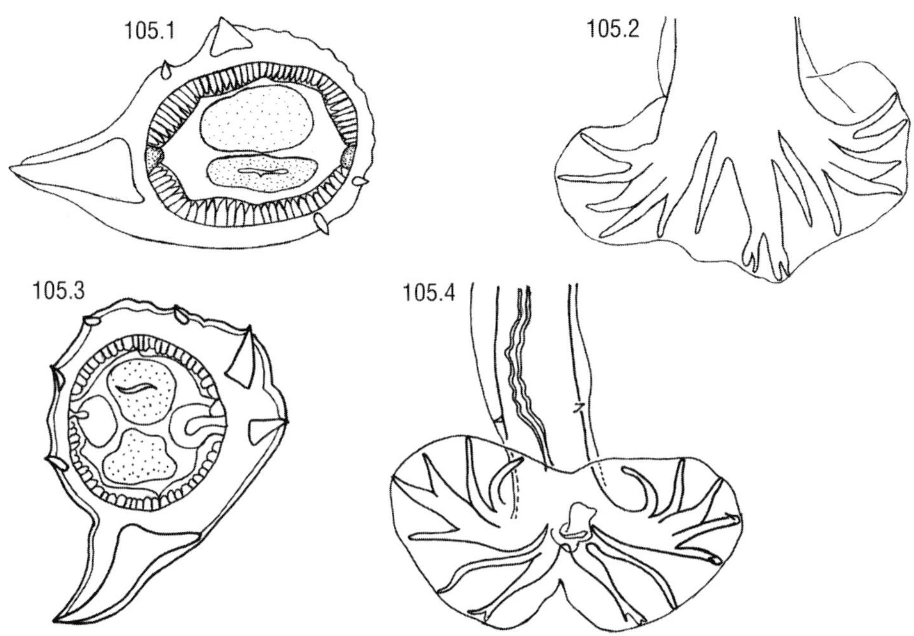

FIGS 105.1 and 105.2 *Durettestrongylus* 105.1 Synlophe in anterior third of body; 105.2 Bursa, ventral view. (Redrawn after Guerrero, 1982.)

FIGS 105.3 and 105.4 *Justinema* 105.3 Synlophe, just posterior to cephalic region; 105.4 Bursa, ventral view. (Redrawn after Durette-Desset, 1970b.)

Dorsal ray divided for half its length, distal tips of branches divided so that ray 9 is shorter than ray 10. *Female*: Monodelphic. Parasites in stomach of Dasyproctidae.

Type species: *Freitastrongylus angelae* Gonçalves, Pinto and Durette-Desset, 2007

Brevistriatinae Durette-Desset, 1971

[The subgenus *Paraheligmonina* Ortlepp, 1939 of *Heligmonina* Baylis, 1928, listed by Spencer Jones and Gibson (1987), was elevated to generic status by Durette-Desset (1971). The genus is included in Durette-Desset (1983) as a member of this subfamily.]

***Lagostrongylus* Fukumoto, Kamiya and Ohbayashi, 1986 (Figs 106.3 and 106.4)**

DIAGNOSIS: Synlophe with 12–13 continuous, different-sized ridges of arête-type, 7 dorsal, 6 ventral at mid-body. Axis of orientation of ridges ventral right to dorsal left. Small caréne present in left lateral field. Cephalic vesicle with fine, transverse striations. *Male*: Bursa symmetrical. Ventral rays with common stem, separate at distal tips. Lateral rays separate at distal

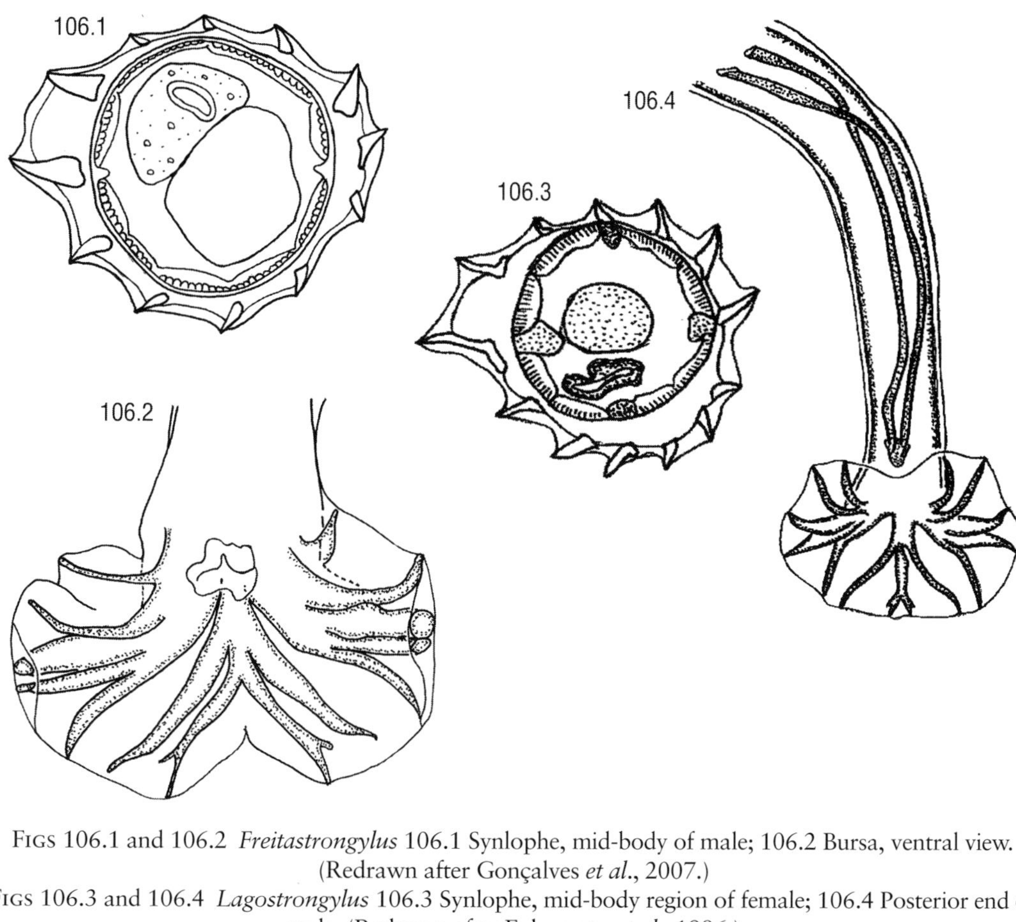

Figs 106.1 and 106.2 *Freitastrongylus* 106.1 Synlophe, mid-body of male; 106.2 Bursa, ventral view. (Redrawn after Gonçalves *et al.*, 2007.)
Figs 106.3 and 106.4 *Lagostrongylus* 106.3 Synlophe, mid-body region of female; 106.4 Posterior end of male. (Redrawn after Fukumoto *et al.*, 1986.)

tips, antero-lateral curving anteriorly. Externo-dorsal rays arise one-third from base of dorsal ray. Dorsal ray thick, divides distally, each branch dividing again. Spicules equal, long and slender, distal end inflated and blunt or simple and sharp. Gubernaculum present. *Female*: Monodelphic. Parasites of *Lepus* spp.

Type species: *Lagostrongylus leporis* (Schulz, 1931) Fukumoto, Kamiya and Ohbayashi, 1986

Macrostrongylus Ow Yang, Durette-Desset and Ohbayashi, 1983 (Fig. 107.1)

DIAGNOSIS: Synlophe absent. *Male*: Bursa: 2-2-1 type, i.e. ventral rays with common stem, separate distally; lateral rays with common stem, antero- and medio-lateral rays close together, separate distally, postero-lateral separate; dorsal ray well developed. Externo-dorsal rays arise in mid-region of main stem of dorsal ray. *Female*: Vestibule of ovejector very long. Parasite in Malaysian rodents, family Muridae (*Rattus* spp.).

Type species: *Macrostrongylus ratti* Ow Yang, Durette-Desset and Ohbayashi, 1983

[*Macrostrongylus* Ow Yang, 1967 and *M. ratti* Ow Yang, 1967 were first described in a PhD thesis (Ow Yang, 1967); the published date and authority of the genus and species are as shown. Listed in Spencer Jones and Gibson (1987).]

Nesomystrongylus Durette-Desset, Lethonen and Haukisalmi, 2002 (Figs 107.2, 107.3, 107.4 and 107.5)

DIAGNOSIS: Synlophe: longitudinal cuticular ridges discontinuous. Ridges terminate in sharp tip in cross-section. *Male*: Bursa: symmetrical; antero-ventral (ray 2) very short, arising separately from other rays; postero-ventral, ray 3, large; length of rays decreasing towards externo-dorsal ray (ray 8), evenly spaced; dorsal ray divided almost to base, each branch divided at distal tip. Genital cone small. Spicules filiform. Gubernaculum present. *Female*: Monodelphic. Vulva opens near anus. Caudal spine present. Parasites of Nesomyinae (Muridae) in Madagascar.

Type species: *Nesomystrongylus fissicauda* Durette-Desset, Lethonen and Haukisalmi, 2002

[Durette-Desset *et al.* (2002) consider this genus to be near *Fissicauda* Durette-Desset and Krishnasamy, 1976.]

Nippostrongylinae Durette-Desset, 1971

Guerrerostrongylus Sutton and Durette-Desset, 1991 (Fig. 108)

DIAGNOSIS: Body length over 8 mm. Synlophe: minimum of 40 longitudinal cuticular ridges, arête type; height of ridges in anterior body unequal, equal from mid-region. *Male*: Bursa with long postero-lateral rays; dorsal lobe and ray long; externo-dorsal rays arise from base of dorsal ray, shorter than dorsal ray. Genital cone not enlarged. *Female*: Posterior end not curved. Vulva opens near anus. Tail tip rounded. Parasites in small intestine of Cricetidae and Caviomorpha.

Type species: *Guerrerostrongylus uruguayensis* Sutton and Durette-Desset, 1991

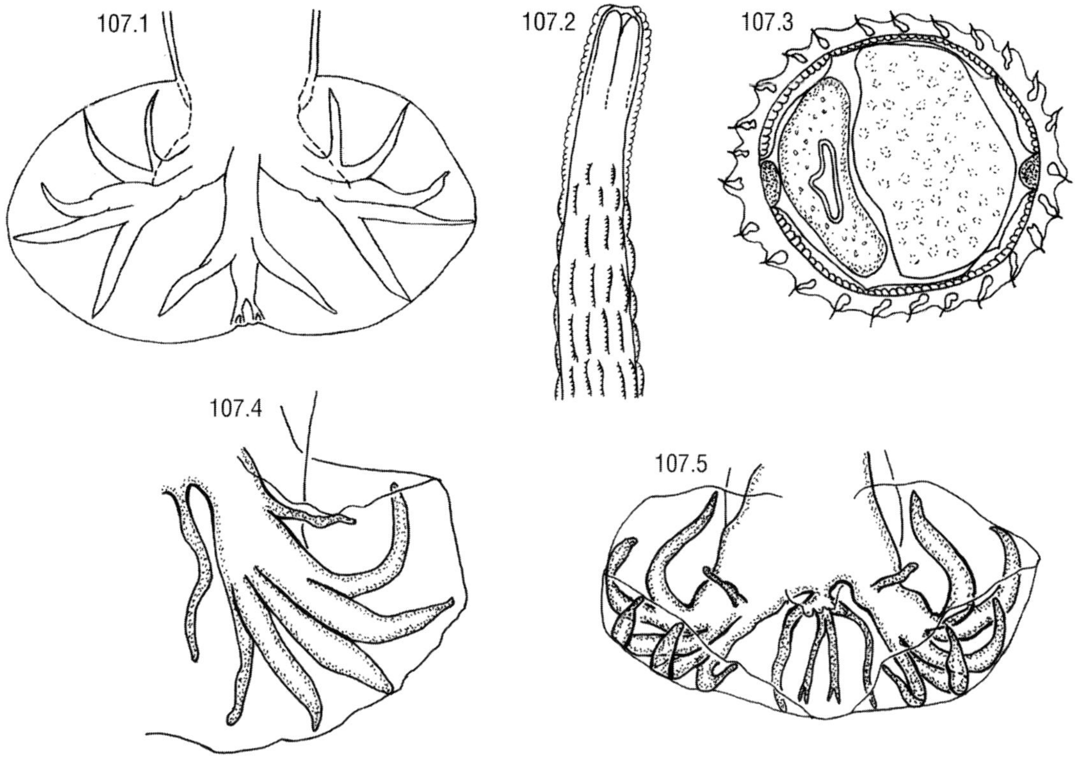

FIG. 107.1 *Macrostrongylus* Posterior end of male, ventral view. (Redrawn after Ow Yang *et al.*, 1983.) FIGS 107.2, 107.3, 107.4 and 107.5 *Nesomyostrongylus* 107.2 Anterior end of female showing cuticular ridges, ventral view; 107.3 Synlophe, mid-body of female; 107.4 Left lobe of bursa, ventral view; 107.5 Bursa, ventral view. (Redrawn after Durette-Desset *et al.*, 2002.)

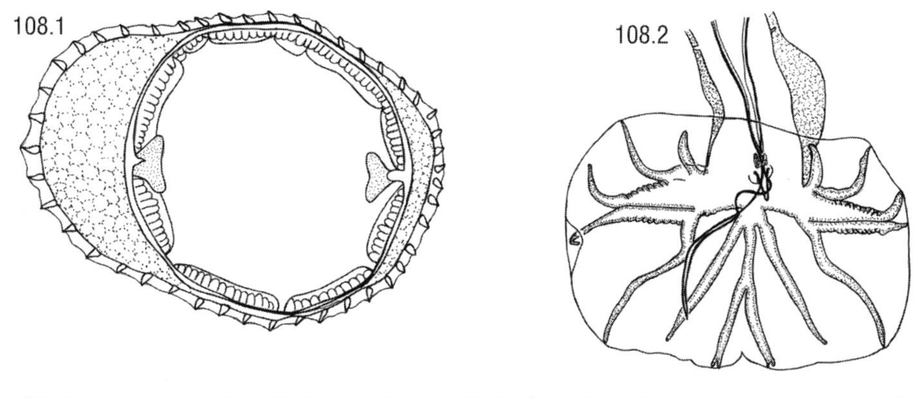

FIG. 108 *Guerrerostrongylus* 108.1 Synlophe, in mid-body region of male; 108.2 Bursa, ventral view. (Redrawn after Sutton and Durette-Desset, 1991.)

Trichofreitasia Sutton and Durette-Desset, 1991 (Fig. 109)

DIAGNOSIS: Body length over 8 mm. Synlophe: longitudinal cuticular ridges of arête type; perpendicular to body surface except anterior, where there is an oblique axis of orientation. *Male*: Bursa with large lateral lobes. Spicules delicate. Genital cone not enlarged. *Female*: Cuticle inflated at level of ovejector. Posterior end strongly curved ventrally. Parasite in small intestine of South American Cricetidae (*Oryzomys flavescens*).

Type species: *Trichofreitasia lenti* Sutton and Durette-Desset, 1991

Spalacina Biserkov, Durette-Desset and Genov, 1995 (Fig. 110)

DIAGNOSIS: Anterior body forms 1–4 spirals. Synlophe: 14–19 ridges; size gradient absent; carène formed of enlarged ventral–left ridge; dorsal left ridge absent; axis of orientation oblique, 70–80° from sagittal axis. *Male*: Bursa slightly asymmetrical; antero-ventral ray separate; postero-ventral, antero- and medio-lateral rays close together diverging at distal tips; postero-lateral separate; arrangement 1-3-1 type; externo-dorsal ray arises at base of dorsal, long; dorsal ray divides in distal third. *Female*: Monodelphic. Vulva opens near anus. Parasites in small intestine of spalacid rodents in Palaearctic region.

Type species: *Spalacina yanchevi* Biserkov, Durette-Desset and Genov, 1995

[Biserkov *et al.* (1995) believe this genus is derived from *Heligmonina*.]

Bunomystrongylus Hasegawa and Mangali, 1996 (Fig. 111)

DIAGNOSIS: Cephalic vesicle present. Mouth opening circular. Cephalic sensory organs: 6 outer labial papillae; 4 submedian cephalic papillae; amphids near labial papillae. Synlophe: in

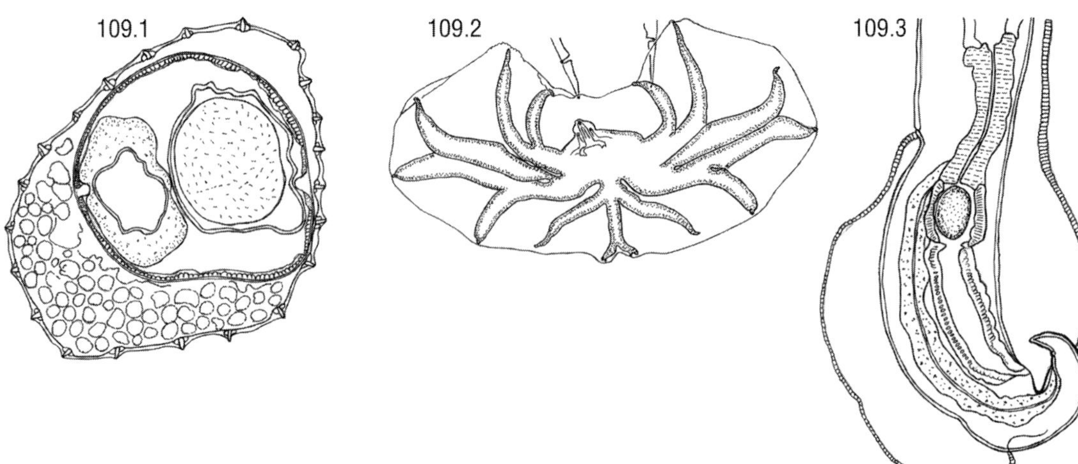

FIG. 109 *Trichofreitasia* 109.1 Synlophe, female at oesophago-intestinal junction; 109.2 Bursa, ventral view
FIG. 109.3 Posterior end of female, lateral view. (Redrawn after Sutton and Durette-Desset, 1991.)

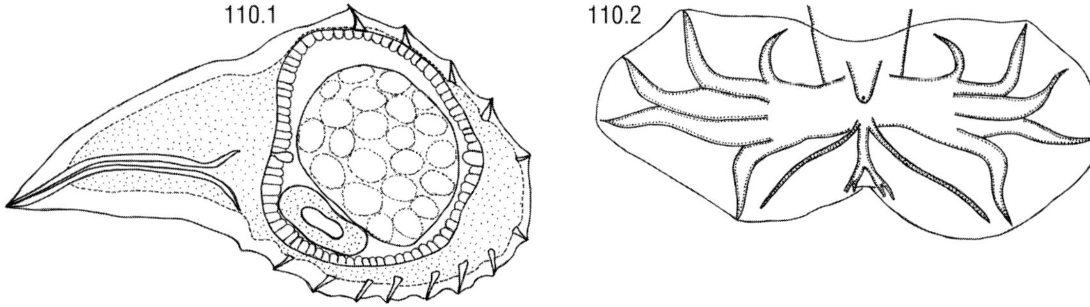

FIG. 110 *Spalacina* 110.1 Synlophe, mid-body of male; 110.2 Bursa, ventral view. (Redrawn after Biserkov *et al.*, 1995.)

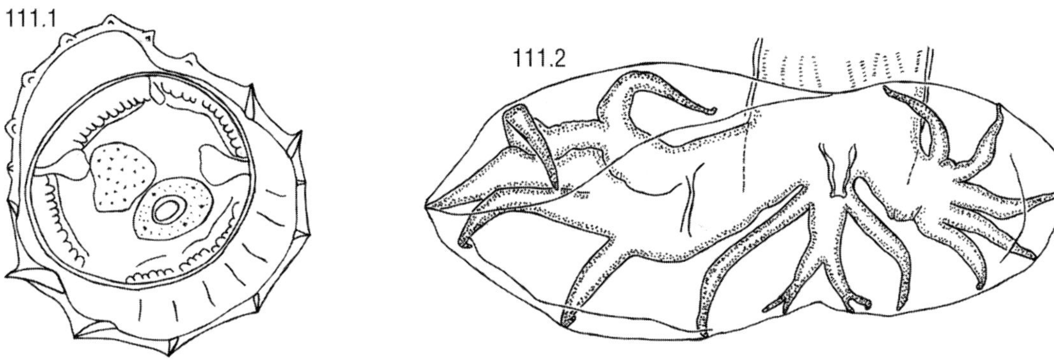

FIG. 111 *Bunomystrongylus* 111.1 Synlophe, mid-body of male; 111.2 Bursa, ventral view. (Redrawn after Hasegawa and Mangali, 1996.)

mid-body ridges dorsal to lateral left rounded, without internal support, remainder pointed of arête type, in cross-section; ridges decrease in height from dorsal right to dorsal left, increase in height from ventral right to ventral left; axis of orientation oblique 65° from sagittal axis in mid-body. *Male*: Bursa asymmetrical; ventral and lateral rays separated at distal tips; externo-dorsal rays longer than dorsal ray, arise from base of dorsal ray. Genital cone prominent: papillae 0 (ventral raylet) at tip of cone; papillae 7 (dorsal raylets) small. Spicules equal, alate, filiform. Gubernaculum present. *Female*: Monodelphic. Vulva opens near anus. Ovejector well developed, vestibule long, coiled. Parasites of *Bunomys* spp. (Rodentia, Muridae) in Sulawesi, Indonesia.

Type species: *Bunomystrongylus abadii* Hasegawa and Mangali, 1996

Hasanuddinia Hasegawa and Syafruddin, 1994 (Fig. 112)

DIAGNOSIS: Body in sinistral coil or irregular curves. Cephalic vesicle present. Mouth opening triangular. Cephalic sensory organs: 6 outer labial papillae; 4 submedian cephalic papillae; amphids lateral to outer labial papillae. Synlophe: longitudinal cuticular ridges of the arête type; in mid-body axis of orientation from ventral right to dorsal left, angle of 70° from sagittal axis; ridges reduce in height from dorsal right to dorsal left, increase in height from ventral right to ventral left, last ventral left ridge less developed. *Male*: Bursa asymmetrical, right lobe larger; ventral rays with common stem, diverge at distal tips; lateral rays with common stem, antero- and medio-lateral

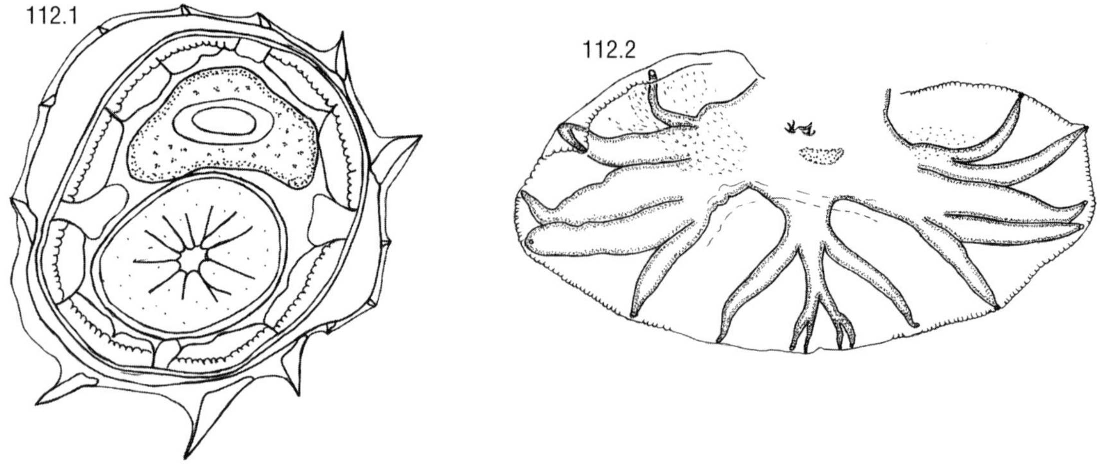

FIG. 112 *Hasanuddinia* 112.1 Synlophe, mid-body of male; 112.2 Bursa, ventral view. (Redrawn after Hasegawa and Syafruddin, 1994a.)

rays parallel, close together, postero-lateral ray separate; externo-dorsal ray arises from main stem of dorsal ray, not longer than dorsal ray, which bifurcates in distal half, each branch dividing again. Genital cone: papilla 0 (ventral raylet) on short cone, papilla 7 (dorsal raylet) single. Spicules equal, alate. Gubernaculum present. *Female*: Monodelphic. Vulva opens near anus. Tail short with pointed tip. Parasites in small intestine of murine rodents in Sulawesi, Indonesia.

Type species: *Hasanuddinia maxomyos* Hasegawa and Syafruddin, 1994

Maxomystrongylus Hasegawa and Syafruddin, 1997 (Fig. 113)

DIAGNOSIS: Body coiled. Cephalic vesicle present. Mouth opening triangular. Cephalic sensory organs: 6 outer labial papillae; 4 submedian cephalic papillae; 2 lateral amphids. Synlophe: axis of orientation 70° from sagittal axis in mid-body, ventral right to dorsal left; ridges pointed in cross-section; carène in lateral field; dorsal ridges decrease in size right to left, increase towards carène; ventral ridges increase in size right to left, decrease towards carène. *Male*: Bursa asymmetrical, left lobe larger; ventral rays with common stem, diverge; lateral rays with common stem, antero- and medio-laterals thick, diverge at distal tips, postero-lateral short, separate at distal tip; externo-dorsal arises from stem of dorsal ray; dorsal ray divides in distal half, each branch dividing again at distal tip. Genital cone slightly elevated. *Female*: Ovejector with unilateral diverticulum. Parasites of small intestine of murines.

Type species: *Maxomystrongylus yasumai* Hasegawa and Syafruddin, 1997

Malaistrongylus Ow Yang, Durette-Desset and Ohbayashi, 1983 (Fig. 114)

DIAGNOSIS: Synlophe: 30-plus longitudinal cuticular ridges of arête type; dorsal left and right ridges most developed. *Male*: Bursa with dorsal ray divided in its anterior half; antero- and medio-lateral rays fused. Parasites in Muridae in Malaysia (*Rattus* spp. and *Chiropodomys gliroides*) and Thailand (*Rattus surifer*).

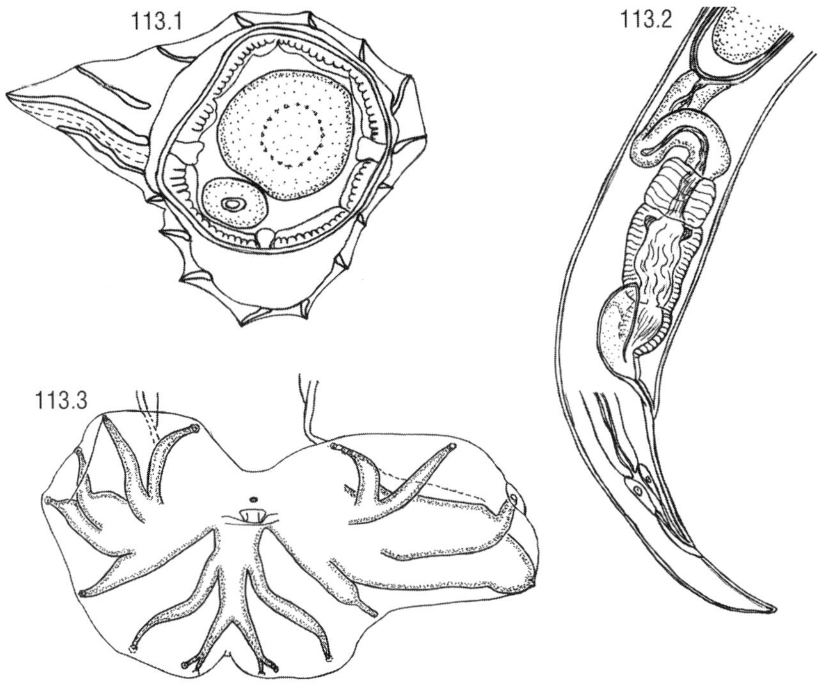

FIG. 113 *Maxomystrongylus* 113.1 Synlophe, mid-body of male; 113.2 Posterior end of female, right lateral view; 113.3 Bursa, ventral view. (Redrawn after Hasegawa and Syafruddin, 1997.)

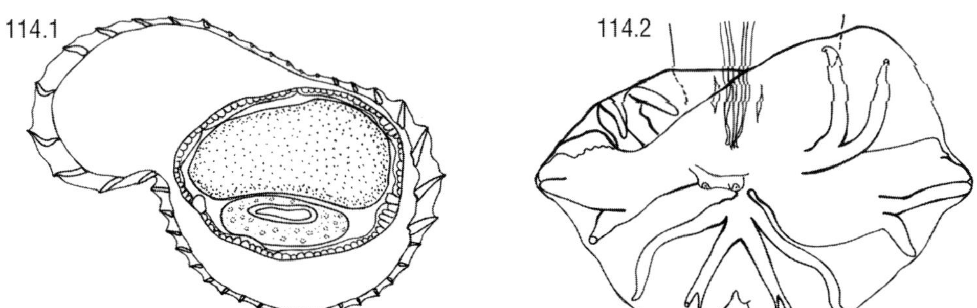

FIG. 114 *Malaistrongylus* 114.1 Synlophe, mid-body region of female; 114.2 Posterior end of male, ventral view. (Redrawn after Ow Yang *et al.*, 1983.)

Type and only species: *Malaistrongylus odontospicularis* Ow Yang, Durette-Desset and Ohbayashi, 1983

[*Malaistrongylus* Ow Yang, 1967 and *M. odontospicularis* Ow Yang, 1967 were first described in a PhD thesis (Ow Yang, 1967); the published date and authority of the genus and species are as shown. Listed in Spencer Jones and Gibson (1987).]

Rattustrongylus Ow Yang, Durette-Desset and Ohbayashi, 1983 (Fig. 115)

DIAGNOSIS: Synlophe: longitudinal cuticular ridges of arête type; dorsal left ridges equal, less developed than other ridges; ventral left and dorsal right ridges large. *Male*: Bursa with left lobe slightly larger than right lobe; dorsal ray divided from its anterior half. Parasites in Malaysian Muridae (*Rattus* spp.).

Type species: *Rattustrongylus odontoconus* Ow Yang, Durette-Desset and Ohbayashi, 1983

[*Longistriata odontoconus* Ow Yang, 1967 was first described in a PhD thesis (Ow Yang, 1967); the species is listed as a synonym of *Rattustrongylus odontoconus* by Ow Yang *et al.* (1983). Listed in Spencer Jones and Gibson (1987).]

Sabanema Ow Yang, Durette-Desset and Ohbayashi, 1983 (Fig. 116)

DIAGNOSIS: Synlophe: numerous longitudinal cuticular ridges (25-plus at level of excretory pore) of arête type; ridges at dorsal left and ventral right subequal, smaller than other ridges. *Male*: Bursa with left lobe larger than right; dorsal ray very long, divided for at least two-thirds of its length. Parasites in Muridae in Malaysia (*Rattus* spp. and *Chiropodomys gliroides*) and Thailand (*Rattus fluvescens*).

Type species: *Sabanema sabana* Ow Yang, Durette-Desset and Ohbayashi, 1983

[Listed in Spencer Jones and Gibson (1987).]

Euzetoda Elias and Durette-Desset, 2003 (Fig. 117)

DIAGNOSIS: Synlophe: carène absent; ridges unequal of arête or crête type; left lateral ala formed of 2 closely adjacent, unequal ridges, lack obvious gradient. *Male*: Bursa subsymmetrical, ray pattern of 1-4 type, antero-ventral ray separated from postero-ventral and

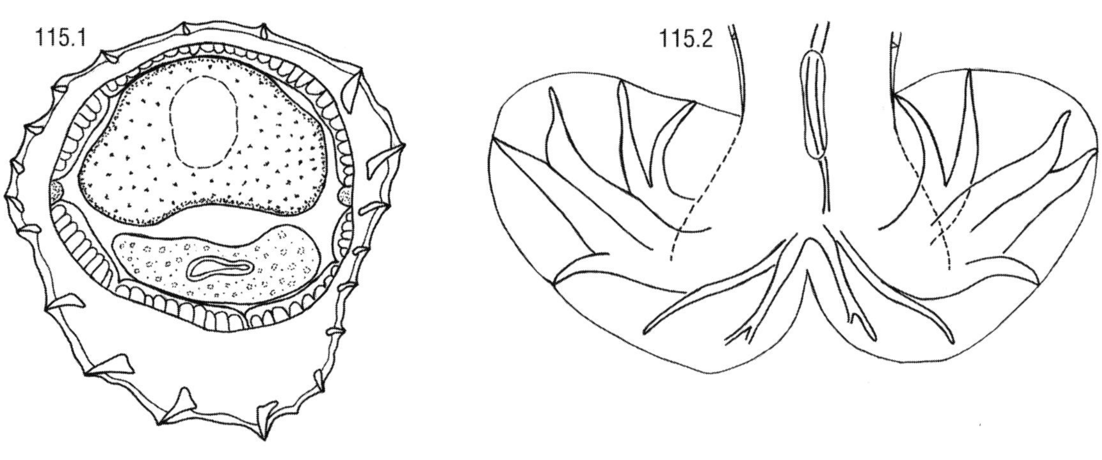

FIG. 115 *Rattustrongylus* 115.1 Synlophe, mid-body region of female; 115.2 Posterior end of male, ventral view. (Redrawn after Ow Yang *et al.*, 1983.)

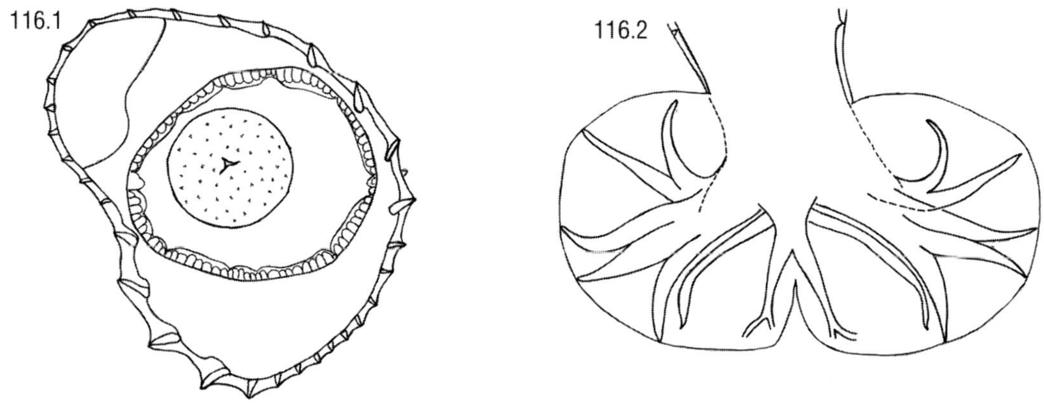

FIG. 116 *Sabanema* 116.1 Synlophe, at level of excretory pore; 116.2 Posterior end of male, ventral view. (Redrawn after Ow Yang *et al.*, 1983.)

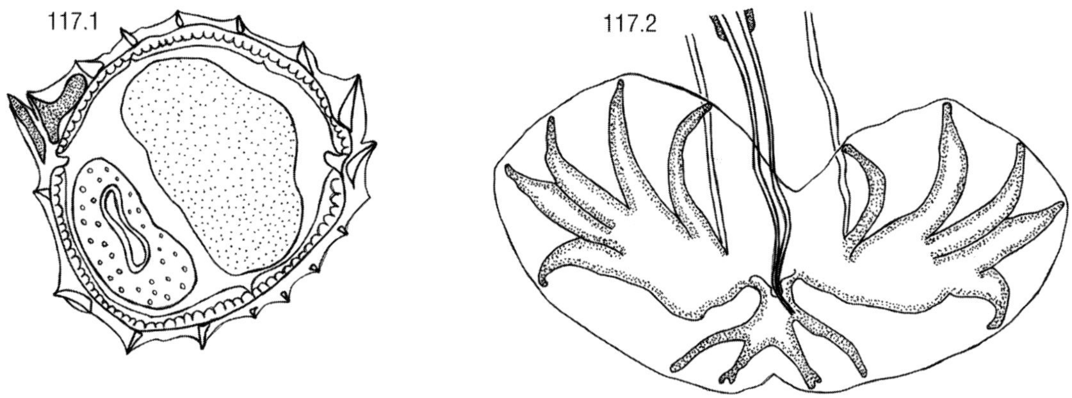

FIG. 117 *Euzetoda* 117.1 Synlophe, mid-body of female; 117.2 Bursa, ventral view. (Redrawn after Elias and Durette-Desset, 2003.)

lateral rays; dorsal lobe reduced, ray wide. Spicules short, thin. *Female*: Ovejector and uterus short. Parasite of Chinese Arvicolinae (Muridae).

Type species: *Euzetoda ludovici* Elias and Durette-Desset, 2003

Suttonema Digiani and Durette-Desset, 2003 (Figs 118.1 and 118.2)

DIAGNOSIS: Cephalic vesicle present. External labial and cephalic papillae linked in pairs by U-shaped cuticular ridge. Synlophe: few uninterrupted ridges; ridges on dorsal surface rounded without supports; ridges on lateral and ventral surfaces with supports; 'comarêtes in left ventral and ventral fields'; axis of orientation ventral right to dorsal left, approximately '45° to sagittal axis at mid-body'. *Male*: Bursa asymmetrical, right lobe larger than left, ventral rays and lateral rays with common trunk. Ray pattern of 1-4 type. Antero-ventral ray separate, shorter than postero-ventral ray. Antero- and medio-lateral rays diverge sharply at distal tips, externo-dorsals asymmetrical, dorsal ray divides in mid-region, each branch dividing at distal tips. Genital cone well developed. *Female*: Monodelphic. Vulva opens near anus. Tail tapers sharply posterior to anus, spine absent. Parasites in intestine of neotropical Sigmodontinae, Muridae (*Oxymycterus rufus*).

Type species: *Suttonema delta* Digiani and Durette-Desset, 2003

[Digiani and Durette-Desset (2003a) place this genus in the Heligmosomoidea in the suborder Trichostrongylina.]

Malvinema Digiani, Sutton and Durette-Desset, 2003 (Figs 118.3 and 118.4)

DIAGNOSIS: Synlophe: longitudinal cuticular ridges continuous; axis of orientation ventral right to dorsal left, inclination 55° to sagittal axis; largest ridges ventral left, dorsal right. *Male*: Bursa asymmetrical, right lobe hypertrophied, 1-4 type, i.e. antero-ventral ray separate from other rays. Genital cone well developed. *Female*: Monodelphic. Vulva opens near anus. Tail sharply pointed. Parasites in small intestine of neoptropical Sigmodontinae (Muridae).

Type species: *Malvinema frederici* Digiani, Sutton and Durette-Desset, 2003

[Digiani *et al.* (2003) place this genus in the Heligmosomoidea in the suborder Trichostrongylina.]

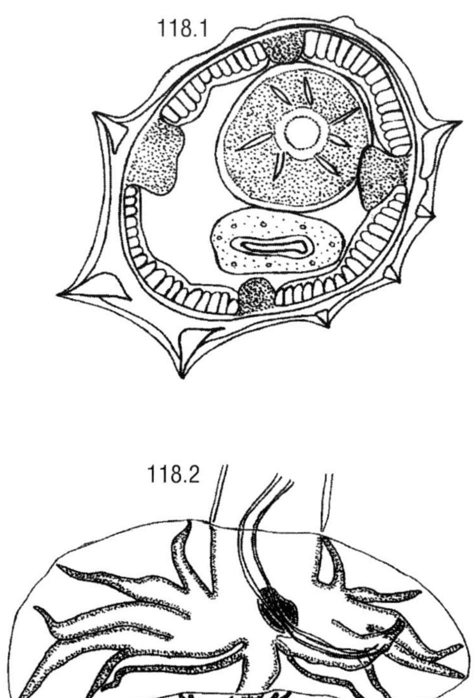

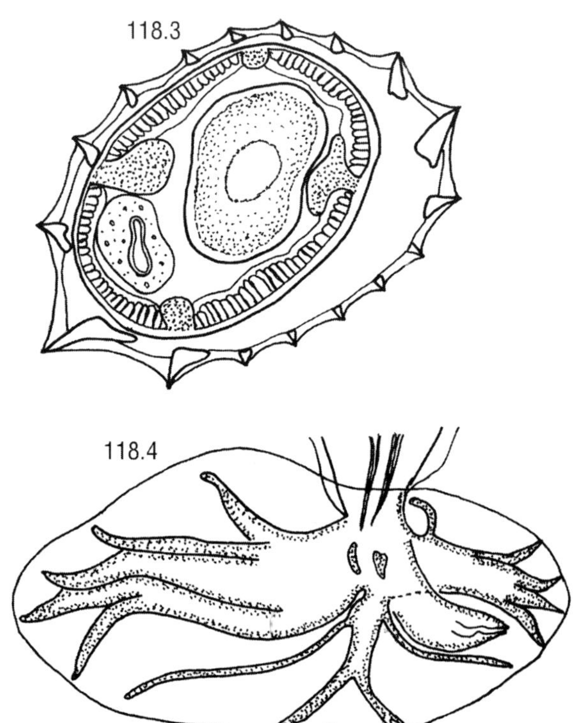

FIGS 118.1 and 118.2 *Suttonema* 118.1 Synlophe, mid-body of male; 118.2 Bursa, ventral view. (Redrawn after Digiani *et al.*, 2003a.)
FIGS 118.3 and 118.4 *Malvinema* 118.3 Synlophe, 900 μm from anterior (mid-body) of female; 118.4 Bursa, ventral view. (Redrawn after Digiani *et al.*, 2003b.)

Ichthyostrongylidae Yamaguti, 1961

Neoichthyostrongylus Naqvi and Gupta, 1987 (publ. 1989) (Fig. 119)

DIAGNOSIS: Cuticle with transverse striations. Head broad. Mouth with 3 rudimentary lips. Cephalic sensory organs: 2 papillae per lip. Buccal capsule absent. *Male*: Bursa bell-shaped, symmetrical, dorsal lobe well developed; ventral rays with common stem, diverge at distal tips; lateral rays with common stem, widely separated at distal tips, medio-lateral divided in distal half, branches stay close together; externo-dorsal ray slender, arises separately from dorsal ray near its base; dorsal ray bifurcates for two-thirds of its length, each branch dividing again forming 4 distinct branches. Spicules equal, similar, slender, up to one-fifth body length. Gubernaculum rod-shaped if present. *Female*: Vulva opens subterminally. Eggs oval, thin-shelled. Anus terminal. Parasites in intestinal mesentery of the marine fish, *Equula daura*, from Panjim, Goa, India.

Type species: *Neoichthyostrongylus thapari* Naqvi and Gupta, 1987 (publ. 1989)

[The genus is separated from *Ichthyostrongylus* Mawson, 1954 on the basis of the branching of the dorsal ray and the bifurcation of the medio-lateral ray. Durette-Desset (1983) did not include *Ichthyostrongylus* in her key, on the grounds that the species are 'insufficiently known to

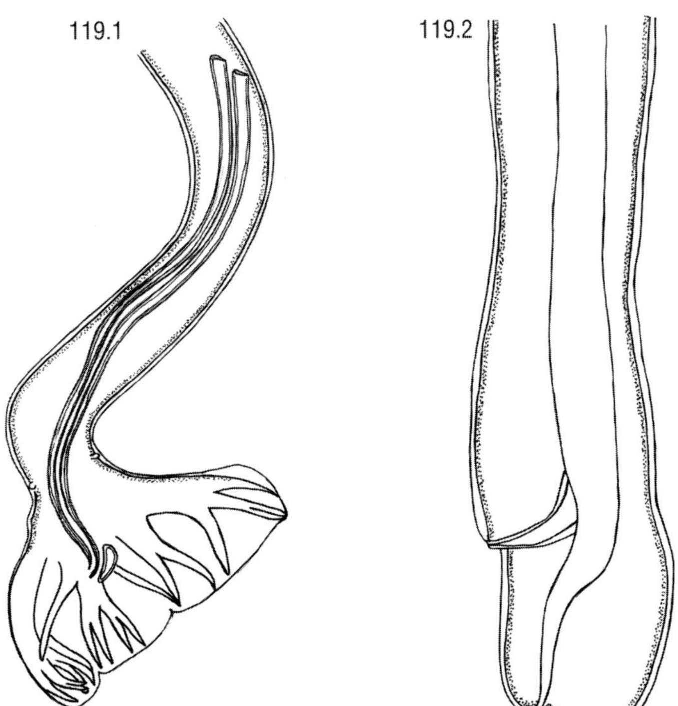

FIG. 119 *Neoichthyostrongylus* 119.1 Posterior end of male; 119.2 Posterior end of female. (Redrawn after Naqvi and Gupta, 1987 (publ. 1989).)

be classified'. Mawson suggested they are probably marine nematodes. *Neoichthyostrongylus* is provisionally included in this superfamily and family Ichthyostrongylidae, to which the authors had assigned their new genus until further studies can establish the true position of these two genera.]

Cylicostrongylidae Yamaguti, 1961

Neocylicostrongylus Arya and Johnson, 1977 (Fig. 120)

DIAGNOSIS: Cephalic region swollen but narrower than cervical region. Cephalic sensory organs: 3 pairs. 'Lips shallow'. Buccal capsule with '7 longitudinal rays' without 'spines or horn-like thickenings'. Single pair of cervical papillae. Oesophagus undivided, muscular. *Male*: Bursa bell-shaped, not clearly divided into separate lobes, rounded at distal margin with 5 simple rays, 11 pairs prebursal caudal papillae pedunculate, 3 pairs bursal papillae sessile. Spicules: equal, stout, similar. Gubernaculum present. Parasites in the intestine of fish (*Cybium guttatum*) in India.

Type species: *Neocylicostrongylus surmani* Arya and Johnson, 1977

[Listed in Spencer Jones and Gibson (1987) in the Camallanoidea.]

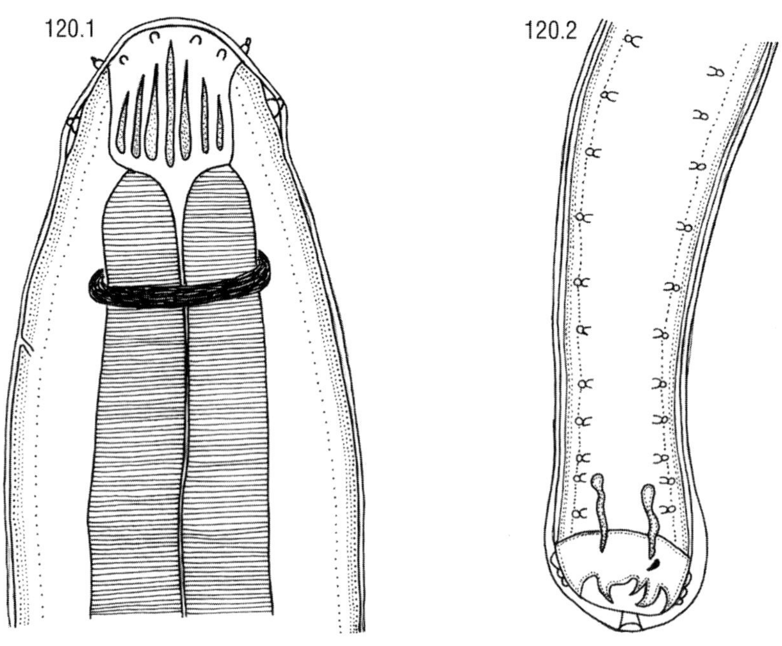

FIG. 120 *Neocylicostrongylus* 120.1 Anterior end of male; 120.2 Posterior end of male. (Redrawn after Arya and Johnson, 1977.)

METASTRONGYLOIDEA

[Durette-Desset and Chabaud (1993) propose that this superfamily should be included in a new suborder Metastrongylina in the Strongylida. No differential diagnosis was given.]

Genera not designated to a family

Micronematodum Castelli, 1926

DIAGNOSIS: Microscopic adults 0.025–0.030 mm long, 0.002 mm wide, intestine 'granuloso'; larvae 0.012 mm long, 0.002–0.003 mm wide; eggs 0.008–0.009 mm × 0.004–0.005 mm. Parasite in lungs of sheep.

Type species: *Micronematodum ovis* Castelli, 1926

[The 'adults' are possibly larvae of *Dictyocaulus*. Description based on Castelli (1926). Listed in Spencer Jones and Gibson (1987).]

Angiostrongylidae (Boehm and Gebauer, 1934 subfamily) Anderson, 1978

Angiostrongylus Kamensky, 1905 (Fig. 121)

DIAGNOSIS: Mouth without lips. 'Peristyles' present. Six cephalic sensory papillae surround mouth; 4 double submedian cephalic sensory papillae present. *Male*: Bursal lobes joined ventrally, overlapping anterior to cloaca, with dorsal median notch. Ventral rays with common stem, separate only near distal tip; antero-lateral ray separate from other laterals; medio- and postero-laterals with common stem, separate in distal half; externo-dorsal ray longer than dorsal, arises separately from dorsal ray; dorsal ray reduced, divided at distal tip. Spicules long. Gubernaculum present, little chitinized. *Female*: Vulva opens near anus. Ovoviviparous. Ovejector prominent. Anus subterminal. Parasites in the heart and arterial circulation of carnivores.

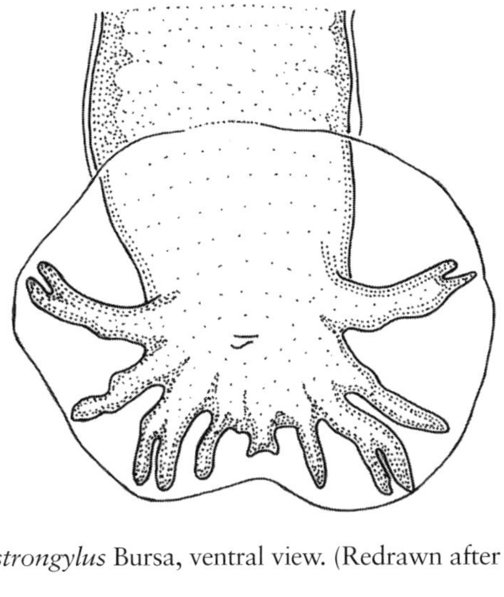

FIG. 121 *Angiostrongylus* Bursa, ventral view. (Redrawn after Ubelaker, 1986.)

Type species: *Angiostrongylus vasorum* (Baillet, 1866) Kamensky, 1905.

[Emended diagnosis based on Ubelaker (1986), who confirms *Haemostrongylus* Railliet and Henry, 1907 as a synonym of the genus.]

Parastrongylus Baylis, 1928 (Fig. 122)

DIAGNOSIS: Mouth without lips. 'Peristyles' present. Six cephalic sensory papillae surround mouth; 4 double submedian cephalic sensory papillae present. *Male*: Bursal lobes joined ventrally, overlapping anterior to cloaca with dorsal median notch. Ventrals with common stem, separate in distal half; lateral rays with common stem, separate in distal half; externo-dorsal ray longer than dorsal, arises separately from dorsal ray; dorsal ray reduced, divided at distal tip. Spicules long. Gubernaculum present, generally little chitinized. *Female*: Vulva opens near anus. Oviparous. Ovejector prominent. Anus subterminal. Parasites in the arterial circulatory system of rodents.

Type species: *Parastrongylus tateronae* (Baylis, 1928)

[Anderson (1978) included this as a subgenus of *Angiostrongylus*. Ubelaker (1986) restored it to generic status and the emended diagnosis is based on his description. Ubelaker (1986) includes *Pulmonema* Chen, 1935, *Rattostrongylus* Schulz, 1951, *Morerastrongylus* Chabaud, 1972 and *Chabaudistrongylus* Kontrimavichus, 1979 (in: Kontrimavichus and Delyamure, 1979) as synonyms.]

Angiocaulus Schulz, 1951 (Fig, 123)

DIAGNOSIS: *Male*: Bursa well developed; ventral rays with common stem only separate near distal ends; lateral rays with common stem, antero-lateral separates first, diverges from other

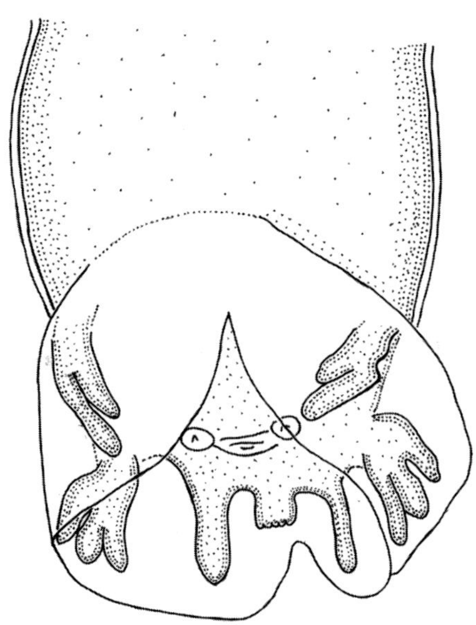

FIG. 122 *Parastrongylus* Bursa, ventral view. (Redrawn from Ubelaker, 1986 after Baylis, 1928.)

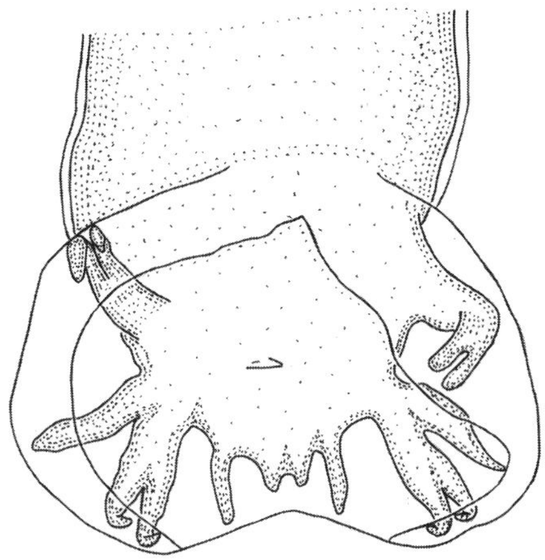

FIG. 123 *Angiocaulus* Bursa, ventral view. (Redrawn from Ubelaker, 1986 after Dougherty, 1946.)

laterals, medio- and postero-laterals only separate in distal half; dorsal ray short, divided for distal half; externo-dorsal ray longer than dorsal ray and separate. Spicules long, filiform, alate. Gubernaculum present, proximal end bifurcated. *Female*: Vulva opens near anus. Oviparous. Parasites in vascular system of mustelids.

Type species: *Angiocaulus gubernaculatus* (Dougherty, 1946) Schulz, 1951

[Anderson (1978) included this genus and *Cardionema* as a synonym of *Angiostrongylus*. Ubelaker (1986) restored the validity of *Angiocaulus* and considers *Cardionema* Yamaguti, 1941 a synonym. The emended diagnosis is based on Ubelaker (1986).]

Gallegostrongylus Mas-Coma, 1977

[Ubelaker (1986) considers *Thaistrongylus* Ohbayashi, Kamiya and Bhaibulaya, 1979 a synonym of *Gallegostrongylus*.]

Procyonostrongylus Anderson, Prestwood and Strelive, 1979 (Fig. 124)

DIAGNOSIS: Delicate worms, tegumental sheath well developed, lips absent. 'Peristyles' present, laterals with 2 papillae and amphid, subdorsals and subventrals each with 3 papillae. *Male*: Bursa well developed, lobes joined ventrally, deep dorsal incision. Ventral rays with common stem, equal or subequal, rounded distal tips; lateral rays with common stem, medio-lateral longer then other laterals; dorsal ray highly developed, larger than other rays. Spicules equal. Gubernaculum well developed. *Female*: Vulva and anus terminal. Ovejector prominent with well-developed distal sphincter. Oviparous. Parasites under the epithelium of the epiglottis of mammals (*Procyon lotor lotor*) in Glades, Highlands and Hillsborough counties, Florida, USA.

Type species: *Procyonostrongylus muelleri* Anderson, Prestwood and Strelive, 1979

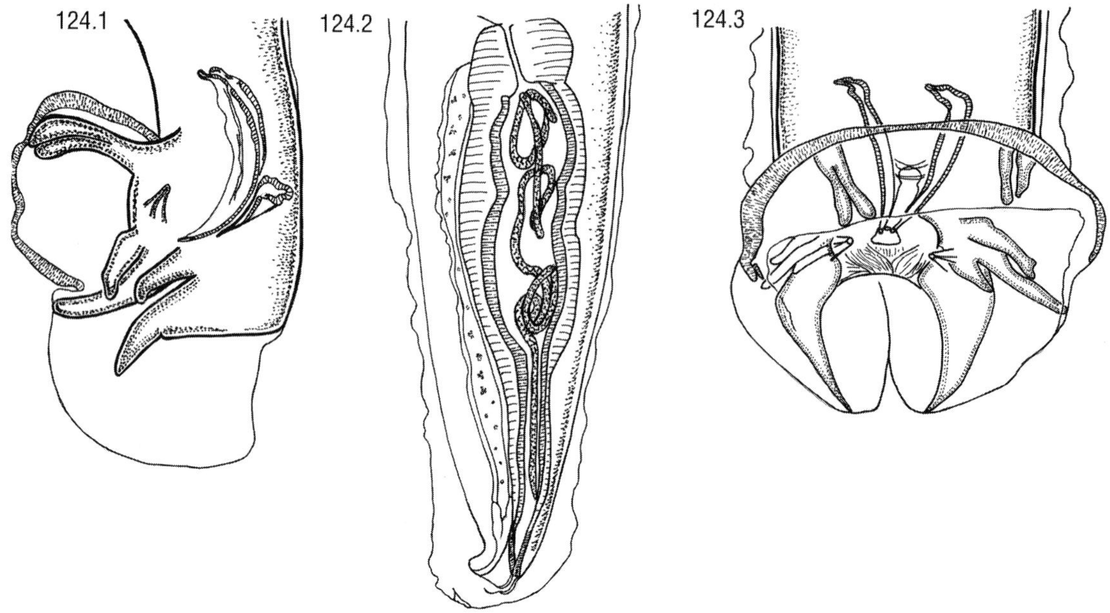

FIG. 124 *Procyonostrongylus* 124.1 Posterior end of male, lateral view; 124.2 Posterior end of female, lateral view; 124.3 Bursa, ventral view. (Redrawn after Anderson *et al.*, 1979.)

[Listed in Spencer Jones and Gibson (1987).]

Chabaudistrongylus Kontrimavichus, 1979 (in: Kontrimavichus and Delyamure, 1979) (= *Angiocaulus* Schulz, 1951 (in part); *Rattostrongylus* Schulz, 1951 (in part))

DIAGNOSIS: Translation not available. See original publication.

Type species: *Chabaudistrongylus ryjikovi* (Jushov, 1971) Kontrimavichus, 1979

[Kontrimavichus, 1979 places the genus in the subfamily Angiostrongylinae of the Filaroididae. Anderson (1978) put all the genera of the subfamily Angiostrongylinae into the family Angiostrongylidae. Anderson (1978) synonymizes *Angiocaulus* Schulz, 1951 with *Angiostrongylus* subgenus, and *Rattostrongylus* Schulz, 1951 with *Parastrongylus* subgenus. Ubelaker (1986) restored the validity of *Angiocaulus* (see above) and synonymizes *Chabaudistrongylus* with *Parastrongylus*, which he restores to generic rank. Listed in Spencer Jones and Gibson (1987).]

Rauschivingylus Kontrimavichus, 1979 (in: Kontrimavichus and Delyamure, 1979)

DIAGNOSIS: Translation not available. See original publication.

Type species: *Rauschivingylus microti* (Rausch and Rausch, 1969) Kontrimavichus, 1979

[Kontrimavichus (1979) placed the genus in the subfamily Sobolevingylinae of the Filaroididae. The type genus of this subfamily is included in the Angiostrongylidae in Anderson (1978), and therefore this genus is listed in the same family. Listed in Spencer Jones and Gibson (1987).]

Malayometastrongylus Gibbons and Krishnasamy, 1986 (Fig. 125)

DIAGNOSIS: Head region with prominent elevations. Buccal capsule small with weakly cuticularized walls. *Male*: Bursa reduced, terminal, undivided, oval in dorso-ventral view. Bursal rays reduced, ventrals and laterals each with common stem. Externo-dorsal arises near base of dorsal ray, dorsal ray divides distally, combined width of branches greater than main stem width. Spicules short, arcuate, sclerotized striations support alae. Gubernaculum present with complex branching. *Female*: Ovoviviparous. Vulva opens near anus. Tail conical with subterminal papillae. Parasite in the lungs of *Rattus rattus diardii* in Malaysia.

Type species: *Malayometastrongylus diardinematus* Gibbons and Krishnasamy, 1986

[Listed in Spencer Jones and Gibson (1987).]

Thaistrongylus Ohbayashi, Kamiya and Bhaibulaya, 1979 (Fig. 126)

DIAGNOSIS: Cuticle smooth. Mouth with 4 lips. *Male*: Bursa with dorsal notch. Bursal rays reduced, short, stout. Dorsal ray reduced to pair of subventral papillae. Spicules short, arcuate. Gubernaculum present. *Female*: Vulva and anus open close together. Tail short, rounded. Parasite in lungs of *Rattus berdmorei* in Mae Sariang, Thailand.

Type species: *Thaistrongylus harinasutai* Ohbayashi, Kamiya and Bhaibulaya, 1979

[Considered a synonym of *Gallegostrongylus* Mas-Coma, 1977 by Ubelaker (1986). Listed in Spencer Jones and Gibson (1987).]

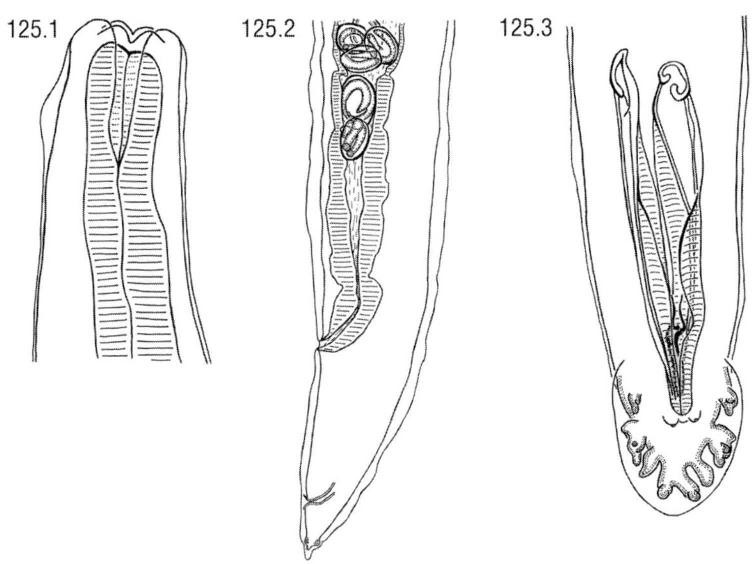

FIG. 125 *Malayometastrongylus* 125.1 Cephalic region; 125.2 Posterior end of female, lateral view; 125.3 Posterior end of male, ventral view. (Redrawn after Gibbons and Krishnasamy, 1986.)

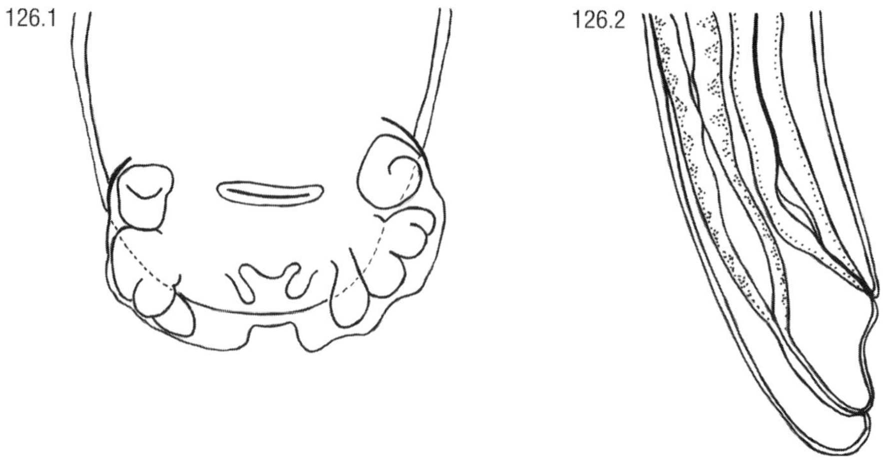

FIG. 126 *Thaistrongylus* 126.1 Bursa, ventral view; 126.2 Posterior end of female, lateral view. (Redrawn after Ohbayashi *et al.*, 1979.)

Cercogylus Petter and Brochier, 1989 (Fig. 127)

DIAGNOSIS: Mouth triangular without lips. Four double submedian cephalic papillae. Buccal cavity very reduced, walls not sclerotized. *Male*: Bursa reduced, lobes not separated, longer than wide. Dorsal ray completely divided, branches elongate and separated from each other with 2 papillae at distal tips. Spicules short, sturdy. Gubernaculum present. *Female*: Vulva opens near anus. Anus subterminal. Viviparous. Tail rounded. Parasite in primates, Cercopithecidae, in Africa.

Type species: *Cercogylus africanus* Petter and Brochier, 1989

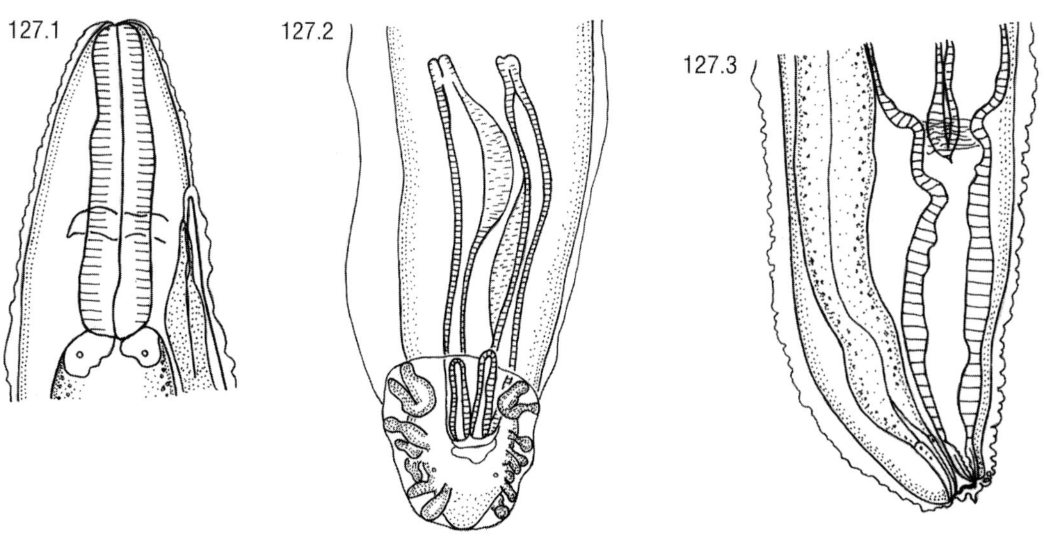

FIG. 127 *Cercoglyus* 127.1 Anterior end, lateral view; 127.2 Posterior end of male, ventral view; 127.3 Posterior end of female, lateral view. (Redrawn after Petter and Brochier, 1989.)

Antechinostrongylus Spratt, 1981 (Fig. 128)

DIAGNOSIS: Tegumental sheath present. Cuticle smooth. Mouth with 3 reduced lips each with a pair of papillae. Four submedian cephalic papillae, 2 lateral amphids. Oesophagus claviform. Excretory pore opens at level of middle third of oesophagus. *Male*: Bursal lobes reduced, not joined anterior to cloaca. Bursal rays reduced; ventral rays fused, form single papilla; lateral rays common bulbous stem, form 3 short processes; externo-dorsal ray small, separate; dorsal ray absent. Spicules equal, short, arcuate. Gubernaculum with corpus attached to distal, spoon-like plate. *Female*: Vulva opens close to anus. Ovoviviparous. Tail short, bluntly rounded. Larvae without dorsal spine. Parasites in 'deep lung parenchyma' of Australian marsupials, Dasyuridae (*Antechinus swainsonii*).

Type species: *Antechinostrongylus disgubernaculus* Spratt, 1981

[Listed in Spencer Jones and Gibson (1987).]

Cosmostrongylus Spratt, 1984 (Fig. 129)

DIAGNOSIS: 'Voluminous' tegumental sheath present, generally formed into folds. Cuticle thin with longitudinal striae. Mouth with 3 small lips, each with 2 papillae of inner circle. Four large submedian cephalic papillae, 2 lateral amphids. Buccal capsule very reduced, walls poorly sclerotized. Oesophagus short, claviform. *Male*: Posterior end curved ventrally. Bursa reduced. Ventral rays with common stem, parallel, separate; laterals with common stem, postero-lateral maybe shorter and thicker than other laterals; externo-dorsal ray separate; dorsal ray forms 2 small papillae, which may be joined at base, surrounded by tiny lobe.

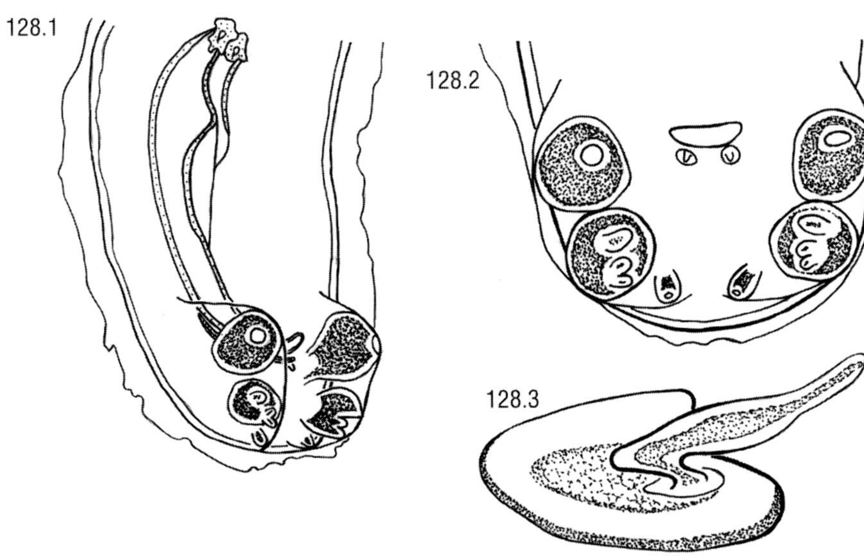

FIG. 128 *Antechinostrongylus* 128.1 Posterior end of male, lateral view; 128.2 Bursa, ventral view; 128.3 Gubernaculum, latero-ventral view. (Redrawn after Spratt, 1981.)

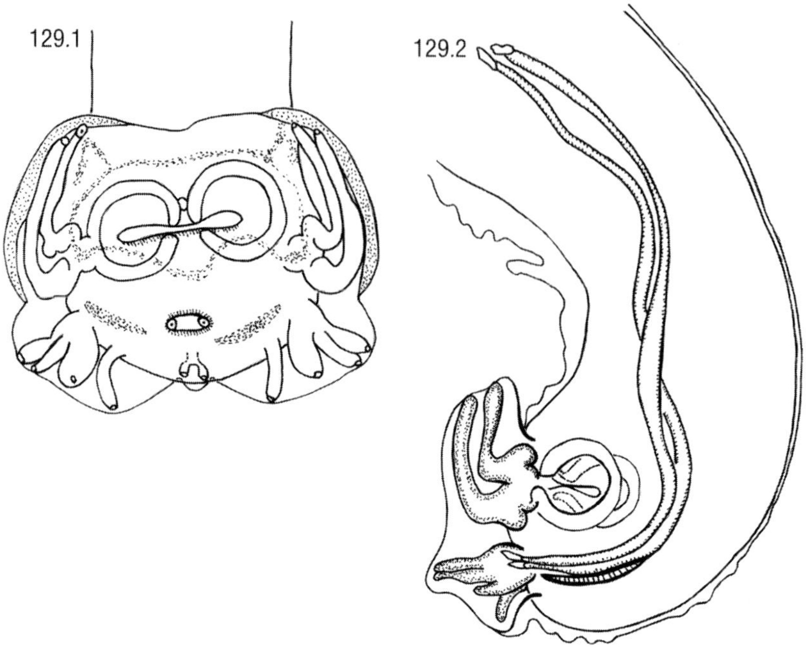

FIG. 129 *Cosmostrongylus* 129.1 Bursa, ventral view; 129.2 Posterior end of male, lateral view. (Redrawn after Spratt, 1984.)

Spicules equal, similar. Gubernaculum present. Telamon well developed, forming 2 rings and sclerotized. *Female*: Vulva opens near anus. Vagina short, muscular ovejector. Didelphic, prodelphic, ovoviviparous. Parasite in lungs of macropodid marsupials (*Thylogale stigmatica*) in Australia.

Type: *Cosmostrongylus conspectus* Spratt, 1984

[Listed in Spencer Jones and Gibson (1987).]

Viverrostrongylus Asakawa, Ohbayashi and Ow Yang, 1986 (Fig. 130)

DIAGNOSIS: Mouth with 3 lips. *Male*: Free in bronchioles and alveoli. Bursa with 2 lateral lobes with separate dorsal lobe. Ventral rays fused, only separated at distal tips; lateral rays fused, only separated at distal tips; externo-dorsal ray short, separate from dorsal ray; dorsal ray thick, bifurcating halfway along its length, each branch dividing distally. Spicules short with wide, striated alae. Gubernaculum reduced. *Female*: Encapsulated in nodules, several females present. Vulva opens near subterminal anus. Ovoviviparous. Parasites of lungs of a large spotted civet, *Viverra megaspila*, in Sungai Petani, Malaysia.

Type species: *Viverrostrongylus brauni* (von Linstow, 1897) Asakawa, Ohbayashi and Ow Yang, 1986

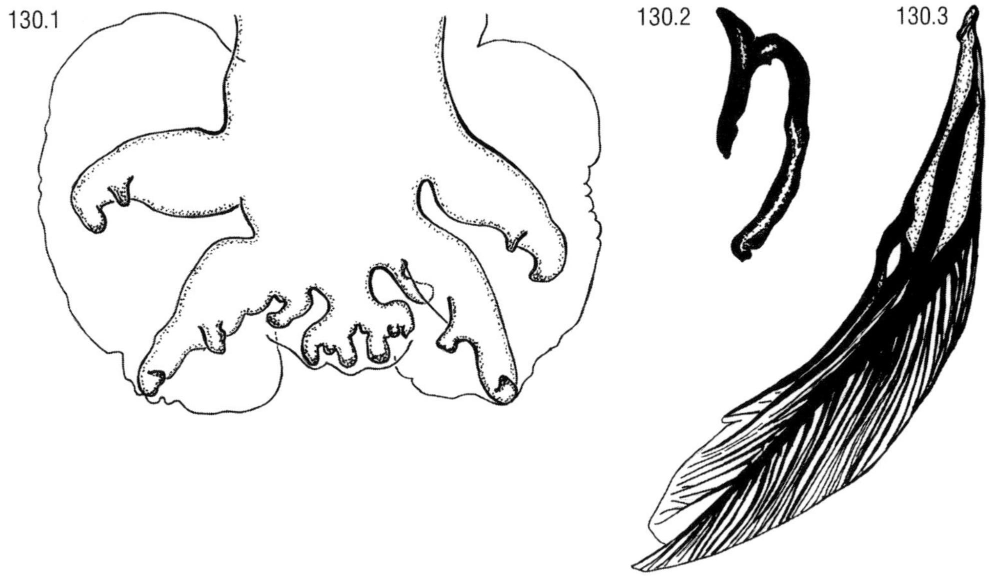

FIG. 130 *Viverrostrongylus* 130.1 Bursa, ventral view; 130.2 Gubernaculum; 130.3 Left spicule. (Redrawn after Asakawa *et al.*, 1986.)

Protostrongylidae Leiper, 1926

***Dukerostrongylus* Dinnik and Boev, 1982 (Fig. 131)**

DIAGNOSIS: Body filiform. *Male*: Telemon present. Bursa with dorsal ray ovoid with 3 branches, 2 outer branches divided. Spicules equal, split distally to form a broad main stem and narrow, pointed alate branch. Gubernaculum complex, divided into colourless rudimentary corpus and well-developed crura. *Female*: Not studied. Larvae have spine on tail. Parasites in the lung alveoli of *Sylvicapra grimmia*, common bush duiker in Kenya.

Type species: *Dukerostrongylus kenyae* Dinnik and Boev, 1982

[The authors place this genus in the subfamily Muelleriinae Skrjabin, 1933 on the basis of the trifid dorsal ray, the spicules, the gubernaculum and the larvae with a spine on the tail. *Cystocaulus* Schulz, Orlov, and Kutass, 1933 and *Muellerius* Cameron, 1927 are also included in the subfamily. Anderson (1978) did not divide the family into subfamilies. Listed in Spencer Jones and Gibson (1987).]

***Umingmakstrongylus* Hoberg, Polley, Gunn and Nishi, 1995 (Fig. 132)**

DIAGNOSIS: Body with very large measurements. *Male*: Up to 171 mm long. Bursa bilobed, generally asymmetrical. Externo-dorsal ray separate from dorsal ray. Dorsal ray thick, trilobate with 6 papillae. Telamon forms distal transverse plate. Gubernaculum with paired arcuate capitulum, ventral prongs curve around spicules, dorsal prongs directed over corpus, crura paired, lack

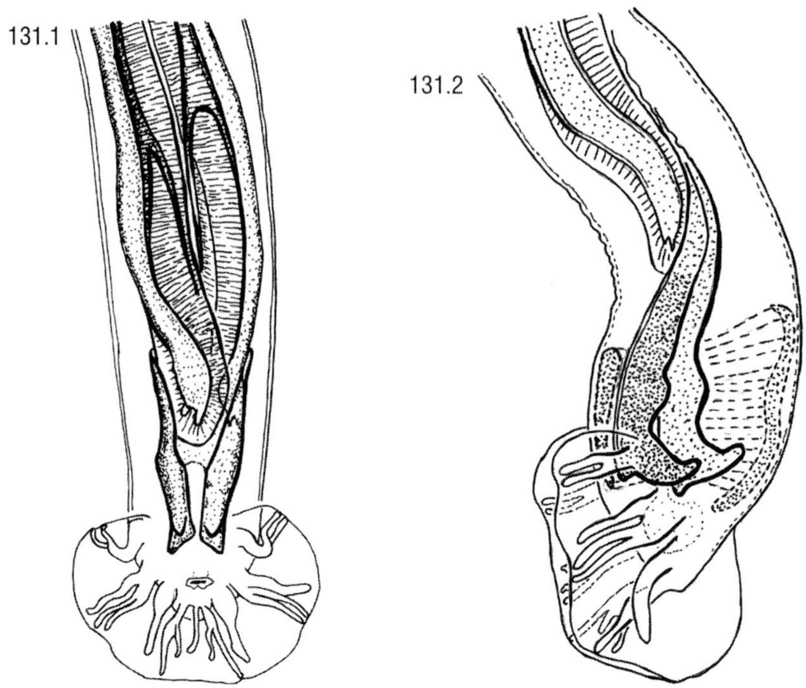

FIG. 131 *Dukerostrongylus* 131.1 Posterior end of male, ventral view; 131.2 Posterior end of male, lateral view. (Redrawn after Dinnik and Boev, 1982.)

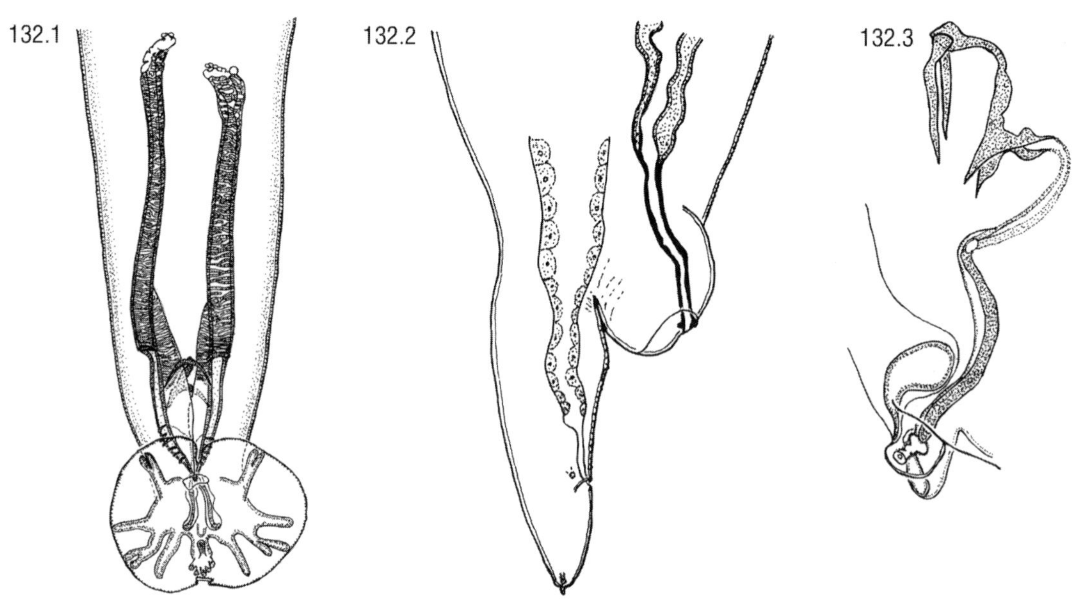

FIG. 132 *Umingmakstrongylus* 132.1 Posterior end of male, ventral view; 132.2 Posterior end of female, lateral view; 132.3 Gubernaculum and distal transverse plate of telamon, lateral view. (Redrawn after Hoberg *et al.*, 1995.)

prominent falcate dorsal feet. Spicules equal to subequal, similar, left up to 0.431 mm long, right up to 0.463 mm long, not split distally, distinctive joint in distal third. *Female*: Up to 468 mm long. Vulva opens near anus, provagina absent. Tail conical, tapers to blunt point. L_1 with prominent dorsal spine on tail followed by 3 cuticular folds. Parasites in cysts in lung tissue of musk ox, *Ovibos moschatus moschatus*, in Central Arctic of the Northwest Territories, Canada.

Type species: *Umingmakstrongylus pallikuukensis* Hoberg, Polley, Gunn and Nishi, 1995

[The authors use the subfamily Muelleriinae for this genus, one of the six subfamilies used by Boev (1975) but indicate that, as yet, a phylogenetic basis has not been established to confirm the validity of these subfamilies. Anderson (1978) declined to use the subfamilies in Boev's monograph, as the number of subgenera and genera was so small. The brief diagnosis is based on Hoberg *et al.* (1995).]

Imparispiculus Luo, Duo and Chen, 1988 (Fig. 133)

DIAGNOSIS: Cuticle smooth, striations and longitudinal ridges absent. Four submedian cephalic papillae, 2 lateral amphids. Mouth with 3 rudimentary lips. Oesophagus slightly wider posteriorly. Nerve ring anterior to middle region of oesophagus. Excretory pore opens at level of posterior end of oesophagus. Cervical papillae near oesophago-intestinal junction. *Male*: Bursa bilobed with median notch. Bursal rays well developed. Dorsal ray rounded with 5 sessile papillae. Spicules unequal, dissimilar. Gubernaculum with capitulum, corpus and crura. Crural border smooth. Telamon with basal plate, 2 ventral and 2 lateral plates. *Female*: Vulva opens near anus. Vagina short. Provagina present. Oviparous. Tail conical. L_1 tail acute without spine. Parasite in pulmonary alveoli and bronchiole of hare, *Lepus oiostolus*, in Jiuzhi County, Qinghai Province, China.

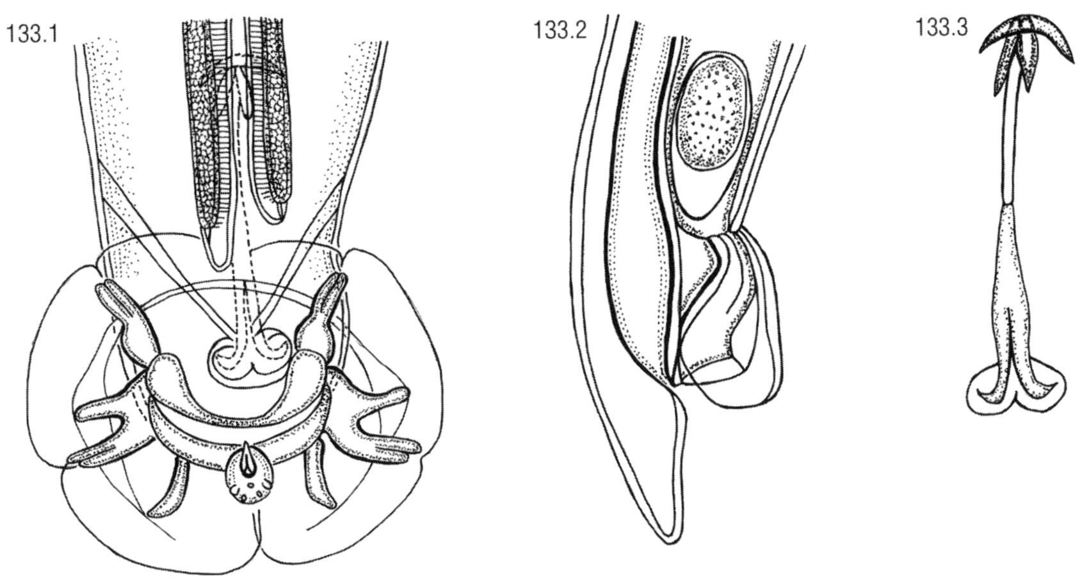

FIG. 133 *Imparispiculus* 133.1 Bursa ventral view; 133.2 Posterior end of female, lateral view; 133.3 Gubernaculum. (Redrawn after Luo *et al.*, 1988.)

Type species: *Imparispiculus jiuzhiensis* Luo, Duo and Chen, 1988

Mariostrongylus Teixeira de Freitas and Franco, 1967 (Fig. 134)

DIAGNOSIS: Cuticle with transverse striations. *Male*: Bursa symmetrical with lobes joined, rays short. Ventral rays with common stem; lateral rays with common stem; externo-dorsal ray separate from dorsal ray; dorsal ray divided into 2 branches, distal tips simple. Spicules alate in distal half. Gubernaculum boat-shaped. Telamon present. *Female*: Didelphic. Prodelphic. Vulva opens near anus. Tail tapers with rounded tip. Parasite in lungs of rodents in the State of São Paulo, Brazil.

Type species: *Mariostrongylus pessoai* Freitas and Franco, 1967

[Texeira de Freitas and Negrão de Souza Franco (1967) placed their genus in the subfamily Elaphostrongylinae. Anderson (1978) declined to use subfamilies in his key (see above). Vicente *et al.* (1997) placed the genus in the family Protostrongylidae. Listed in Spencer Jones and Gibson (1987).]

Pulmostrongylus Hsü, 1935 (Fig. 135)

DIAGNOSIS: Lateral alae if present narrow, extend along length of body. Mouth simple. Inner circle of 6 cephalic papillae, outer circle with 4 submedian cephalic papillae and 2 lateral amphids. Buccal capsule reduced or absent. Oesophagus clavate. *Male*: Bursa bilobed. Dorsal lobe reduced or absent. Ventral rays only divided distally; antero-lateral ray separate; medio- and postero-lateral rays only divided distally; externo-dorsal ray separate to base; dorsal ray if present divides in distal half, each branch dividing again. Spicules variable in form. Gubernaculum large or reduced. Telamon present or absent. *Female*: Vulva opens from posterior to mid-body or near anus. Vulvar appendages present or absent. Uteri parallel.

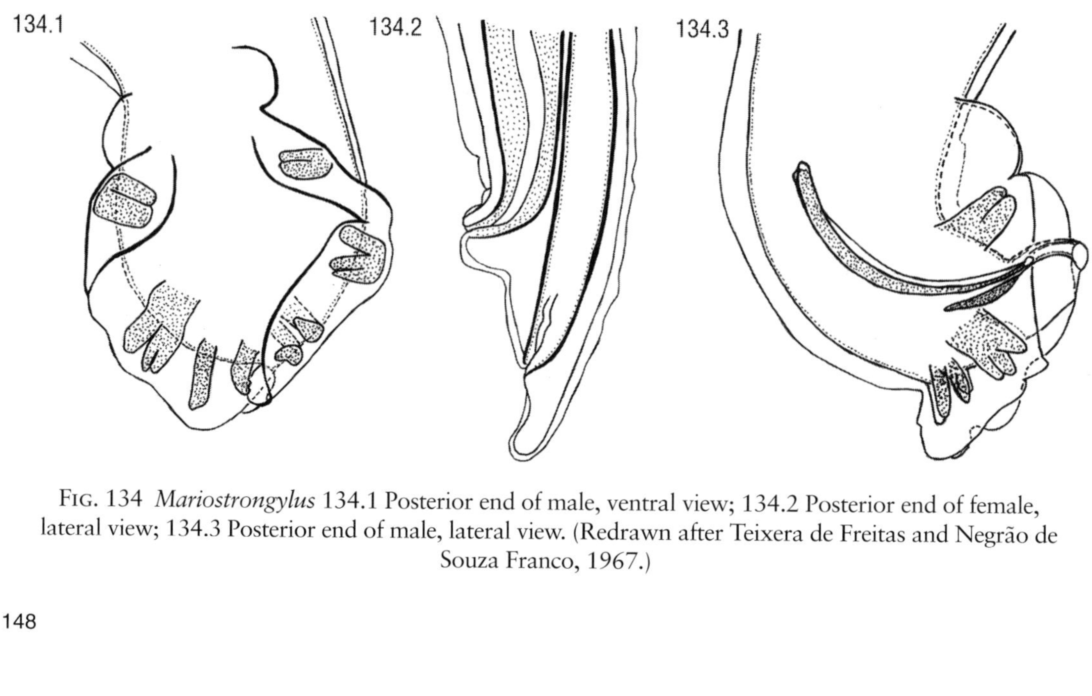

FIG. 134 *Mariostrongylus* 134.1 Posterior end of male, ventral view; 134.2 Posterior end of female, lateral view; 134.3 Posterior end of male, lateral view. (Redrawn after Teixera de Freitas and Negrão de Souza Franco, 1967.)

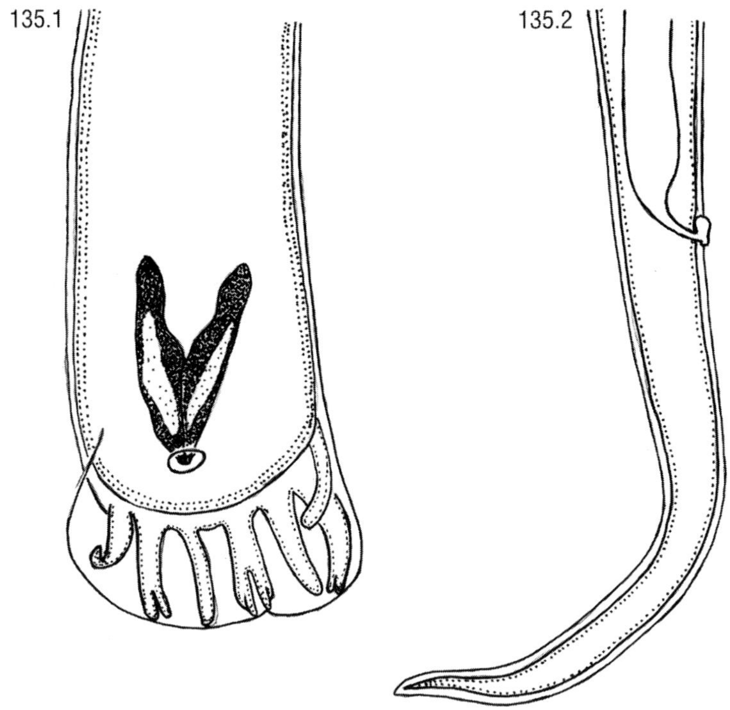

FIG. 135 *Pulmostrongylus* 135.1 Posterior end of male, ventral view; 135.2 Posterior end of female, lateral view. (Redrawn after Trivedi, 1987, publ. 1989.)

Oviparous. Tail slender. Parasites in lung, plural cavities, peritoneum, gall bladder and body cavity of carnivores.

Type species: *Pulmostrongylus fengi* Hsü, 1935

[Anderson (1978) considers this genus a subgenus of *Protostrongylus* Kamensky, 1905. The emended diagnosis is based on Trivedi (1987, publ. 1989), who validates the genus.]

Skrjabingylidae (Skrjabin, 1933 subfamily) Delamure and Kontrimavichus, 1976

Skrjabingylus Petrov, 1927 (Fig. 136)

DIAGNOSIS: Tegumental sheath with very distinct, widely spaced annulations. Buccal capsule very short, walls strongly sclerotized. Mouth opening terminal. Cephalic sensory organs: 6 papillae in inner circle; 10 papillae in external circle. *Male*: Bursa reduced to '2 fleshy lateral lobes', rays reduced within lobes to 6 pedunculated papillae. Spicules equal, alate. Gubernaculum reduced to corpus. *Female*: Vulva opens near middle of body. Amphidelphic. Ovejector with prominent sphincter. Viviparous or ovoviviparous. Parasites of the frontal sinuses of mustelids.

Type species: *Skrjabingylus nasicola* (Leuckart, 1842).

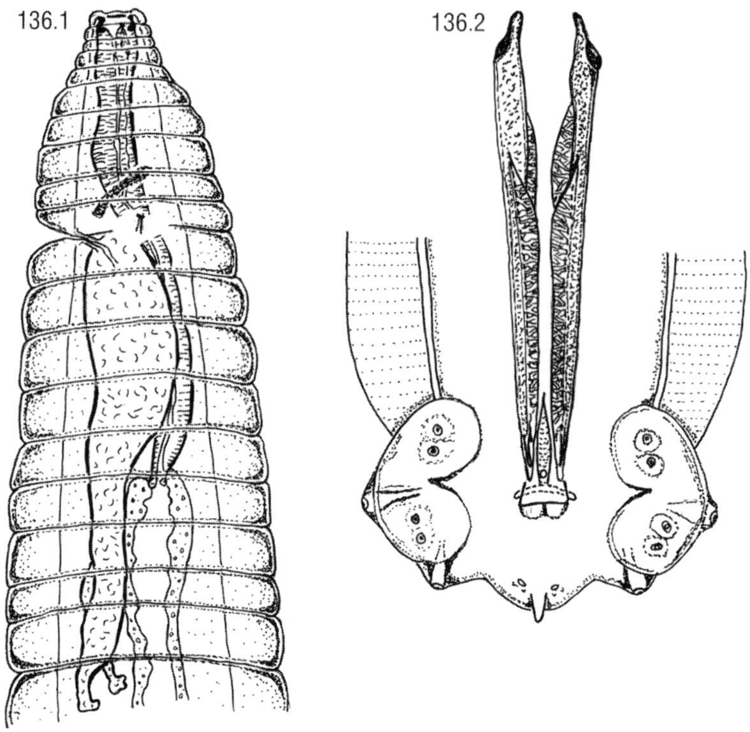

FIG. 136 *Skrjabingylus* 136.1 Anterior end of female, lateral view; 136.2 Posterior end of male, ventral view. (Redrawn after Lankester, 1983.)

[The emended diagnosis is based on Lankester (1983).]

Pseudaliidae Railliet and Henry, 1909

Halocercus Baylis and Daubney, 1925

[Anderson (1978) indicated that 'little [was] to be gained by recognizing subgenera of *Halocercus* (see Skrjabin, 1942)'. The subgenera are therefore not referred to in Anderson's key. The descriptions below are based on Skrjabin (1942), and the subgenera *Prohalocercus* and *Posthalocercus* are listed in Spencer Jones and Gibson (1987).]

Subgenus *Halocercus* (Baylis and Daubney, 1925 genus)

DIAGNOSIS: *Male*: Bursa present, not divided into lobes. Bursal rays rudimentary.

Type species: *Halocercus* (*Halocercus*) *delphini* Baylis and Daubney, 1925

[Type species as listed in Kontrimavichus *et al.* (1976).]

Subgenus *Prohalocercus* Skrjabin, 1942

DIAGNOSIS: *Male*: Bursa absent.

Type species: *Halocercus* (*Prohalocercus*) *invaginatus* (Quekett, 1841) Dougherty, 1943

[Type species as listed in Kontrimavichus *et al.* (1976).]

Subgenus *Posthalocercus* Delamure, 1942 (in Skrjabin, 1942)

DIAGNOSIS: Male: Bursal lobes more developed. Bursal rays poorly developed.

Type species: *Halocercus* (*Posthalocercus*) *taurica* Delamure, 1942 (in Skrjabin, 1942)

[Type species as listed in Kontrimavichus *et al.* (1976).]

4 OXYURIDA

OXYUROIDEA

Genera not assigned to a family

Paraleptonema Wang, 1980 (Fig. 137)

DIAGNOSIS: Lateral alae absent. Oesophagus with posteriorly dilated corpus, isthmus and bulb. *Male*: Spicules equal, similar. Gubernaculum present. One pair precloacal, 6 pairs postcloacal papillae. Tail sharply pointed. *Female*: Vulva opens in posterior half of body. Tail sharply pointed. Parasite in intestine of *Rana spinosa* in Beifeng, Fuzhou, Fujiang Province, China.

Type species: *Paraleptonema ranae* Wang, 1980

[Wang (1980) considers the genus close to *Cosmoxynemoides* Travassos, 1948. Petter (1976) considered *Cosmoxynemoides* of doubtful status as the males are unknown, and did not include the genus in her key. Translation by Professor Jian Heng. Listed in Spencer Jones and Gibson (1987).]

Archeonema Ricci, 1987 (publ.1988)

DIAGNOSIS: *Male*: Unknown. *Female*: Unknown. *Pre-adult*: Buccal capsule present. Oesophagus with corpus and distal bulb. Excretory pore opens near anterior end, passes into large excretory vesicle, oval in sagittal section. Parasite in *Speleomantes supramontis*, Amphibia (Urodela, Plethodontidae) in Italy.

Type species: *Archeonema lanzai* Ricci, 1987 (publ. 1988)

[Ricci (1987, publ. 1988) provisionally places the genus in the Oxyurida. The reproductive system is not obvious and the author regards the specimen as pre-adult. As the genus is insufficiently described it is regarded as *incertae sedis*.]

Oxyuridae Cobbold, 1864

[Erkulov and Moldopiyazova (1975) proposed the subfamily Skrjabinemiinae (Vereschagin, 1926) within the family Syphaciidae for *Citellina* and *Skrjabinema*, transferring them from the Syphaciinae.

Hugot (1981) reinstates the subfamily Syphaciinae for genera characterized by: mouth with 3 lips (except *Hilgertia*); 4 submedian cephalic papillae, arranged in a square more or less near the amphids; inner labial papillae generally visible; cephalic vesicle, cervical alae and lateral alae present or absent. *Male*: Single spicule, gubernaculum with hook (morphology varies), *area*

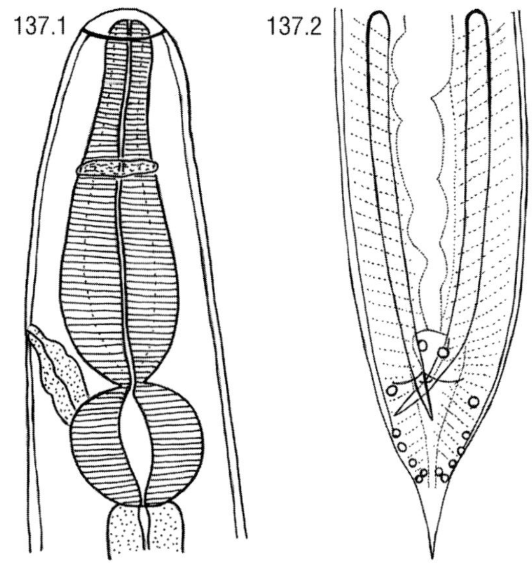

FIG. 137 *Paraleptonema* 137.1 Anterior end, lateral view; 137.2 Posterior end of male, ventral view. (Redrawn after Wang, 1980.)

rugosa with a series of parallel transverse grooves either on all posterior part of body or grouped at level of mamelons, which number 1–4, 3 pairs of caudal papillae (2 pairs surrounding cloaca, 1 pair postcloacal associated with phasmids), caudal appendix well developed. *Female*: Oblong eggs, embryonated or not, asymmetrical, operculum present or absent.

Hugot (1981) includes *Syphacia*, *Hilgertia*, *Sypharista*, *Syphatineria* and *Syphabulea* in the subfamily parasites from rodents (Ctendactylomorphs, Sciuromorphs, Myomorphs, rarely Bathyergidae). Hugot (1981) includes a key to species of the subfamily. *Syphatineria* and *Syphabulea* are subgenera of *Syphacia* in Petter and Quentin (1976).

Petter and Quentin (1976) incorporated all these genera within the Oxyuridae and did not divide the family into subfamilies.

Hugot (1983a), using cladistic analysis, considers the Oxyuridae group parasites of rodents (except *Citellina* Prendel, 1948) and lagomorphs to form a monophyletic group. Using cladistics, Hugot was unable to distinguish the genera *Helminthoxys* Freitas, Lent and Almeida, 1937 and *Octodonthoxys* Quentin, Courtin and Fontecilla, 1975. Petter and Quentin (1976) separate them in their key on the basis of the cephalic extremity, i.e. cephalic region without cuticular shields (*Helminthoxys*) and cephalic region with dorsal cuticular shield ending in 2 lobes (*Octodonthoxys*). Later, Hugot (1988) applied numerical and cladistic methods to the subfamily Syphaciinae and created five tribes: Syphaciini (Railliet, 1916 subfamily), Hilgertiini Hugot, 1988, Passulurini Hugot, 1988, Acanthyoxyurini Schulz, 1948 and Protozoophagini Hugot, 1988. For full details of the analysis and taxa proposed, Hugot (1981, 1983a, 1985, 1988, 1991) should be consulted. As a result of these studies the following new genera and subgenera (within *Sypharista* and *Syphacia*) are established by Hugot: *Sypharista* (*Sypharista* (Quentin, 1971 genus)), type species *S.* (*Sypharista*) *kamegaii*; *S.* (*Quentinema* Hugot, 1985), type species *S.* (*Quentinema*) *callosciuri* Quentin, 1977; *Sypharista* (*Petauxyuris* Hugot, 1985), type species *S.* (*Petauxyuris*) *tridentata* Quentin and Krishnasamy, 1975; *Syphacia* (*Syphacia* (Seurat, 1916 genus)), type species *S.* (*Syphacia*) *obvelata* (Rudolphi, 1802); *Syphacia* (*Seuratoxyuris* Hugot, 1988), type species *S.* (*Seuratoxyuris*) *pahangi* Ow Yang, 1971; *Syphacia* (*Cricetoxyuris* Hugot,

1988), type species *S.* (*Cricetoxyuris*) *okuensis* Hugot and Quentin, 1985; *Idiuoxyuris* Hugot, 1988, type species *I. quentini* (Hugot, 1982).

Hugot *et al.* (1996) create the subfamily Enterobiinae within the family Oxyuridae for parasites of primates and squirrels. Hugot *et al.* (1996) point out that recent redescriptions show that members of the Oxyuridae from primates share derived characters that show it to be a monophyletic group, which they classify as a subfamily, namely Enterobiinae. The authors include *Enterobius* Leach, 1853, *Trypanoxyuris* Vevers, 1923, *Xeroxyuris* Hugot, 1995 and *Lemuricola* Chabaud and Petter, 1959 (except *Lemuricola* (*Ingloxyuris*) *inglisi* Chabaud and Petter, 1959). This single species, previously included in a subgenus, does not share the derived characters of the others included in the subfamily, and Hugot *et al.* (1996) place it in the family Oxyuridae *sensu lato*. For a full analysis of the characters assigned to this subfamily, Hugot *et al.* (1996) should be consulted.]

Enterobius Leach, 1853

DIAGNOSIS: Parasites of Cercopithecidae, Pongidae and Hominidae.

Type species: *Enterobius vermicularis* (Linnaeus, 1758)

[Hugot *et al.* (1996) place this genus and subgenera in their subfamily Enterobiinae.]

Subgenus *Enterobius* (Leach, 1853 genus)

DIAGNOSIS: Lateral alae arise immediately behind cephalic vesicle, well developed. Buccal cavity without teeth or cuticular folds. *Male*: Spicule generally short. *Female*: Lateral alae formed by 2 longitudinal parallel crests. Parasites in humans, catarrhine primates and rodents (Sciuridae).

Type species: *Enterobius* (*Enterobius*) *vermicularis* (Linnaeus, 1758)

[Emended diagnosis based on Hugot (1987).]

Subgenus *Colobenterobius* Quentin, Betterton and Krishnasamy, 1979 (Fig. 138)

DIAGNOSIS: Cephalic vesicle distinctly striated. Oesophageal teeth with distinct projections. Lateral alae start behind level of oesophageal bulb, height reduced. *Male*: Spicule generally long. *Female*: Lateral alae formed by single crest. Parasites of Catarrhini primates, Colobinae and orang-utan.

Type species: *Enterobius* (*Colobenterobius*) *longispiculum* Quentin, Betterton and Krishnasamy, 1979

[Emended diagnosis based on Hugot (1987). Following the description of the species *E.* (*C.*) *paraguerezae*, Hugot found that the spicule length (88 μm) in this subgenus had a greater range than described by Quentin *et al.* (1979) (described by them as larger than 100 μm), and thus no longer constituted a suitable character to define the subgenus. Differences in the morphological features of the lateral alae described by Hugot (1987), however, confirmed the validity of the subgenus.]

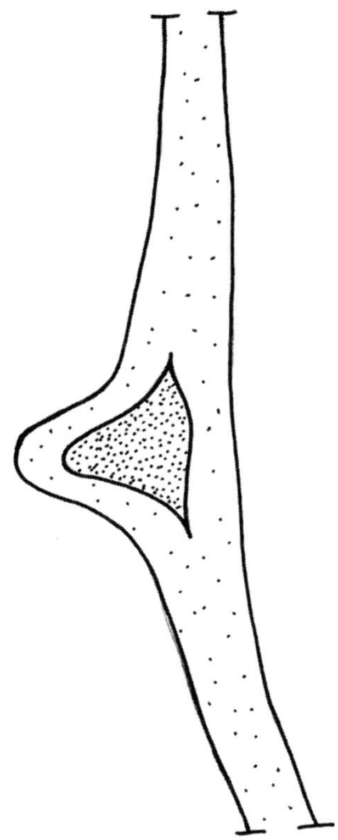

FIG. 138 *Colobenterobius* Cross-section of lateral ala in mid-region of female body. (Redrawn after Hugot, 1987.)

Wellcomia Sambon, 1907

DIAGNOSIS: Anterior features (except *W. mitchelli* and *W. decorata*) range from peribuccal region triangular; 3 pseudolabia reduced; interlabial processes and apical crests simple to peribuccal region round; 3 pseudolabia well developed; interlabial processes and apical crests complex (in end-on view each process with lateral denticles arranged symmetrically either side of a median, more developed denticle). *Male*: Posterior region ornamented, mamelons present. Gubernaculum without capitulum; caudal papillae weakly pedunculate; caudal appendix short and thick or gubernaculum with capitulum; caudal papillae strongly pedunculate; caudal appendix elongate and slender. *Female*: *Vagina vera* ranging from short, not extended beyond body wall, to long, extended beyond body wall. Parasites of New and Old World rodents.

Type species: *Wellcomia mitchelli* Sambon, 1907

[The genus was considered as *genus inquirendus* provisionally by Petter and Quentin (1976), and the species dispersed throughout several genera including *Evaginuris* Skrjabin and Schikhobalova, 1951. Hugot (1982) reinstated the genus *Wellcomia* and regarded *Evaginuris* Skrjabin and Schikhobalova, 1951 (*sensu* Quentin, 1973), type species *E. evaginata* (Smyth, 1908), a synonym of the genus. A thorough analysis of the characters of the species assigned to the genus is given by Hugot (1982), and the emended diagnosis is based on this analysis. For full details of this analysis and a full list of synonyms, Hugot (1982) should be consulted.]

Trypanoxyuris Vevers, 1923

Type species: *Trypanoxyuris trypanuris* Vevers, 1923

[Petter and Quentin (1976) include two subgenera, *Trypanoxyuris* and *Hapaloxyuris*. Hugot *et al.* (1996) revalidate and include as a third subgenus *Paraoxyuronema* Artigas, 1936 for parasites of Atelidae in the neotropical region. Hugot *et al.* (1996) include *Buckleyenterobius* Sandosham, 1950 and *Odontorobius* Skrjabin and Schikhobalova, 1951 (not included in Petter and Quentin, 1976) as synonyms of the subgenus *Paraoxyuronema*.

Paraoxyuronema Artigas, 1936 was not included by Petter and Quentin (1976), as doubts about its validity had been raised by Inglis and Cosgrove (1965). The type species *P. brachytelesi* Artigas, 1936 was known only by female specimens, and Inglis and Cosgrove (1965) considered the genus *incertae sedis* until males were available for study. Hugot *et al.* (1996) consider the original description of the female by Artigas (1936) 'allows the recognition of the characteristic structure of the female genital [system] that unites the family'. Sandosham (1950) overlooked the description by Artigas (1936) and described a new genus, *Buckleyenterobius*, for parasites from primates of the family Atelidae, and Hugot *et al.* (1996) regard the genus as a synonym of *Paraoxyuronema* Kreis (1932).

The subgenus *Rodentoxyuris* Quentin and Tenora, 1974 (publ. 1975) is transferred to this genus by Hugot, 1984 (referred to in Hugot *et al.* 1996) from *Lemuricola* Chabaud and Petter, 1959.

Hugot *et al.* (1996) place *Trypanoxyuris* in their new subfamily Enterobiinae.]

Skrjabinema Werestchagin, 1926

Parasites of ruminants.

Type species: *Skrjabinema ovis* (Skrjabin, 1915) Werestchagin, 1926

[Mönnig (1932) created two subgenera, namely *S.* (*Skrjabinema*) Werestchagin, 1926 genus for species with lips that project forward and are not surrounded by the wall of the mouth capsule (*S. ovis* (Skrjabin, 1915); *S. tarandi* Skrjabin and Mizkevitsch, 1930) and *S.* (*Chilocrypta*) Mönnig, 1932 for species with lips that project towards the centre of mouth opening and are surrounded by the wall of the mouth capsule (*S. rupicaprae* Bohm and Gebauer, 1930; *S. africana* Mönnig, 1932; *S. alata* Mönnig, 1932). The subgenus *Chilocrypta* is listed in Spencer Jones and Gibson (1987). Skrjabin and Shikhobalova (1951) did not consider the division of the genus justified, and Petter (1976) did not include them in her key.]

Acanthoxyurus Sandground, 1928

Subgenus *Acanthoxyurus* (Sandground, 1928 genus) Hugot, 1983

DIAGNOSIS: Mouth with 3 lips absent or 3 well-developed lips present. Cephalic sensory organs not grouped on latero-dorsal lobes. *Male*: Cephalic region without hooks. Cervical alae absent. *Female*: Cephalic shield with dorsal cephalic hooks. Cervical alae present, latero-dorsal.

Vagina short, simple. Eggs rounded. Tail rarely longer than 1 cm. Parasites of African rodents, genus *Anomalurus*.

Type species: *Acanthoxyurus* (*Acanthoxyurus*) *anomaluri* Sandground, 1928

Subgenus *Petronema* Hugot, 1983 (Fig. 139)

DIAGNOSIS: Mouth with 6 lips. Cephalic sensory organs arranged on 2 latero-dorsal lobes. Cephalic shield in both sexes prolonged into 2 strong dorsal hooks, 2 bilobed cervical alae attached and directed dorsally. Cuticle ornamented at base of each cervical ala. *Male*: Lateral alae present, one stops abruptly at first mamelon. Ventral cuticle ornamented with longitudinal parallel crests from excretory pore to cloaca. Two mamelons present. Ornamentation asymmetric to sagittal axis. *Female*: Lateral alae present, one stops abruptly anterior to vulvar opening. Cuticle ornamented posterior to vulva. Body narrows posterior to vulva in young adults. Vagina very long, marked by longitudinal crenellated folds near vulva, directed anteriorly. Gravid females with thick tail, up to 3 cm long, 4 distinct annules towards distal end. Egg triangular with double shell, not operculate, not embryonated. Parasites of African rodents, family Thryonomyidae.

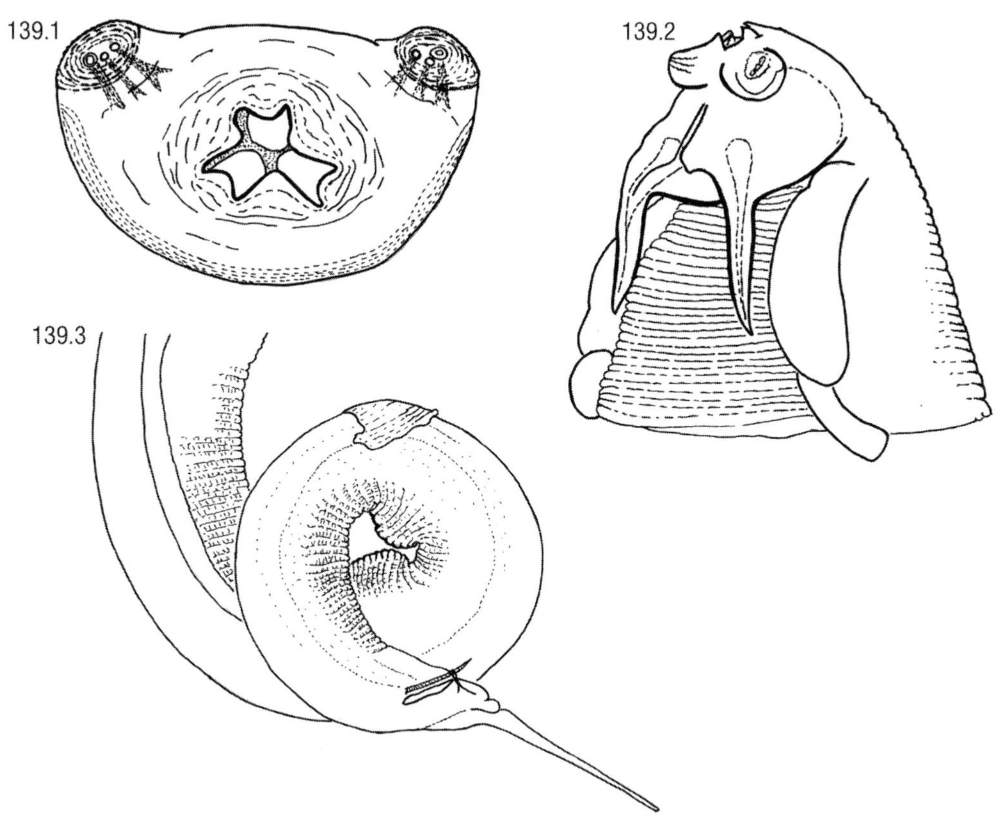

FIG. 139 *Petronema* 139.1 Cephalic region, *en face* view of male; 139.2 Anterior end of male, dorso-lateral view; 139.3 Posterior end of male, lateral view. (Redrawn after Hugot, 1983b.)

Type species: *Acanthoxyurus* (*Petronema*) *shortridgei* Mönnig, 1931

[Hugot (1988), in his numerical and cladistic study, raises the rank of this subgenus to genus. Listed in Spencer Jones and Gibson (1987).]

Hoepplius Chu, 1931 (Fig. 140)

DIAGNOSIS: Mouth circular with 6 lips. Cephalic sensory organs: 4 submedian papillae; 2 amphids. Buccal capsule absent. Oesophagus: corpus; slight isthmus; bulb. *Male*: Single spicule. Caudal alae present. Caudal papillae: single, large precloacal papilla; 1 pair precloacal papillae; 1 pair elongate slender adcloacal papillae. Slender appendix on tail with terminal papilla. *Female*: Each lip with single tooth. Vulva opens in anterior region of body. Two ovaries

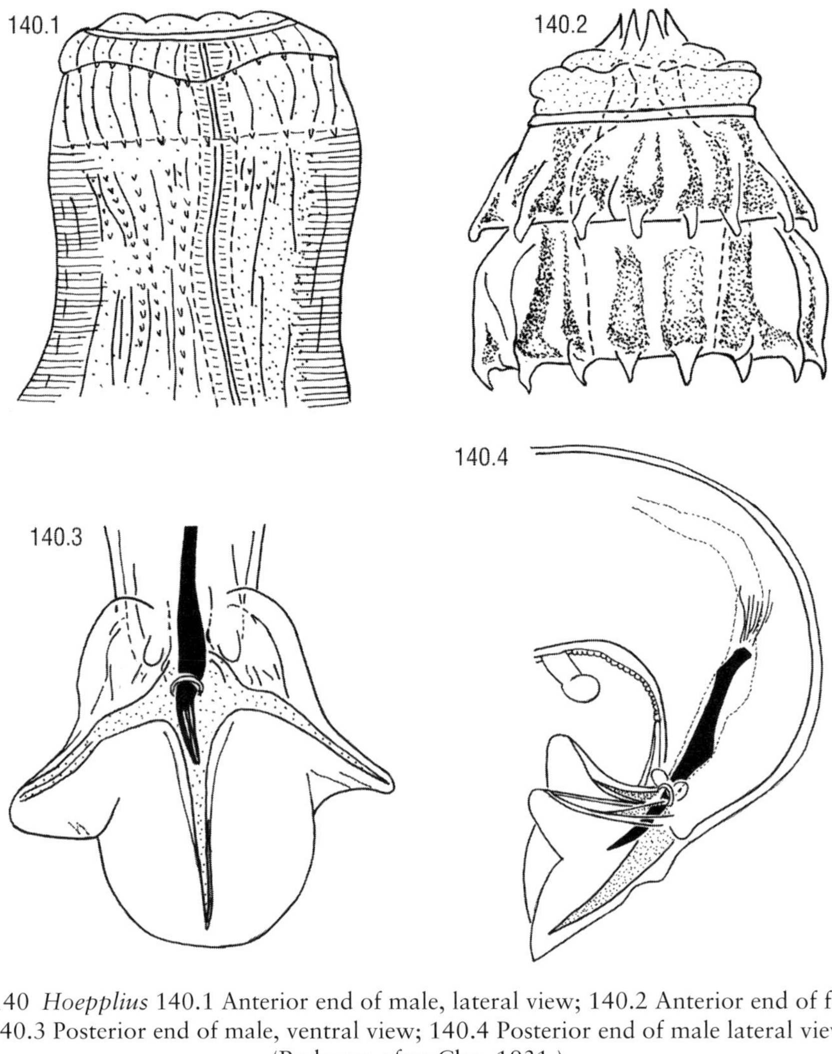

FIG. 140 *Hoepplius* 140.1 Anterior end of male, lateral view; 140.2 Anterior end of female; 140.3 Posterior end of male, ventral view; 140.4 Posterior end of male lateral view. (Redrawn after Chu, 1931.)

present. Eggs asymmetrical. Tail tapers to fine point. Parasite in large intestine and caecum of *Cynocephalus volans* in Bilar, Bohol, Philippine Islands.

Type species: Not designated? *Hoepplius spinosus* Chu, 1931

[Listed in Spencer Jones and Gibson (1987). Petter and Quentin (1976) did not include this genus in their key.]

Oxyuronema Kreis, 1932

DIAGNOSIS: Cuticle with transverse striations. Mouth with 3 lips. Amphids open with oval pores. Buccal capsule present. Oesophagus with 2 bulbs, 3 spears at proximal end. *Male*: Single spicule. Gubernaculum not seen. Single testis reflected. Four pairs of papillae on posterior end. Tail roundly truncated. Terminal appendix. *Female*: Vulva opens in anterior third of body. Monodelphic. Ovary reflexed anteriorly. Eggs numerous, shell double. Tail elongate, tapers to point. Parasite in rectum of red spider monkey, *Ateles geoffroyi*, Panama.

Type species: *Oxyuronema atelophora* Kreis, 1932

[Kreis (1932) differentiated the type species from *Enterobius*. Spencer Jones and Gibson (1987) list the genus in the Oxyuroidea. *Oxyuronema atelophora* Kreis, 1932 is considered by Hugot *et al.* (1996) as likely to have been described from material containing several species without designating a single type specimen. As a result of their studies, *Oxyuronema* is considered a *genus inquirendum*.]

Lobatorobius Skrjabin and Schikhobalova, 1951 (Fig. 141)

DIAGNOSIS: Cephalic vesicle present with 3 asymmetrical parts. Lateral alae present. Three lips, each with 2 papillae. Subventral lips each with small inner tooth. Oesophagus with proximal swelling, distinct distal bulb. *Male*: Four pairs caudal papillae: anterior pair pedunculate,

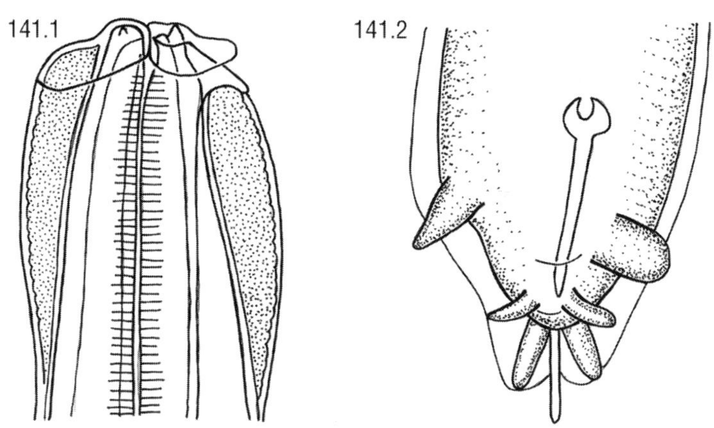

FIG. 141 *Lobatorobius* 141.1 Anterior end of body; 141.2 Posterior end of male, ventral view. (Redrawn after Travassos, 1925 in Skrjabin *et al.*, 1960 (English translation 1974).)

directed anteriorly; second pair sessile at base of anterior pair (not shown in diagram); third and fourth pairs close, third pair directed laterally. Spicule straight, long. Gubernaculum absent. Appendix present. *Female*: Vulva opens in anterior half of body. Two uteri. Egg slightly pitted. Tail long, pointed. Parasites in monkeys.

Type species: *Lobatorobius scleratus* (Travassos, 1925) Skrjabin and Schikhobalova, 1951

[Listed in Spencer Jones and Gibson (1987).]

Acanthoxynema Khera, 1954 (publ. 1956)

DIAGNOSIS: *Male*: Unknown. *Female*: Body small, cuticle striated. Narrow lateral alae present. Cephalic region with distinct distal margin. Mouth with 3 papillae and 2 pairs of hooks. Oesophagus with distal bulb. Nerve ring just anterior to bulb. Vulva opens anterior to middle of body. Vagina directed anteriorly. Didelphic. Eggs oxyurid type. Parasite of alimentary canal of goat in India.

Type species: *Acanthoxynema lucknowensis* Khera, 1954 (publ. 1956)

[Spencer Jones and Gibson (1987) regard this genus as *incertae cedis*. The type species may represent a damaged specimen of *Skrjabinema ovis*, a cosmopolitan parasite of the large intestine of goats and sheep.]

Lemuricola Chabaud and Petter, 1959

[Hugot, 1984 (referred to in Hugot *et al.* 1996) transfers the subgenus *Rodentoxyuris* Quentin and Tenora, 1974 (publ. 1975) from *Lemuricola* to *Trypanoxyuris* Vevers, 1923. Hugot *et al.* (1996) place these two genera in their new subfamily Enterobiinae. Hugot *et al.* (1996) consider that, as the species *Lemuricola* (*Inglisoxyuris*) *inglisi* Chabaud and Petter, 1959 does not share the derived characters of other members of their new subfamily, the genus *Inglisoxyuris* formerly regarded as a subgenus is not included in the Enterobiinae but is retained in the Oxyuridae *sensu lato*.]

Gynaecometra Araujo, 1977 (Fig. 142)

DIAGNOSIS: Cuticle with fine transverse striations. Anterior region with sexual dimorphism. Lateral alae absent. Oesophagus: posterior bulb with valve. *Male*: Anterior cuticular swelling present. Lips absent. Buccal capsule absent. Excretory pore opens posterior to oesophageal bulb. Tail dorso-ventrally rectangular. Spicules robust. Gubernaculum present, well chitinized. *Female*: Six conical lips. Buccal capsule present. Anterior cuticular swelling absent. Excretory pore opens at level of oesophageal bulb. Vulva opens anterior to mid-body. Opisthodelphic. Two short ovaries, uterus single, ovejector short. Viviparous. Tail slender, pointed. Parasites of reptiles in Estado Bahia, Brazil.

Type species: *Gynaecometra bahiensis* Araujo, 1977

[The author places the genus in the subfamily Syphaciinae. Petter and Quentin (1976) chose not to divide the family Oxyuridae into subfamilies in their key. Listed in Spencer Jones and Gibson (1987).]

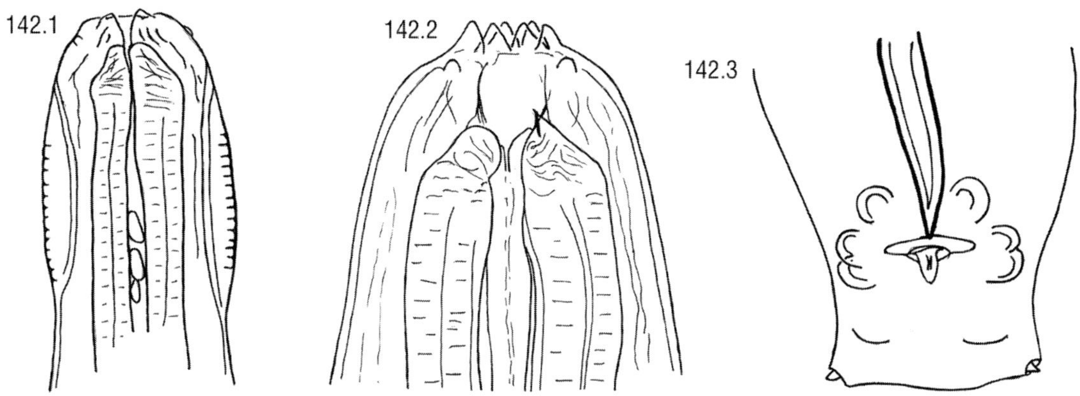

FIG. 142 *Gynaecometra* 142.1 Anterior end of male, lateral view; 142.2 Anterior end of female, lateral view; 142.3 Posterior end of male, ventral view. (Redrawn after Araujo, 1977.)

Adelonema Mawson, 1978 (Fig. 143)

DIAGNOSIS: Lateral alae present. Distinct annular constriction separates cephalic region from body. Mouth opening circular. Four submedian cephalic papillae, 2 lateral amphids. Buccal capsule, circular in cross-section, with 6 longitudinal ridges projecting anteriorly as 3 large and 3 small 'pseudolabia'.

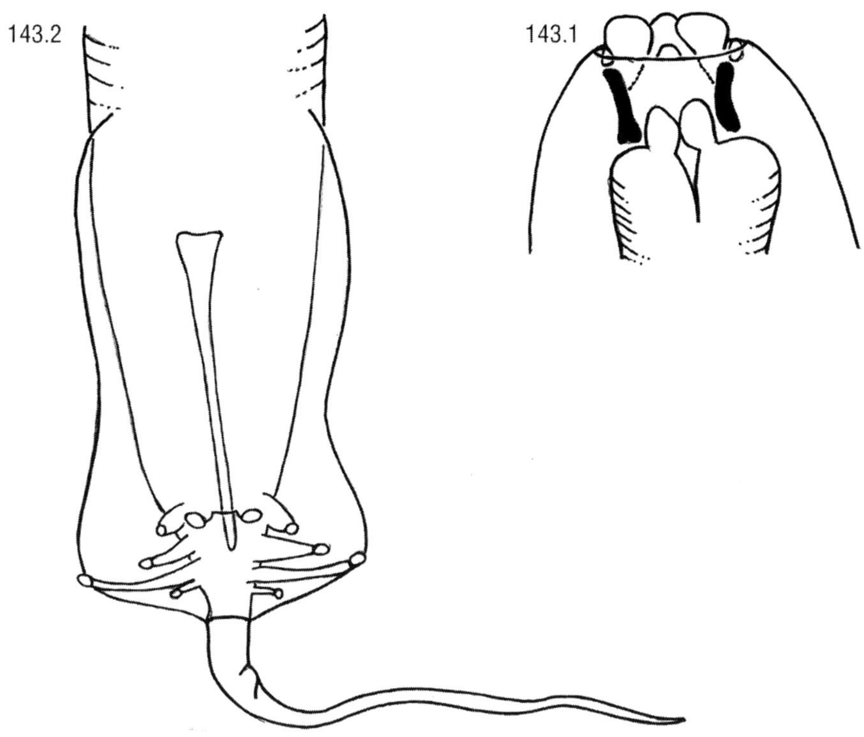

FIG. 143 *Adelonema* 143.1 Anterior end of female, lateral view; 143.2 Posterior end of male, ventral view. (Redrawn after Mawson, 1978.)

Proximal end of oesophagus forms 3 lobes extending into buccal cavity. Oesophagus with corpus, isthmus and bulb. Excretory pore opens posterior to oesophagus. *Male*: Lateral alae extend from distal end of oesophagus to just anterior to caudal alae. Caudal alae extend to just posterior to cloaca. Four pairs of pedunculate, 1 pair sessile, caudal papillae. Single spicule. Caudal appendix present. *Female*: Lateral alae extend to anus. Vulva opens near mid-body. Uteri opposed. Eggs without operculum. Tail long, tapers to sharp point. Parasite of Australian phalangerid marsupials.

Type species: *Adelonema trichosuri* (Johnston and Mawson, 1938) Mawson, 1978

[Listed in Spencer Jones and Gibson (1987).]

Rauschtineria Hugot, 1980 (Fig. 144)

DIAGNOSIS: Dimorphism in cephalic features. *Male*: Mouth triangular with 3 equal lips. Three strong oesophageal teeth, each with a median denticle. Four submedian cephalic papillae. Two ventral mamelons. Three pairs of caudal papillae, first pair postcloacal surrounded by cuticular ring. Gubernaculum with capitulum and corpus partially fused. Caudal appendix present. *Female*: Mouth triangular. Three oesophageal teeth. Three interlabia between oesophageal teeth forming true labial teeth, each covered by a tongue-like pseudolip. Eggs non-operculate, without embryos. Parasites of Sciurinae.

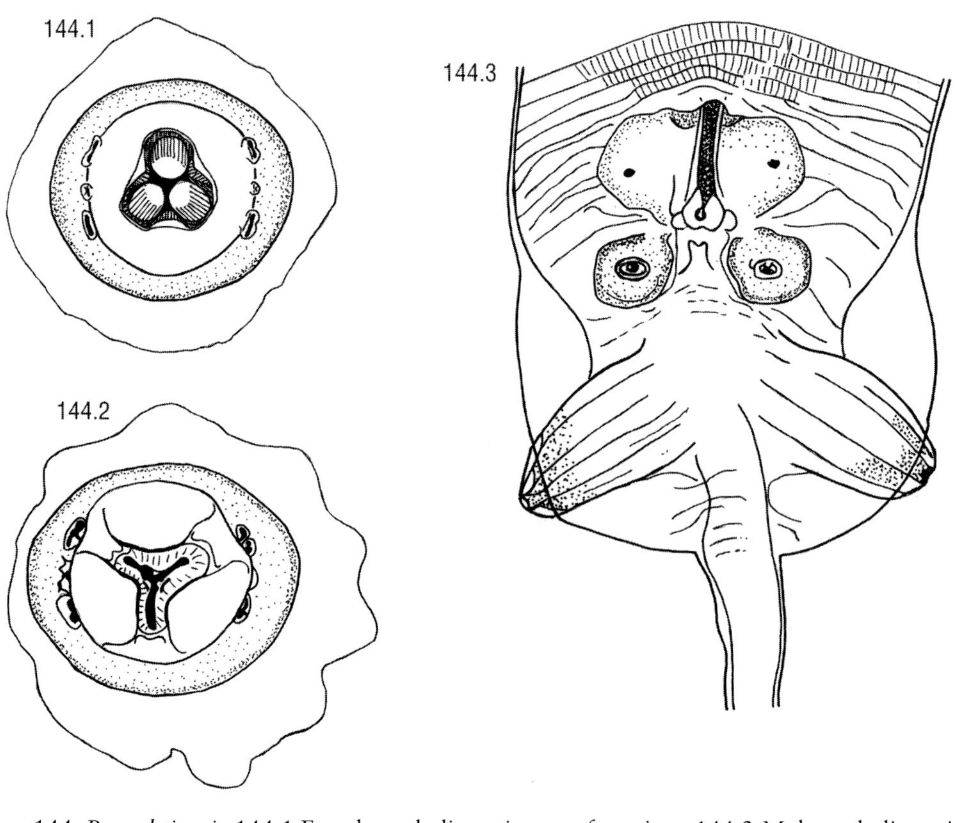

FIG. 144 *Rauschtineria* 144.1 Female cephalic region, *en face* view; 144.2 Male cephalic region, *en face* view; 144.3 Posterior end of male, ventral view. (Redrawn after Hugot, 1980.)

Type species: *Rauschtineria citelli* (Tiner and Rausch, 1950) Hugot, 1980

[Hugot (1980) puts his genus in the subfamily Oxyurinae. Petter and Quentin (1976) chose not to divide the family Oxyuridae into subfamilies for their key. Listed in Spencer Jones and Gibson (1987).]

Syphatineria (Chabaud and Biocca, 1955 subgenus) Hugot, 1981

DIAGNOSIS: Peribuccal area round, features generally well developed. Cephalic papillae close to amphids. Cephalic vesicle and lateral alae well developed (except *S. pallaryi* and *S. antiqua*). *Male*: Two to three mamelons. Spicule delicate. Accessory hook of gubernaculum elongated ventrally by 2 slender prongs ornamented with chitinous projections. *Female*: Muscular ovejector directed posteriorly. Eggs elongate, asymmetric with one side concave, the other convex, small operculum, non-embryonated. Parasites of Sciurinae in Africa and the Oriental region.

Type species: *Syphatineria pallaryi* (Seurat, 1915)

[Hugot (1981) elevates *Syphatineria* to generic rank in the subfamily Syphacinae. The genus was formerly a subgenus of *Syphacia*, as listed by Petter and Quentin (1976). Hugot (1981) divides the genus into four subgenera, and the emended diagnosis is based on his publication.]

Subgenus *Syphatineria* Chabaud and Biocca, 1955 (Fig. 145)

DIAGNOSIS: Edges of lips reinforced with chitinous rim. Peribuccal features little developed. Lips not completely separated. Cephalic papillae arranged in square or near to amphids.

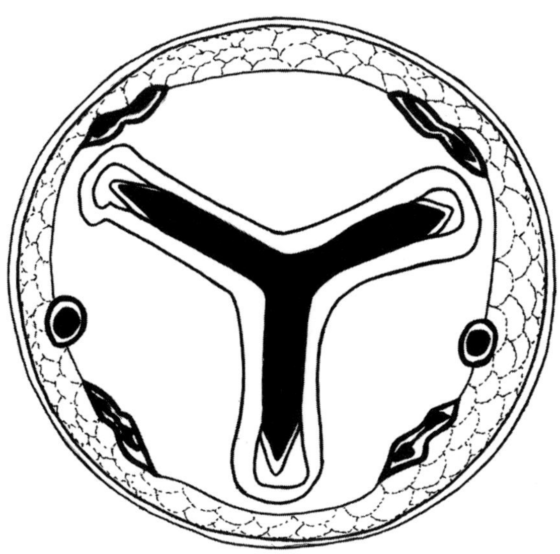

FIG. 145 *Syphatineria* Female, *en face* view. (Redrawn after Hugot, 1981.)

Cephalic vesicle absent or poorly developed. Lateral alae arise posterior to oesophagus. Two mamelons present. Parasites of Xerini in Africa.

Type species: *Syphatineria* (*Syphatineria*) *pallaryi* (Seurat, 1915) Hugot, 1981

[The emended diagnosis is based on Hugot (1981).]

Subgenus *Africanoxys* Hugot, 1981 (Fig. 146)

DIAGNOSIS: Edges of lips reinforced with a chitinous rim. Peribuccal features well developed. Lips completely separate. Cephalic papillae near amphids. Cephalic vesicle well developed. Lateral alae arise posterior to cephalic vesicle. *Male*: Two to three mamelons. Parasites of Protoxerini and Funambulini in the Ethiopian region.

Type species: *Syphatineria* (*Africanoxys*) *adami* (Quentin, 1971) Hugot, 1981

[Listed in Spencer Jones and Gibson (1987).]

Subgenus *Orientoxys* Hugot, 1981 (Fig. 147)

DIAGNOSIS: Peribuccal features well developed. Cephalic papillae always close to amphids. *Male*: Two to three mamelons. Accessory hook of gubernaculum ornamented with chitinous projections. Parasites of Callosciurini in the Indochina region and Malaysia.

Type species: *Syphatineria* (*Orientoxys*) *owyangi* (Quentin and Krishnasamy, 1975) Hugot, 1981

[Listed in Spencer Jones and Gibson (1987).]

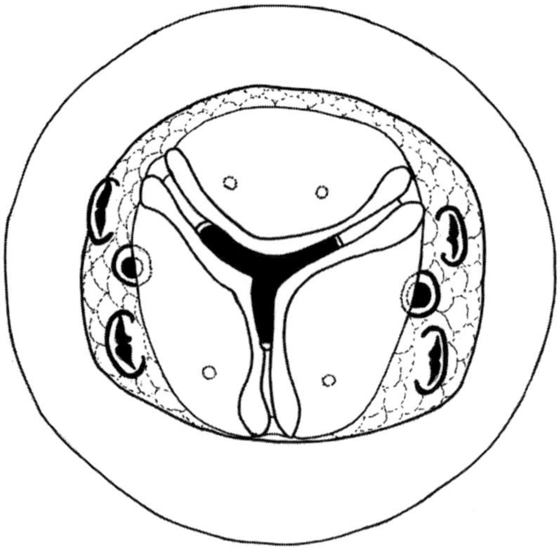

FIG. 146 *Africanoxys* Female, *en face* view. (Redrawn after Hugot, 1981.)

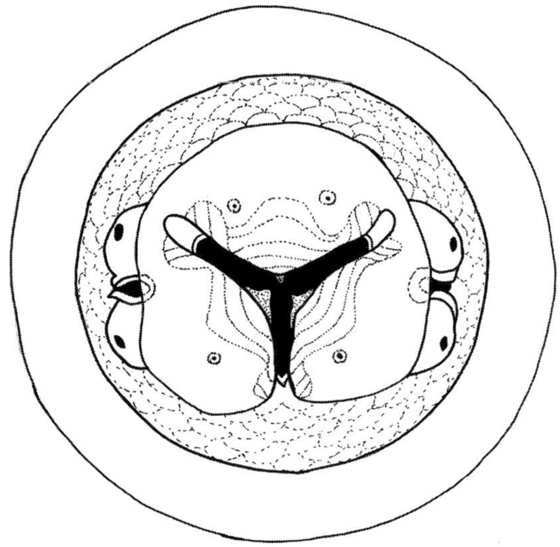

FIG. 147 *Orientoxys* Female, *en face* view. (Redrawn after Hugot, 1981.)

Subgenus *Quentenora* Hugot, 1981 (Fig. 148)

DIAGNOSIS: Peribuccal features well developed. Lips only partially separated. Cephalic papillae near amphids. *Male*: Two to three mamelons. Accessory hook of gubernaculum with plain ornamentation. Parasites of Funambulini in the Indian region.

Type species: *Syphatineria* (*Quentenora*) *funambuli* (Johnson, 1967) Hugot, 1981

[Listed in Spencer Jones and Gibson (1987).]

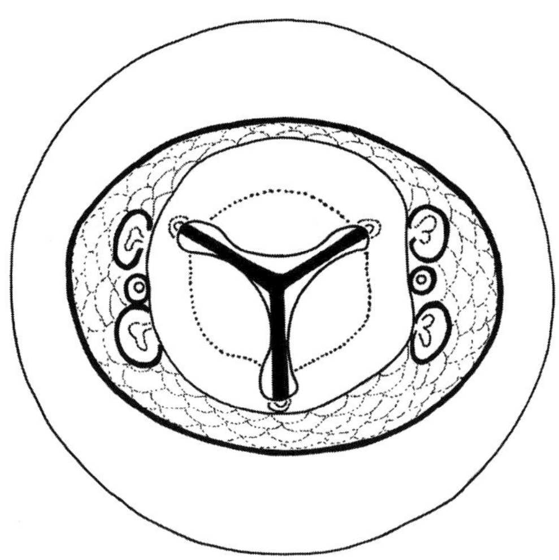

FIG. 148 *Quentenora* Female, *en face* view. (Redrawn after Hugot, 1981.)

Syphabulea (Gubanov, 1964 subgenus) Hugot, 1981 (Fig. 149)

DIAGNOSIS: Peribuccal region round or triangular with rounded angles. Lips well developed. Labial denticles strengthen lips in some species. Cephalic papillae near amphids. *Male*: Three to four mamelons. Accessory hook of gubernaculum with complex ornamentation formed of ventral chitinous projections. *Female*: Muscular ovejector directed posteriorly. Eggs long, asymmetrical, non-embryonated, provided with one large operculum occupying whole of convex side of shell. Parasites of Petauristinae of the Oriental (under canopy species) and Holarctic regions and of the Sciurinae.

Type species: *Syphabulea sobolevi* (Gubanov, 1964) Hugot, 1981

[Hugot (1981) elevated the subgenus to generic rank in the subfamily Syphaciinae, and the emended diagnosis is based on his publication. The genus was formerly a subgenus of *Syphacia* Seurat, 1916 and is listed as such in the key of Petter and Quentin (1976).]

Neohilgertia Navone, Suriano and Pujol, 1990 (Fig. 150)

DIAGNOSIS: Cephalic vesicle present. Cuticle with transverse striations. Four submedian cephalic papillae, 2 lateral amphids. Buccal capsule absent. Four oesophageal teeth: 2 dorsal; 2 latero-ventral. *Male*: Longitudinal mid-ventral swelling present. Eight to nine pairs ventro-lateral papillae on posterior end, 4 pairs cloacal papillae (1 sessile, 3 pedunculate). Spicule single. Gubernaculum without accessory hook. Tail truncated. *Female*: Monodelphic. Vagina muscular, non-protuberant. Eggs operculate. Parasite in caecum of *Thylamys venustus cinderellus*, Marsupialia, Didelphidae in Burruyacu, Tucumán Province, Argentina.

Type species: *Neohilgertia venusti* Navone, Suriano and Pujol, 1990

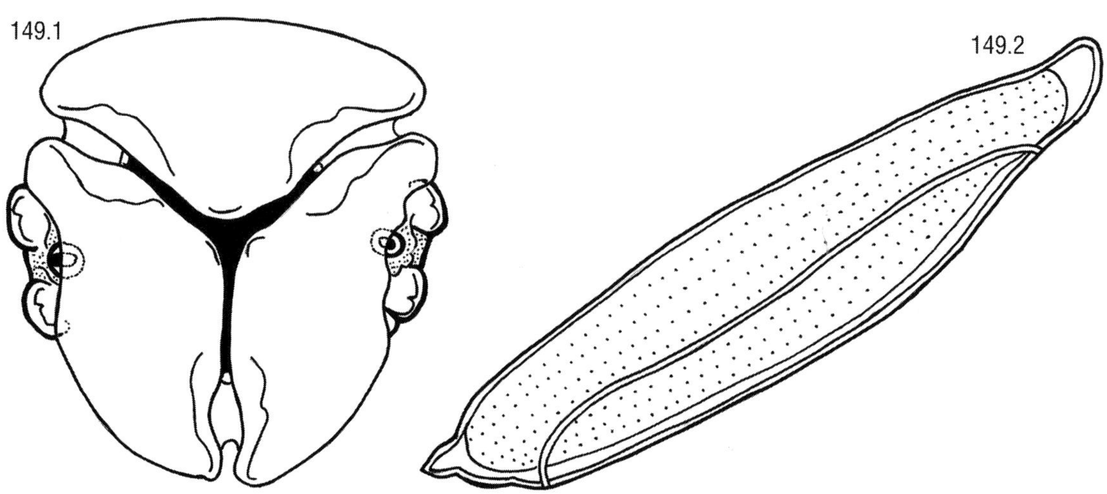

FIG. 149 *Syphabulea* 149.1 Female, *en face* view; 149.2 Egg showing large operculum. (Redrawn after Hugot, 1981.)

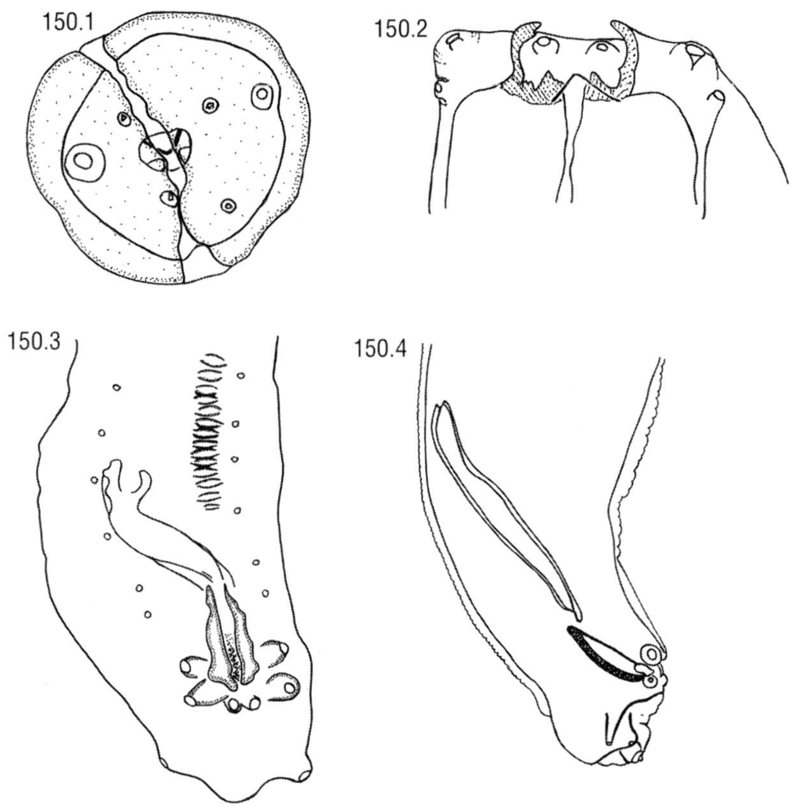

FIG. 150 *Neohilgertia* 150.1 Female *en face* view; 150.2 Anterior end of female, dorsal view; 150.3 Posterior end of male, ventral view; 150.4 Posterior end of male, lateral view. (Redrawn after Navone *et al.*, 1990.)

Lepusius Farooq and Khatoon, 1995 (Fig. 151)

DIAGNOSIS: Absence of cephalic vesicle or lateral flanges. Lateral alae present. Hexagonal mouth with 6 small lips, each with small papillae. Prominent buccal cavity without teeth. Typical oxyuroid oesophagus. *Male*: Long, pointed tail with narrow alae. Papillae not seen. Short, single spicule, spade-like. Gubernaculum absent. *Female*: Vulva opens in mid-body. Oviparous. Long, pointed tail. Eggs without filaments. Parasites of rabbits, *Lepus nigricollis dayanus*, in Pakistan.

Type species: *Lepusius nigricollisi* Farooq and Khatoon, 1995

[The authors consider this genus close to *Protozoophaga* Travassos, 1923 (Farooq and Khatoon, 1995).]

Xeroxyuris Hugot, 1995 (Fig. 152)

DIAGNOSIS: Mouth with 3 lips, each partially overlapping an oesophageal tooth. Oesophageal teeth triangular each with sharp denticle at proximal tip. Submedian cephalic

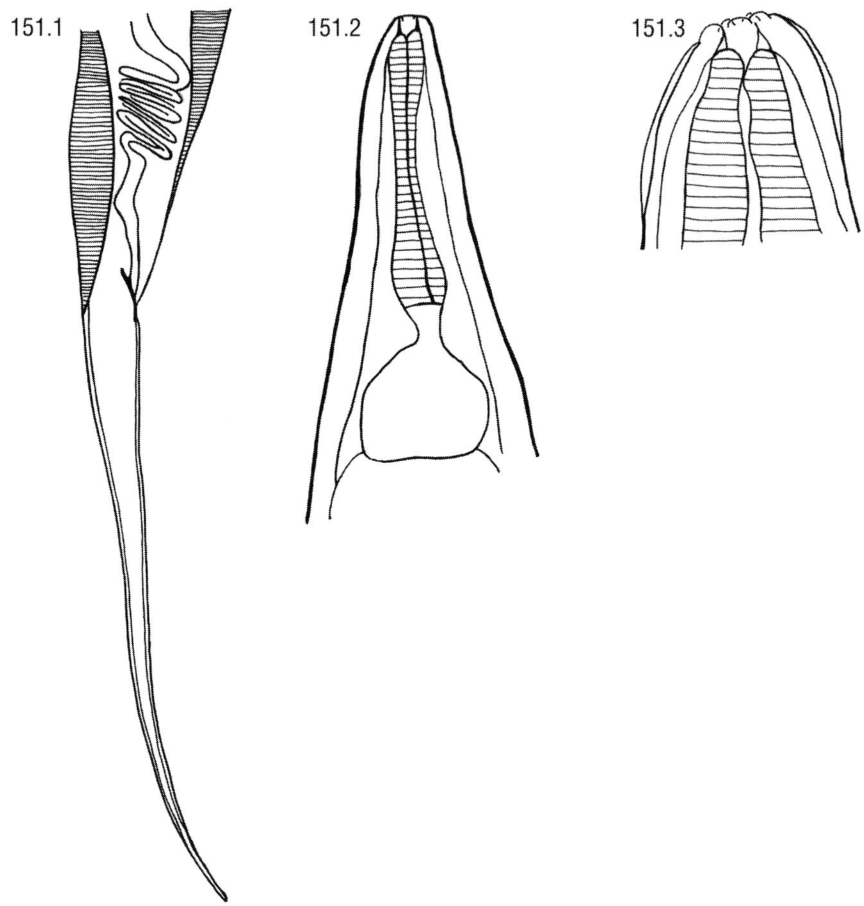

FIG. 151 *Lepusius* 151.1 Posterior end of male, lateral view; 151.2 Anterior end of male; 151.3 Cephalic region of male. (Redrawn after Farooq and Khatoon, 1995.)

papillae arranged with 2 dorsal papillae associated with dorsal lip, 1 ventral papilla and amphid associated with each subventral lip. Cephalic vesicle present. Oesophageal bulb spherical. *Male*: Lateral alae formed by triangular crest arising halfway between anterior extremity and nerve ring, extends to caudal papillae. Posterior to excretory pore ventral cuticle forms *area rugosa*, each annule split in ventral region. Four pairs of caudal papillae, first and fourth pedunculated, second and third sessile with weakly chitinized rings. Two spicules with dorsal angular bend between median and posterior thirds. Cloaca with hood-shaped cuticular fold. Gubernaculum ring-shaped with longitudinal projections. *Female*: First part of lateral alae formed by triangular crest, arises immediately posterior to nerve ring and extends to middle of oesophagus. Second part of alae separate, formed by 2 fused parallel crests, extends to middle of tail. Vulva prominent, opens between first and middle thirds of body. Vagina directed posteriorly. Uterine tube divided by cellular diaphragm. Beyond diaphragm uterus forms 2 prodelphic branches. Eggs oblong, weakly asymmetrical, thin-shelled, non-embryonated, non-operculated. Parasite of the caecum and large intestine of *Xerus* (*Geosciurus*) *inauris* in southern Africa.

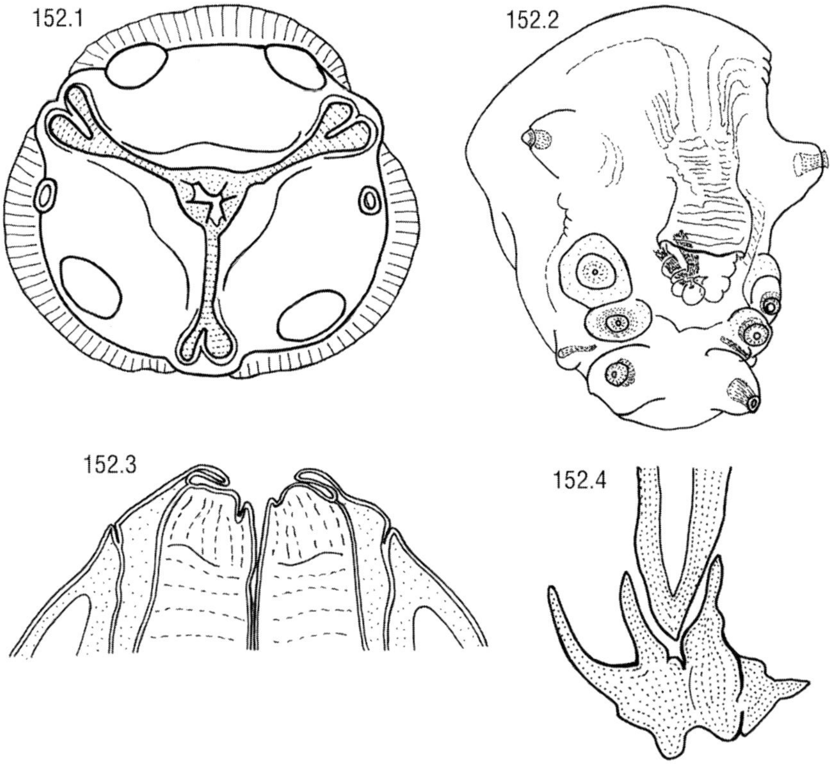

FIG. 152 *Xeroxyuris* 152.1 Female cephalic region, *en face* view; 152.2 Posterior end of male, ventro-lateral view; 152.3 Cephalic region of female, optical section through amphids, ventral view; 152.4 Gubernaculum and distal tip of spicule, left lateral view. (Redrawn after Hugot, 1995.)

Type species: *Xeroxyuris parallela* (Linstow, 1908) Hugot, 1995

[Hugot *et al.* (1996) include this genus in their new subfamily Enterobiinae.]

Didelphoxyuris Gardner and Hugot, 1995 (Fig. 153)

DIAGNOSIS: Mouth opening terminal, at base of depression with 3 lips. Four submedian cephalic papillae, 2 lateral amphids. Lateral alae present with 2 longitudinal ridges in each lateral field. *Male*: Ventral median crest present. Four pairs of caudal papillae: 2 pairs sessile, adcloacal; 1 pair postcloacal; 1 posterior pair pedunculate. Phasmids lateral and dorsal to posterior pair of papillae. Spicule and gubernaculum well developed. Tail truncated. *Female*: Cephalic vesicle well developed. Excretory pore opens just posterior to oesophagus. Vulva opens immediately posterior to excretory pore. Prodelphic. Oviduct S-shaped near muscular vagina and muscular uterine junction. Vagina thick, muscular wall. Eggs oval without operculum, non-embryonated. Parasite in caecum and colon of *Thylamys elegans* (Marsupialia, Didelphidae) in Bolivia (eastern region of Andes).

Type species: *Didelphoxyuris thylamisis* Gardner and Hugot, 1995

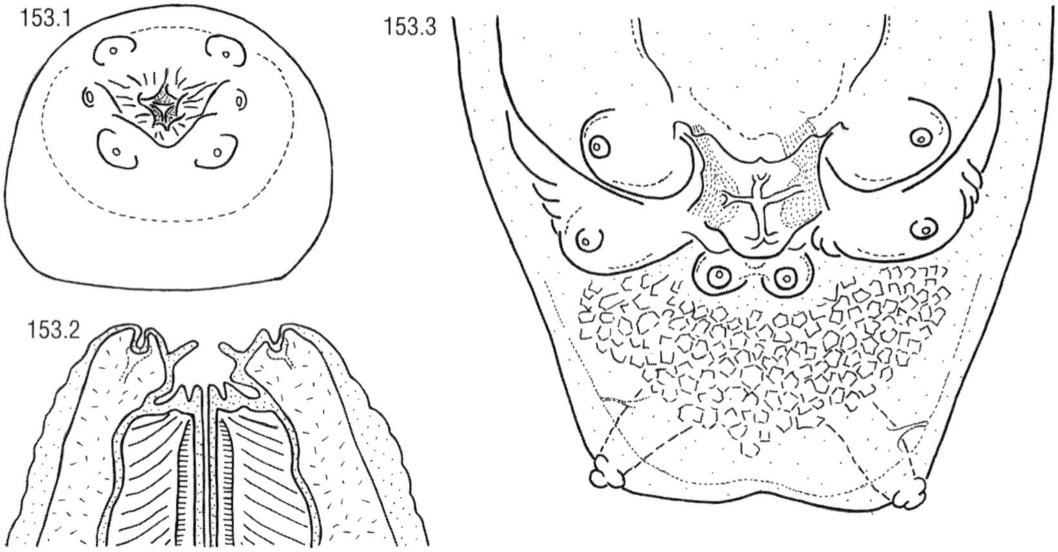

FIG. 153 *Didelphoxyuris* 153.1 Cephalic region of male, *en face* view; 153.2 Optical section of male cephalic region through amphids; 153.3 Posterior end of male, ventral view. (Redrawn after Gardner and Hugot, 1995.)

Monodelphoxyuris Guerrero and Hugot, 2003 (Fig. 154)

DIAGNOSIS: Mouth opens in depression. Three poorly developed lips. Cephalic sensory organs: 4 submedian papillae; 2 amphids, slightly pedunculated. Cephalic vesicle poorly developed. Lateral alae present with 2 longitudinal ridges in each lateral field. *Male*: Two groups of 3 mamelons. Four pairs of caudal papillae: 2 pairs sessile, adcloacal; 1 pair just posterior to cloaca; fourth posteriormost pair pedunculate; phasmids just anterior to fourth pair of papillae. Spicule 'well sclerotized'. Gubernaculum present. *Female*: Vulva opens near excretory pore in anterior

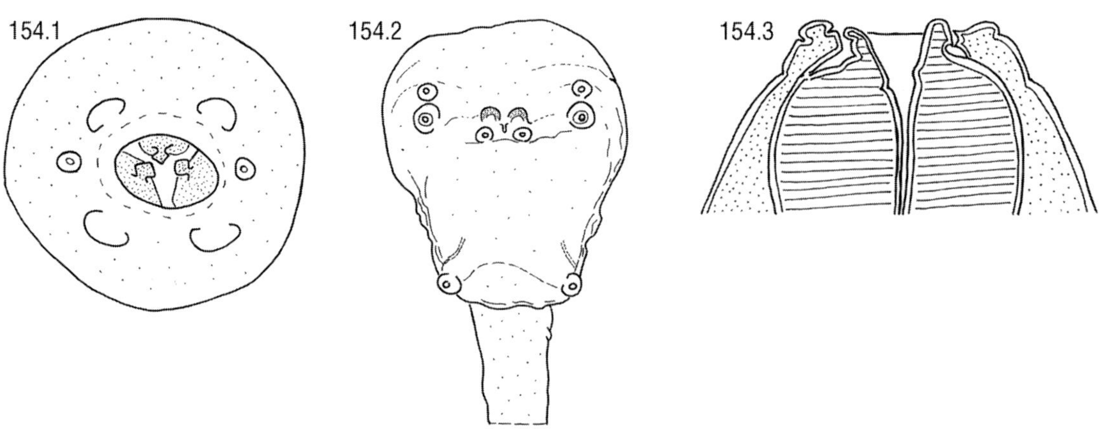

FIG. 154 *Monodelphoxyuris* 154.1 Cephalic region of female, *en face* view; 154.2 Posterior end of male, ventral view; 154.3 Cephalic region of female in optical section, ventral view. (Redrawn after Guerrero and Hugot, 2003.)

fourth of body. Prodelphic. Eggs operculate. Tail long, tapering to point. Parasite in caecum of *Monodelphis emiliae* (Marsupialia, Monodelphidae) in Peru (eastern region of Andes).

Type species: *Monodelphoxyuris dollmeiri* Guerrero and Hugot, 2003

Pharyngodonidae Travassos, 1919

Genera not designated to a subfamily

Skrjabineandros Markov, 1962

DIAGNOSIS: Translation of description not available – see original publication.

Type species: *Skrjabineandros popovi* (Markov and Bogdanov, 1962) Markov, 1962

[The type species was formerly placed in *Thelandros*. Listed in Spencer Jones and Gibson (1987).]

Pharyngodon Diesing, 1861

Subgenus *Pharyngodon* (Diesing, 1861 genus) Calvente, 1948

DIAGNOSIS: *Male*: Spicules visible with little chitinization.

Type species: *Pharyngodon* (*Pharyngodon*) *spinicauda* (Dujardin, 1845)

[Petter and Quentin (1976) did not divide the genus into subgenera.]

Subgenus *Neyrapharyngodon* Calvente, 1948

DIAGNOSIS: *Male*: Spicules invisible or absent.

Type species: *Pharyngodon* (*Neyrapharyngodon*) *neyrae* Calvente, 1948

[Petter and Quentin (1976) did not divide the genus into subgenera. Listed in Spencer Jones and Gibson (1987).]

Avilandros Skrjabin and Schikhobalova, 1951 (Fig. 155)

DIAGNOSIS: Cuticle with distinct transverse striations. Mouth opening round. Six cephalic sensory papillae. Buccal capsule short. Oesophagus: corpus; isthmus; bulb. Excretory pore opens posterior to oesophagus. *Male*: Posterior end of body expanded. Two pairs of adanal papillae, large; 1 pair ventral papillae middle region of tail. Single spicule pointed. Gubernaculum absent. *Female*: Vulva opens posterior to mid-body, anterior lip projects above body surface. Muscular ovejector, directed anteriorly before curving posteriorly. Egg asymmetrical, single polar operculum. Tail short, tapers to distal point. Parasite in intestine of *Tringa hypoleucos* in Calcutta Zoo, India.

Type species: *Avilandros avis* (Maplestone, 1940) Skrjabin and Schikhobalova, 1951

[Listed in Spencer Jones and Gibson (1987). Petter and Quentin (1976) did not include this genus in their key.]

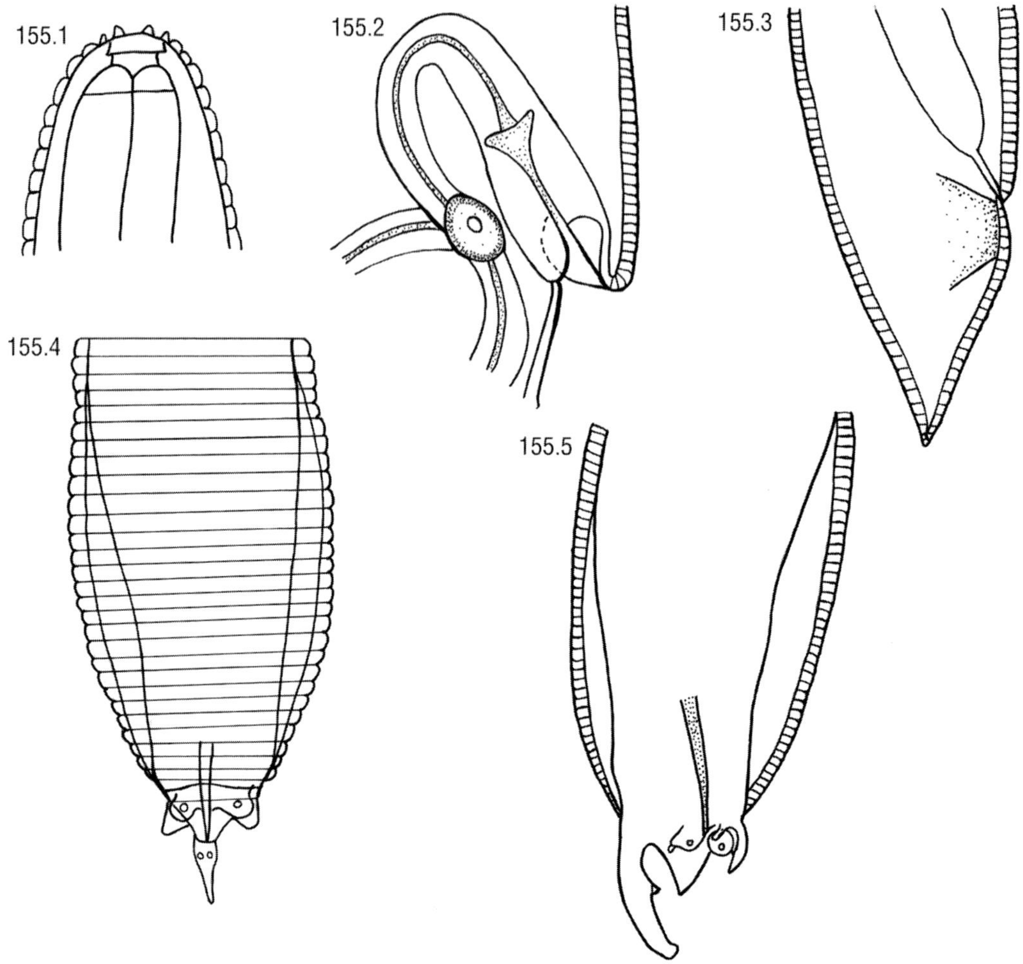

FIG. 155 *Avilandros* 155.1 Cephalic region; 155.2 Female vulvar region, lateral view; 155.3 Posterior end of female, lateral view; 155.4 Posterior end of male, ventral view; 155.5 Posterior end of male, lateral view. (Redrawn after Maplestone, 1940 from Skrjabin *et al.*, 1960 (translation 1974).)

Gopheruris Petter and Douglass, 1976 (Fig. 156)

DIAGNOSIS: *Male*: Mamelon extends to lower lip of short cloaca and is directed perpendicular to body surface, supported by a poorly developed chitinous structure. Caudal alae without papillae, extend posteriorly from in front of cloaca. Two pairs of papillae around cloaca, 1 pair posteriorly on tail. Spicule absent. *Female*: Vulva opens in middle of body. Parasite in colon of Testudinidae (*Gopherus* sp.) in Sonora, Mexico.

Type species: *Gopheruris aspicula* Petter and Douglass, 1976

[Listed in Spencer Jones and Gibson (1987).]

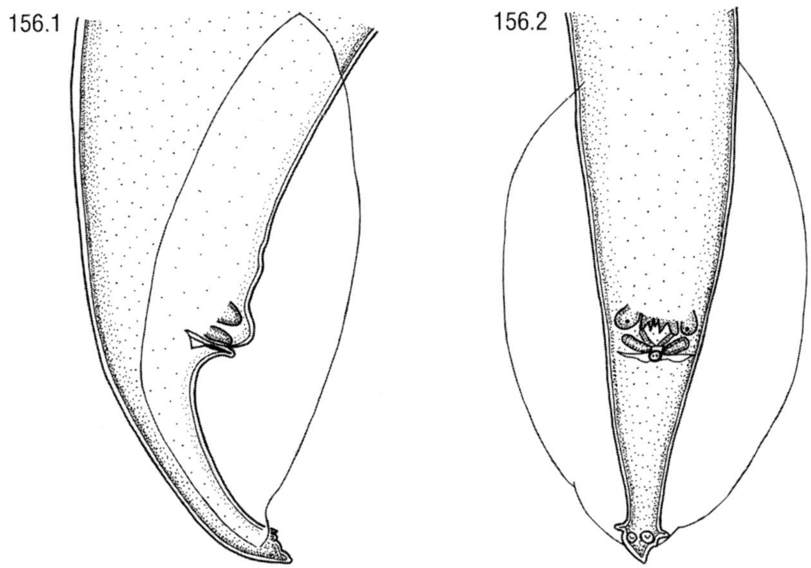

FIG. 156 *Gopheruris* 156.1 Posterior end of male, lateral view; 156.2 Posterior end of male, ventral view. (Redrawn after Petter and Douglass, 1976.)

Parathelandros Magzoub, Kasim and Shawa, 1980 (Fig. 157)

DIAGNOSIS: Cuticle with fine transverse striations. Mouth with 6 small lips. Buccal capsule short. Oesophagus short with corpus and distal bulb. *Male*: Posterior end of body truncate. One pair of caudal alae present. Caudal 'papillae absent'. Caudal appendix short. Spicule short, 'needle-like'. *Female*: Vulva opens slightly posterior to mid-body. Parasite in the caecum of Dabb lizard, *Uromastyx aegyptia*, in Najd Area, Saudi Arabia.

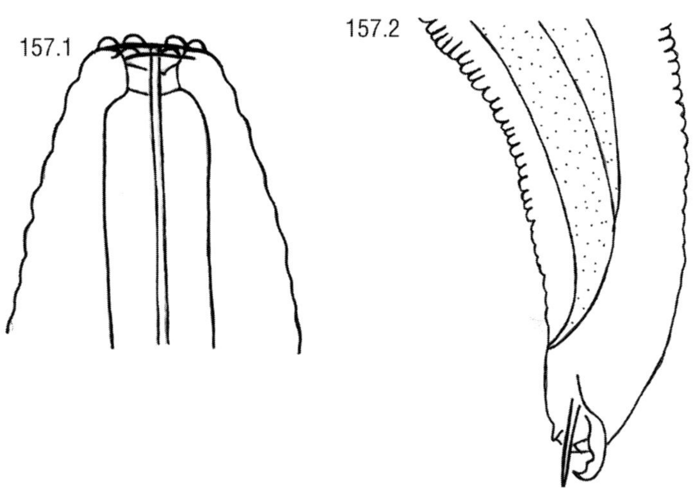

FIG. 157 *Parathelandros* 157.1 Anterior end of body; 157.2 Posterior end of male, lateral view. (Redrawn after Magzoub *et al.*, 1980.)

Type species: *Parathelandros kanistica* Magzoub, Kasim and Shawa, 1980

[The generic name is preoccupied, having already been given to nematode parasites of Australian amphibians, namely *Parathelandros* Baylis, 1930. The authors distinguish their genus from *Thelandros* Wedl, 1862 in the shape and arrangement of the lips, the presence of transverse striations on the cuticle, the absence of male caudal papillae and the presence of 1 pair of caudal alae. They also distinguish the genus from *Tachygonetria* Wedl, 1862 by these features, as well as by the length of the oesophagus, the truncate male tail and the presence of a caudal appendix. Hornero and Roca (1992) consider this genus a junior homonym of *Parathelandros* Baylis, 1930. Listed in Spencer Jones and Gibson (1987).]

Brasilnema Moravec, Kohn and Fernandes, 1992 (Fig. 158)

DIAGNOSIS: Cuticle with distinct transverse striations. Lateral alae present. Mouth hexagonal with 6 triangular lamellae raised above body surface. Buccal capsule well developed.

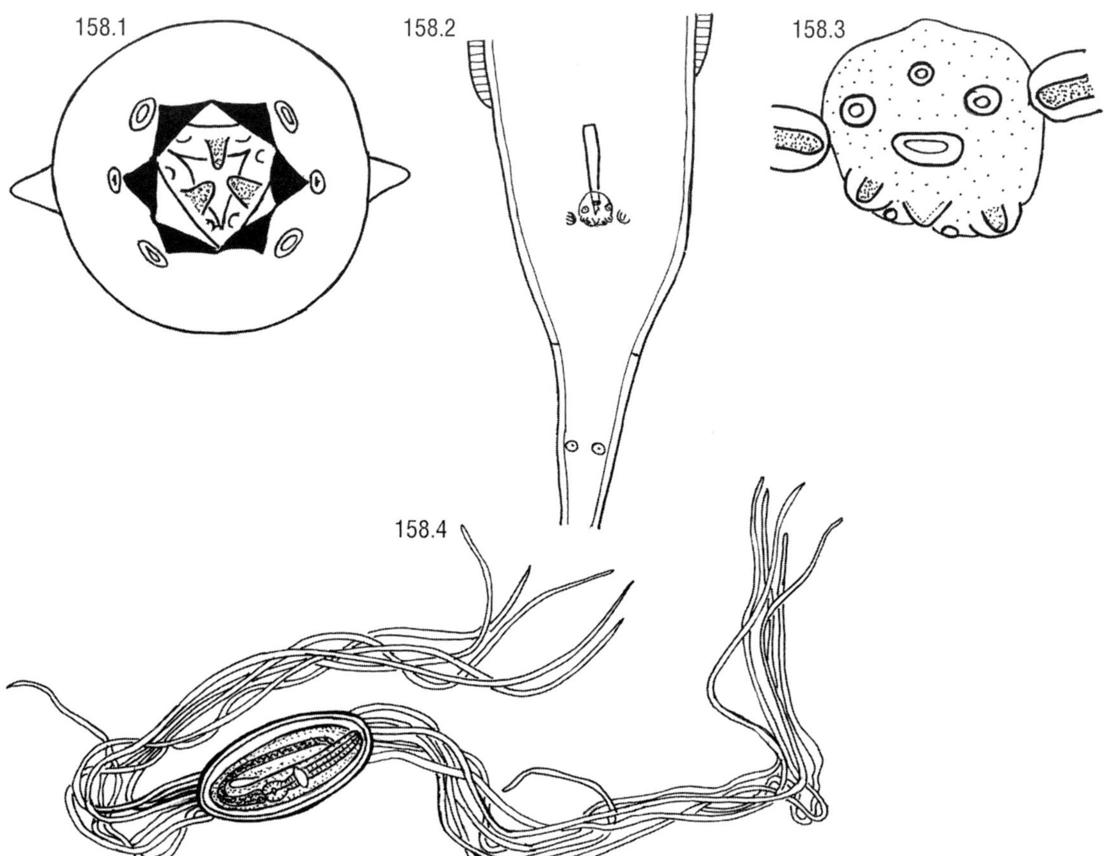

FIG. 158 *Brasilnema* 158.1 *En face* view; 158.2 Posterior end of male showing papillae, ventral view; 158.3 Cloacal region of male; 158.4 Mature egg. (Redrawn after Moravec *et al.*, 1992a.)

Tail narrow, long and pointed. *Male*: Spicule simple. Gubernaculum absent. Cluster of papillae surrounding cloaca on protruding cone, additional pair posterior to cloaca. *Female*: Vulva opens just anterior to mid-body. Eggs with long polar filaments, embryonated. Parasites in intestine of fat catfish, *Pimelodella lateristriga* in Guaira, Paraná State, Paraná River, Brazil.

Type species: *Brasilnema pimelodellae* Moravec, Kohn and Fernandes, 1992

Parasynodontisia Moravec, Kohn and Fernandes, 1992 (Fig. 159)

DIAGNOSIS: Cuticle with fine transverse striations. Lateral alae absent. Mouth triangular with 6 triangular lamellae raised above body surface. Buccal capsule absent. Tail long, tapers to point. *Male*: Spicule simple. Gubernaculum absent. Cluster of papillae surround cloaca on protruding cone, additional pair on mid-region of tail. *Female*: Vulva opens just anterior to mid-body. Eggs without polar filaments, embryonated. Parasite in intestine of armoured catfish, *Rhinelepis aspera* in Guaira, Paraná State, Paraná River, Brazil.

Type species: *Parasynodontisia petterae* Moravec, Kohn and Fernandes, 1992

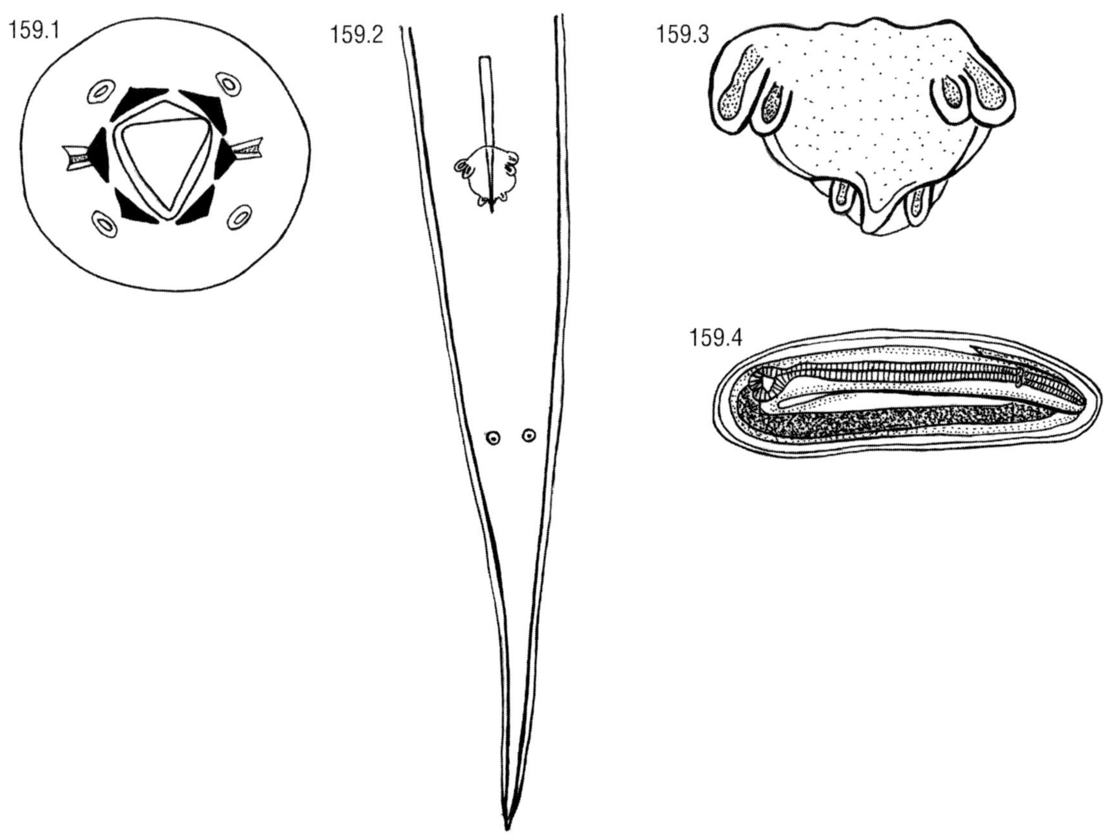

FIG. 159 *Parasynodontisia* 159.1 Female, *en face* view; 159.2 Posterior end of male, ventral view; 159.3 Cloacal region of male; 159.4 Mature egg. (Redrawn after Moravec *et al.*, 1992a.)

Spinoxyuris Petter, 1994 (Fig. 160)

DIAGNOSIS: Mouth bordered by 6 lips. Four oval submedian cephalic papillae, 2 amphids. Buccal capsule small, without teeth. Oesophagus with elongate corpus, short isthmus, bulb with sclerotized valve. Lateral alae present, extend to tail, terminate in a spine. *Male*: Lateral alae terminate posterior to cloaca. Genital cone well developed. One pair precloacal papillae on genital cone; 1 double postcloacal papilla, posterior to genital cone. Spicule simple. Gubernaculum absent. *Female*: Lateral alae terminate posterior to anus. Vulva opens anterior to mid-body. Uterus prodelphic. Eggs embryonated at deposition, armed with filaments. Parasites in fish in neotropical region, Nacunday, Rio Parana, Paraguay.

Type species: *Spinoxyuris oxydoras* Petter, 1994

Royandersonia Moravec and Van As, 2004 (Fig. 161)

DIAGNOSIS: Mouth hexagonal with '6 small oral lamellae'. Cephalic sensory organs: 4 submedian papillae; 2 lateral amphids. Buccal cavity 'well developed'. Oesophagus: corpus long, 'uniform in width'; isthmus short; bulb present. Lateral alae absent. *Male*: Genital cone with 3 pairs of small papillae around cloaca. One pair of large papillae on tail. Spicule absent. Tail long, slender, pointed. *Female*: Vulva opens in posterior half of body. Tail long, slender, pointed. Parasites in intestine of Asian freshwater fish.

Type species: *Royandersonia moraveci* (Anderson and Lim, 1996) Moravec and Van As, 2004

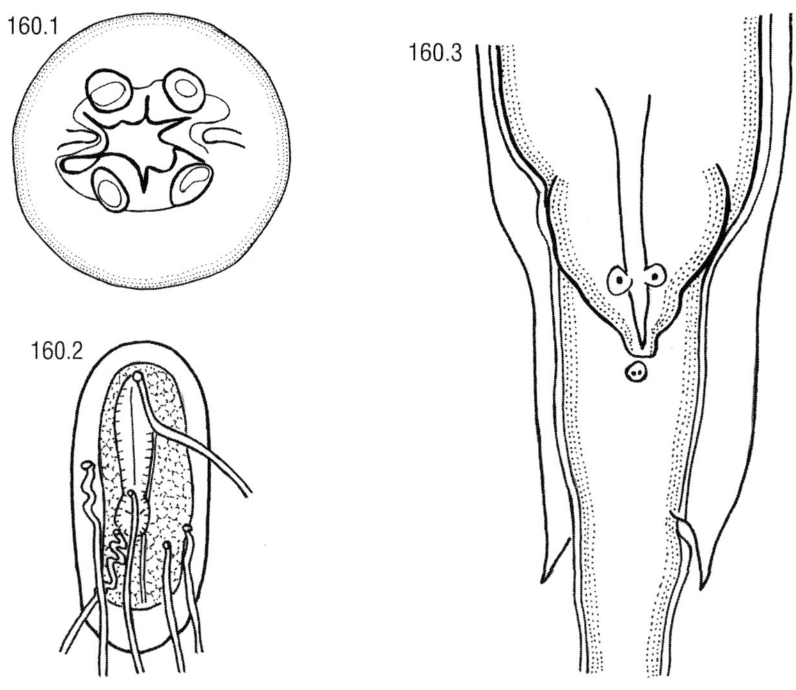

FIG. 160 *Spinoxyuris* 160.1 Female, *en face* view; 160.2 Egg; 160.3 Male cloacal region, ventral view. (Redrawn after Petter, 1994.)

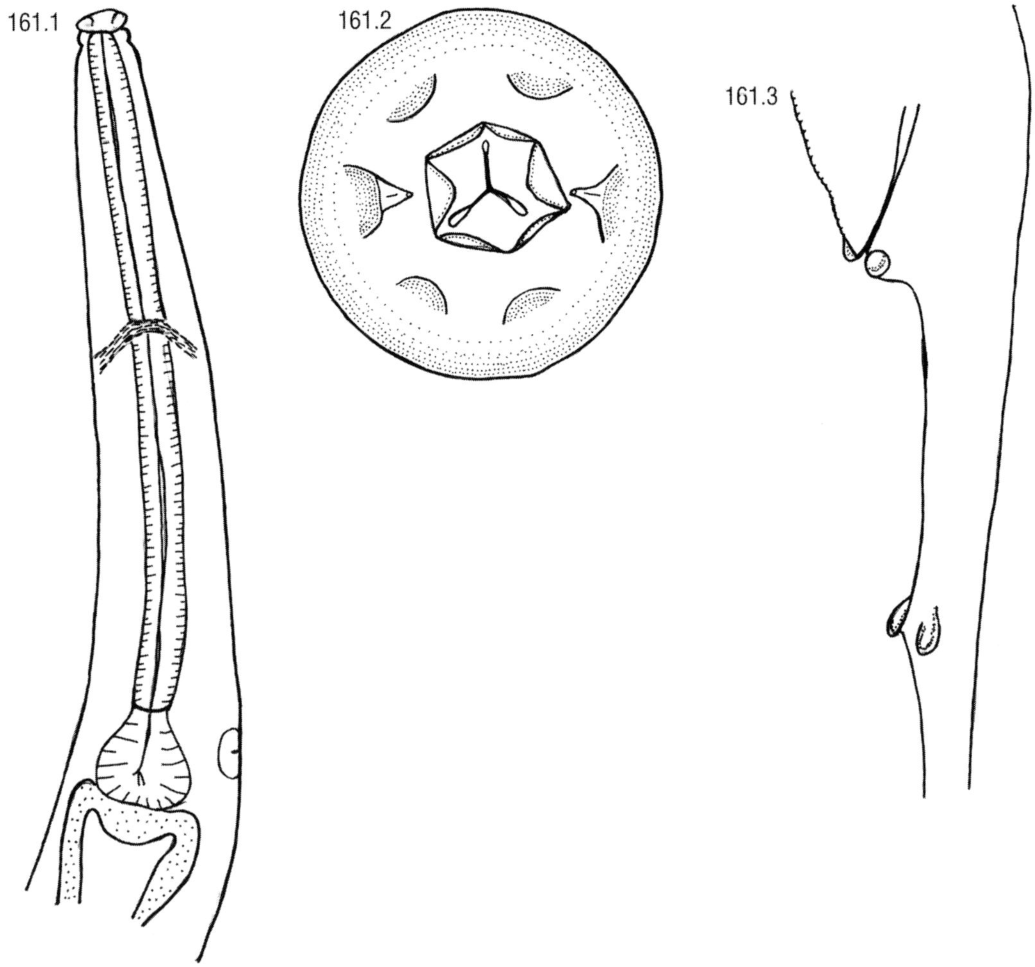

FIG. 161 *Royandersonia* 161.1 Anterior end of female, lateral view; 161.2 Female, *en face* view; 161.3 Posterior end of male, lateral view. (Redrawn after Anderson and Lim, 1996.)

Travnematinae Skrjabin, Shikhobalova and Lagodovskaya, 1963 (family)

Type genus: *Travnema* Pereira, 1938

[Petter and Quentin (1976) place the genus *Travnema* in the family Pharyngodonidae and do not divide the family into subfamilies. Moravec *et al.* (1994) reinstate Travnematidae as the subfamily Travnematinae on the basis of the morphology of the cephalic region and oesophagus, which they consider differs from other members of the Pharyngodonidae from fish. They also include *Hakynema* Moravec and Sey, 1988 in this subfamily.]

Hakynema Moravec and Sey, 1988 (Fig. 162)

DIAGNOSIS: Mouth circular with 3 weakly developed lips, each with distal cuticular point. Four large submedian cephalic papillae, pair of lateral amphids. Buccal capsule absent. Oesophagus

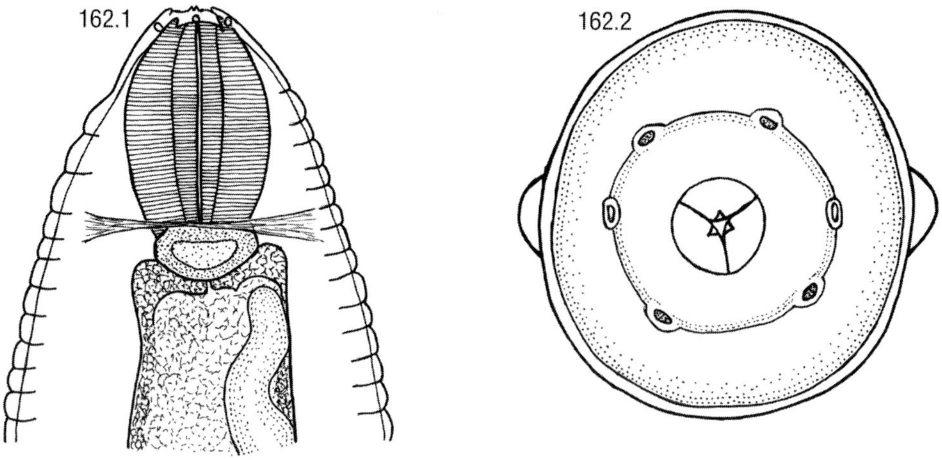

FIG. 162 *Hakynema* 162.1 Female, anterior end of body; 162.2 Female *en face* view showing inflated cervical cuticle. (Redrawn after Moravec and Sey, 1988c.)

with large, oval-shaped corpus, poorly developed bulb without sclerotized valve, isthmus absent. Excretory pore opens posterior to oesophagus. *Male*: Unknown. *Female*: Cervical alae present in oesophageal region. Vulva opens posterior to mid-body. Eggs ovoid. Tail sharply narrows to a point posterior to anus. Parasite of intestine of fish (*Spinibarbichthys denticulatus*, Cyprinidae) in Red River near Hanoi, Vietnam.

Type species: *Hakynema vietnamensis* Moravec and Sey, 1988

[Moravec and Sey (1988c) originally placed this genus in the Travnematidae Skrjabin, Shikhobalova and Lagodovskaya, 1963 in the superfamily Atractoidea. Petter and Quentin (1976) placed the type genus *Travnema* in the Pharyngodonidae in the Oxyuroidea. Later Moravec *et al.* (1994) reinstate the family Travnematidae as the subfamily Travnematinae within the Pharyngodonidae.]

Heteroxynematidae Skrjabin and Shikhobalova, 1948

Ivaschkinonema Erkulov, 1975

DIAGNOSIS: Four labia present. *Male*: Caudal annules present on tail. Precloacal cuticular appendages absent. Parasite of *Alticola argentatus*.

Type species: *Ivaschkinonema alticola* Erkulov, 1975

[Erkulov (1975) places this genus in the subfamily Aspiculurinae with *Cephaluris* and *Aspiculuris*. Petter and Quentin (1976) do not accept this subfamily and place *Aspiculuris* in the subfamily Heteroxynematinae, and *Cephaluris* Akhtar, 1947 in the subfamily Labiostomatinae. The subfamily position of this genus is therefore uncertain. Listed in Spencer Jones and Gibson (1987).]

Heteroxynematinae (Skrjabin and Shikhobalova, 1948 family)

Heteroxynema Hall, 1916

[The genus is divided into three subgenera by Petter and Quentin (1976), namely *Heteroxynema* – parasites of neoarctic rodents, family Sciuridae, type species *Heteroxynema* (*Heteroxynema*) *cucullatum* Hall, 1916; *Cavioxyura* Quentin, 1975 – parasites of neotropical Caviomorpha, type species *Heteroxynema* (*Cavioxyura*) *chilensis* Quentin, 1975; and *Proxyuronema* Quentin, 1975 – parasites of palaeartic Sciuridae, type species *Heteroxynema* (*Proxyuronema*) *getula* (Seurat, 1915). Hugot (1983b) adds a fourth subgenus, *Nadinema*.]

Subgenus *Nadinema* Hugot, 1983 (Fig. 163)

DIAGNOSIS: Mouth opening hexagonal with 6 rudimentary lips, each with a cephalic sensory papilla of internal circle, 6 additional digitiform processes appear to extend to each labial papilla, project into buccal cavity. Four submedian cephalic papillae near amphids frame mouth opening. Three large oesophageal teeth each terminate proximally with a small median denticle. Cervical and lateral alae absent. In cervical region enlarged annules ornamented with slightly overlapping striations forming a sequence of junctions on each border of lateral field. *Female*:

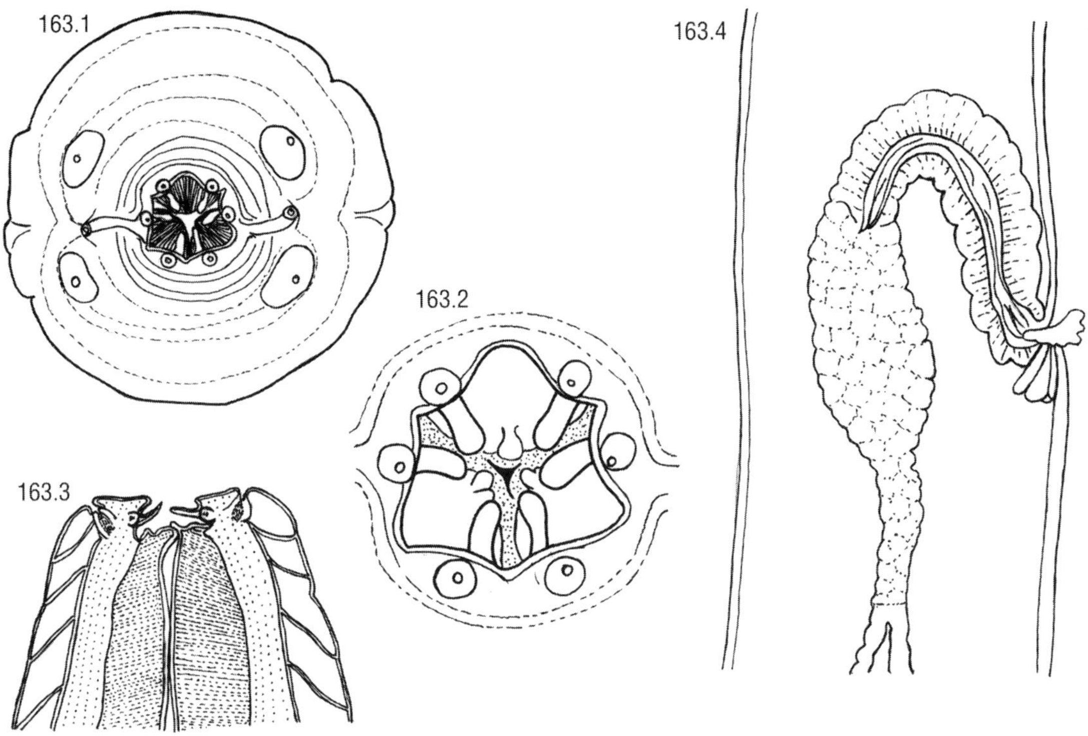

FIG. 163 *Nadinema* 163.1 Female, *en face* view; 163.2 Female, details of mouth opening; 163.3 Optical section through cephalic region at level of amphids; 163.4 Female vulvar region showing glandular region of ovejector. (Redrawn after Hugot, 1983b.)

Ovejector with 3 parts: infundibulum; glandular segment; strongly muscular vagina. Parasite of African rodents, family Thryononyidae.

Type species: *Heteroxynema* (*Nadinema*) *cafer* Hugot, 1983

[Listed in Spencer Jones and Gibson (1987).]

Aspiculuris Schulz, 1924

[Petter and Quentin (1976) did not divide this genus into subgenera. Akhtar (1955) placed the genus in the subfamily Aspiculurinae Skrjabin and Shikhobalova, 1950, a subfamily not accepted by Petter and Quentin (1976), who place the genus in the Heteroxynematinae.]

Subgenus *Aspiculuris* (Schulz, 1924 genus) Akhtar, 1955

DIAGNOSIS: Cephalic vesicle present. Cervical alae with distal, sickle-shaped border, discontinuous with lateral alae.

Type species: *A.* (*Aspiculuris*) *tetraptera* (Nitzch, 1821) Schulz, 1924

Subgenus *Pseudaspiculuris* Akhtar, 1955

DIAGNOSIS: Cephalic vesicle present. Cervical alae without distal, sickle-shaped border, continuous with lateral alae.

Type species: *A.* (*Pseudaspiculuris*) *asiatica* Schulz, 1927

[Listed in Spencer Jones and Gibson (1987).]

Subgenus *Paraspiculuris* Akhtar, 1955

DIAGNOSIS: Cephalic vesicle present. Cervical alae with distal, sickle-shaped border, continuous with lateral alae.

Type species: *A.* (*Paraspiculuris*) *pakistanica* Akhtar, 1955

[Listed in Spencer Jones and Gibson (1987).]

Subgenus *Anaspiculuris* Akhtar, 1955

DIAGNOSIS: Cephalic vesicle present. Cervical alae with distal, sickle-shaped border, lateral alae absent.

Type species: *A.* (*Anaspiculuris*) *americana* Erickson, 1938

[Listed in Spencer Jones and Gibson (1987).]

Subgenus *Subaspiculuris* Akhtar, 1955

DIAGNOSIS: Cephalic vesicle absent. Cervical alae with distal, sickle-shaped border, lateral alae absent.

Type species: *A.* (*Subaspiculuris*) *caviellae* Freitas, Lent and Almeida, 1937

[Listed in Spencer Jones and Gibson (1987).]

Kahmannia Mas-Coma and Esteban, 1982 (Fig. 164)

DIAGNOSIS: Cuticle with marked transverse striations. Cervical and lateral alae absent. Cephalic vesicle present. Sinous peribuccal crown with 8 elements. Mouth circular, lips absent. Buccal capsule triangular in cross-section. Small oesophageal teeth present. Oesophagus with corpus, isthmus and valvate bulb. *Male*: Caudal alae absent, *area rugosa* absent, gubernaculum absent. Single spicule. *Female*: Vulva opens just anterior to mid-body. Thin-shelled eggs without operculum. Tail tapers to distal point. Parasites in intestine of *Eliomys quercinus*, Gliridae, Rodentia in Menorca, Balearic Islands.

Type species: *Kahmannia eliomydis* Mas-Coma and Esteban, 1982

[Listed in Spencer Jones and Gibson (1987).]

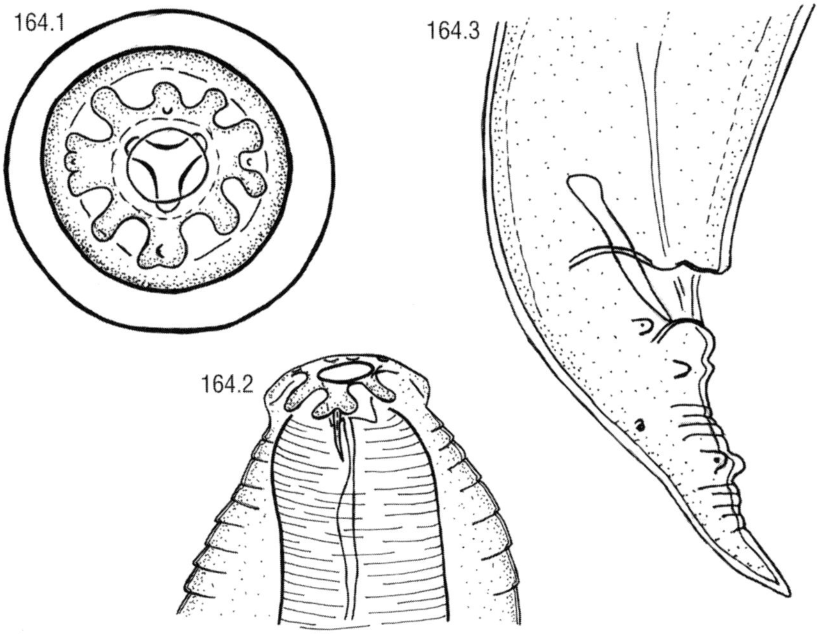

FIG. 164. *Kahmannia* 164.1 Female, *en face* view; 164.2 Female cephalic region, lateral view; 164.3 Male posterior end, latero-ventral view. (Redrawn after Mas-Coma and Esteban, 1982.)

5 ASCARIDIDA

COSMOCERCOIDEA

Genera not assigned to a family

Spiruroides Cameron and Parnell, 1933

DIAGNOSIS: Body slender. Mouth surrounded by 6 sensory papillae. Small, distinct subspherical buccal capsule, 3 teeth at base. Oesophagus: corpus; isthmus; bulb without valve. Nerve ring anterior to bulb. *Male*: Short tail, irregularly trifurcate. Spicules equal, similar, bifurcate distal tips. Gubernaculum poorly chitinized, approximately half length of spicules. Narrow caudal alae present. Single pair of precloacal papillae, 2 pairs of postcloacal papillae, 5 pairs of adcloacal papillae. *Female*: Body short. Vulva opens anterior to mid-body. Buccal capsule shorter than male. Tail irregularly trifurcate. Didelphic. Opposed uteri without ovejectors. Parasites in stomach of hedgehog, *Erinaceous europeus*, in Scotland, UK.

Type species: *Spiruroides erinacei* Cameron and Parnell, 1933

[Cameron and Parnell (1933) refers to this genus as a spirurid, whose most closely related genus appears to be *Seuratum*, which is now placed in the Seuratoidea in the order Ascaridida. The features of the oesophagus, however, suggest it is more likely to be associated with the *Cosmocercoidea* within the same order. Further study is required to assess the validity of this genus and its taxonomic position. Listed in Spencer Jones and Gibson (1987) as *incertae sedis*.]

Oxysomoides Walton, 1928

DIAGNOSIS: Species assigned to *Oxysoma* Schneider, 1866 (which Walton, 1928 refers to as a homonym) or to *Oxysomatium* Railliet and Henry, 1916 that (Walton, 1928) cannot be assigned to the family Kathlaniidae (Lane, 1914 subfamily) Travassos, 1918.

Type species: *Oxysomoides tricirratus* (Linstow, 1901)

[Chabaud (1978) assigns *Oxysomatium* to the Cosmocercidae, not the Kathlaniidae as in Walton's classification. The incomplete diagnosis for this genus and the classification of Chabaud (1978) puts its validity in doubt. Baker (1987) synonymizes *Oxysomoides* with *Falcaustra* Lane, 1915, which is in the subfamily Kathlaniinae. Listed in Spencer Jones and Gibson (1987).]

Cosmocercidae (Railliet, 1916 subfamily) Travassos, 1925

DIAGNOSIS: Slit-like excretory pore present. *Male*: Presence of 2 or more paired papillae in addition to unpaired papilla on anterior lip of cloaca. Presence of 2 pairs of subventral postcloacal papillae grouped on anterior half of tail.

[Emended diagnosis based on Adamson and Baccam (1988) and the basis for moving *Maracaya*, *Schrankiana*, *Schrankianella*, *Labeonema* and *Ibrahimia* from the Atractidae *sensu* Chabaud (1978) to the Cosmocercidae.]

Genera not assigned to a subfamily

Odontogeton Allgén, 1921

DIAGNOSIS: Mouth opens in depression. Cuticular 'cap' present, ring of posteriorly directed hooks outside and teeth inside buccal capsule. Oesophagus with broad corpus, narrow isthmus, well-developed bulb incorporating valve. *Male*: Spicules equal. One pair precloacal, 3 pairs postcloacal papillae. *Female*: Vulva opens behind mid-body. Didelphic. Parasite in intestine of *Phacochoerus aethiopicus* in Natal, South Africa.

Type species: *Odontogeton phacochoeri* Allgén, 1921

[Skrjabin *et al.* (1961) place this genus in the Cosmocercidae. Spencer Jones and Gibson (1987) list it in the Oxyuroidea. Originally described by Allgén (1921). Diagnosis based on Allgén (1925) and Skrjabin *et al.* (1961).]

Blanusia Zapatero, Castaño, Lopez and Martin, 1990 (Fig. 165)

DIAGNOSIS: Lateral alae present. Mouth terminal, with 2 lips. Buccal capsule without teeth. Oesophagus with muscular corpus, metacorpus dilated without valve, large isthmus, bulb with valve. Excretory pore round, simple. *Male*: Spicules equal, joined at distal tips. Gubernaculum present. Ten pairs papillae, distal 2 pairs close, subventral. Plectanes and rosette papillae absent. Tail with sharply pointed distal tip. *Female*: Vulva opens in mid-body. Didelphic. Amphidelphic. Parasite in intestine of *Blanus cinereus*, Amphisbaenidae, in Spain.

Type species: *Blanusia pseudorhabditis* Zapatero, Castaño, Lopez and Martin, 1990

Cosmocercinae Railliet, 1916

Cosmocerca Diesing, 1861

DIAGNOSIS: *Male*: 'Two spicules present, usually weakly sclerotized or rudimentary'.

Type species: *Cosmocerca trispinosa* Railliet and Henry, 1916

[Moravec and Kaiser (1994) made the above emendment to the generic diagnosis to include their species *Cosmocerca longispicula* from Dendrobatid frogs in Martinique, French Antilles.]

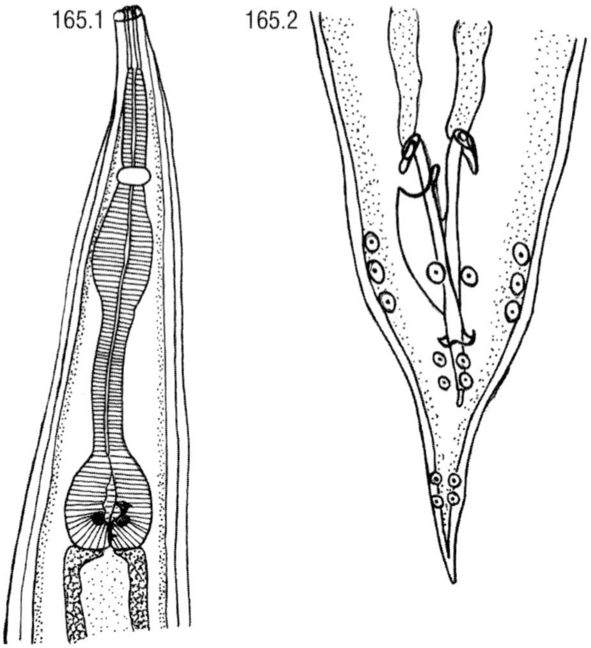

Fig. 165 *Blanusia* 165.1 Anterior end of female; 165.2 Posterior end of male. (Redrawn after Zapatero *et al.*, 1990 (publ. 1991).)

Oxysomatium Railliet and Henry, 1916

DIAGNOSIS: Lateral alae, somatic papillae present. *Male*: Rosettes or plectanes absent from tail. *Female*: Numerous thin-shelled eggs in uterus. Ovary of anterior uterus just posterior to vulva, ovary of posterior uterus well anterior to vulva. Parasites of amphibians and reptiles.

Type species: *Oxysomatium brevicaudatum* (Zeder, 1800) Railliet and Henry, 1916

[Emended diagnosis based on Baker (1980c).]

Neoxysomatium Ballasteros-Márquez, 1945

[Baker (1980c) synonymizes this genus with *Oxysomatium* (see emended diagnosis above), as the type species of the two genera are considered to be synonyms of the same species.]

Aplectana Railliet and Henry, 1916

DIAGNOSIS: Lateral alae, somatic papillae present. *Male*: Rosettes or plectanes absent from tail. *Female*: Numerous thin-shelled eggs in uteri. Two ovaries anterior to vulva.

Type species: *Aplectana acuminata* (Schrank, 1788) Railliet and Henry, 1916

[Emended diagnosis based on Baker (1980d). For a full list of species, Baker (1980d) should be consulted. Baker (1987) synonymizes *Paraleptonema* Wang, 1980, *Neosomatiana* Islam,

Farooq and Khanum, 1979 and *Paraplectana* Rhyzikov, Sharpilo and Shevchenko, 1980 with *Aplectana*.]

Neoraillietnema Ballestros-Márquez, 1945

DIAGNOSIS: Distinct sexual dimorphism. Mouth opening with 3 lips. Oesophagus with valvate posterior bulb. *Male*: Two spicules, 'reduced'. Gubernaculum present, well developed. Precloacal papillae in 2 rows, postcloacal papillae smaller, more irregularly arranged. Plectanes absent. Tail pointed. *Female*: Vulva opens in mid-body. Prodelphic. Ovoviviparous. Tail pointed. Parasites of African toads.

Type species: *Neoraillietnema praeputiale* (Skrjabin, 1916) nec *Raillietnema praeputiale* Semenow, 1929

[Chabaud (1978) considered this genus a synonym of *Aplectana* Railliet and Henry, 1916. Grabda-Kazubska (1985) reinstates and emends the generic diagnosis.]

Luzonema De Oliveira Rodrigues, Carvalho Varela, Sodré Rodrigues and Cristófaro, 1973 (Fig. 166)

DIAGNOSIS: Mouth with 2 small lips. Buccal cavity globular with small teeth on proximal part; divided by 2 laminae in median region. Two chitinous, disc-like formations situated laterally to oesophagus at base of buccal capsule. Oesophagus: short; muscular; bulb present. *Male*: Spicules short. Gubernaculum present. Three pairs precloacal papillae, 4 pairs postcloacal papillae. Tail with 2 pairs of papillae, point rounded. *Female*: Unknown. Parasites of intestine of marine fish, *Lethrinus atlanticus*.

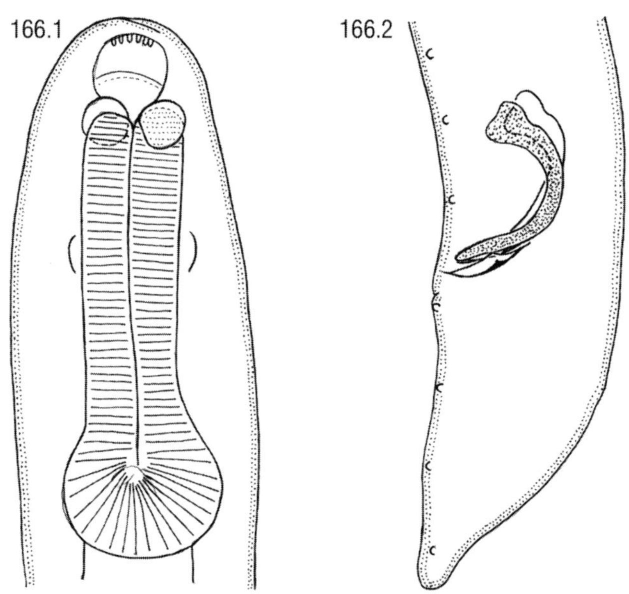

FIG. 166 *Luzonema* 166.1 Anterior end of male, lateral view; 166.2 Posterior end of male, lateral view. (Redrawn after De Oliveira Rodrigues *et al.*, 1973.)

Type species: *Luzonema cruzi* De Oliveira Rodrigues, Carvalho Varela, Sodré Rodrigues and Cristófaro, 1973

[The authors place the genus and subfamily Cosmocercinae in the family Oxyuridae. The subfamily is now in the Cosmocercidae in the superfamily Cosmocercoidea.]

Neosomatiana Islam, Farooq and Khanum, 1979 (Fig. 167)

DIAGNOSIS: Cephalic collar present. Mouth with lips, each lip with 2 papillae. Additional cephalic papillae present. Buccal cavity absent. Oesophagus with distal bulb. Pedunculate somatic papillae present. *Male*: Numerous pedunculate caudal papillae present. Spicules unequal, thin. Gubernaculum absent. Tail tapers to a sharp point. *Female*: Vulva opens anterior to middle of body. Amphidelphic. Eggs embryonated when laid. Tail tapers to sharp point. Parasite in intestine of toad, *Bufo andersoni*, in Pakistan.

Type species: *Neosomatiana akrami* Islam, Farooq and Khanum, 1979

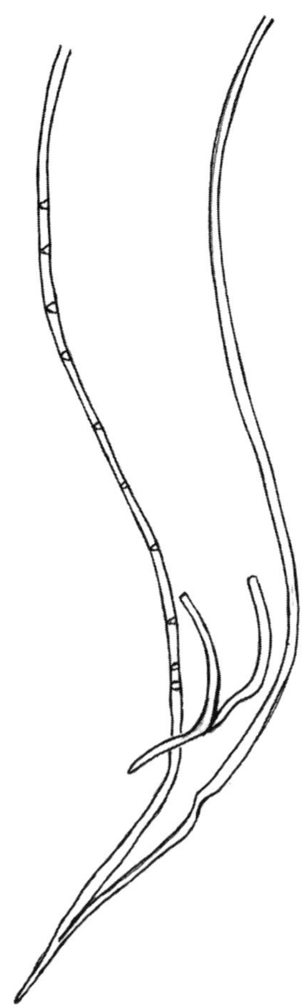

FIG. 167 *Neosomatiana* Posterior end of male. (Redrawn after Islam *et al*., 1979.)

[The authors place the genus in the subfamily Oxysomatiinae of the family Oxyuridae. The genus is distinguished from *Oxysomatium* Railliet and Henry, 1916 and other genera originally placed in this subfamily on the basis of the morphology of the spicules and position of the vulva. Chabaud (1978) places the genus *Oxysomatium* in the subfamily Cosmocercinae in the superfamily Cosmocercoidea; consequently, this genus is now listed in the same subfamily. Baker (1987), however, synonymizes this genus with *Aplectana* Railliet and Henry, 1916. Listed in Spencer Jones and Gibson (1987).]

Parasomatium Islam, Farooq and Khanum, 1979 (Fig. 168)

DIAGNOSIS: Cuticle smooth. Lateral alae absent. Mouth with indistinct lips. Buccal capsule absent. Oesophagus with distal bulb. *Male*: Caudal alae absent. Eleven pairs of pedunculate precloacal papillae. Spicules equal, alate, thick. Gubernaculum absent. Tail curved ventrally, tapers to point. *Female*: Vulva opens at level of oesophagus. Amphidelphic. Eggs embryonated when laid. Tail tapers to sharp point. Parasite in the intestine of the toad, *Bufo andersoni*, in Pakistan.

Type species: *Parasomatium ishaqi* Islam, Farooq and Khanum, 1979

[The authors place the genus in the subfamily Oxysomatiinae of the family Oxyuridae. The genus is distinguished from the genus *Oxysomatium* Railliet and Henry, 1916 on the basis of the pedunculated caudal papillae and the position of the vulva. Chabaud (1978) places the genus *Oxysomatium* in the subfamily Cosmocercinae in the superfamily Cosmocercoidea; consequently, this genus is now listed in the same subfamily. Baker (1987), however, synonymizes this genus with *Cosmocerca*. Listed in Spencer Jones and Gibson (1987).]

Dollfusnema Baker, 1981

DIAGNOSIS: Cuticle with small number of somatic papillae present. Mouth with 3 lips, interlabia present. Lateral alae present. *Male*: Caudal papillae simple, rosettes and plectanes absent. *Female*: Didelphic. Uteri opposed, ovaries anterior to vulva. Small number of large, thin-shelled eggs in uteri. Parasites of the lizard *Leposternon phocaena*, in Brazil.

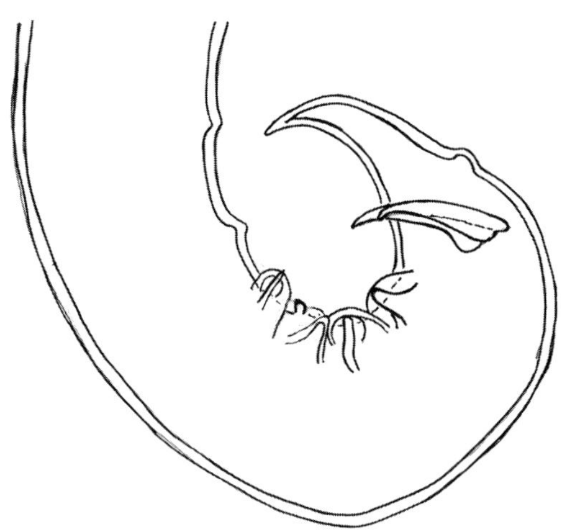

FIG. 168 *Parasomatium* Posterior end of male. (Redrawn after Islam *et al.*, 1979.)

Type species: *Dollfusnema amphisbaenia* Baker, 1981

[Baker (1982b) renames this genus *Paradollfusnema*, as the name *Dollfusnema* is preoccupied by the spirurid genus *Dollfusnema* Caballero, 1974. Listed in Spencer Jones and Gibson (1987).]

Paradollfusnema Baker, 1982 (Fig. 169)

[*Nom. nov.* for *Dollfusnema* Baker, 1981. See *Dollfusnema* above. Listed in Spencer Jones and Gibson (1987).]

Neocosmocercella Baker and Vaucher, 1983 (Fig. 170)

DIAGNOSIS: Cuticle with prominent transverse striations. Lateral alae well developed. Somatic papillae not seen. Mouth hexagonal with 3 bilobed lips, each lobe with anteriorly directed cuticular flange. Cheilostom triradiate. Cephalic sensory organs: 4 large submedian papillae. Oesophagus: corpus; isthmus; bulb. *Male*: Large, vesiculated precloacal papillae present. Prominent muscle cells present, extend from proximal large papillae to cloaca. Spicules long, pointed. Gubernaculum present. Posterior half of body coiled. *Female*: Body straight. Didelphic. Vulva opens anterior to mid-body. Vagina elongate (over 1.5 mm long), uterine sac present, uteri directed anteriorly, ovaries anterior to vulva. Eggs over 150 µm long. Tail long, conical. Parasite of *Phyllomedusa hypochondrialis* (Amphibia) in Paraguay.

Type species: *Neocosmocercella paraguayensis* Baker and Vaucher, 1983

[Listed in Spencer Jones and Gibson (1987).]

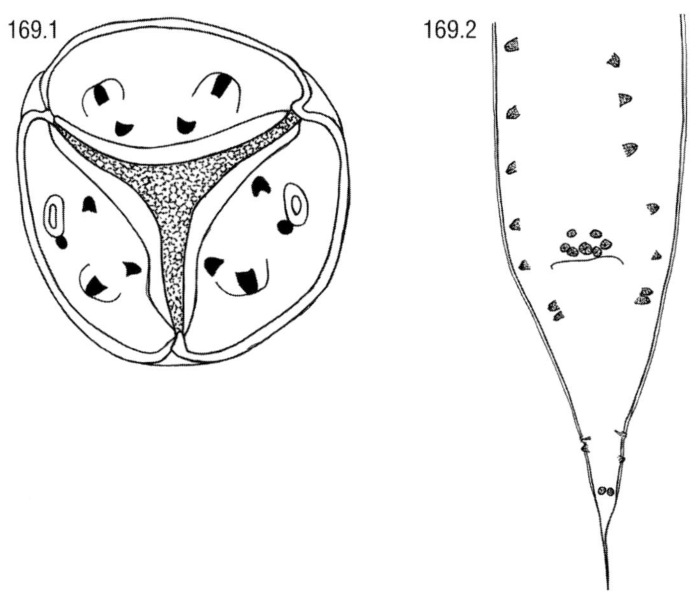

FIG. 169 *Paradollfusnema* (= *Dollfusnema* preoccupied) 169.1 Female cephalic region, *en face* view; 169.2 Posterior end of male, ventral view. (Redrawn after Baker, 1981b.)

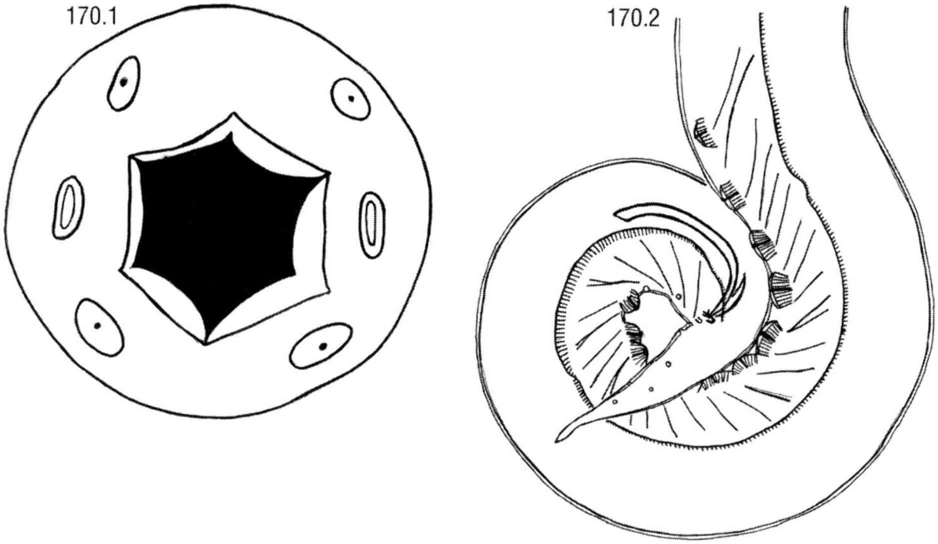

FIG. 170 *Neocosmocercella* 170.1 Cephalic region, *en face* view; 170.2 Posterior end of male, lateral view. (Redrawn after Baker and Vaucher, 1983.)

Furconema Gairola and Malhotra, 1988 (Fig. 171)

DIAGNOSIS: Lateral alae along body. Mouth with 3 lips, each with single papilla. Large, chitinous buccal capsule with 4 teeth at base. Oesophagus claviform with valve at distal end. *Male*: Tail alate with 10 pairs of precloacal papillae, 2 pairs adcloacal papillae, 5 pairs postcloacal papillae. Spicules unequal, dissimilar. Gubernaculum present. *Female*: Vulva opens posterior to mid-body. Tail short, bifurcates distally. Viviparous. Parasite of Indian freshwater teleosts (*Clarias batrachus*).

Type species: *Furconema viviparous* Gairola and Malhotra, 1988

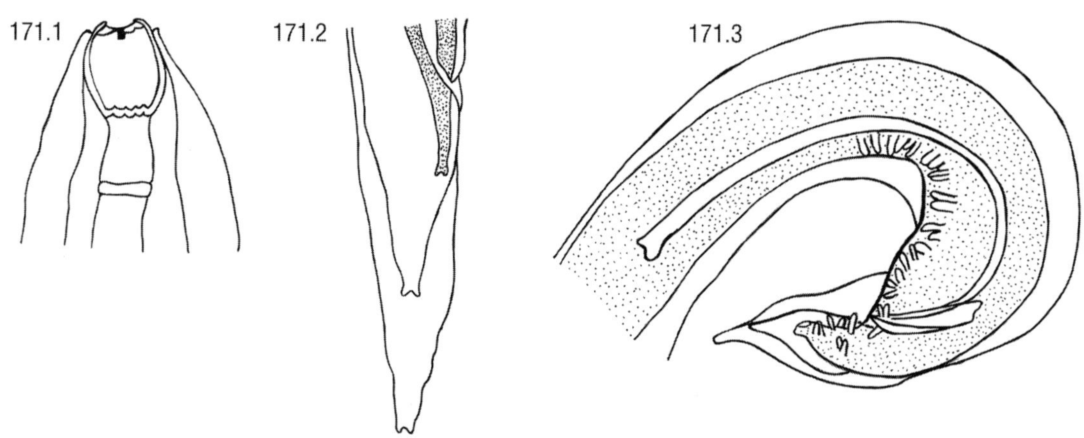

FIG. 171 *Furconema* 171.1 Female, anterior end; 171.2 Female, posterior end, lateral view; 171.3 Male, posterior end, lateral view. (Redrawn after Gairola and Malhotra, 1988.)

Stewartia Rao, 1989 (Fig. 172)

DIAGNOSIS: Lateral alae present. Sexes not dimorphic. Rosette-type ornamentation or 'false' plectanes along length of body. Mouth with 3 lips. Proximal end of pharynx with 3 teeth. *Male*: Spicules equal. Gubernaculum present. *Female*: Prodelphic. Parasites of amphibians.

Type species: *Stewartia macintoshi* (Stewart, 1914) Rao, 1989

[Rao (1989) created a new subfamily Stewartiinae for the new genus and placed it in the Oxyuridae. Rao suggested that the subfamily is intermediate between the Aplectaninae and the Cosmocercinae. The genus *Aplectana* is now in the Cosmocercinae in the Cosmocercoidea. The presence of the rosette-type ornamentation, the shape of the oesophagus (corpus, isthmus and bulb) and the presence of numerous papillae on the male tail (12–14 pairs) place the genus *Stewartia* in the Cosmocercinae, and the family Stewartiinae is considered a synonym. Rao (1977) initially described the new subfamily and genus in the *Abstracts of the 1st National Congress of Parasitology*, Baroda, India, 24–26 February 1977, Indian Society for Parasitology. Listed in Spencer Jones and Gibson (1987).]

Dentinema Moravec, Chara and Shinn, 2004 (Fig. 173)

DIAGNOSIS: Mouth opening triangular with 3 reduced lips. Each lip associated internally with cuticularized 'lamella-like' margins of buccal capsule. Four submedian cephalic papillae,

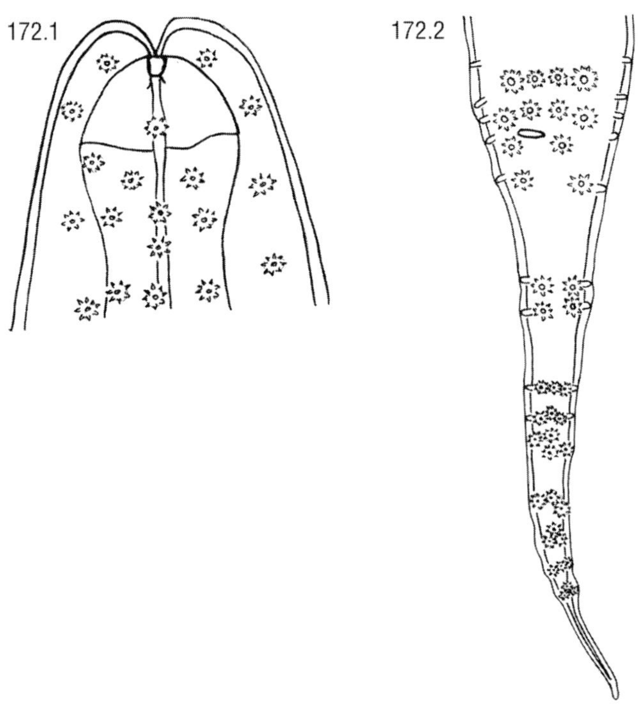

FIG. 172 *Stewartia* 172.1 Male, anterior end; 172.2 Male, posterior end, ventral view. (Redrawn after Rao, 1989.)

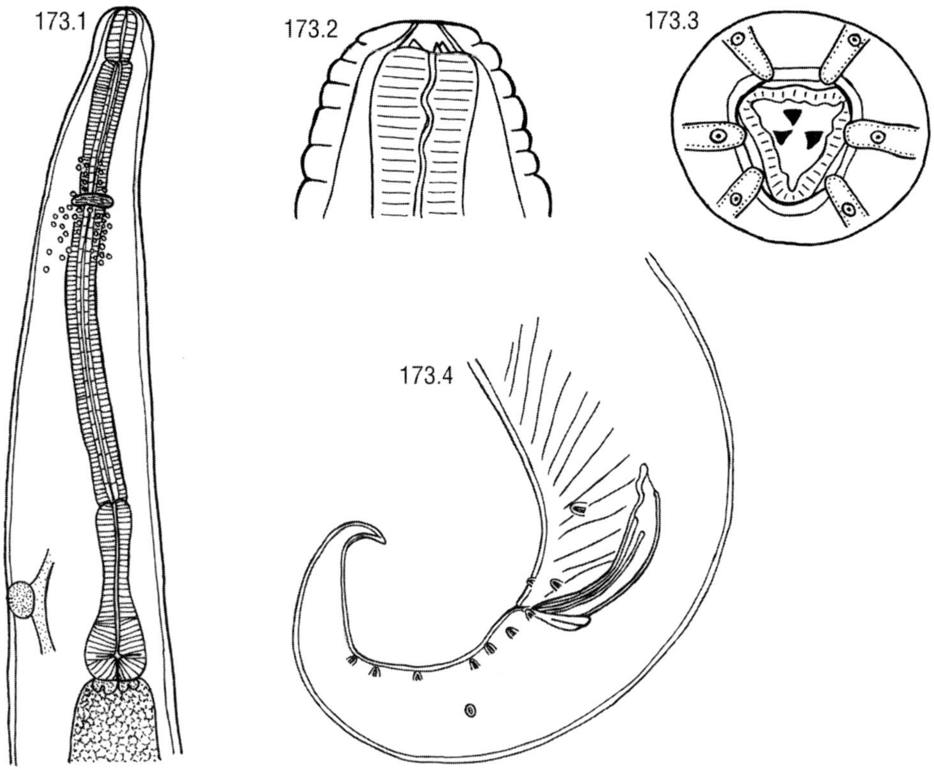

FIG. 173 *Dentinema* 173.1 Anterior end of male, lateral view; 173.2 Cephalic region, lateral view; 173.3 Cephalic region, *en face* view; 173.4 Posterior end of male, lateral view. (Redrawn after Moravec *et al.*, 2004.)

2 lateral amphids associated with 6 radiating ridges, which extend to mouth margin. Three oesophageal teeth protrude into buccal cavity. Oesophagus with anterior 'pharyngeal region', corpus, long isthmus, 'spherical' valved bulb. Lateral alae absent. *Male*: Precloacal region with paired oblique muscles. Precloacal sucker absent. Spicules equal, wide. Gubernaculum present, triangular in ventral view. Precloacal, adcloacal, postcloacal papillae several pairs present. Caudal alae absent. Tail conical. *Female*: Vulva opens in posterior part. Amphidelphic. Eggs oval, thin-walled. Parasites of intestine of neotropical fish.

Type species: *Dentinema trichomycteri* Moravec, Chara and Shinn, 2004

Austraplectaninae Baker, 1981

DIAGNOSIS: Two nuclei in each oesophageal gland. Somatic musculature coelomyarian. *Female*: Vulva opens near middle of body. Ovejector with single, short, muscular anterior region and paired, thin-walled region extending almost to posterior end of body. Uterus directed anteriorly. Ovaries anterior. Oviparous, eggs without filaments. Tail conical, relatively short.

Type genus *Austraplectana* Baker, 1981 (Fig. 174)

DIAGNOSIS: Cuticle with somatic papillae. Mouth with 3 small lips each with a small anterior cuticular extension. Lateral alae present. *Male*: Caudal alae thick, numerous papillae present. Precloacal region with few pairs of latero-ventral muscles. Sucker absent. Parasites of Australian amphibians (Myobatrachidae, Pelodryadidae).

Type species: *Austraplectana kartanum* (Johnston and Mawson, 1941) Baker, 1981

[Listed in Spencer Jones and Gibson (1987).]

Kathlaniidae (Lane, 1914 subfamily) Travassos, 1918

Kathlaniinae Lane, 1914

Megalobatrachonema Yamaguti, 1941

Subgenus *Megalobatrachonema* (Yamaguti, 1941 genus) Baker, 1980 (Fig. 175)

DIAGNOSIS: Cephalic inflation absent. Mouth with 3 small or reduced lips. Four submedian cephalic papillae, '?double with short, thick peduncles'. Buccal cavity reduced, teeth absent.

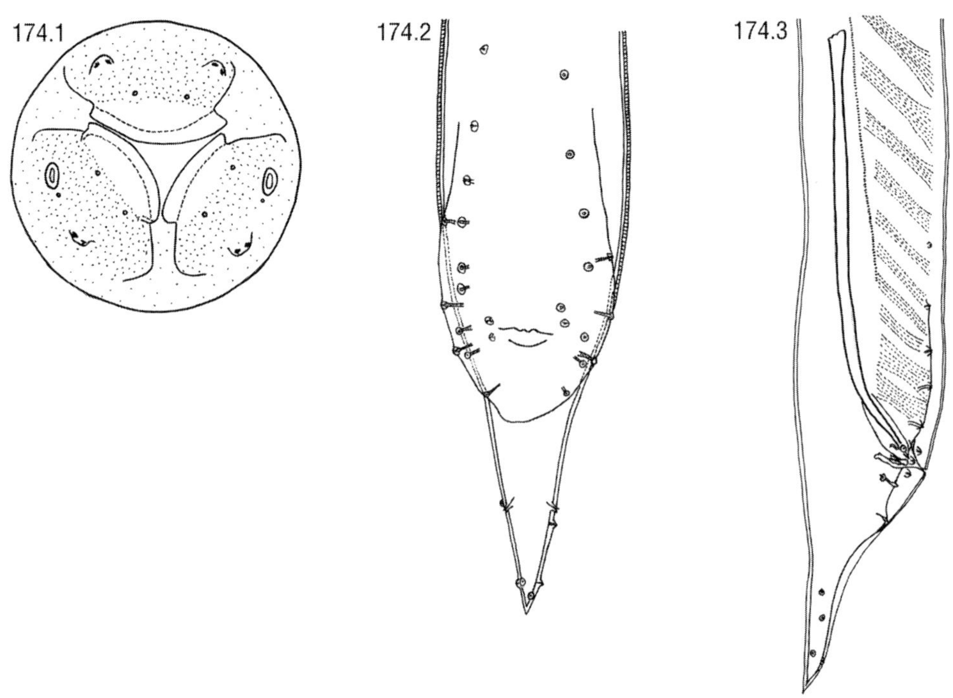

Fig. 174 *Austraplectana* 174.1 Male, *en face* view; 174.2 Posterior end of male, ventral view; 174.3 Posterior end of male, lateral view. (Redrawn after Baker, 1981d.)

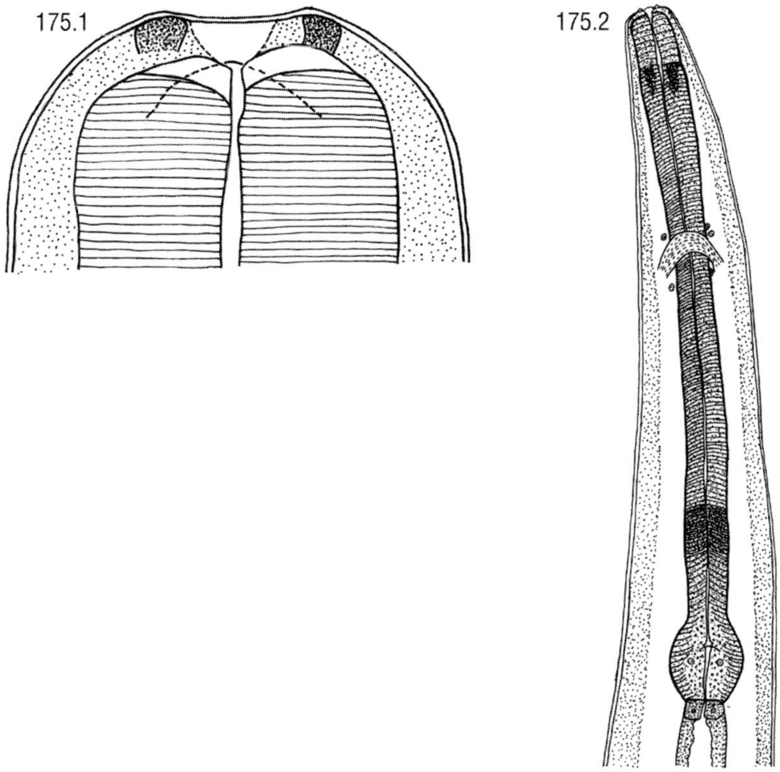

FIG. 175 *Megalobatrachonema* 175.1 Cephalic region of male, dorsal view; 175.2 Anterior end of male, dorsal view. (Redrawn after Baker, 1980b.)

Oesophagus: Proximal end of oesophagus with 3 blunt lobes, covered by thick cheilostomal ring; short, muscular 'pharyngeal' region; elongate corpus; isthmus narrow, poorly defined; bulb with small valves. *Male*: Pseudosucker present or absent. Parasites of urodeles and anurans.

Type species: *Megalobatrachonema* (*Megalobatrachonema*) *nipponicum* Yamaguti, 1941

Subgenus *Chabaudgolvania* (Freitas, 1958 genus) Baker, 1980 (Fig. 176)

DIAGNOSIS: Cephalic inflation absent. Mouth with 3 lips. Four double submedian cephalic papillae with thick, short peduncles. Buccal cavity reduced, teeth absent. Oesophagus: club-shaped; proximal end with 3 tooth-like lobes; cheilostomal ring absent; distinct muscular 'pharyngeal' region; corpus, isthmus and bulb not defined; valves absent at distal end. *Male*: Pseudosucker absent. Parasites of *Triturus* spp. in Western Europe.

Type species: *Megalobatrachonema* (*Chabaudgolvania*) *campanae* Chabaud and Golvan, 1957

[Chabaud (1978) considered *Chabaudgolvania* a synonym of *Megalobatrachonema*. Richardson and Adamson (1988) support Baker's classification of this genus. Later, Richardson and Adamson (1990) reinstate *Chabaudgolvania* and transfer it to the Seuratoidea.]

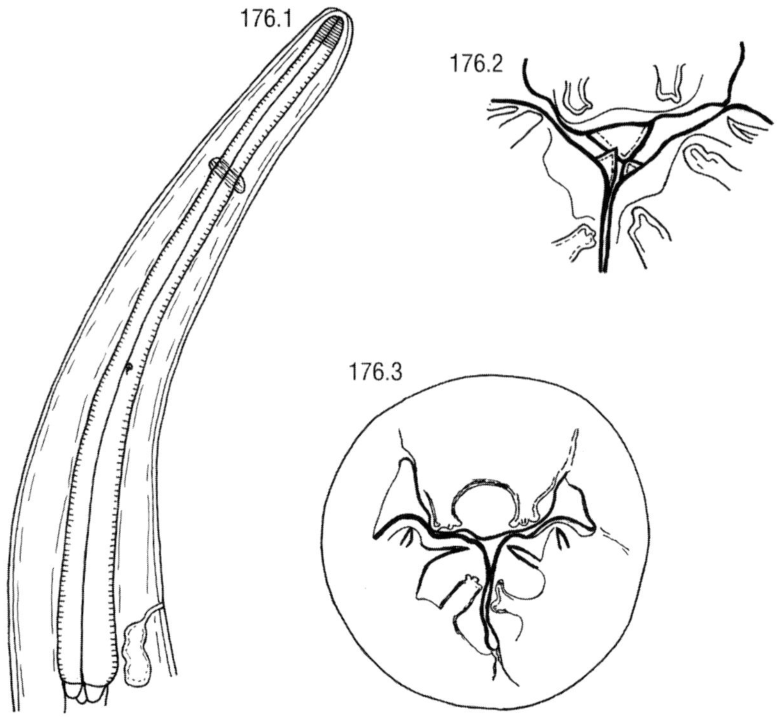

Fig. 176 *Chabaudgolvania* 176.1 Anterior end of male, lateral view; 176.2 *En face* view of female showing oesophageal teeth; 176.3 *En face* view of female. (Redrawn after Chabaud and Golvan, 1957.)

Cordonema Baker, 1981

DIAGNOSIS: See *Urodelnema* Baker, 1981 below.

Type species: *Cordonema mackini* (Walton, 1941) Baker, 1981

[The generic name *Cordonema* is preoccupied – see below.]

Urodelnema Baker, 1981 (Fig. 177)

DIAGNOSIS: Mouth with 3 large, equal lips, interlabia absent. Cephalic region with 3 large, semi-circular cordons. Oesophagus: short, proximal muscular 'pharyngeal' region; elongate corpus; swollen isthmus; bulb with valves. Parasites of amphibians.

Type species: *Urodelnema mackini* (Walton, 1941) Baker, 1981

[Baker (1981c) proposes the name *Urodelnema* for *Cordonema* Baker, 1981, which was preoccupied. Listed in Spencer Jones and Gibson (1987).]

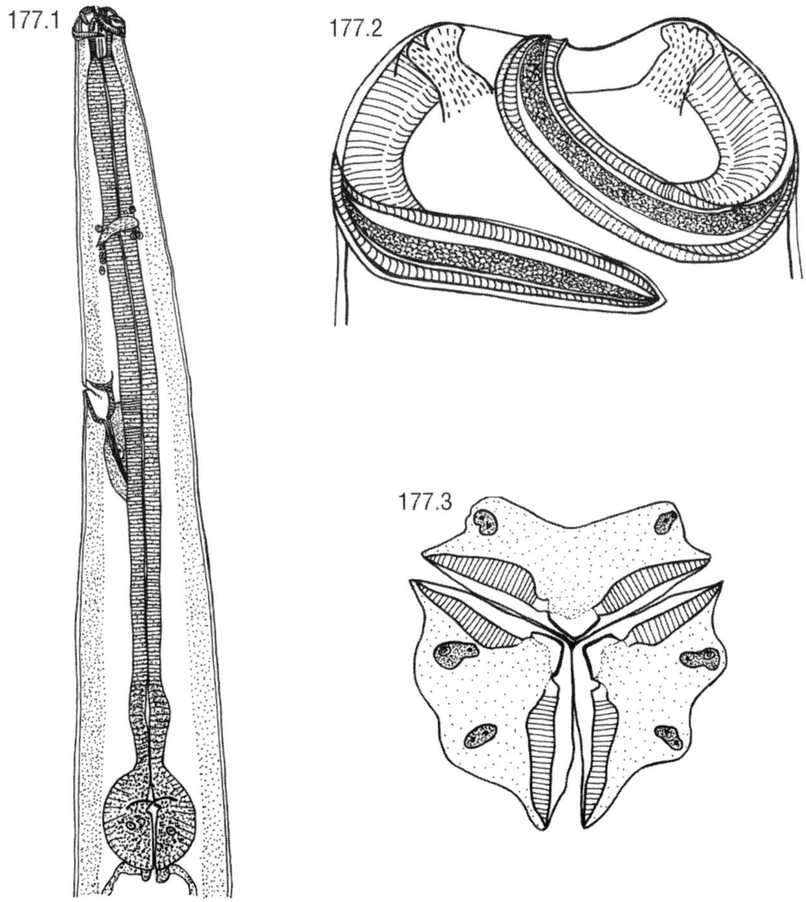

FIG. 177 *Urodelnema* 177.1 Anterior end of male, lateral view; 177.2 Cephalic region of male, dorsal view; 177.3 Cephalic region of male, *en face* view. (Redrawn after Baker, 1981a.)

Myleusnema Moravec and Thatcher, 1996 (Fig. 178)

DIAGNOSIS: Cephalic region separated by transverse fold, near nerve ring, from remainder of body. Mouth triangular with 3 small lips. Buccal capsule poorly developed, lamella-like formation associated with lips form margins. Oesophagus without proximal pharyngeal portion, distal end with elongate isthmus and valvate bulb. *Male*: Precloacal sucker present. Posterior margin of cloaca with 2 projections associated with gubernaculum. Spicules equal, alate, similar. Numerous paired pre- and postcloacal papillae present. Caudal alae absent. *Female*: Vulva opens posterior to mid-body. Prodelphic. Eggs oval, thin-walled, embryonated. Parasites in intestine of characoid freshwater fish, *Myleus ternetzi* (Cypriniformes, Serrasalmidae) in French Guiana.

Type species: *Myleusnema bicornis* Moravec and Thatcher, 1996

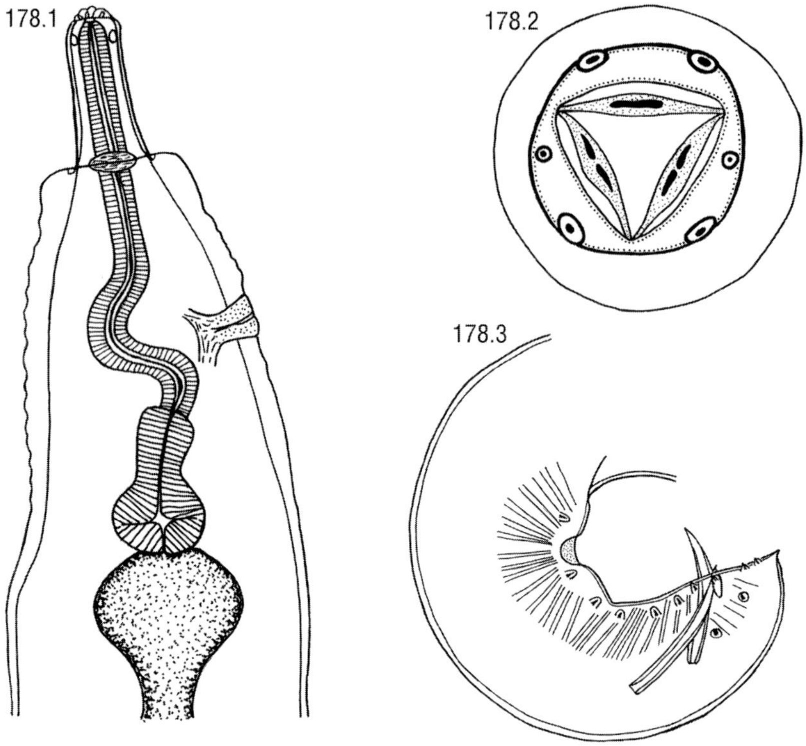

FIG. 178 *Myleusnema* 178.1 Anterior end of female, lateral view; 178.2 *En face* view of female; 178.3 Posterior end of male, lateral view. (Redrawn after Moravec and Thatcher, 1996.)

Spectatus Travassos, 1923

[Baker (1980b) synonymizes *Chabaudinema* Díaz-Ungría, 1968 with *Spectatus*, considered two valid genera in Chabaud (1978).]

Oxyascaridinae Freitas, 1958

[Baker and Vaucher (1985) synonymize this subfamily with Cosmocercinae Railliet, 1916. These are separate subfamilies in Chabaud (1978).]

Oxyascaris Travassos, 1920

DIAGNOSIS: Mouth opening triangular. Lips not bilobed. *Male*: Mature worms less than half size of females. Plectanes or rosettes absent on tail. *Female*: Mature worms more than double size of males; larger than 15 mm in length. Uteri with many small eggs or larvae. Didelphic. Uteri divergent posterior to vulva or with one uterus just anterior to vulva, both ovaries never end a long way anterior to vulva. Parasite in intestine of neotropical anurans and snakes.

Type species: *Oxyascaris oxyascaris* Travassos, 1920

[The emended diagnosis is based on Baker and Vaucher (1985). Baker and Vaucher (1985) synonymize *Pteroxyascaris* Freitas, 1958 with *Oxyascaris* and transfer *Oxyascaris* to the Cosmocercinae.]

Pteroxyascaris Freitas, 1958

[Baker and Vaucher (1985) synonymize this genus with *Oxyascaris* Travassos, 1920. This genus was considered valid in Chabaud (1978).]

Paraoxyascaris De Oliveira Rodrigues and Sodré Rodrigues, 1971 (Fig. 179)

DIAGNOSIS: Lateral alae absent. *Male*: Posterior end with 6 pairs of papillae, 3 pairs precloacal, 3 pairs postcloacal. Spicules equal. Gubernaculum present. *Female*: Didelphic. Viviparous. Parasitic in amphibians in Brazil.

Type species: *Paraoxyascaris travassosi* De Oliveira Rodrigues and Sodré Rodrigues, 1971

[Listed in Spencer Jones and Gibson (1987).]

Pseudoxyascaris Uchida and Itagaki, 1979 (Fig. 180)

DIAGNOSIS: Lateral alae absent. Mouth with 3 reduced lips. Oesophagus with anterior muscular 'pharyngeal' portion, corpus and elongate bulb. Excretory pore opens near distal end of corpus. *Male*: Pseudosucker absent. Spicules equal, distal tips pointed. Four pairs precloacal (only 3 pairs shown in Fig. 180.2), 6 pairs postcloacal papillae. Tail tapers to sharp point. *Female*:

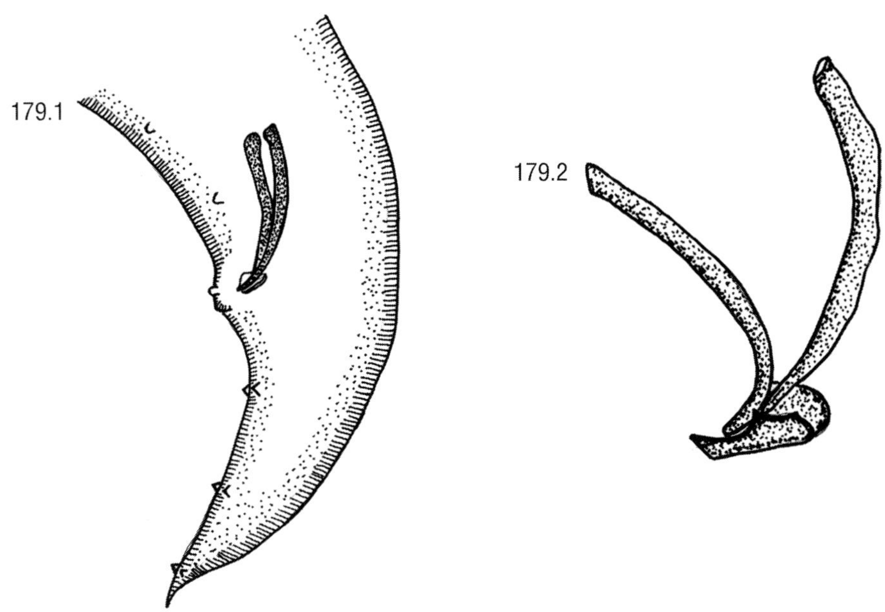

FIG. 179 *Paraoxyascaris* 179.1 Posterior end of male, lateral view; 179.2 Spicules and gubernaculum. (Redrawn after De Oliveira Rodrigues and Sodré Rodrigues, 1971.)

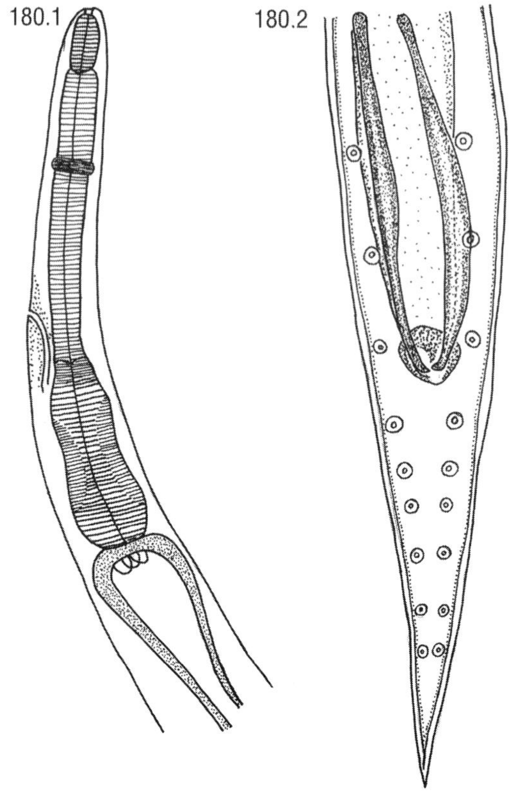

FIG. 180 *Pseudoxyascaris* 180.1 Anterior end, lateral view; 180.2 Posterior end of male, ventral view. (Redrawn after Uchida and Itagaki, 1979.)

Vulva opens in posterior half of body. Parasite in small intestine of amphibians (salamander, *Onchyodactylus japonicus*), in Japan.

Type species: *Pseudoxyascaris japonicus* Uchida and Itagaki, 1979

[The authors place their genus in the family Oxyascarididae. Chabaud (1978) considers this a subfamily of the Kathlaniidae. Listed in Spencer Jones and Gibson (1987).]

Cruziinae (Travassos, 1917 family) Ortlepp, 1924

Cruzia Travassos, 1917

[Baker (1987) synonymizes *Schizobucca* Schuurman-Stekhoven, 1950 with *Cruzia*.]

Schizobucca Schuurman-Stekhoven, 1950

[Baker (1987) synonymizes this genus with *Cruzia* Travassos, 1917. The genus is considered valid by Chabaud (1978).]

Amphibiogoezia Rao, 1979

[Baker (1987) transfers this genus from the Goeziinae in the superfamily Ascaridoidea (see description under Ascaridoidea) to the subfamily Cruziinae, Cosmocercoidea. Listed in Spencer Jones and Gibson (1987).]

Atractidae (Railliet, 1917 subfamily) Travassos, 1919

DIAGNOSIS: Oesophagus distinctly divided at junction of corpus and isthmus.

[Baker (1982b) retains the classification of Chabaud (1978), in which the family Atractidae is separated from the family Cosmocercidae on the basis of their biology, i.e. the genera of the Atractidae are viviparous and the genera of the Cosmocercidae are oviparous. Using 3 main characters, namely monodelphy or didelphy, the cephalic region and the oesophagus, Baker (1982b) identifies four groups within the family Atractidae: (i) monodelphic, presence of 3 lips and modified oesophagus, includes *Schrankiella*; (ii) monodelphic, presence of 3 lips, oesophagus Cosmocercinae-like, includes includes *Schrankiana*, *Labeonema*, *Ibrahimia*; (iii) didelphic, modified cephalic region, oesophagus Cosmocercinae-like, includes *Probstmayria*, *Fitzsimmonsnema*; (iv) monodelphic, 3 lips present, oesophagus modified, includes *Atractis*, *Paratractis*, *Cyrtosomum*, *Monhysterides*, *Proatractis*, *Maracaya*, *Orientatractis*, *Labiduris*, *Nouvelnema*, *Grassenema*, *Rondonia*, *Leiperenia*, *Cobboldina*, *Crossocephalus*.

Adamson and Baccam (1988) revised this group, and the emended diagnosis is based on their study. They split the family on the basis of the morphology of the oesophagus and reproductive system as follows: (i) didelphic genera with an oesophagus which has a long, narrow pharyngeal portion, include *Fitzsimmonsnema* and *Probstmayria*; (ii) monodelphic genera with an oesophagus of variable corpus, include *Atractis*, *Paratractis*, *Cyrtosomum*, *Rondonia*, *Monhysterides*, *Proatractis*, *Cobboldina*, *Orientatractis*, *Labiduris*, *Grassenema*, *Leiperenia*, *Crossocephalus*. Adamson and Baccam (1988) transferred a number of genera to the Cosmocercidae from the family Atractidae *sensu* Chabaud (1978), and these are listed under the family Cosmocercidae.]

Atractis Dujardin, 1945

[Baker (1987) synonymizes *Cyrtosomum* Gedoelst, 1919 and *Pseudatractis* Yamaguti, 1961 with *Atractis*.]

Cyrtosomum Gedoelst, 1919

[Baker (1987) synonymizes this genus with *Atractis* Dujardin, 1945. It was considered a valid genus in Chabaud (1978).]

Schrankianella Freitas, 1959

Type species: *Schrankianella brasili* (Travassos, 1927)

[Baker and Vaucher (1988) synonymize *Schrankianella* with *Schrankiana* Strand, 1942 on the basis of the morphology of the oesophagus. The type species *S. brasili* is therefore transferred to *Schrankiana*, whose species are found in the intestine of amphibians. Chabaud (1978) includes *Schrankianella* as a valid genus.]

Pseudatractis Yamaguti, 1961

[Baker (1987) synonymizes this genus with *Atractis* Dujardin, 1945. Chabaud (1978) lists this genus as a synonym of *Monhysterides* Baylis and Daubney, 1922.]

Klossiella Gonçalves da Costa, 1960

DIAGNOSIS: Mouth with 2 bilobed lips with y-shaped interlabia. Oesophagus divided with both parts of a 'cellular' structure. *Male*: Two spicules. Gubernaculum present. *Female*: Vulva opens near anus. Viviparous. Parasites of fish and reptiles.

Type species: *Klossiella iheringi* (Travassos, Artigas and Pereira, 1928) Gonçalves da Costa, 1960

[Gonçalves da Costa (1960) separates this genus from *Monhysterides* Baylis and Daubney, 1922. Gonçalves da Costa (1961) changes the name to *Klossinemella* as *Klossiella* is preoccupied, see below. Listed in Spencer Jones and Gibson (1987).]

Klossinemella Gonçalves da Costa, 1961

DIAGNOSIS: See *Klossiella*.

Type species: *Klossinemella iheringi* (Travassos, Artigas and Pereira, 1928) Costa, 1961

[*Klossinemella* Gonçalves da Costa, 1961 is the new name for *Klossiella*, which is preoccupied (*Klossiella* Johansen, 1902, Protozoa). Gonçalves da Costa, 1961 places the renamed genus in the family Cobboldinidae Skrjabin, 1948, a family not accepted by Chabaud (1978). As the genus is distinguished from *Monhysterides* and the type genus of Cobboldinidae is placed in the Atractidae in Chabaud (1978), *Klossinemella* is also placed in the Atractidae. Listed in Spencer Jones and Gibson (1987).]

Pseudocyrtosomum Gupta and Johri, 1987 (Fig. 181)

DIAGNOSIS: Lateral alae absent. Mouth with 3 distinct lips each with 2 outer and 2 inner labial papillae. Buccal capsule absent. Oesophagus with anterior muscular 'pharyngeal' region; corpus; isthmus; valvate bulb. *Male*: Posterior end curved, sharply tapering from cloaca forming sharp, pointed tail. Caudal papillae sessile: 2 pairs precloacal; 5 pairs postcloacal. Spicules subequal, similar, alate. Gubernaculum absent. *Female*: Vulva opens anterior or posterior to mid-body. Monodelphic. Oviparous. Tail tapers to sharp point. Parasite in the intestine of the tortoise, *Kachuga dhongoka*.

Type species: *Pseudocyrtosomum lucknowensis* Gupta and Johri, 1987

[The authors place their genus in the subfamily Labidurinae. This subfamily is not included in Chabaud (1978).]

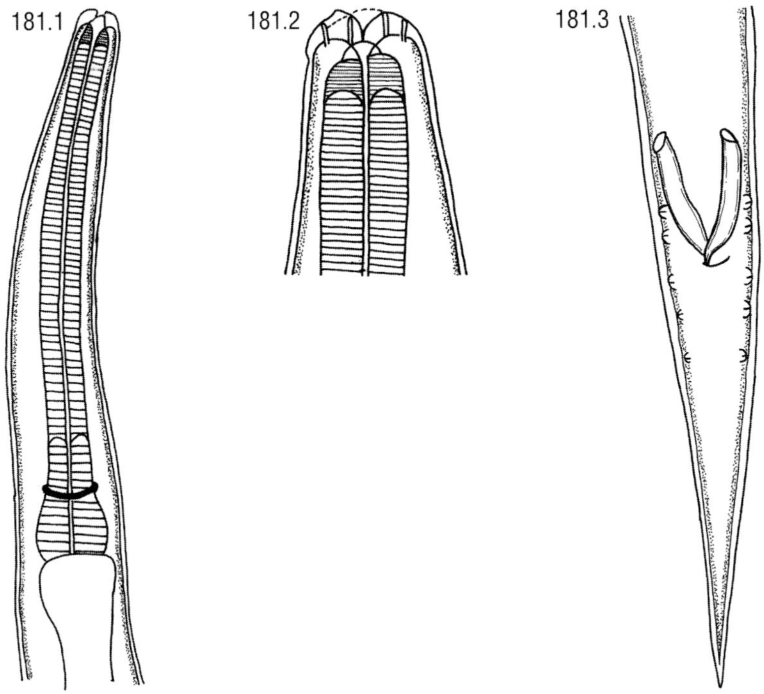

FIG. 181 *Pseudocyrtosomum* 181.1 Male anterior end, lateral view; 181.2 Male cephalic region, lateral view; 181.3 Male posterior end, ventral view. (Redrawn after Gupta and Johri, 1987a.)

Buckleyatractis Khalil and Gibbons, 1988 (Fig. 182)

DIAGNOSIS: Anterior body curved ventrally, with dorsal crest of 3 longitudinal rows of spines, restricted to cervical region. Spines absent on remainder of dorsal surface. Spines of varying shapes and sizes found on other parts of body, arrangement exhibits sexual dimorphism. Ventral surface with 2 large pairs of spines in cervical region. Smaller rows of spines extend to near posterior end. Mouth with 3 lips, margin bilobed. Dorsal lip with 2 papillae, subventral lips with 1 papilla and 1 amphid. Each papilla with 2 projections. Cervical papillae project above body surface with median constriction. Oesophagus divided with distal bulb. Excretory pore opens posterior to oesophagus, surrounded by ornamentation. *Male*: Posterior end with 10 pairs of ventral papillae. Spicules unequal. Gubernaculum present. *Female*: Vulva opens near anus. Viviparous. Parasites in large intestine of freshwater turtles, *Podocnemis unifillis* in Colombia.

Type species: *Buckleyatractis marinkelli* Khalil and Gibbons, 1988

Paratractis Gupta and Naiyer, 1989 (publ. 1990) (Fig. 183)

DIAGNOSIS: Mouth with 3 lips: 1 dorsal, 2 subventral. Buccal capsule short. Oesophagus: elongate corpus; posterior valvate bulb. Excretory pore opens anterior to oesophageal bulb.

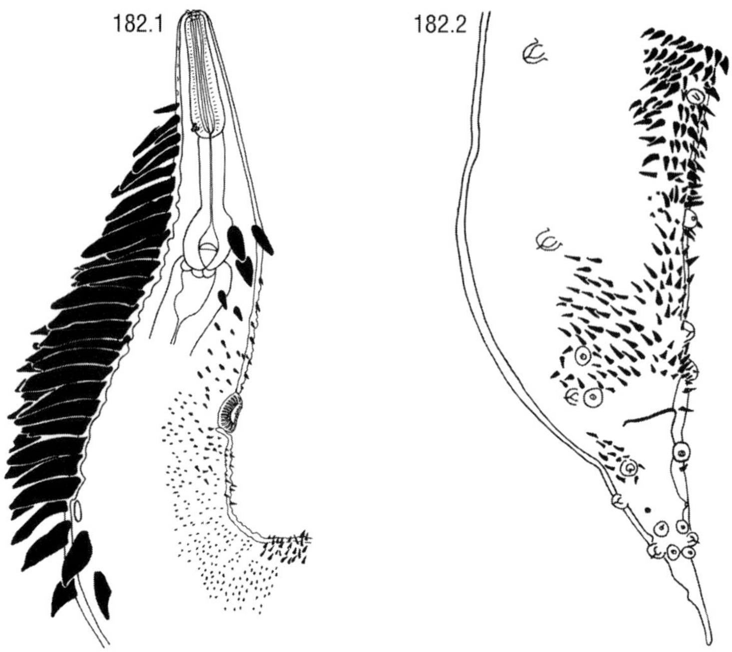

FIG. 182 *Buckleyatractis* 182.1 Anterior end, lateral view; 182.2 Posterior end of male, latero-ventral view. (Redrawn after Khalil and Gibbons, 1988.)

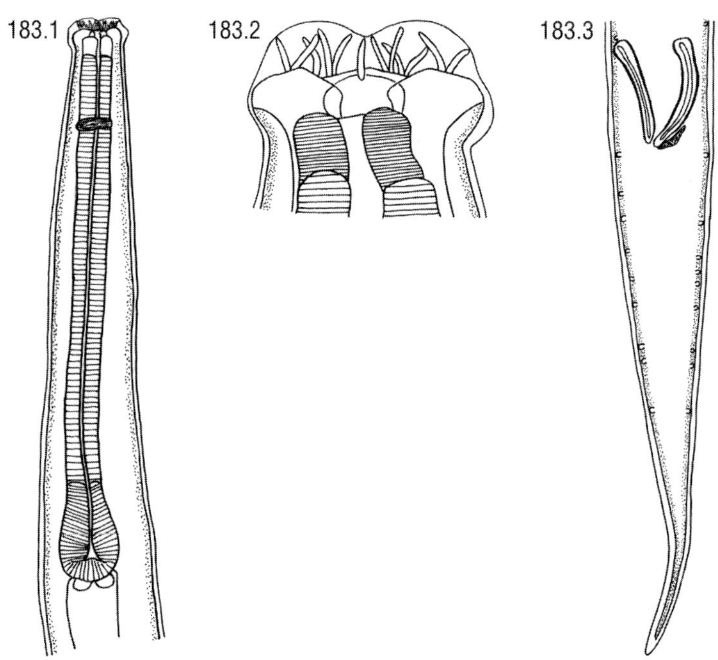

FIG. 183 *Paratractis* 183.1 Anterior end of female, lateral view; 183.2 Cephalic region of female, lateral view; 183.3 Posterior end of male, ventral view. (Gupta and Naiyer, 1989 (publ. 1990).)

Intestinal diverticula absent. *Male*: Body coiled posteriorly. Caudal alae absent. One to two pairs of precloacal papillae; 8–11 pairs of postcloacal papillae. Spicules similar, alate, equal or subequal. Gubernaculum present. Tail elongate, sharply pointed. *Female*: Vulva opens in posterior part of body. Monodelphic. Viviparous. Tail elongate, sharply pointed. Parasite of the intestine of the tortoise *Hardella thurgi* in India.

Type species: *Paratractis indica* Gupta and Naiyer, 1989

[The generic name is preoccupied by *Paratractis* Sarmiento, 1959 in the family Atractidae from the turtle *Podocnemis dumeriliana*.]

Podocnematractis Gibbons, Khalil and Marinkelle, 1995 (Fig. 184)

DIAGNOSIS: Body coiled, transversely striated. Body spines present or absent. Lateral alae well developed. Mouth opening terminal with 6 lips: 2 dorso-lateral; 2 ventro-lateral; 2 lateral. Four cephalic cuticular shields present. Cephalic papillae arranged with 4 filamentous outer labial papillae, 4 submedian cephalic papillae, 2 lateral amphids. Cervical papillae dome-shaped. Oesophagus distinctly divided. Excretory pore large, opens just anterior to posterior bulb of oesophagus into bulbous chamber. *Male*: Nine pairs of dome-shaped caudal papillae

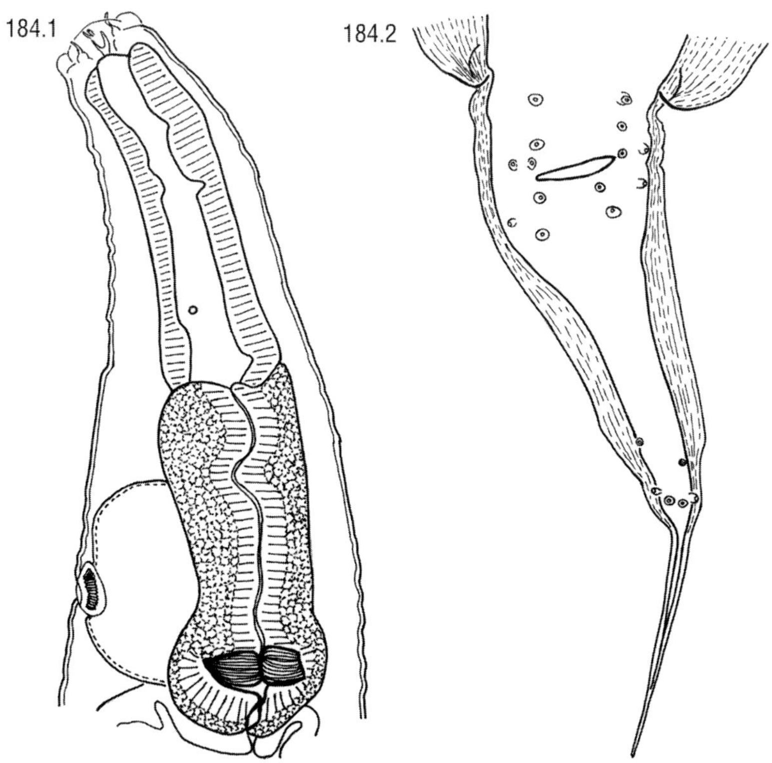

FIG. 184 *Podocnematractis* 184.1 Anterior end of female, lateral view; 184.2 Posterior end of male, ventral view. (Redrawn from Gibbons *et al.*, 1995.)

and pair of phasmids. Spicules slender, unequal, left long and pointed, right short with bifid tip. Gubernaculum present. *Female*: Vulva and anus close together. Viviparous. Adults in large intestine of freshwater turtles, *Podocnemis* spp. in Colombia.

Type species: *Podocnematractis ortleppi* (Thapar, 1925) Gibbons, Khalil and Marinkelle, 1995

Diceronema Gibbons, Knapp and Krecek, 1996 (Figs 185.1, 185.2 and 185.3)

DIAGNOSIS: Cuticle transversely striated. Mouth with rudimentary dorsal and ventral lips. Two pairs of submedian cephalic papillae lateral to mouth. Cup-shaped buccal capsule present. Cheilorhabdion divided dorso-ventrally, overlapping rim of mouth opening. Prorhabdion formed by dorsal and ventral curved semi-lunar plates. Cervical papillae divided into 4 pointed prongs. Oesophagus divided, corpus with distal bulb, isthmus and bulb absent from posterior portion. Nerve ring extends posteriorly from corpus of oesophagus. Excretory pore opens in region of nerve ring. *Male*: Tail coils ventrally. Five pairs of pedunculated papillae asymmetrically arranged; 3 pairs sessile papillae near distal end; additional pair adjacent to distal pair, possibly phasmids. Ventral cuticular flap with 2 small projections at distal tip. Spicules unequal, right spicule with barb at distal tip. Slender caudal alae present. *Female*: Vulva opens just anterior to anus. Viviparous. Adults parasites in stomach of black rhinoceros, *Diceros bicornis bicornis*, in South Africa.

Type species: *Diceronema versterae* Gibbons, Knapp and Krecek, 1996

Paraorientatractis Gibbons, Khalil and Marinkelle, 1997 (Figs 185.4, 185.5 and 185.6)

DIAGNOSIS: Body coiled with large semiannules on dorsal surface. Lateral alae well developed, longitudinally ridged. Caudal appendix present. Mouth with 4 lips, each lip with single papilla, pair of recurved spines, single median spine near distal margin. Two additional smaller spines posterior to large lateral amphids. Cervical papillae small, rounded. Oesophagus divided. Excretory pore opens into elongate chamber. *Male*: Eight pairs of caudal papillae, 4 precloacal, 4 postcloacal. Phasmids open at level of posterior end of lateral alae. Spicules unequal, pointed distal tips. Gubernaculums present. *Female*: Vulva opens close to anus. Viviparous. Parasites in large intestine of freshwater turtles.

Type species: *Paraorientatractis semiannulata* Gibbons, Khalil and Marinkelle, 1997

Rhinoclemmysnema Gibbons and Platt, 2006 (Fig. 186)

DIAGNOSIS: Mouth with 6 lips, subdorsal and subventral lips divided. Four submedian cephalic papillae, 2 lateral cephalic papillae associated with amphids. Oesophagus distinctly divided, with corpus forming more than one-quarter of total length, isthmus present, distal bulb with lobes of valve striated. *Male*: Seven pairs of papillae on posterior end. Spicules unequal. Gubernaculum barbed at distal end. *Female*: Vulva opens near anus. Monodelphic. Viviparous. Parasite in small and large intestine of *Rhinoclemmys pulcherrima* in Costa Rica.

Type species: *Rhinoclemmysnema multilabiatum* Gibbons and Platt, 2006

185.1

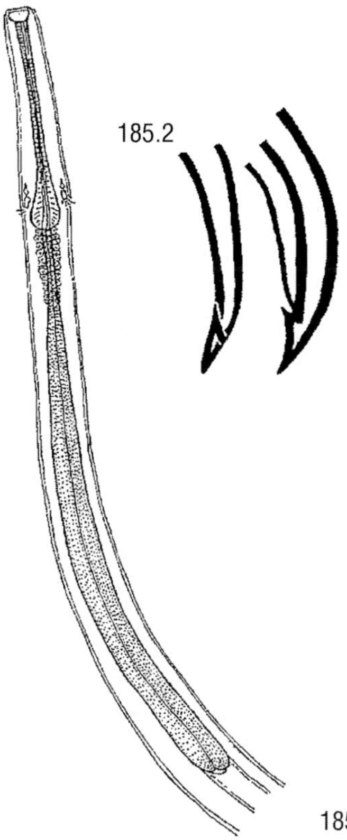

185.2

185.3

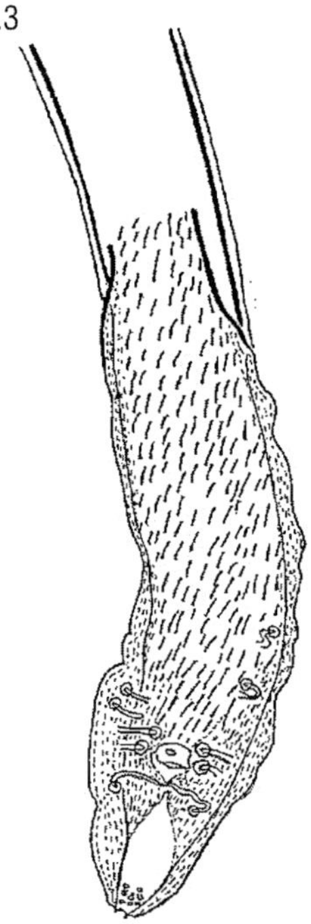

185.4

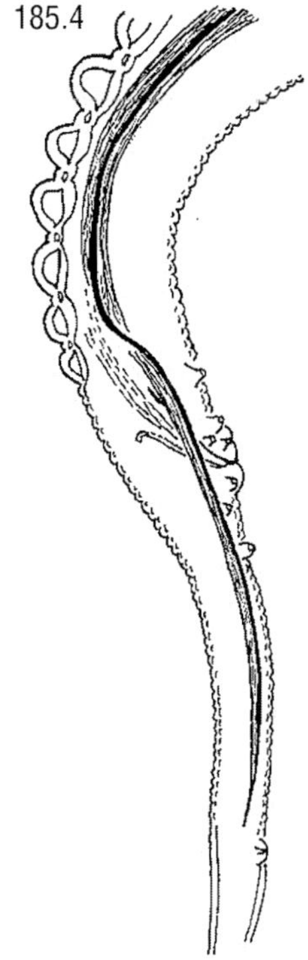

185.5

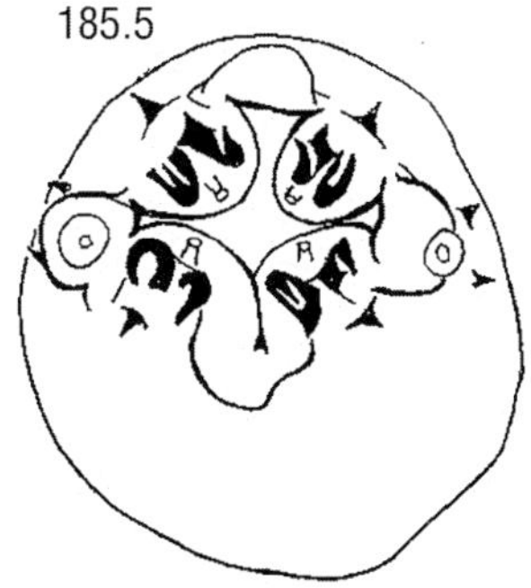

185.6

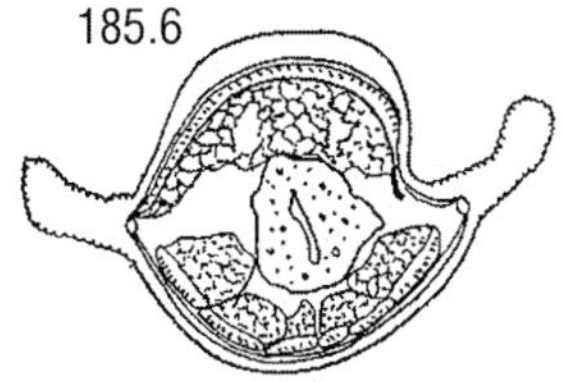

Figs 185.1, 185.2 and 185.3 *Diceronema* 185.1 Anterior end of female, dorso-ventral view; 185.2 Distal tip of right spicule; 185.3 Posterior end of male, ventral view. (Redrawn after Gibbons, Knapp and Krecek, 1996.)

Figs 185.4, 185.5 and 185.6 *Paraorientatractis* 185.4 Male posterior end, lateral view (caudal appendix not shown); 185.5 Cephalic region of male, *en face* view; 185.6 Cross-section of male, mid-body region arranged dorso-ventrally. (Redrawn after Gibbons, Khalil and Marinkelle, 1997.)

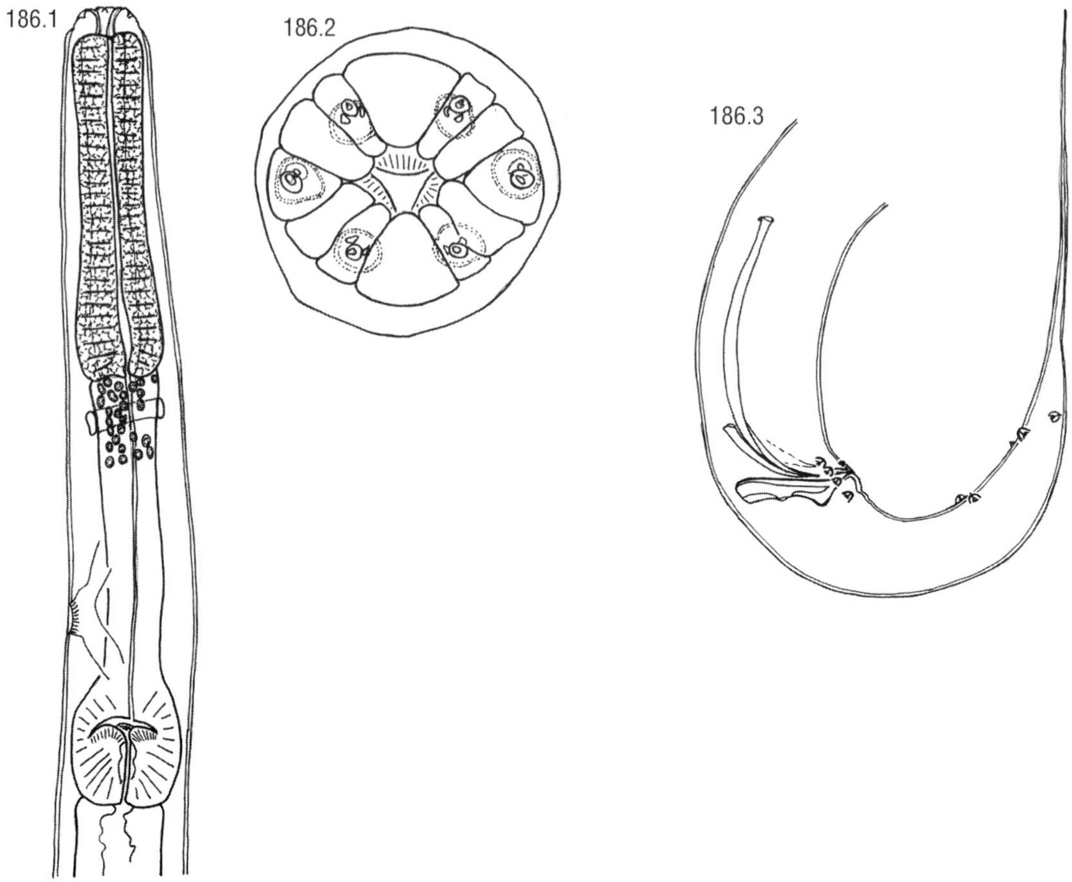

FIG. 186 *Rhinoclemmysnema* 186.1 Anterior end of female, dorso-ventral view; 186.2 *En face* view of male; 186.3 Posterior end of male, lateral view. (Redrawn after Gibbons and Platt, 2006.)

SEURATOIDEA

Seuratidae (Hall, 1916) Railliet, 1906

Seuratinae Hall, 1916

Pseudospinura Wang, Zhao, Wang, Zhang and Chang, 1979 (Fig. 187)

DIAGNOSIS: Translation of description not available. Parasite of *Ninox s. scutulata* in Fuzhou, China.

Type species: *Pseudospinura mackinus* Wang, Zhao, Wang, Zhang and Chang, 1979

[The authors place their new genus in Skrjabinuridae. This family is not included by Chabaud (1978) in his key. The genus *Skrjabinura* is placed in the Seuratinae, and therefore *Pseudospinura* has been included in the same subfamily. Listed in Spencer Jones and Gibson (1987).]

Skrjabinelaziinae Chabaud, Campana-Rouget and Brygoo, 1959

Skrjabinelazia Sypliaxov, 1930

DIAGNOSIS: Mouth with delicate small leaf crown, tiny elements (except *S. hoffmanni*). *Male*: Caudal alae absent. Spicules present or absent. Gubernaculum present, well developed. *Female*: Vulva opens in anterior part of body, at level of oesophagus. Viviparous or oviparous (eggs embryonated). Parasites of intestine of reptiles.

Type species: *Skrjabinelazia taurica* Sypliaxov, 1930

[The emended diagnosis is based on Roca and Garcia-Adell (1988).]

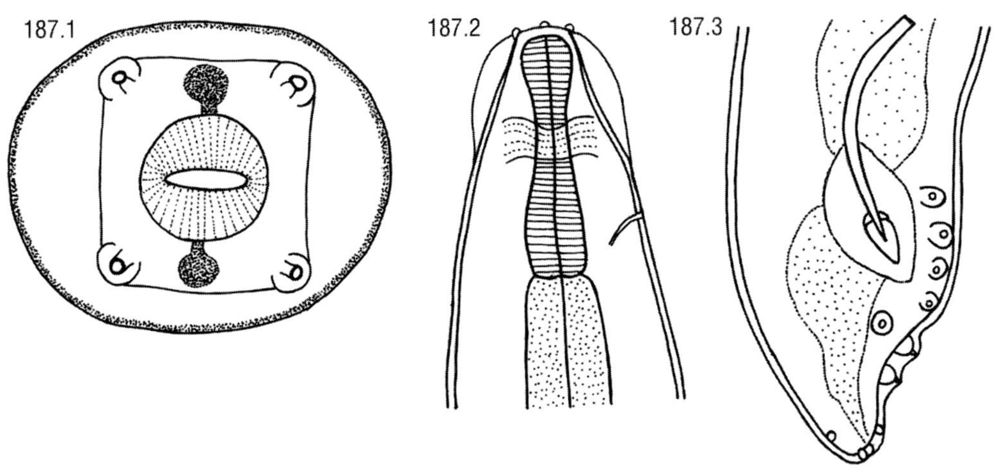

FIG. 187 *Pseudospinura* 187.1 Cephalic region, *en face* view; 187.2 Anterior end, lateral view; 187.3 Posterior end of male, latero-ventral view. (Redrawn after Wang *et al.*, 1979.)

Echinonematinae Inglis, 1967

Seurechina Chabaud, Seureau, Beveridge, Bain and Durette-Desset, 1980 (Fig. 188)

DIAGNOSIS: Body, apart from cephalic region, covered in spines, aligned transversely in each annule. Head region with oval plateau aligned laterally. Cephalic sensory organs: 4 submedian cephalic papillae, 2 lateral amphids. Mouth opens into chitinous capsule, enlarged dorso-ventrally. Oesophagus claviform. Four sublateral plates under cephalic hypodermis, appear to be dilations from lateral fields, possibly assist fixation of cervical hooks. *Male*: Spicules equal. Gubernaculum slender. Anterior fifth of body with 50 rings of large hooks. Remainder of somatic spines small, end in 2 lateral bands. Anterior lip of cloaca swollen, pair of papillae present. Caudal alae present. Caudal papillae: 2 pairs precloacal; 3 pairs adcloacal; cluster on tail tip. *Female*: Anterior fifth of body with 50 rings of large hooks. Remainder of body spines small except anterior lip of vulva with large spines. Vulva opens posterior to mid-body. Tail pointed. Parasites of Dasyuridae (*Dasyurus hallucatus*) in West Australia.

Type species: *Seurechina chaneeti* Chabaud, Seureau, Beveridge, Bain and Durette-Desset, 1980

[Listed in Spencer Jones and Gibson (1987).]

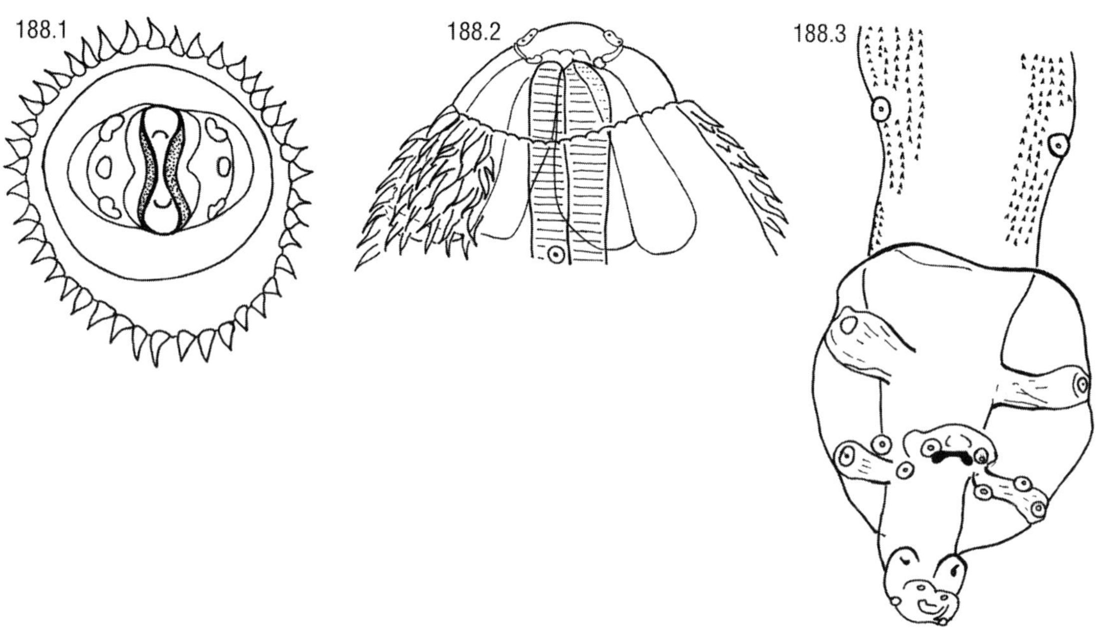

FIG. 188 *Seurechina* 188.1 Female cephalic region, *en face* view; 188.2 Female cephalic region, lateral view showing four sublateral plates; 188.3 Posterior end of male, ventral view. (Redrawn after Chabaud *et al.*, 1980.)

Inglechina Chabaud, Seureau, Beveridge, Bain and Durette-Desset, 1980 (Fig. 189)

DIAGNOSIS: Cephalic region with 3 rows of hooks. Cervical region not inflated. Spines simple, similar along body. Mouth opening triangular. *Male*: Spines continue to just anterior to caudal papillae, gradually increase in size. *Female*: Vulva opens anterior to mid-body. Parasite in large intestine of *Sminthopsis crassicaudata* in South Australia.

Type species: *Inglechina australis* (Inglis and Mawson, 1967) Chabaud, Seureau, Beveridge, Bain and Durette-Desset, 1980

[Listed in Spencer Jones and Gibson (1987).]

Linstowinema Smales, 1997 (Fig. 190)

DIAGNOSIS: Cephalic bulb with 3 transverse rows of 14–16 large hooks. Mouth triangular. Cephalic sensory organs: inner circle of papillae present on mouth margin; 4 pairs of submedian cephalic papillae; 2 lateral amphids. Anterior cervical region 2–11 rows of small hooks, posterior cervical region 8–18 rows of 14–16 hooks. Body with numerous transverse rows of hooks, increase in number mid-body, decrease posteriorly. Oesophagus clavate. Nerve ring at level of cephalic bulb. Cervical papillae located at first row of cervical hooks. *Male*: Body hooks finish anterior to cloaca (ventral surface), at anterior margin of posterior quarter (dorsal surface). Spicules long, equal, similar. Gubernaculum small. *Female*: Body hooks extend to tail tip. Vulva opens in mid-region. Monodelphic. Ovejector directed anteriorly. Parasites in gastrointestinal tract of dasyurid and paramelid marsupials in Australia.

Type species: *Linstowinema cinctum* (Linstow, 1898) Smales, 1997

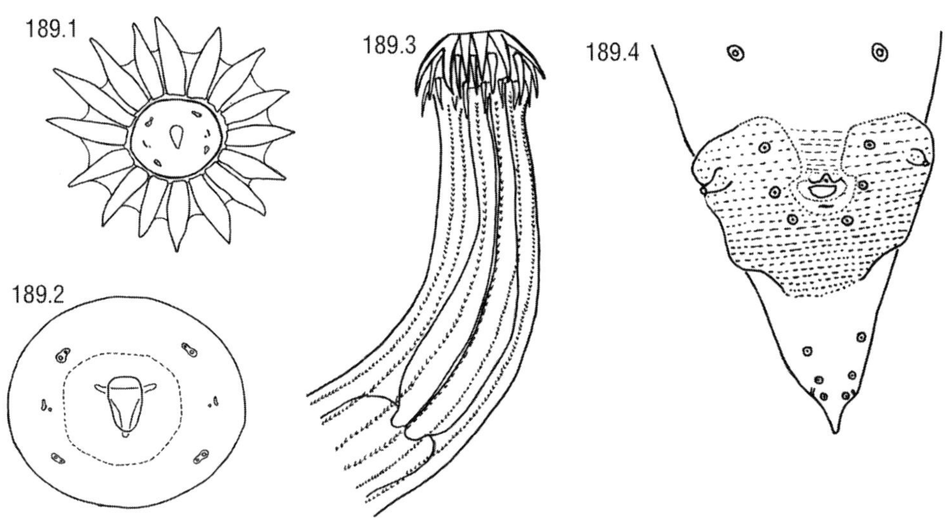

Fig. 189 *Inglechina* 189.1 Cephalic region, *en face* view; 189.2 Cephalic region, higher magnification of *en face* view; 189.3 Anterior end; 189.4 Posterior end of male, ventral view. (Redrawn after Inglis and Mawson, 1967.)

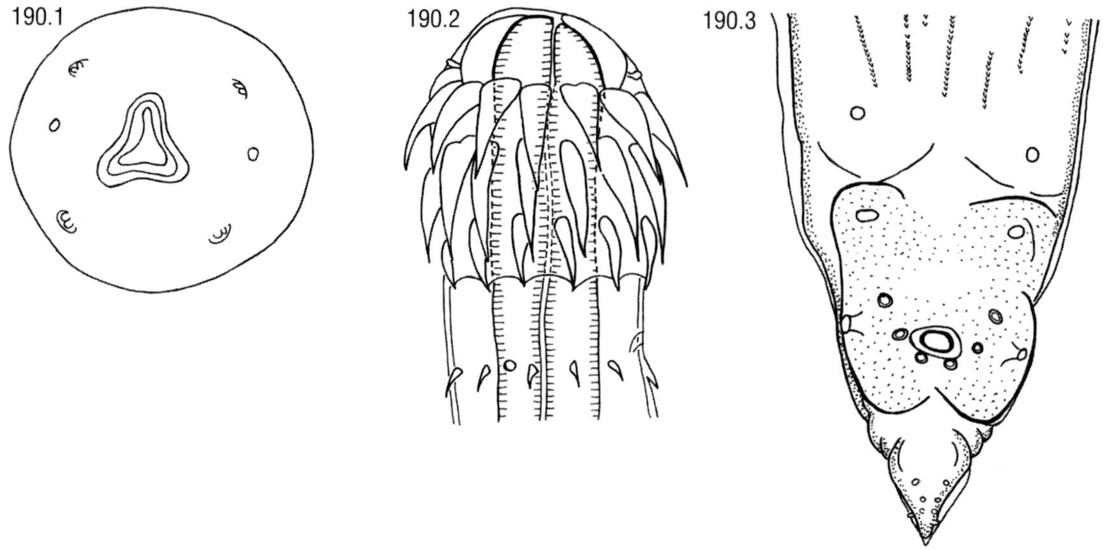

FIG. 190 *Linstowinema* 190.1 Cephalic region, *en face* view; 190.2 Cephalic region, lateral view; 190.3 Posterior end of male, ventral view. (Redrawn after Smales, 1997b.)

[Smales (1997b) revises the family Echinonematinae and reveals that the generic name *Echinonema* Linstow, 1898 is preoccupied by a genus of sponges named by Carter, 1881. The name *Linstowinema* is proposed to replace it, and the emended diagnosis is based on Smales (1997b).]

Chabaudechina Smales, 1999 (Fig. 191)

DIAGNOSIS: Mouth opening triangular, simple without lips. Four pairs submedian cephalic papillae, 2 lateral amphids. Cephalic bulb with 5 rows of hooks, decrease in size posteriorly. Body with numerous rows of hooks and spines, increasing in size to rows 24–28 in cervical region, then decreasing posteriorly. Oesophagus short, claviform. Excretory pore and cervical papillae at same level posterior to nerve ring. *Male*: Spines on ventral surface become more scale-like posteriorly, end anterior to cloaca. Spicules equal, similar, alae absent. Gubernaculum absent. *Female*: Spines extend over whole body. Vulva opens at mid-body. Vagina directed anteriorly. Monodelphic. Parasites in stomach of Australian dasyurid marsupials (*Sminthopsis* spp.).

Type species: *Chabaudechina presidentei* Smales, 1999

Bainechina Smales, 1999 (Fig. 192)

DIAGNOSIS: Mouth opening triangular, simple without lips. Four double submedian cephalic papillae, 1 pair lateral amphids. Cephalic region unarmed. Body covered in numerous rows of hooks or spines. Hooks decrease in size posteriorly forming spines. Oesophagus short, claviform. Four pairs of laminae surround proximal end. Cervical papillae posterior to nerve ring. *Male*: Body spines end anterior to cloaca. Spicules equal, simple, alae absent. *Female*: Spines extend over entire body. Vulva opens in mid-body. Vagina directed posteriorly. Papillae surround body

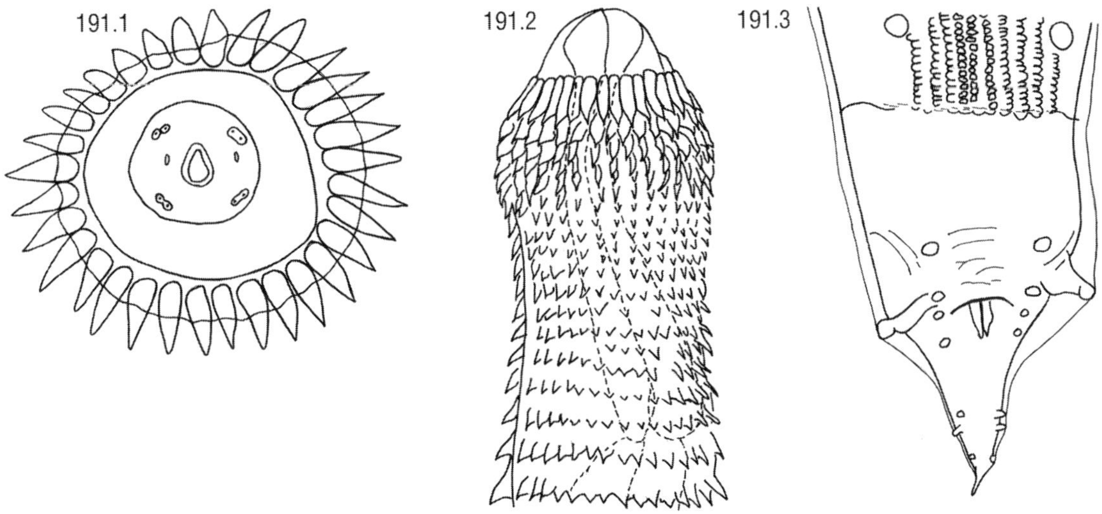

FIG. 191 *Chabaudechina* 191.1 Cephalic region, *en face* view; 191.2 Anterior end, lateral view; 191.3 Posterior end of male, ventral view. (Redrawn after Smales, 1999a.)

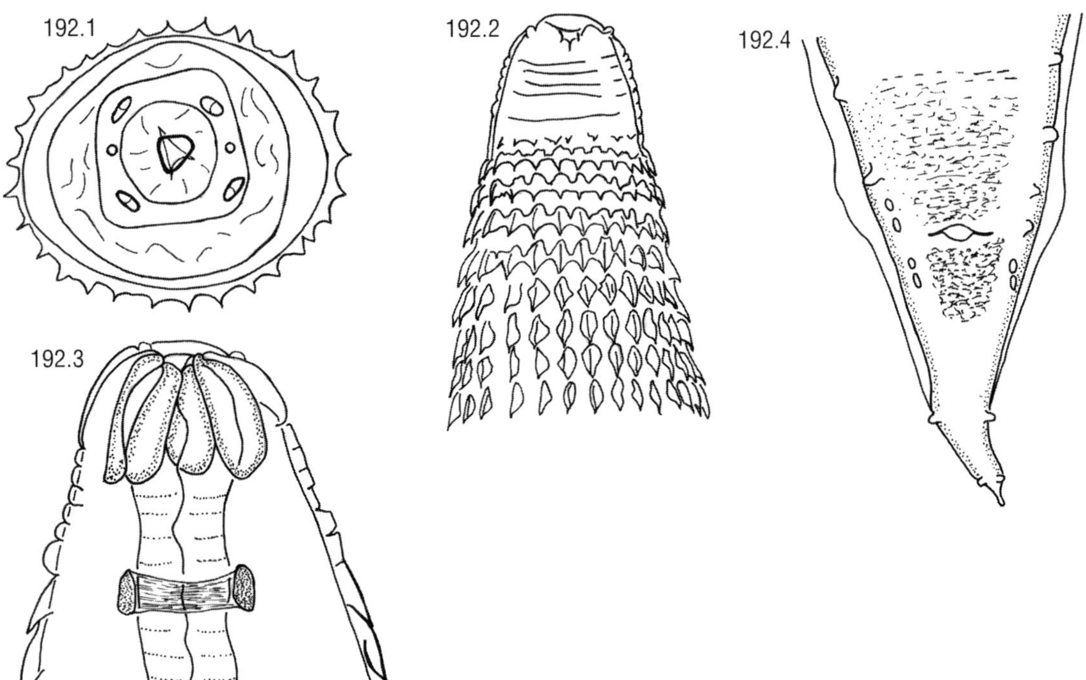

FIG. 192 *Bainechina* 192.1 Cephalic region, *en face* view; 192.2 Anterior end, lateral view; 192.3 Anterior end, optical section showing laminae, lateral view; 192.4 Posterior end of male, ventral view. (Redrawn after Smales, 1999b.)

at level of vulva. Monodelphic. Parasite in stomach and small intestine of Australian dasyurid marsupials (*Planigale* spp.).

Type species: *Bainechina rossiae* Smales, 1999

Durettechina Smales, 2000 (Fig. 193)

DIAGNOSIS: Cuticle with fine annulations. Mouth opening triangular, simple, without lips. Four double submedian cephalic papillae, 1 pair lateral amphids. Cephalic bulb armed with 4 rows of 26 hooks decreasing in size posteriorly. Cervical region with 1 row of spines followed by 15 rows of hooks, remainder of body covered in rows of spines. Oesophagus short, claviform. Nerve ring at level of cephalic bulb. Cervical papillae at level of fourth row of hooks on cephalic bulb. *Male*: Hooks or spines extend over two-thirds of dorsal side of body and to anterior of cloaca on ventral side, decrease in size posteriorly, become scale-like on ventral surface. Spicules long, equal, similar. *Female*: Posterior to cervical region, spines extend over entire body. Vulva opens anterior to mid-body. Vagina directed anteriorly. Monodelphic. Parasites in intestine of Australian dasyurid marsupials (*Antechinus* spp.).

Type species: *Durettechina beveridgei* Smales, 2000

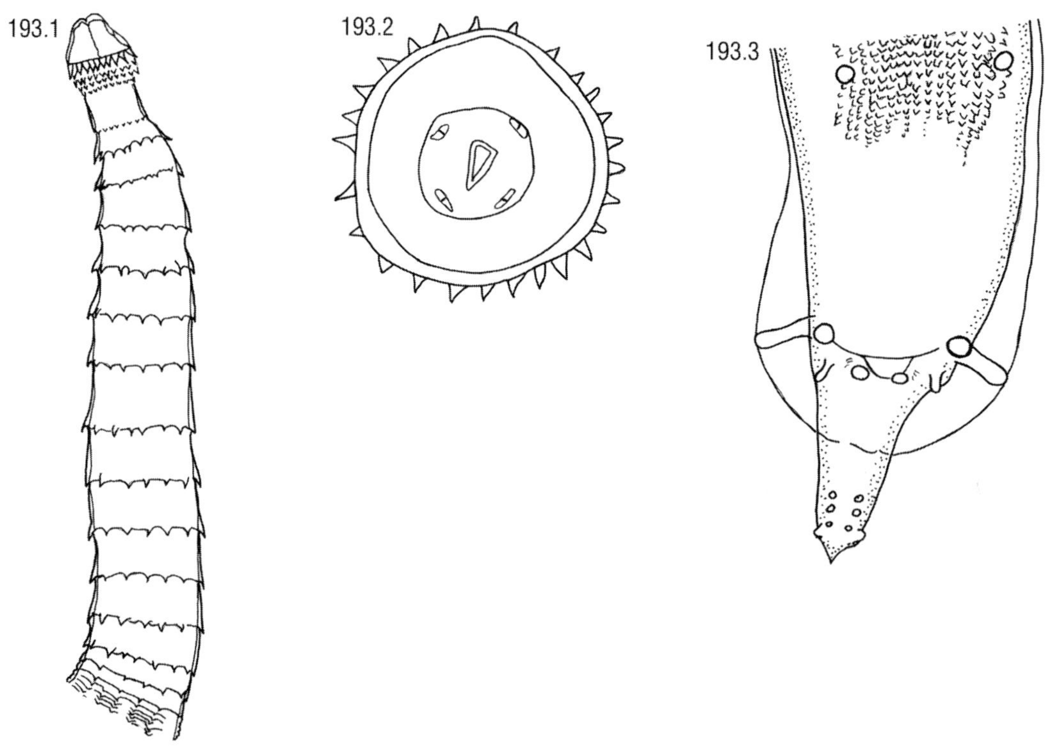

FIG. 193 *Durettechina* 193.1 Anterior end, dorsal view; 193.2 Cephalic region, *en face* view; 193.3 Posterior end of male, ventral view. (Redrawn after Smales, 2000.)

Schneidernematidae Freitas, 1956

Inglisonematinac Mawson, 1968

[Baker (1984), following his studies of the cephalic region, oesophagus and female reproductive system, upholds the proposal by Schmidt and Kuntz (1971) to promote Inglisonematinae to family rank. Baker (1984) proposes that the family should be included in the Heterakoidea and not the Seuratoidea, as in Chabaud (1978).]

Quimperiidae (Gendre, 1928 subfamily) Baylis, 1928

Genera not assigned to a subfamily

***Neoomeia* Yin and Zhang, 1984 (Fig. 194)**

DIAGNOSIS: Mouth opening triangular with 6 reduced lips, 2 dorsal, 2 subventral teeth at base of lips. Four submedian pairs, 1 dorsal pair, 1 ventral pair cephalic papillae; 2 lateral amphids. Buccal capsule present. Two dorsal, 2 subventral teeth attached to anterior margin of oesophagus, protrude into buccal capsule. Oesophagus divided, anterior muscular, posterior glandular

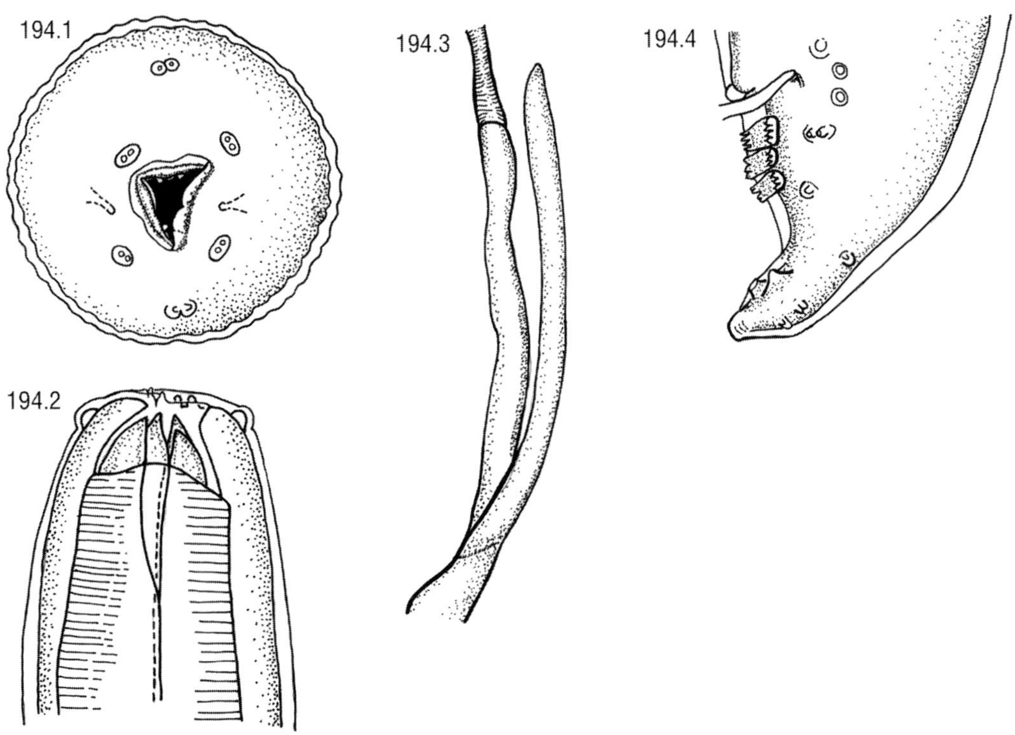

FIG. 194 *Neoomeia* 194.1 Cephalic region, *en face* view; 194.2 Anterior end showing teeth protruding into buccal capsule; 194.3 Intestinal caecum; 194.4 Posterior end of male showing plectanes. (Redrawn after Yin and Zhang, 1984.)

regions. Intestinal caecum present. *Male*: 58–62 pairs of caudal papillae. Caudal alae present. Three plectanes postcloacal with denticulate lateral margin. Spicules equal, similar. Gubernaculum present. *Female*: Vulva opens in anterior half of body. Tail pointed with alae. Parasite in stomach of *Sphyrna lewini*, hammerhead shark, Guangdong and Fujiang Provinces, China.

Type species: *Neoomeia sphyrnae* Yin and Zhang, 1984

[The type species has some similarities to species of *Pulchrascaris*, Ascaridoidea, as described by Deardorff (1987) and Bruce and Cannon (1990), who record *Pulchrascaris chiloscylli* from the same host, *Sphyrna lewini*. Further studies may reveal these species as synonyms. Listed in Spencer Jones and Gibson (1987).]

Quimperiinae Gendre, 1928

Pseudohaplonema Wang, Zhao and Chen, 1978 (Fig. 195)

DIAGNOSIS: Cephalic region with 6 cephalic sensory papillae. Buccal capsule absent. Oesophagus divided, anterior muscular and posterior glandular regions. *Male*: Spicules equal, similar. Gubernaculum present. Four pairs precloacal, 7 pairs postcloacal papillae: 3 pairs just posterior to cloaca, 4 pairs lateral. *Female*: Vulva opens in middle of body. Parasite in intestine of turtle, *Chinemys reevesii* (= *Geochemys reevesii*) in China.

Type species: *Pseudohaplonema geoclemydis* Wang, Zhao and Chen, 1978

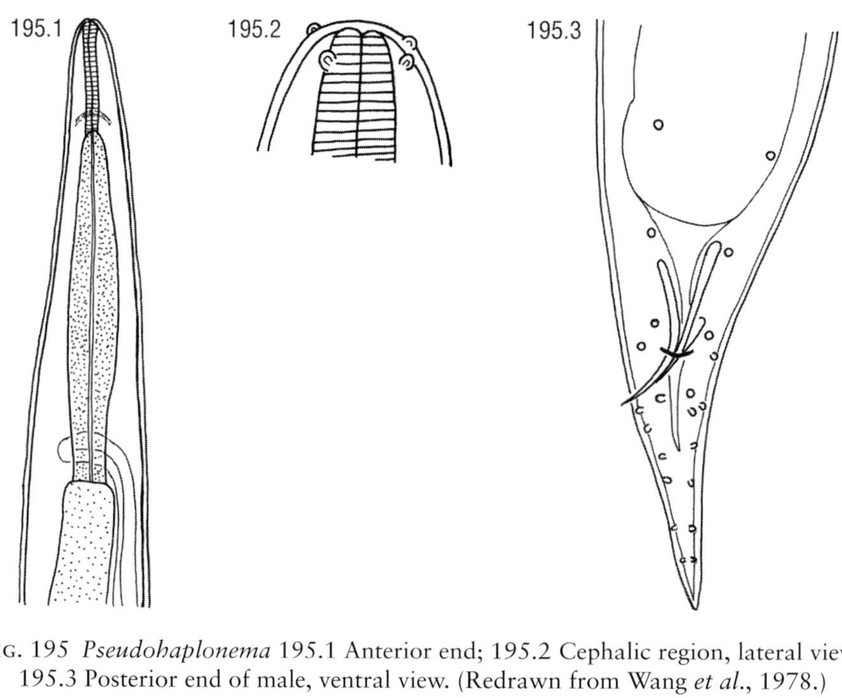

Fig. 195 *Pseudohaplonema* 195.1 Anterior end; 195.2 Cephalic region, lateral view; 195.3 Posterior end of male, ventral view. (Redrawn from Wang *et al*., 1978.)

[The authors consider the genus is closely related to *Haplonema* Ward and Magath, 1917 and, consequently, *Pseudohaplonema* is placed in the subfamily Quimperiinae. Listed in Spencer Jones and Gibson (1987). Translation by Professor Jian Heng.]

Neoquimperia Wang, Zhao, Wang, Zhang and Chang, 1979 (Fig. 196)

DIAGNOSIS: Translation of description not available. Parasite in *Crossostoma davidi* in Fujian, China.

Type species: *Neoquimperia robustum* Wang, Zhao, Wang, Zhang and Chang, 1979

[Moravec *et al.* (2002a) consider this genus a synonym of *Ezonema* Boyce, 1971. Listed in Spencer Jones and Gibson (1987).]

Haplodidentus Naidu and Thakare, 1981 (Fig. 197)

DIAGNOSIS: Cuticle with transverse striations. Mouth without lips. Pair of lateral cephalic papillae and 2 amphids. Cephalic region with pair of didents in dorso-ventral fields. Oesophagus with anterior muscular, posterior glandular regions. Nerve ring surrounds muscular region. *Male*: Unknown. *Female*: Vulva opens on cuticular disc with radiating striations in anterior third of body. Eggs 3–6 µm × 3–4 µm. Adults parasitic in intestine of *Saccobranchus fossilis* and *Mastacembelus*

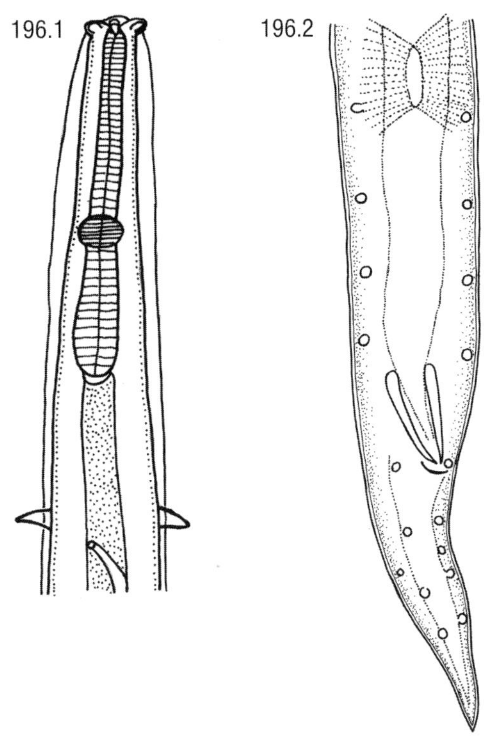

FIG. 196 *Neoquimperia* 196.1 Anterior end, ventral view; 196.2 Posterior end of male, ventral view. (Redrawn after Wang *et al.*, 1979.)

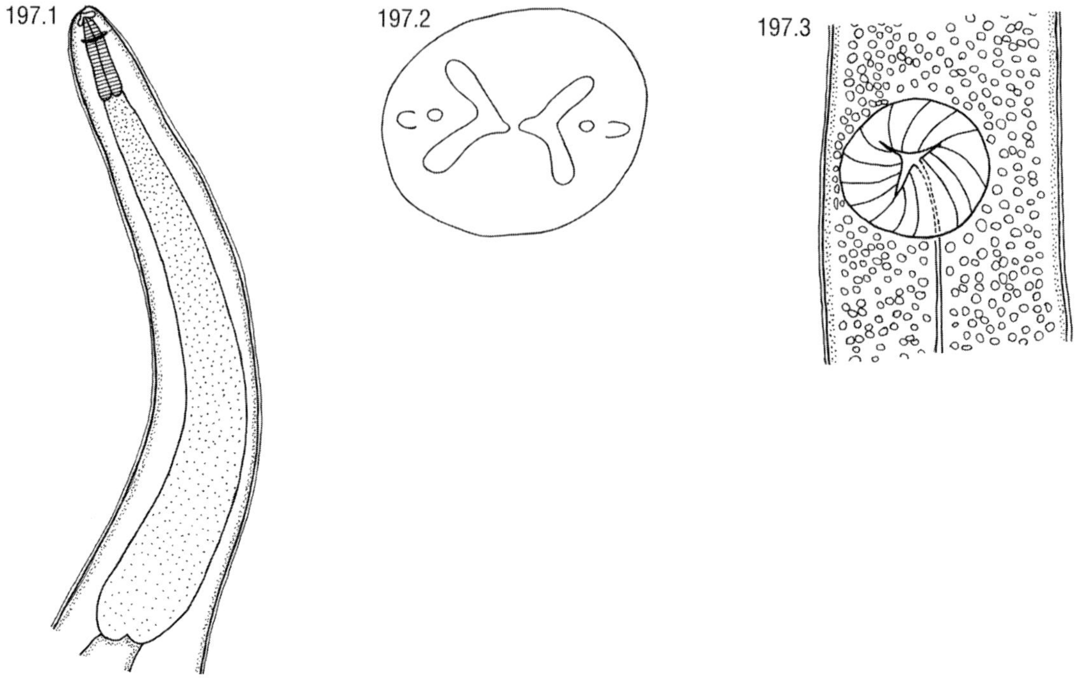

FIG. 197 *Haplodidentus* 197.1 Anterior end of female, dorsal view; 197.2 Cephalic region, *en face* view; 197.3 Vulvar region, ventral view. (Redrawn after Naidu and Thakare, 1981.)

armatus in River Vainganga, India. Immature stages in mesentery, subcutaneous tissue and intestine of *S. fossilis*, *M. armatus*, *Clarias batrachus* and *Rana tigrina* in Tumsar and Nagpur, India.

Type species: *Haplodidentus indicus* Naidu and Thakare, 1981

[The authors place their genus in the family Haplonematidae Sudarikov and Ryzhikov, 1952. They distinguish the genus from genera assigned by Chabaud (1978) to the family Quimperiidae and subfamily Quimperiinae. For these reasons the genus is assigned to the same subfamily, namely Quimperiinae.]

Paraseuratoides Wang, 1984 (Fig. 198)

DIAGNOSIS: Cuticle smooth. Cephalic region bluntly rounded. Mouth encircled by 6 cephalic sensory organs. Buccal capsule small. Teeth absent. Oesophagus not distinctly divided. Cervical papillae distinct, located towards distal end of oesophagus. *Male*: Spicules unequal. Gubernaculum present. Narrow caudal alae. Four pairs of precloacal, 7 pairs of postcloacal pedunculate papillae. Tail narrows to terminal appendix. *Female*: Vulva opens in posterior half of body. Uteri opposed. Tail tapers to terminal appendix. Parasite in intestine of freshwater fish in Fujian province, China.

Type species: *Paraseuratoides acuminicauda* Wang, 1984

[The author places the genus in the family Haplonematidae Sudarikov and Ryzhikov, 1952 and considers it closely related to *Paraseuratum* Johnston and Mawson, 1940. Chabaud (1978)

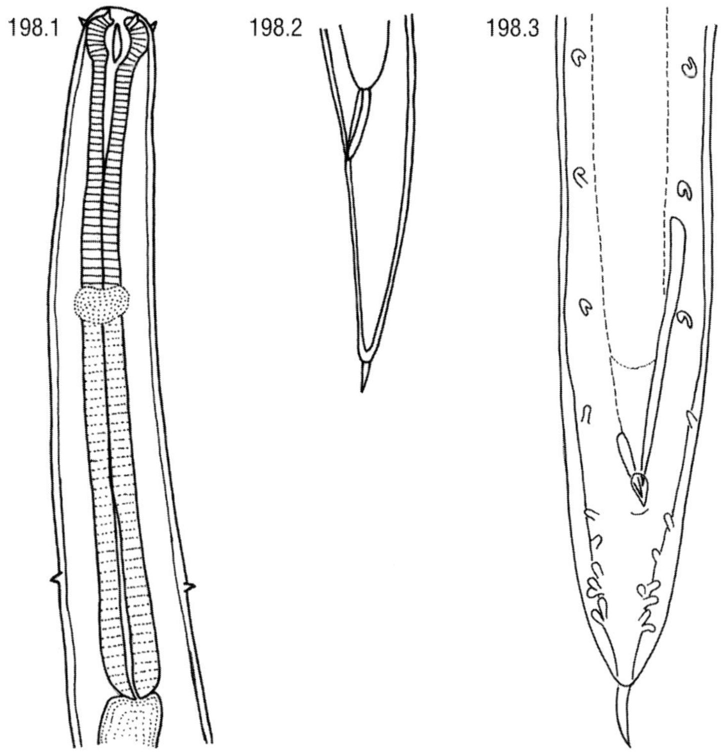

FIG. 198 *Paraseuratoides* 198.1 Anterior end of male, ventral view; 198.2 Posterior end of female, lateral view; 198.3 Posterior end of male, ventral view. (Redrawn after Wang, 1984.)

places *Paraseuratum* in the subfamily Quimperiinae and, consequently, Wang's genus is placed in the same subfamily. Listed in Spencer Jones and Gibson (1987).]

Desmognathinema Baker, Goater and Esch, 1987 (Fig. 199)

DIAGNOSIS: Cephalic vesicle slightly developed or absent. Cervical alae absent. Cuticle with irregular annulations and prominent transverse striations, particularly near anterior end. Mouth opening triangular. Buccal capsule thin-walled, reduced. Oesophageal teeth, small, present or absent. Oesophagus, long, proximal muscular 'pharyngeal' region, remainder divided into muscular and glandular regions. *Male*: Caudal papillae ventral or subventral. Precloacal region with oblique muscle bands. Pseudosucker absent. Spicules equal, alate. Gubernaculum present. *Female*: Vulva opens in posterior region of body. Tail conical, sharply pointed. Parasite in small intestine of amphibians, including plethodontid salamanders (black-bellied salamander, *Desmognathus quadramaculatus*) in south-eastern USA.

Type species: *Desmognathinema nantahalaensis* Baker, Goater and Esch, 1987

[Diagnosis based on Baker *et al.* (1987) and Moravec and Sey (1990).]

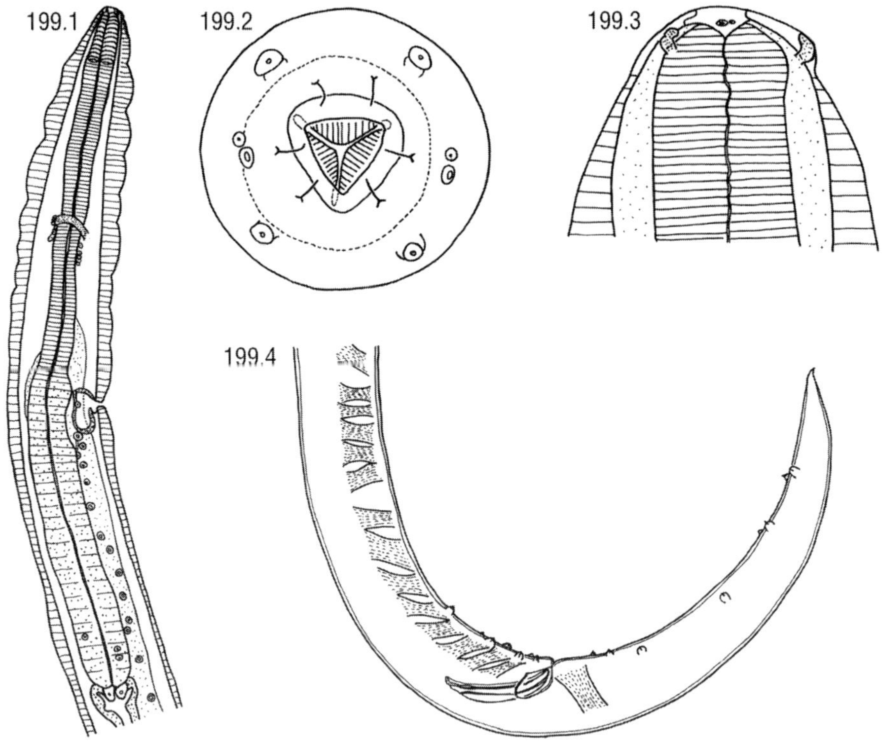

FIG. 199 *Desmognathinema* 199.1 Anterior end of male, lateral view; 199.2 Male cephalic region, *en face* view; 199.3 Male cephalic region, lateral view; 199.4 Posterior end of male, lateral view. (Redrawn after Baker *et al.*, 1987.)

Touzeta Petter, 1987 (Fig. 200)

DIAGNOSIS: Mouth opening round without lips. Buccal cavity without sclerotized walls or teeth. Oesophageal funnel present (oesophagostome). Oesophagus short with spherical swelling at proximal end, divided in mid-region, 2 parts with different diameters. *Male*: Pseudosucker and oblique muscular bands present anterior to cloaca. Spicules equal. Gubernaculum small. *Female*: Vulva opens posterior to mid-body. Uteri opposed. Eggs in uterus not embryonated. Parasites in South American freshwater fish.

Type species: *Touzeta ecuadoris* Petter, 1987

Wakubitinema Hasegawa, 1988 (Fig. 201)

DIAGNOSIS: Cervical and lateral alae absent. Mouth opening triangular. Cephalic papillae: pedunculate; amphids prominent. Buccal capsule present, poorly developed. 'Pharyngeal' portion of oesophagus short with 3 teeth. Remainder of oesophagus divided into distinct anterior muscular and posterior glandular regions. Intestinal caecum absent. Nerve ring surrounds middle of muscular oesophagus. Excretory pore opens posterior to oesophagus, system symmetrical. Cervical papillae very small. *Male*: Spicules alate. Gubernaculum present, bifurcate. Papillae on posterior end sessile, precloacal papilla unpaired. Oblique muscle bands present

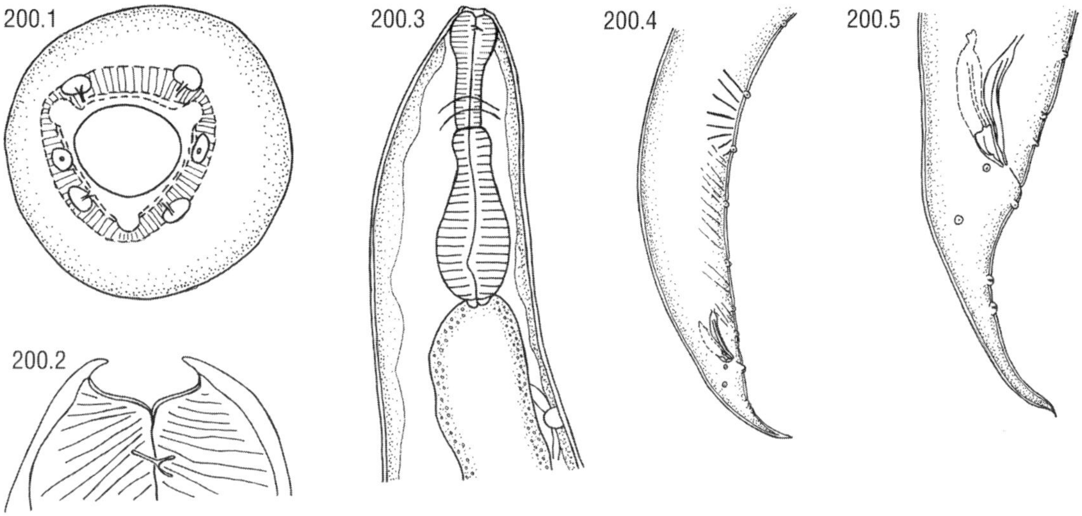

FIG. 200 *Touzeta* 200.1 Cephalic region, *en face* view; 200.2 Anterior showing buccal capsule and oesophageal funnel; 200.3 Anterior end, lateral view; 200.4 Posterior end of male, lateral view; 200.5 Posterior extremity of male, lateral view. (Redrawn after Petter, 1987.)

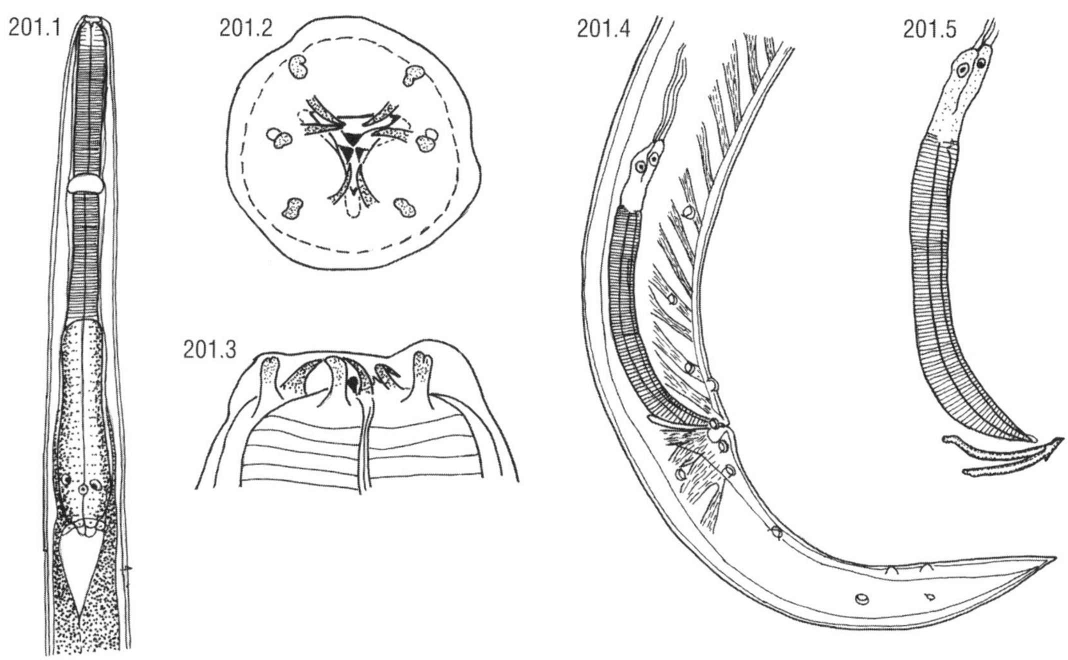

FIG. 201 *Wakubitinema* 201.1 Anterior end, ventral view; 201.2 Cephalic region, *en face* view; 201.3 Cephalic region, lateral view; 201.4 Posterior end of male, lateral view; 201.5 Spicules and gubernaculum. (Redrawn after Hasegawa, 1988.)

anterior to cloaca. Precloacal sucker and caudal alae absent. *Female*: Vulva opens in posterior third of body. Vagina short, directed anteriorly. Amphidelphic. Parasite in small intestine of Ranidae (*Rana* (*Limnonectes*) *namiyei*) in Japan.

Type species: *Wakubitinema toyamai* Hasegawa, 1988

[Moravec and Sey (1990) describe a species *Desmognathinema papuensis* representing a link between the genera *Wakubitinema* and *Desmognathinema*, and consider *Wakubitinema* a synonym of *Desmognathinema*, transferring the type species as *Desmognathinema toyamai* (Hasegawa, 1988) Moravec and Sey, 1990.]

Chabaudgolvania Texeira de Freitas, 1988

DIAGNOSIS: Elevated circumoral ring present, nervous tissue passes through ring. Oesophagus with anterior pharyngeal region only slightly apparent, distal bulb elongate without valves. *Male*: Spicules robust, width to length ratio greater than 6%. Nine to 12 pairs of sensory papillae on posterior end, 1–2 pairs of these lateral. Parasites of Salamandridae and Ambystomatidae.

Type species: *Chabaudgolvania terdentata* Linstow, 1890

[The emended diagnosis is based on Richardson and Adamson (1990). Chabaud (1978) considered the genus a synonym of *Megalobatrachonema* in the Cosmocercoidea. Baker (1980c) considered the genus a subgenus of *Megalobatrachonema*. Richardson and Adamson (1990) now reinstate the genus and transfer it to this superfamily.]

Wuinema Yu and Wang, 1992 (Fig. 202)

DIAGNOSIS: Mouth triangular, lips absent. Cephalic papillae absent. Lateral alae well developed in anterior part of body. Cervical papillae present towards posterior end of oesophagus. Oesophagus with long anterior muscular region, short posterior glandular region. Excretory pore opens at level of anterior oesophagus. *Male*: Spicules equal, similar. Gubernaculum present. Single medium papilla anterior to cloaca. Precloacal sucker present, muscular. Caudal alae absent. Papillae sessile. *Female*: Vulva opens in posterior half of body. Oviparous. Parasite in stomach and anterior intestine of freshwater fish, *Mystus macropterus*, in Youshui River, Guizhou Province and Lishui River, Hunan Province, China.

Type species: *Wuinema mysti* Yu and Wang, 1992

[Moravec *et al.* (2002a) consider this genus a synonym of *Paragendria* Baylis, 1939.]

Neoparaseuratum Moravec, Kohn and Fernandes, 1992 (Fig. 203)

DIAGNOSIS: Head rounded. Mouth hexagonal, 6 ridges radiate from margin. Four submedian cephalic papillae, 2 lateral amphids. Buccal capsule weakly developed. Oesophagus long, undivided, increasing in width posteriorly, 3 very small teeth at proximal end. Cervical region with numerous longitudinal bands of inflated cuticle. Cervical papillae at level of posterior oesophagus. Excretory pore opens posterior to oesophagus. *Male*: Caudal alae absent. Oblique muscles bands present anterior to cloaca. Spicules equal. Gubernaculum short. *Female*: Vulva opens

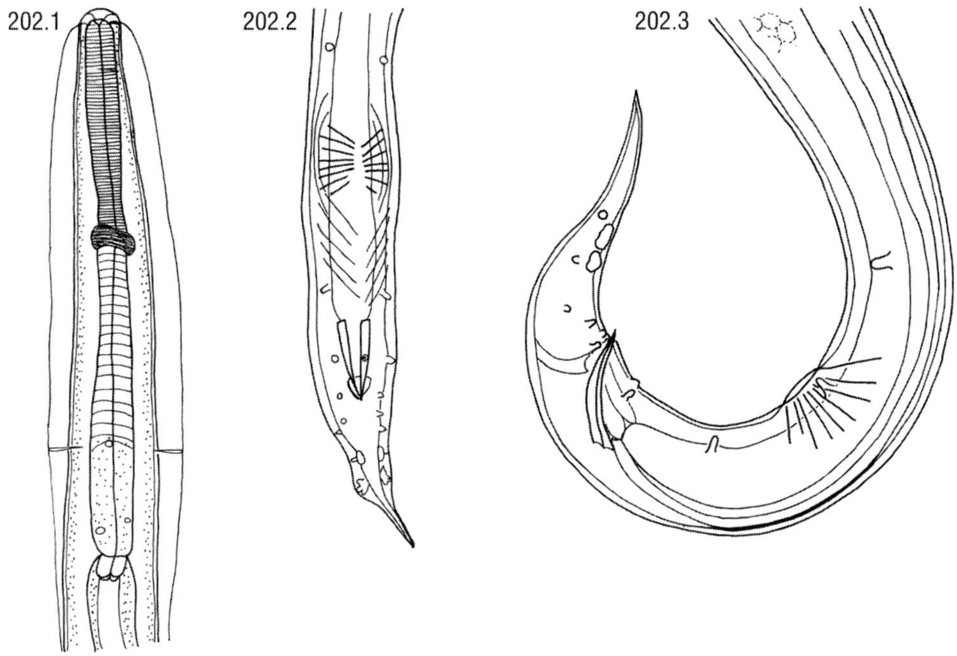

FIG. 202 *Wuinema* 202.1 Anterior end of body, dorsal view; 202.2 Posterior end of male, ventral view; 202.3 Posterior end of male, lateral view. (Redrawn after Yu and Wang, 1992.)

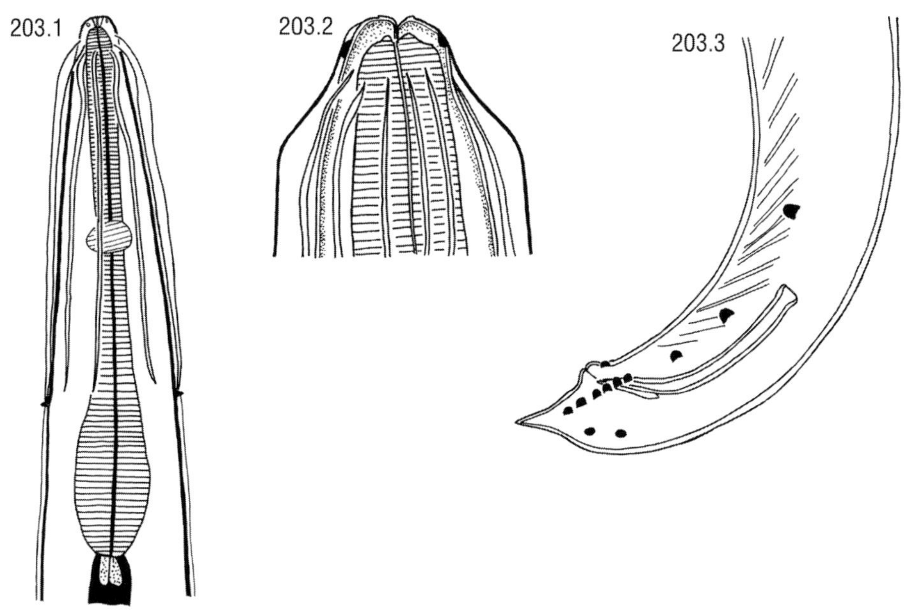

FIG. 203 *Neoparaseuratum* 203.1 Anterior part of male, dorso-ventral view; 203.2 Cephalic region of male, lateral view; 203.3 Posterior end of male, lateral view. (Redrawn after Moravec *et al.*, 1992b.)

posterior to mid-body. Uteri opposed. Eggs in uteri non-embryonated. Parasite of intestine of South American (Brazil) freshwater fish (thorny catfish, *Pterododas granulosus*).

Type species: *Neoparaseuratum travassosi* Moravec, Kohn and Fernandes, 1992

Paraseuratoides Moravec, Salgado-Maldonado and Aguilar-Aguilar, 2002

DIAGNOSIS: Mouth simple, lips absent, 6 spines present. Four submedian cephalic papillae, 2 lateral amphids. Buccal capsule present. Six oesophageal teeth extend into buccal cavity. Lateral alae absent. Cervical papillae, excretory pore in oesophageal region. Oesophagus elongate, undivided, widens posteriorly. *Male*: Caudal alae, precloacal sucker absent. Oblique muscle bands present anterior to cloaca. Spicules equal. Gubernaculum present. *Female*: Vulva opens posterior to mid-body. Parasites in intestine of freshwater fish (swamp eels).

Type species: *Paraseuratoides ophisterni* Moravec, Salgado-Maldonado and Aguilar-Aguilar, 2002

[Moravec *et al.* (2002b) record that the generic name *Paraseuratoides* is a junior homonym to *Paraseuratoides* Wang, 1984. The authors rename their genus *Gibsonnema*. The genus is placed in the Quimperiinae, as it is related to *Paraseuratum*.]

Gibsonnema Moravec, Salgado-Maldonado and Aguilar-Aguilar, 2002 (Fig. 204)

DIAGNOSIS: see *Paraseuratoides* Moravec, Salgado-Maldonado and Aguilar-Aguilar, 2002.

Type species: *Gibsonnema ophisterni* Moravec, Salgado-Maldonado and Aguilar-Aguilar, 2002

[Moravec *et al.* (2002b) designated *Gibsonnema nom. nov.* for *Paraseuratoides* Moravec, Salgado-Maldonado and Aguilar-Aguilar, 2002, which is preoccupied.]

Cucullanidae Cobbold, 1864

Cucullaninae (Cobbold, 1864 family)

Cucullanus Mueller, 1777

Subgenus *Stelmioides* Antonucci, 1929

DIAGNOSIS: Three lips. Eggs embryonated.

Type species: *Cucullanus* (*Stelmioides*) *ammocoetis* Linstow, 1897

[Listed in Spencer Jones and Gibson (1987).]

Dacnitoides Ward and Magath, 1916

DIAGNOSIS: Cephalic region straight. Lips not chitinized. Oesophagostome present. Intestinal caecum present. *Male*: Precloacal sucker without chitinous ring. Spicules equal. Gubernaculum

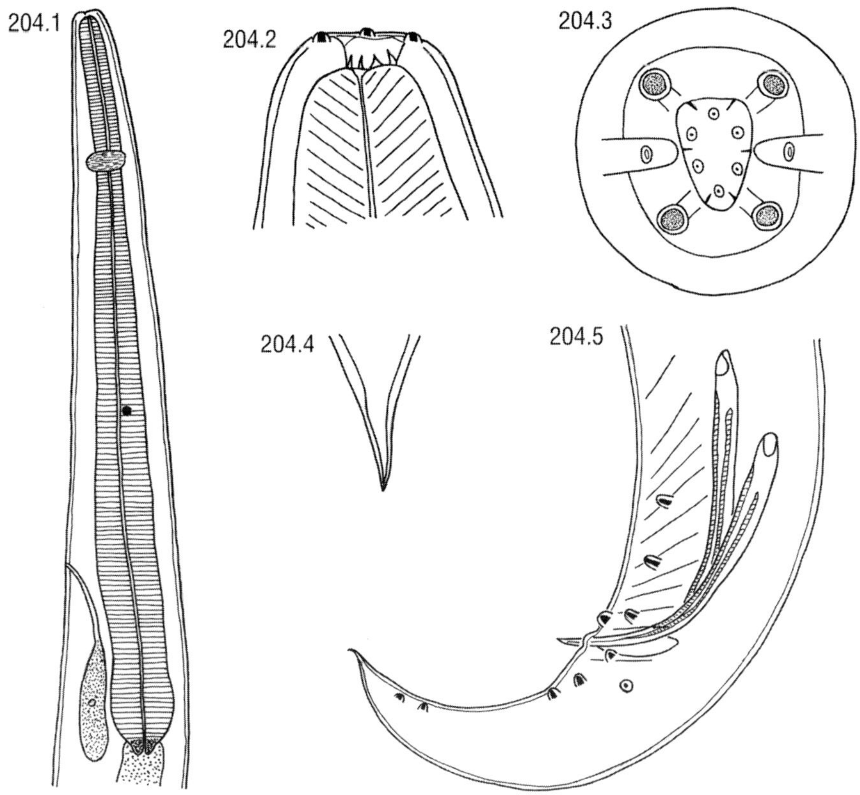

FIG. 204 *Gibsonnema* 204.1 Anterior end of female, lateral view; 204.2 Cephalic region, lateral view; 204.3 Cephalic region, *en face* view; 204.4 Distal tip of female tail; 204.5 Posterior end of male, lateral view. (Redrawn after Moravec *et al.*, 2002b.)

absent. *Female*: Vulva opens just posterior to mid-body. One functional uterus, posterior branch ends in blind sac. Parasite in intestine of North American freshwater fish, *Perca flavescens*, *Stizostedion vitreum* (rarely).

Type species: *Dacnitoides cotylophora* Ward and Magath, 1916

[Diagnosis based on Ward and Magath (1916) and Yorke and Maplestone (1926). The authors consider the genus close to *Cucullanus*. Yorke and Maplestone (1926) create a new subfamily Dacnitoididae for the genus within the Cucullanidae, as it has an anteriorly directed caecum. Ivaschkin and Khromova (1976) list the type species in the genus *Dichelyne* as *Dichelyne cotylophora* (Ward and Magath, 1916) Gendre, 1927. The genus is listed in Spencer Jones and Gibson (1987).]

Neocucullanus Travassos, Artigas and Pereira, 1928 (Fig. 205)

DIAGNOSIS: Mouth elongated dorso-ventrally, bordered by a denticulate membraneous collarette, supported by peribuccal sclerotized frame attached to sclerotized parts which extend into underlying oesophageal muscles. Four large submedian cephalic papillae, 2 lateral amphids,

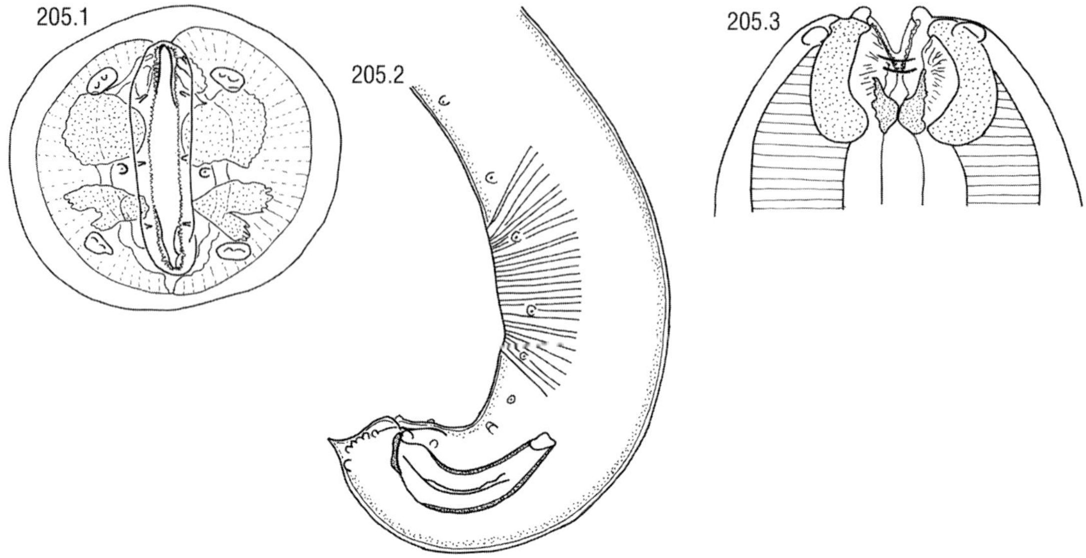

FIG. 205 *Neocucullanus* 205.1 Cephalic region, *en face* view; 205.2 Posterior end of male, lateral view; 205.3 Cephalic region, ventral view. (Redrawn after Petter, 1989.)

6 papillae of internal circle. Oesophagus claviform, internal wall thickened with sclerotized rods dilated anteriorly into an oesophagostome. Intestinal caecum absent. *Male*: Spicules equal, short, wide, distal tips rounded. Precloacal sucker present. Numerous pairs of precloacal papillae present. Tail short and rounded. *Female*: Vulva opens slightly posterior to mid-body. Ovejector short. Uteri opposed. Parasites in Characidae, Pisces in South America.

Type species: *Neocucullanus neocucullanus* Travassos, Artigas and Pereira, 1928

[The emended diagnosis is based on Petter (1989)].

Paracucullanellus Agrawal, 1965

DIAGNOSIS: Anterior extremity straight, not bent dorsally. Mouth triangular with 'short, transverse cuticular ridge'. Oesophagus club-shaped, muscular without glandular portion, dilated anteriorly to form 'pseudobuccal capsule'. Intestinal caecum absent. Cephalic glands present, extend from nerve ring to just anterior to posterior end of oesophagus. *Male*: Precloacal sucker present without chitinous rim. Caudal alae well developed. Spicules equal. Gubernaculum present. Caudal papillae: 4 pairs precloacal; 1 pair adcloacal; 6 pairs postcloacal. Tail pointed. *Female*: Vulva opens posterior to mid-body. Didelphic, uteri opposed. Oviparous. Tail pointed with lateral pair of papillae in mid-region. Parasitic in the intestine of freshwater fish, *Rita rita*.

Type species: *Paracucullanellus indica* Agrawal, 1965

[De (1989) considers there are no substantial morphological and size differences between *Cucullanus ritai* Karve, 1952 and *Paracucullanellus indica*, and that the two species are therefore conspecific. De regards *Paracucullanellus* as a synonym of *Cucullanus*. Gupta and Masoodi (1982) also consider this genus a synonym of *Cucullanus* Mueller, 1777.]

Pseudocucullanus Kalyankar, 1971

[Chabaud (1978) considers this genus insufficiently described to include in his key. Gupta and Masoodi (1982) consider the genus is a synonym of *Cucullanus* Mueller, 1777.]

Bacudacnitis Ferraz and Thatcher, 1988 (Fig. 206)

DIAGNOSIS: Cephalic region bulbous. Mouth not markedly orientated dorsally. Complex, cuticularized thickenings absent from oesophagostome lining. Prominent buccal support present. Two pairs of cephalic papillae present. Excretory pore with rounded protruberance projecting above body surface. Lateral, single, rounded cuticular vesicle near cephalic region. Intestinal caecum absent. *Male*: Precloacal sucker absent. Spicules subequal. Gubernaculum present, y-shaped. *Female*: Amphidelphic. Vulva opens in posterior third of body. Parasite in intestine of Amazonian catfish, *Pseudodoras niger*, in Brazilian Amazon.

Type species: *Bacudacnitis grandistomis* Ferraz and Thatcher, 1988

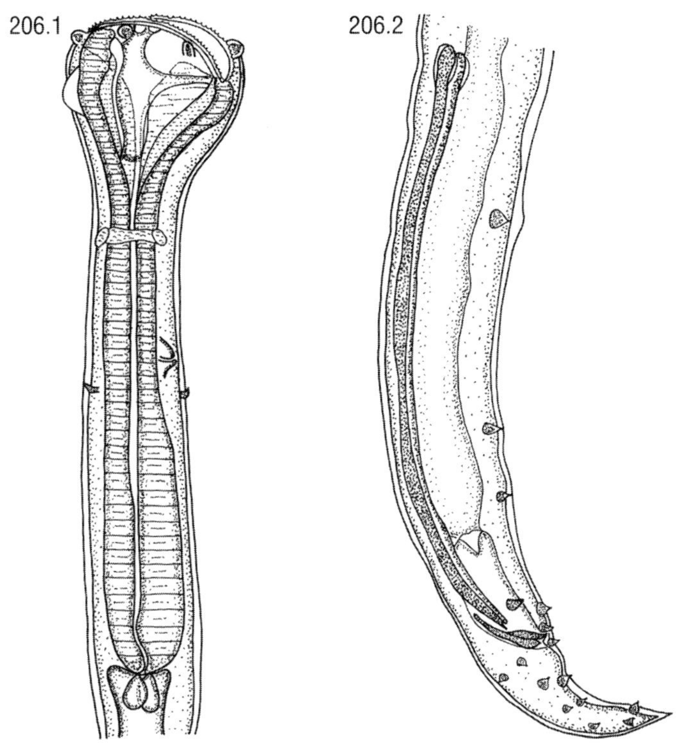

FIG. 206 *Bacudacnitis* 206.1 Anterior end of male, ventral view; 206.2 Posterior end of male, lateral view. (Redrawn after Ferraz and Thatcher, 1988.)

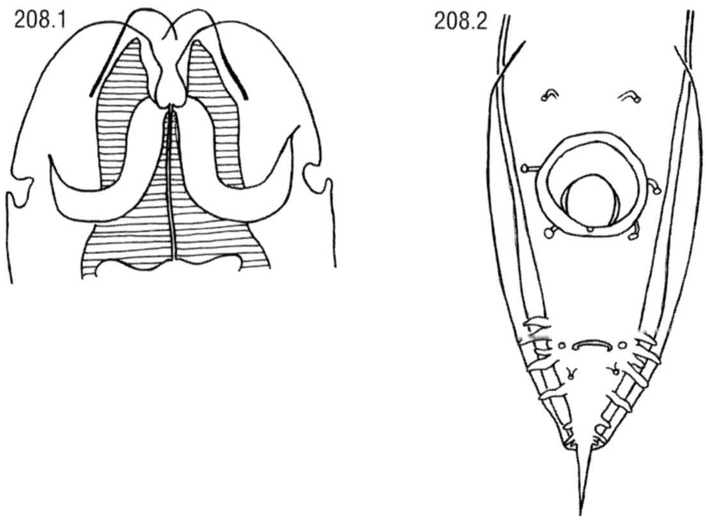

FIG. 208 *Neoheterakis* 208.1 Cephalic region showing arrangement of cordons on lips; 208.2 Posterior end of male. (Redrawn after Kumar and Thienpoint, 1974.)

Type species: *Neoheterakis chapini* Kumar and Thienpoint, 1974

[The authors place their genus in the subfamily Spinaspidoderinae Freitas, 1956. This subfamily is not accepted in Chabaud (1978) and the four genera – *Pseudaspidoderoides* Freitas, 1956, *Pseudaspidoderina* Freitas, 1956, *Spinaspidodera* Skrjabin and Schikhobalova, 1947 and *Pseudaspidoderella* Ali and Deskmukh, 1969 – are considered synonyms of *Pseudaspidodera* Baylis and Daubney, 1922. The authors consider *Neoheterakis* to be intermediary between *Pseudaspidodera* and *Odontoterakis* Skrjabin and Schikhobalova, 1947. Chabaud (1978) includes both these genera in the subfamily Heterakinae in his key and, consequently, *Neoheterakis* is placed in the same subfamily. Listed in Spencer Jones and Gibson (1987).]

Haroldakis Inglis, 1991 (Fig. 209)

DIAGNOSIS: Interlabia modified to form uncurved, non-anastomosing, non-recurrent cordons. Cheilorhabdions not incorporated in lips, project beyond lips. Five teeth (onchia) on each proximal lobe of 'pharyngeostome'. *Male*: Posterior end with 11 pairs of caudal papillae, additional pedunculate pair between sucker group of papillae and cloacal group of papillae absent. Transverse cushion with 2 sessile papillae between sucker and cloaca present. Medial cushion with sessile pair of papillae on posterior lip of cloaca. Parasites in caecum and large intestine of South American tinamous.

Type species: *Haroldakis multidentata* (Baylis, 1944) Inglis, 1991

Meteterakinae Inglis, 1958

DIAGNOSIS: Lips rounded. Interlabia absent. Cordons present or absent. Excretory pore opens into a large vesicle. Lateral alae extend nearly all length of body. *Male*: Spicules equal. Narrow

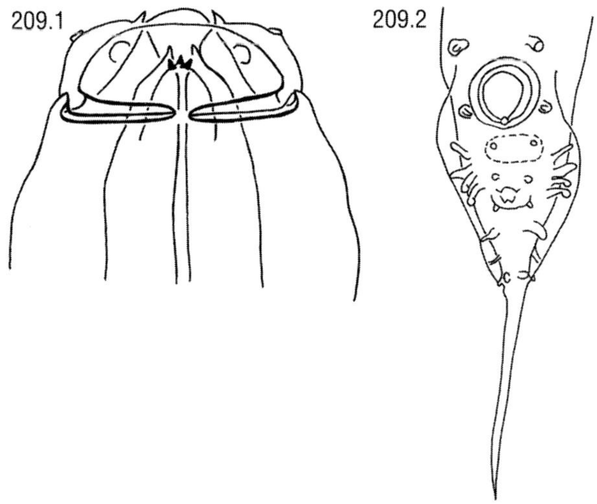

FIG. 209 *Haroldakis* 209.1 Cephalic region, dorsal view; 209.2 Posterior end of male, ventral view. (Redrawn after Inglis, 1991b.)

caudal alae present. Three or four pairs of large papillae pedunculate, short and thick: 1 pair at level of cloaca; 2 pairs, rarely 3 at level of precloacal sucker. Small sessile or pedunculate cloacal papillae also present, variable in number. *Female*: Vulva with prominent anterior lip, except in *Bufonerakis*. Parasites of amphibians, reptiles, mammals (*Hystrix*) in Oriental and South American regions and birds in New Caledonia.

Type genus: *Meteterakis* Karve, 1930

[Emended diagnosis based on Petter *et al.* (1988), who include three genera – *Gireterakis* Lane, 1917, *Bufonerakis* Baker, 1980 and *Cagourakis* Petter, Chermette and Vassart, 1988.]

Bufonerakis Baker, 1980 (Fig. 210)

DIAGNOSIS: Head cordons and interlabia absent. Three lobes at proximal end of oesophagus, project into buccal cavity, each with 'tooth', strengthened with thick cuticle. *Male*: Caudal alae present with numerous papillae, size and shape variable. Spicules equal, simple. Gubernaculum absent. Dorsal wall of cloaca thickened. *Female*: Vulva opening slit-like, flap absent. Short, flexed, muscular vagina, anteriorly and posteriorly directed uteri, ovaries prevulvar. Parasites of amphibians and reptiles in neotropical regions.

Type species: *Bufonerakis andersoni* Baker, 1980

[Listed in Spencer Jones and Gibson (1987).]

Cagourakis Petter, Chermette and Vassart, 1988 (Fig. 211)

DIAGNOSIS: Cephalic cordons absent. Oesophagus short with wide distal bulb. Cervical alae wide, arise medio-dorsally just posterior to cephalic region, curve posteriorly towards lateral

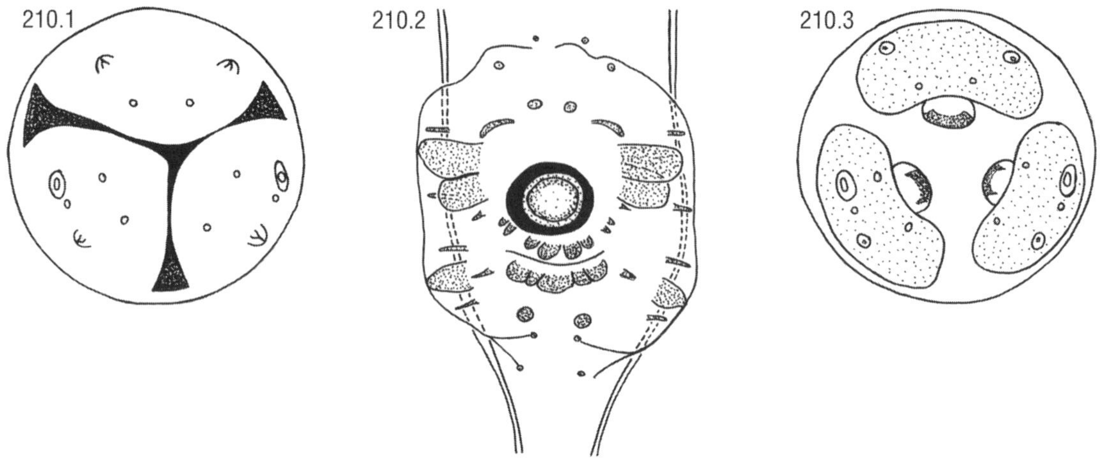

FIG. 210 *Bufonerakis* 210.1 Cephalic region of female, *en face* view; 210.2 Posterior end of male, ventral view; 210.3 Cephalic region of female, optical section through oesophagus just posterior to lips. (Redrawn after Baker, 1980a.)

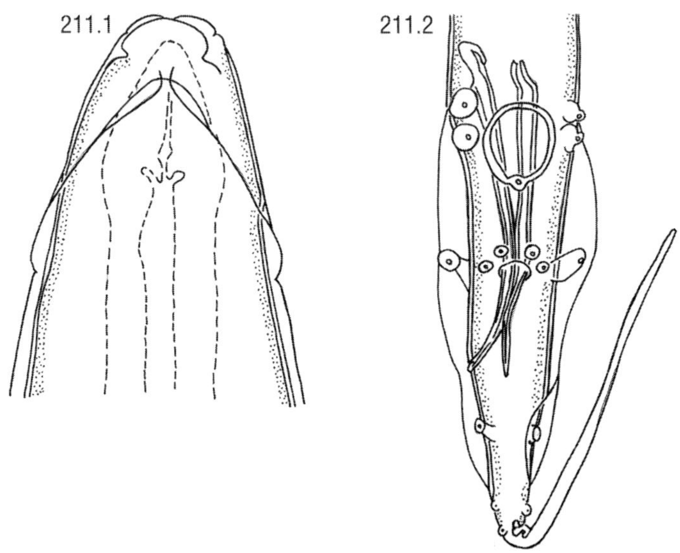

FIG. 211 *Cagourakis* 211.1 Anterior end, dorsal view; 211.2 Posterior end, ventral view. (Redrawn after Petter *et al.*, 1988.)

line. *Male*: Precloacal sucker wide. Nine pairs of caudal papillae: 4 pairs pedunculate, 5 pairs sessile. Spicules equal, similar. Gubernaculum absent. *Female*: Vulva opens between anterior third and middle of body, anterior lip prominent. Tail long, tapered. Parasite in caecum, posterior ileum and rectum of birds (*Rhynchetos jubatus*) in New Caledonia.

Type species: *Cagourakis dorsalata* Petter, Chermette and Vassart, 1988

Aspidoderidae Skrjabin and Schikhobalova, 1947

Lauroiinae Skrjabin and Schikhobalova, 1951

Nematomystes Sutton, Chabaud and Durette-Desset, 1980 (Fig. 212)

DIAGNOSIS: Mouth with 3 lips, each lip extended by a rectangular plate. Lateral alae present. *Male*: Precloacal sucker well developed. Spicules equal. Parasite in South American rodents (Cricetidae, *Oxymycterus misionalis*).

Type species: *Nematomystes rodentiphilus* Sutton, Chabaud and Durette-Desset, 1980

[Listed in Spencer Jones and Gibson (1987).]

Narsingiellinae Rao, 1978

DIAGNOSIS: Cephalic cordons present without plaques. Oesophagus with 2 bulbs at distal end. *Male*: Spicules equal. Parasites of amphibians and reptiles.

Type genus: *Narsingiella* Rao, 1978

Narsingiella Rao, 1978 (Fig. 213)

DIAGNOSIS: Characters of the subfamily. *Male*: Gubernaculum present. Parasite in the intestine of amphibian *Bufo viridis*, Berhampur, India.

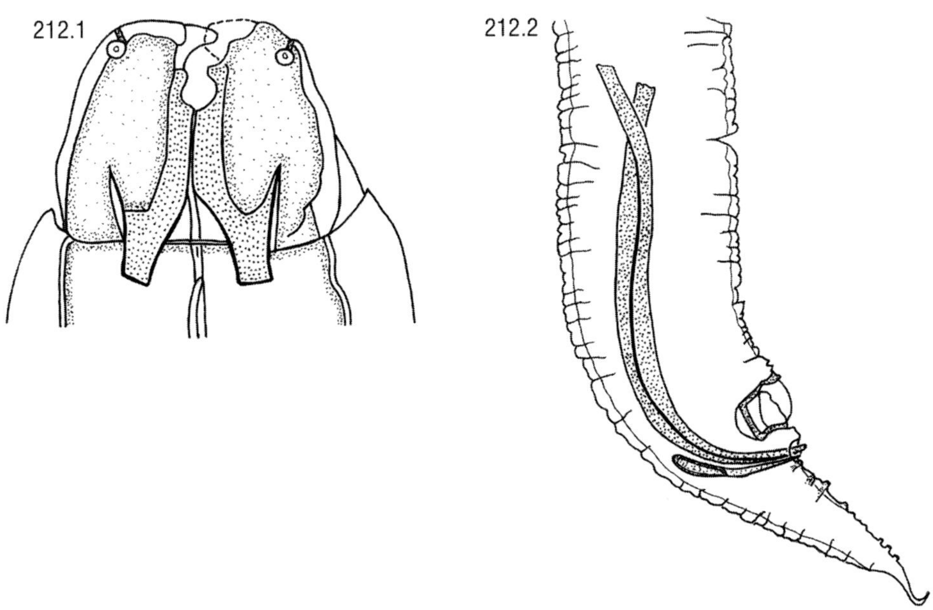

FIG. 212 *Nematomystes* 212.1 Cephalic region, ventral view; 212.2 Posterior end of male, lateral view. (Redrawn after Sutton *et al.*, 1980.)

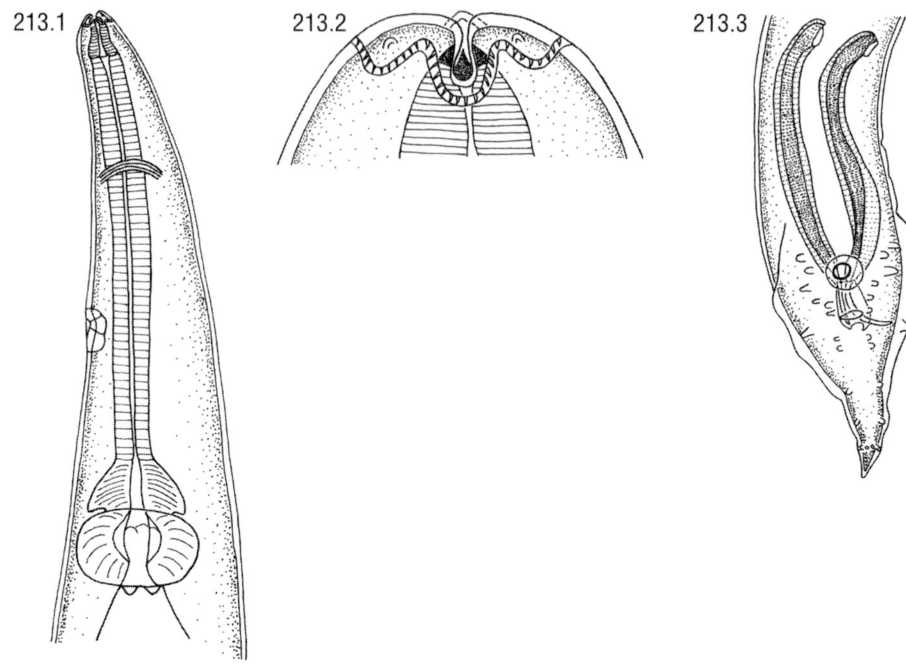

FIG. 213 *Narsingiella* 213.1 Female anterior end, lateral view; 213.2 Cephalic region, ventral view; 213.3 Posterior end of male, ventral view. (Redrawn after Rao, 1978.)

Type species: *Narsingiella narsingi* Rao, 1978

[Rao (1978) also includes *Cheloniheterakis* Yamaguti, 1961, a parasite of reptiles, in the new subfamily. Inglis (1967) did not accept the validity of *Cheloniheterakis* and Chabaud (1978), accepting this, did not include *Cheloniheterakis* in his key.]

Kiwinematidae Inglis and Harris, 1990

DIAGNOSIS: Lips shallow, rounded. Cordons, interconnecting lobes absent. Lips without distinct distal margin separating them from body. Lateral alae present. Cephalic cap absent. Oesophagus with distal trivalvulate bulb; radial tubes present associated with lumen in mid-region. *Male*: Caudal papillae sessile to mammiliform. Caudal alae absent. Spicules equal, identical. Precloacal sucker ventral. *Female*: Eggs with roughened surface, non-embryonated *in utero*.

Type genus: *Kiwinema* Inglis and Harris, 1990

[Emended diagnosis based on Inglis (1991b).]

Kiwinematinae Inglis and Harris, 1990

DIAGNOSIS: Proximal part of cheilorhabdions of cheilostome form cuticular flanges. Oesophagus narrow with distinctly formed distal bulb. *Male*: Expansion of lateral alae adjacent to pre-cloacal sucker and cloaca absent. *Female*: Mature eggs with rough external shell. Parasites of vertebrates indigenous to New Zealand.

[Emended diagnosis based on Inglis (1991b).]

Kiwinema Inglis and Harris, 1990 (Fig. 214)

DIAGNOSIS: Somatic papillae absent. Lateral alae extend along almost entire body. Excretory pore not opening on cuticular plate. Oesophagus without isthmus. *Male*: Eight pairs caudal papillae, 3 prominent and mammiliform. Gubernaculum absent. Tail with distal appendix. *Female*: Ventral postanal cuticular thickening absent. Parasite in caecum of *Apteryx* sp. in South Island, New Zealand.

Type species: *Kiwinema gracilicauda* (Harris, 1975) Inglis and Harris, 1990

[Emended diagnosis based on Inglis (1991b).]

Mammalakinae Inglis, 1991

DIAGNOSIS: Cheilorhabdions not obvious. Cuticular flanges on internal surface of lips absent. Oesophagus stout, club-shaped, isthmus absent, distal bulb not distinctly separated from corpus. *Male*: Expansion of lateral alae adjacent to precloacal sucker and cloaca present, without incorporating caudal papillae. *Female*: Mature eggs with smooth external shell. Parasites of mole rats, African Bathyergidae and palaeartic Spalacidae in South Africa, Eastern Europe and southern former USSR.

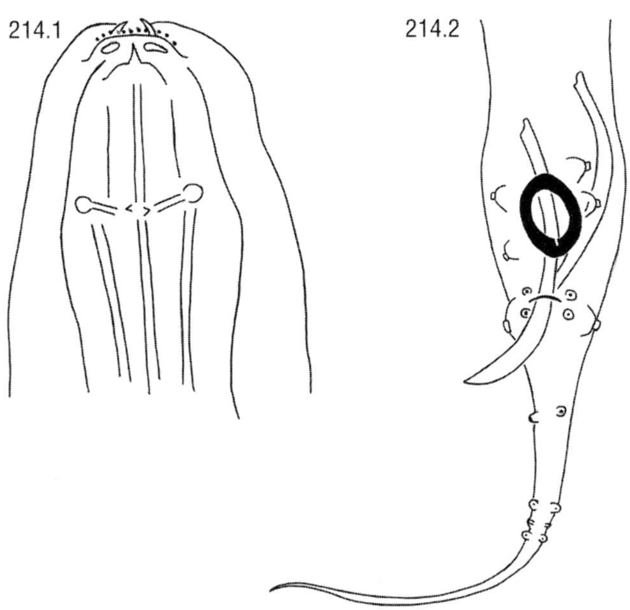

FIG. 214 *Kiwinema* 214.1 Anterior end, dorsal view; 214.2 Posterior end of male, ventral view. (Redrawn after Inglis and Harris, 1990.)

Type genus: *Mammalakis* Inglis, 1991

Mammalakis Inglis, 1991 (Fig. 215)

DIAGNOSIS: Characteristics of the subfamily. Parasite of *Spalax micropthalmus*.

Type species: *Mammalakis macrospiculum* (Ortlepp, 1939) Inglis, 1991

Hatterianema Chabaud and Dollfus, 1966

DIAGNOSIS: Somatic papillae present. Excretory pore opens on cuticular plate. Oesophagus with distinct isthmus. *Male*: Ten pairs of true caudal papillae, sessile or indistinctly mammiform. Gubernaculum present. Tail without caudal appendix. *Female*: Lateral alae stop just posterior to vulva. Postanal plate-like cuticular thickening present. Parasite of intestine of the reptile *Hatteria punctata* in New Zealand.

Type species: *Hatterianema hollandi* Chabaud and Dollfus, 1966

[Emended diagnosis based on Inglis (1991b). This genus was placed in the subfamily Spinicaudinae Travassos, 1920 of the family Heterakidae in Chabaud (1978).]

Ascaridiidae Travassos, 1919

[Skrjabin and Mozgovoi (1973) proposed that a new superfamily, Ascaridioidea, should be erected for this family within the order Ascaridida. Chabaud (1978) did not follow this classification.]

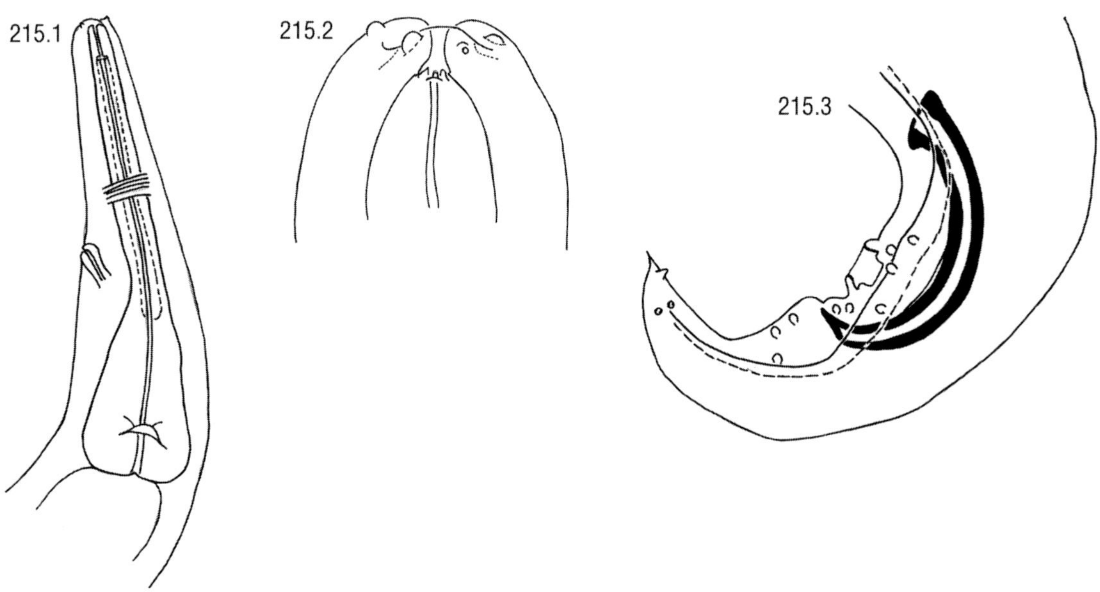

FIG. 215 *Mammalakis* 215.1 Anterior end, lateral view; 215.2 Cephalic region, ventro-lateral view; 215.3 Posterior end of male, lateral view. (Redrawn after Inglis, 1991b.)

ASCARIDOIDEA

[Gibson and Taylor (1976) comment on the taxonomic significance and function of the excretory system of ascaridoid nematodes. Their results indicate that all anisakines and *Sulcascaris* appear to have an excretory pore between the subventral lips and that not all anisakines have a one-sided system, as suggested by Hartwich's key. *Paranisakis* and *Paranisakiopsis* have a two-sided system and *Sulcascaris*, *Raphidascaris* and *Terranova scoliodontis* have a reduced right-hand duct. *Paranisakiopsis* should be transferred to the Anisakinae, leaving all raphidascarines with a pore near the nerve ring. The remaining ascaridoids appear to have a two-sided tubular system except for angusticaenes, multicaecines, acanthocheilid *Paranisakis* and crossophorid *Dartevellenia*. Fagerholm (1991) studied the morphology of the posterior end of male ascaridoids and indicated that there is considerable stability in the arrangement of sensory papillae present. Fagerholm (1991) defines the following terminology to describe the arrangement of these papillae: *proximal papillae* include those situated anterior to the cloaca and, in a few cases, those that extend level and slightly posterior to the cloaca, the latter generally referred to as the adcloacal papillae in other groups; *distal papillae* are those present on the ventral to sublateral surface on the distal half of the tail; *paracloacal papillae* refer to the two pairs of subventral papillae close to the distal lip of the cloaca.]

Genera not assigned to a family

Pseudostrongyloides Kreis, 1940

DIAGNOSIS: Parasite of *Natrix hypomelas*, *Dendrophis calligastra*, *Typhlops subocularis*, Squamata, Reptilia.

Type species: *Pseudostrongyloides ophidia* Kreis, 1940

[Kreis (1940) placed the genus in the superfamily Rhabdiasoidea, family Rhabdiasidae. This superfamily was not accepted in the *CIH Keys* and the family Rhabdiasidae is placed in the Rhabditoidea. Spencer Jones and Gibson (1987) place the genus in the superfamily Ascaridoidea. The genus is based on the description of juvenile specimens with 3 distinct lips. Re-evaluation of the specimens may lead to a synonymy of this genus with one of the established genera recorded from reptiles.]

Pandascaris Porter, 1948

DIAGNOSIS: Parasite of *Ailuropoda melanoleuca*, Giant panda, Zoological Society of London.

Type species: *Pandascaris ailuropoda* Porter, 1948

[The species is listed by Porter in the annual report for 1946 (publ. 1948) of the Zoological Society of London. No description could be traced for this genus or species. The genus is listed in Spencer Jones and Gibson (1987) as a *nomen nudum*.]

Mariporrocaecum Sheenko, 1985 (Fig. 216)

DIAGNOSIS: Three lips with dentigerous ridges. Interlabia present. Cephalic papillae form 2 rings. Ventriculus present without appendix. Intestinal caecum present. Excretory cell H-shaped, excretory pore opens anterior to hemizonid. *Male*: Spicules long. *Female*: Vulva opens in posterior half of body. Parasite in fish (rays).

Type species: *Mariporrocaecum petrovi* (Mozgovoi, 1950) Sheenko, 1985

[Listed in Spencer Jones and Gibson (1987).]

Heterocheilidae (Railliet and Henry, 1912 subfamily) Railliet and Henry, 1915

[Hartwich (1974) maintains the family Heterocheilidae for the single genus *Heterocheilus*. Sprent (1980a) proposes that this family should return to subfamily status. He considers the subfamily has affinities to the Acanthocheilidae and Ascarididae on the basis of their symmetrical excretory systems, but on the basis of labial structure suggests the subfamily is closer to the Ascarididae. Sprent (1980a) therefore proposes the subfamily should join the Ascaridinae, Toxocarinae and Angusticaecinae in the family Ascarididae. He includes the following genera in the subfamily Heterocheilinae: *Heterocheilus* Diesing, 1839, *Paradujardinia* Travassos, 1933, *Dujardinascaris* Baylis, 1927, *Multicaecum* Baylis, 1923, *Gedoelstascaris* Sprent, 1978, *Ortleppascaris* Sprent, 1978, *Brevimulticaecum* Mozgovoi, in Skrjabin, Shikhobalova and

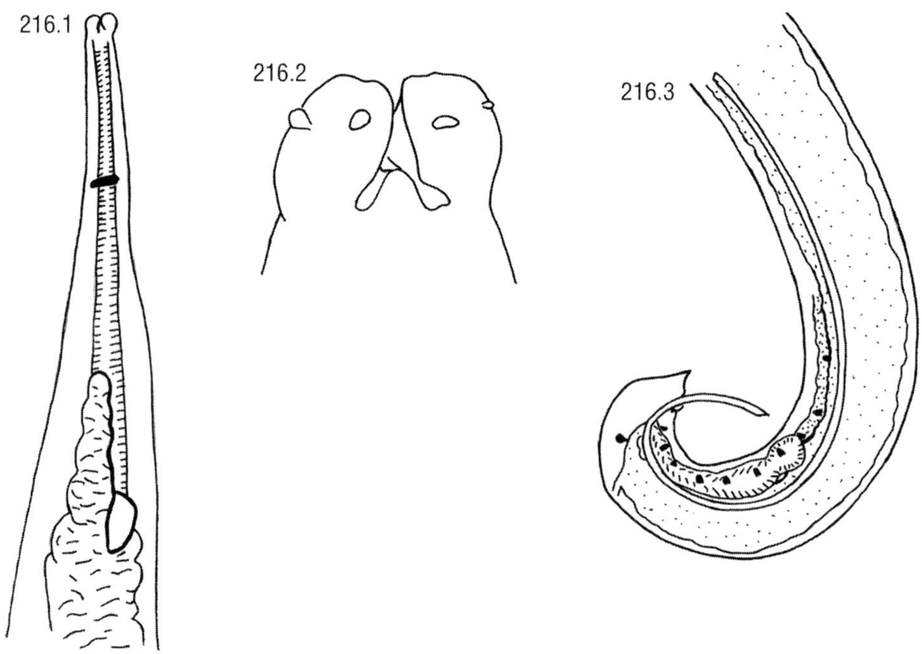

FIG. 216 *Mariporrocaecum* 216.1 Anterior end; 216.2 Cephalic region, ventral view; 216.3 Posterior end of male, lateral view. (Redrawn after Sheenko, 1985.)

Mozgovoi, 1952 and *Krefftascaris* Sprent, 1980. He does away with the subfamily Multicaecinae (Hartwich, 1954 family) Campana-Rouget, 1960. The affinities of *Typhlophorus* von Linstow, 1906, *Trispiculascaris* Skrjabin, 1916 (former synonym of *Dujardinascaris*), *Hartwichia* Chabaud and Bain, 1966 Sprent (1980a) proposes to consider at a later date. *Trispiculascaris* is redescribed by Sprent (1983a).

Sprent (1980a) emends the diagnosis of the subfamily as follows: mouth with 3 lips, generally with notches and interlocking processes, subventral lips with asymmetric pulp. Median lobe of pulp absent (except *Heterocheilus*, *Ortleppascaris*). Dentigerous ridges absent (except *Multicaecum*). Interlabia present, reduced (*Ortleppascaris*) or form interlabial ridges (*Heterocheilus*). Cervical papillae near nerve ring. Cervical alae present or absent. Excretory pore generally opens near nerve ring or anterior, system with 2 posterior filaments that join anteriorly to form commissure containing nucleus, which proceeds forward as excretory duct, anterior filaments present or absent. Ventriculus present with nuclei of oesophageal glands, may be lobed or with appendices (*Multicaecum*, *Brevimulticaecum*). Intestinal caecum present extending over half of length of oesophagus. *Male*: 4 pairs of precloacal, 2 single paracloacal, 2–3 postcloacal papillae near distal tip of tail. Spicules alate in mid-region, alae do not reach tip. Gubernaculum present (except *Paradujardinia* absent, *Heterocheilus* vestigial). *Female*: Vulva opens in anterior half of body. Uterus didelphic, opisthodelphic or amphidelphic. Parasites mainly of freshwater, estuarine or coastal aquatic vertebrates.]

Type genus: *Heterocheilus* Diesing, 1839.

[Fagerholm (1991) restores the subfamily to family rank and lists *Multicaecum* Baylis, 1923, *Brevimulticaecum* Mozogovoi, 1951, *Typhlophorus* Linstow, 1906, *Krefftascaris* Sprent, 1980, *Gedoelstascaris* Sprent, 1978, *Ortleppascaris* Sprent, 1978, *Trispiculascaris* Skrjabin, 1916, *Dujardinascaris* Baylis, 1947, *Heterocheilus* Diesing, 1839, *Paradujardinia* Travassos, 1933 and *Hartwichia* Chabaud and Bain, 1966 in the family.]

Heterocheilus Diesing, 1839

DIAGNOSIS: Body small to medium-sized. Mouth with 3 alate lips, distal margin elongated overlapping 'postlabial region'. Dentigerous ridges, interlabia absent. Longitudinal ridges present between lip extensions. Cervical papillae posterior to nerve ring. Excretory pore opens near nerve ring, excretory filaments bilateral. Cervical alae absent. Ventriculus present. Intestinal caecum present. *Male*: Body slightly smaller than female. Four pairs of precloacal papillae, paracloacal papillae separate. Spicules alate. Gubernaculum present. *Female*: Body larger than male. Vulva opens anterior to mid-body. Didelphic. Parasites in stomach of Sirenians, *Trichechus* spp., in Amazonas, Brazil and Caribbean Sea.

Type species: *Heterocheilus tunicatus* Diesing, 1839

[Emended diagnosis based on Sprent (1980a).]

Trispiculascaris Skrjabin, 1916

DIAGNOSIS: Body small. Mouth with 3 lips, large in relation to body, each with dentigerous ridge composed of triangular denticles. Interlabia present. Excretory pore opens at level of nerve ring. Cervical alae present. Ventriculus, intestinal caecum 'probably present'. *Male*: Spicules equal, alate. Gubernaculum present. Approximately 4 pairs of precloacal papillae. Parasite in stomach of *Crocodylus niloticus* in Central Africa.

Type species: *Trispiculascaris trispiculascaris* Travassos, 1920

[Emended diagnosis based on Sprent (1983a), who had no access to specimens and was only able to examine Skrjabin's description and figures.]

Pelecanascaris Ali and Farooqui, 1970 (Fig. 217)

DIAGNOSIS: Stout worms with thick cuticle. Head with 3 prominent lips, interlabia present, 1 interlabium bifid at distal and proximal ends. Excretory pore opens at apex of one interlabium. Intestinal caecum fairly well developed, appendix present. *Male*: Caudal alae absent. Numerous caudal papillae. Spicules subequal. Gubernaculum absent. *Female*: Smaller than male. Vulva opens at junction of first and middle third of body. Eggs thick-shelled. Parasites of proventriculus of birds (*Pelecanus philippenisis*) in India.

Type species: *Pelecanascaris deodhari* Ali and Farooqui, 1970

[Fagerholm (1991) indicates this genus may be a synonym of *Contracaecum*. Listed in Spencer Jones and Gibson (1987).]

Krefftascaris Sprent, 1980b (Fig. 218)

DIAGNOSIS: Body small. Mouth with 3 trapezoidal, rounded lips. Dentigerous ridges present. Interlabia absent. Excretory pore opens near nerve ring, excretory filaments bilateral. Cervical

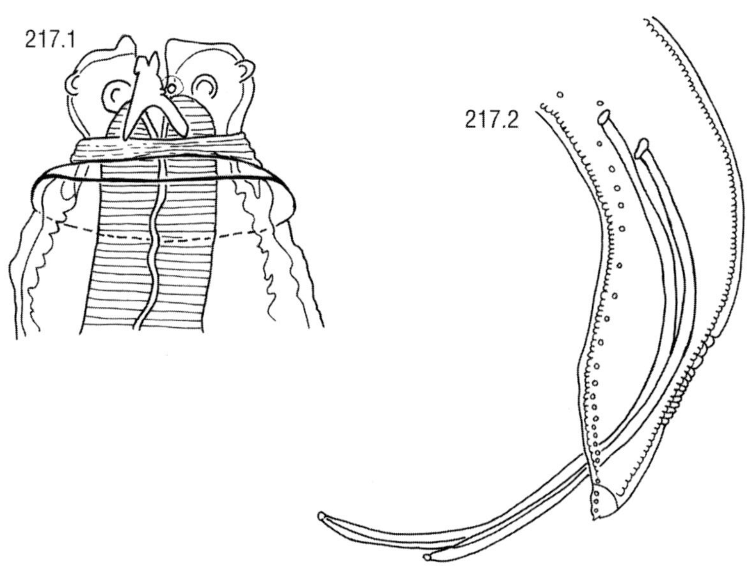

FIG. 217 *Pelecanascaris* 217.1 Female cephalic region, dorso-ventral view; 217.2 Posterior end of male, lateral view. (Redrawn after Ali and Farooqui, 1970.)

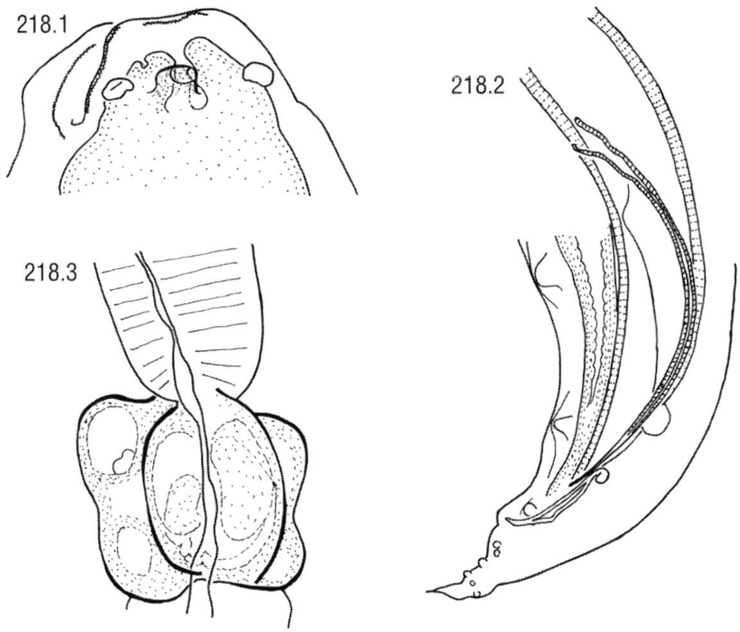

FIG. 218 *Krefftascaris* 218.1 Lips, dorsal view; 218.2 Posterior end of male, ventro-lateral view; 218.3 Oesophago-intestinal junction showing ventriculus. (Redrawn after Sprent, 1980b.)

papillae posterior to nerve ring. Lateral alae present, extend whole length of body. Ventriculus present, lobulate. Intestinal caecum present. *Male*: Body slightly smaller than females. Spicules alate. Gubernaculum present. Four pairs precloacal, 2 pairs separate paracloacal, 3 pairs postcloacal papillae. *Female*: Vulva opens in anterior third of body. Didelphic. Amphidelphic. Parasite in stomach of chelonians (*Chelodina longicollis*, *Elseya dentata*) from Gara River, Northern New South Wales, Australia.

Type species: *Krefftascaris parmenteri* Sprent, 1980b

[Sprent (1980b) included this genus in the subfamily Heterocheilinae in the Ascardidae. Fagerholm (1991) lists the genus in the family Heterocheilidae. Listed in Spencer Jones and Gibson (1987).]

Gedoelstascaris Sprent, 1978 (Fig. 219)

DIAGNOSIS: Body coiled. Mouth with 3 elongate lips, dentigerous ridges and interlocking processes absent. Median lobe of lip pulp inconspicuous, 'anterior rostral plate present'. Interlabia present. Cervical alae absent. Excretory pore opens just anterior to level of nerve ring, system bilateral with anterior extensions of excretory cell. Cervical papillae sessile, just posterior to level of nerve ring. Ventriculus present, short anterior and posterior lobes incorporate nuclei. Dorsal oesophageal gland opens near anterior end of oesophagus, subventral glands open posteriorly. Intestinal caecum present. *Male*: Spicules alate. Gubernaculum slender. Caudal papillae: 4 pairs precloacal; 1 pair lateral; 2 pairs paracloacal; 3 pairs postcloacal papillae. *Female*: Vulva opens anterior to middle of body, 2 uterine branches. Parasites in stomach and duodenum of *Crocodylus* spp. in Africa and Australia.

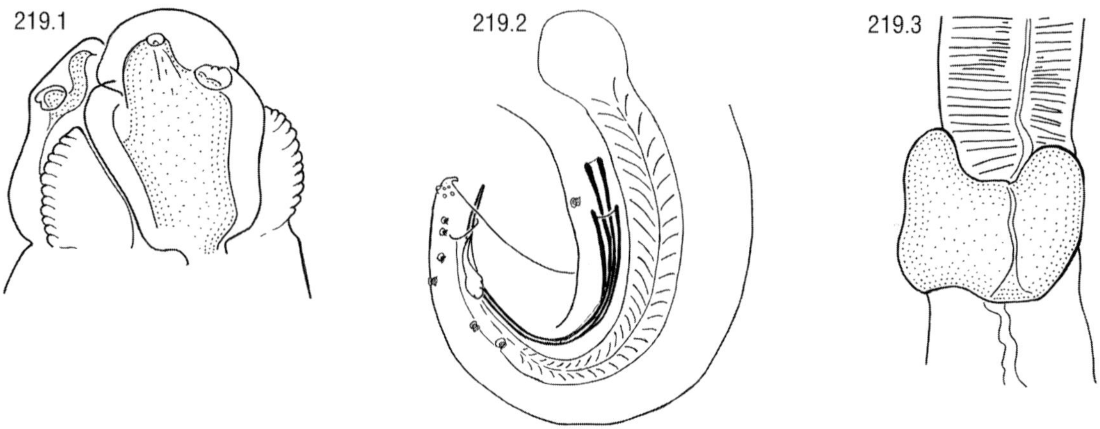

FIG. 219 *Gedoelstascaris* 219.1 Cephalic region showing subventral lip and interlabia; 219.2 Posterior end of male, latero-ventral view; 219.3 Oseophago-ventricular region. (Redrawn after Sprent, 1978b.)

Type species: *Gedoelstascaris vendenbrandeni* (Baylis, 1929) Sprent, 1978

[The species are former members of the genus *Dujardinascaris*. Sprent (1980a) places this genus in the Heterocheilinae. Sprent (1990b) studied species of the Heterocheilinae from fish, which he proposed were secondarily derived from species in crocodiles. The examination of species of *Brevimulticaecum* Mozgovoi, 1951 led him to consider that there were insufficient features for them to be separated from the species assigned to *Gedoelstascaris*. He therefore considered *Gedoelstascaris* a synonym of *Brevimulticaecum*. Fagerholm (1991) lists both genera in the family Heterocheilidae. Listed in Spencer Jones and Gibson (1987).]

Ortleppascaris Sprent, 1978 (Fig. 220)

DIAGNOSIS: Mouth with 3 diamond-shaped lips, margins thin, median pulp distinct. Rostral plate, dentigerous ridges, interlocking processes absent. 'Interlabia lightly cuticularized', ventral interlabium may be absent. Lateral alae or thickening of cuticle in lateral field of oesophageal region present. Excretory pore opens near cephalic region, system bilateral, nucleus in commissure of excretory filaments, anterior prolongations of filaments absent. Ventriculus lobed. Intestinal caecum present. *Male*: Spicules alate. Gubernaculum slender, sharply pointed. Caudal papillae: 4 pairs precloacal; 1 pair lateral; 2 pairs paracloacal; 1 pair subdorsal postcloacal; 2 pairs subventral postcloacal papillae. Tail narrows to a point. *Female*: Vulva opens anterior to mid-body, two uterine branches. Tail long, tapers to distal tip. Parasites in stomach, intestine or body cavity of *Crocodylus*, *Caiman*, alligator, in Zaire, Zimbabwe, Trinidad, Venezuela, Brazil, USA (Florida) and Cuba.

Type species: *Ortleppascaris nigra* (Gedoelst, 1916) Sprent, 1978

[The species are former members of the genus *Dujardinascaris*. Sprent (1980a,b) places this genus in the Heterocheilinae. Fagerholm (1991) does not use the subfamily rank and lists this genus in the family Heterocheilidae. Listed in Spencer Jones and Gibson (1987).]

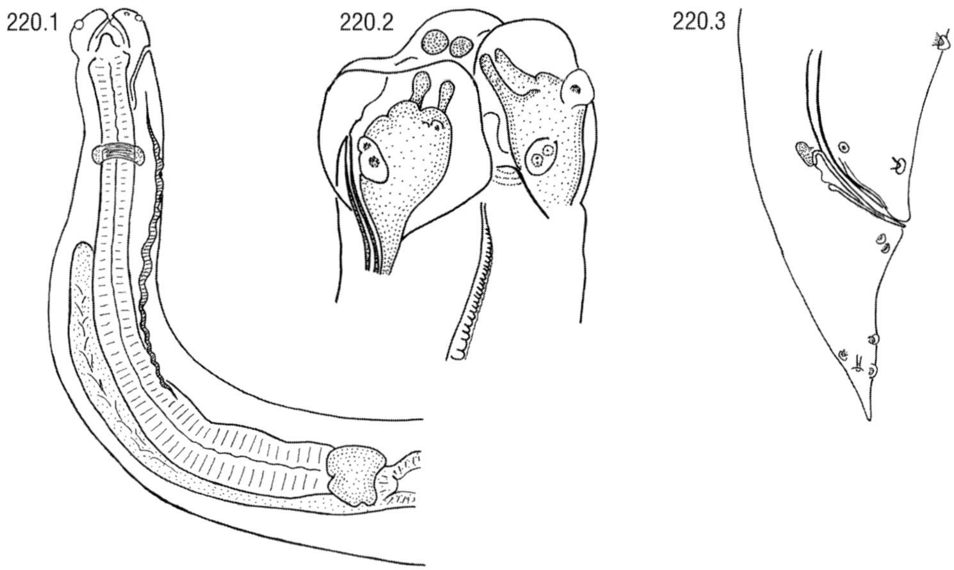

FIG. 220 *Ortleppascaris* 220.1 Anterior end; 220.2 Cephalic region showing excretory pore through subventral lip, lateral view; 220.3 Posterior end of male, lateral view. (Redrawn after Sprent, 1978b.)

Anisakidae (Railliet and Henry, 1912 subfamily) Skrjabin and Karokhin, 1945

[Shakhmatova (1971) erected the subfamily Cloeascarinae for *Cloeoascaris* Baylis, 1923 on the basis of the head structures of the genus. Hartwich (1974) lists *Cloeoascaris* as a synonym of *Galeiceps* Railliet, 1916 in the tribe Anisakinea.

Petter *et al.* (1991) propose three additional subfamilies within the Anisakidae. The family Acanthocheilidae Wülker, 1930 becomes a subfamily, the tribe Paranisakinea Hartwich, 1974 is elevated to a subfamily and a new subfamily, Pseudanisakinae Petter, Paradiznik, Radujkovic and Cassone, 1991, is proposed. Petter *et al.* (1991) define the subfamilies as follows:

Acanthocheilinae is characterized by the presence of small lips, an excretory pore that opens at the level of the nerve ring, a bilateral excretory system (filamentous nuclei in posterior left filament anterior to oesophago-intestinal junction), a rounded ventriculus, oesophageal appendix absent, numerous precloacal papillae on male, simple spicules without alae, the absence of a gubernaculum and parasitic in sharks, Pleurotremata, Selachii. The subfamily is erected for the type and only genus *Acanthocheilus* Molin, 1858.

Paranisakinae is characterized by the presence of well-developed lips, an excretory pore that opens at the level of the nerve ring, a bilateral excretory system (left filament larger than right, nucleus in posterior left filament anterior to oesophago-intestinal junction), an elongate ventriculus, oesophageal appendix absent, fewer than 13 pairs of precloacal papillae on male, alate spicules, the presence of a gubernaculum and parasitic in elasmobranchs, Selachii. The subfamily is erected for *Paranisakis* Baylis, 1923 (type genus) and *Mawsonascaris* Sprent, 1990.

Pseudanisakinae is characterized by the presence of small lips, an excretory pore that opens at the level of the nerve ring, a bilateral excretory system (posterior left filament enlarged into a ribbon, nucleus in posterior left filament at level of oesophago-intestinal junction), an elongate ventriculus, oesophageal appendix absent, fewer than 10 pairs of precloacal papillae on male, alate spicules, the absence of a gubernaculum and parasitic in rays, Hypotremata, Selachii. The subfamily is erected for the type and only genus *Pseudanisakis* Layman and Borovkova, 1926.

Hartwich (1974) considered *Metanisakis* Mozgovoi, 1951 as valid and included the genus in the Acanthocheilidae. Gibson (1973) considers the genus *Metanisakis* a synonym of *Pseudanisakis*. Petter *et al.* (1991) follow Gibson's classificiation.]

Anisakinae Railliet and Henry, 1912

[Considered a subfamily of Ascarididae by Sprent (1983b). Fagerholm (1991) maintains it as a subfamily of the family Anisakidae and lists *Terranova* Leiper and Atkinson, 1915, *Sulcascaris* Hartwich, 1957, *Anisakis* Dujardin, 1845, *Pseudoterranova* Mozogovoi, 1951, *Peritrachelius* Diesing, 1851, *Pulchrascaris* Vicente and Santos, 1972 and *Paranisakiopsis* Yamaguti, 1941 in the subfamily.]

Genera not designated to a tribe

Pulchrascaris Vicente and Dos Santos, 1972 (Fig. 221)

DIAGNOSIS: Lateral alae distinct. Lips reduced, equal in size, cuticular flanges absent, internal pulp indistinct. Dorsal lip with 2 lateral double papillae and 2 tooth-like structures on inner

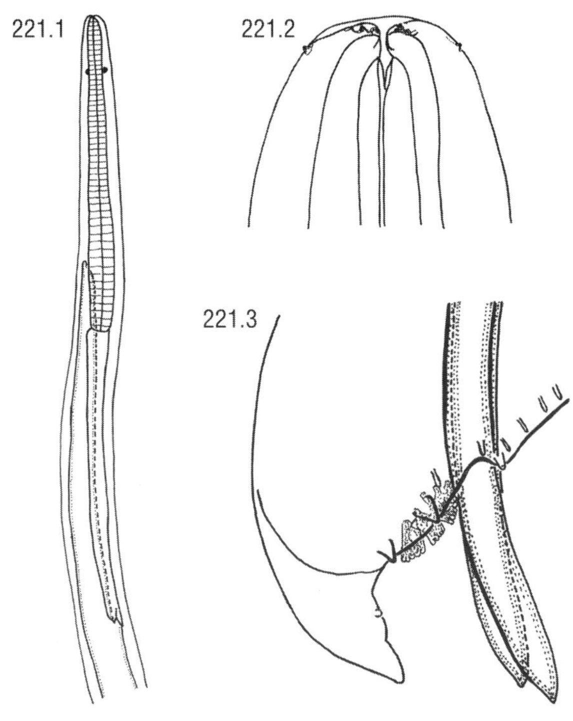

FIG. 221 *Pulchrascaris* 221.1 Anterior end of female; 221.2 Female cephalic region; 221.3 Distal end of male, lateral view. (Redrawn after Vicente and Dos Santos, 1972.)

dorsal surface. Subventral lips each with double medio-lateral papillae, single lateral papilla, an amphid and single tooth-like structure on inner dorsal surface. Dentigerous ridges and interlabia absent. Cervical papillae near level of nerve ring. Excretory pore between base of subventral lips. Excretory duct with nucleus near nerve ring, excretory filament extends posteriorly within left lateral cord to past mid-body, absent in posterior quarter. Ventriculus elongate. Intestinal caecum present. *Male*: Spicules equal to slightly unequal, similar, alate. Gubernaculum absent. Cuticular plates present immediately posterior to cloaca. Modified annules on ventral surface of cuticle anterior to cloaca. *Female*: Vulva opens anterior to mid-body. Uterus didelphic, opisthodelphic. Tail conical. Parasites in stomach of marine elasmobranchs and teleosts.

Type species: *Pulchrascaris caballeroi* Vincente and Dos Santos, 1972

[Emended diagnosis based on Deardorff (1987). Although the type species was thought to be a synonym of *P. chiloscyllii* by Gibson and Colin (1982) from the original description, Deardorff (1987) re-examined the holotype and considered it should remain a separate species. Vicente and Dos Santos (1972) originally placed the genus in the subfamily Filocapsulariinae Yamaguti, 1961. Fagerholm (1991) places the genus in the Anisakinae within the family Anisakidae. Listed in Spencer Jones and Gibson (1987).]

Anisakinea Chabaud, 1965

Terranova Leiper and Atkinson, 1914

DIAGNOSIS: Body reaches maximum width in anterior one-third of length. Cuticle with annulations. Lateral alae generally present. Mouth with 3 short lips. Flanges and interlabia absent. Cervical papillae well developed, project above body surface near nerve ring. Ventriculus cylindrical, 2.6–6.5 times long as wide, without appendix. Intestinal caecum present. Excretory system generally unilateral, ribbon-like, nucleus posterior but close to ventriculus, pore opens between base of subventral lips. *Male*: Spicules filiform, subequal. Precloacal papillae pedunculate, form continuous latero-ventral ridge. Median ventral precloacal papilla present. Median transverse cuticular plates present posterior to cloaca. Tails tapers, ornamentation absent. *Female*: Vulva opens in approximately anterior third of body. Muscular vagina clearly defined from uterus. Didelphic. Opisthodidelphic. Oviducts not extending anterior to vagina. Parasites of fish, crocodiles and snakes.

Type species: *Terranova antartica* Leiper and Atkinson, 1914

[Emended diagnosis based on Bruce and Cannon (1990). The type species is based on a single female from the stomach of a shark, thus creating taxonomic difficulties as the specific characters are generally associated with male specimens. Sprent (1979b), Gibson and Colin (1982) and Bruce and Cannon (1990) should be consulted for detailed reviews of the genus and lists of the species.]

Pseudoterranova Mozgovoi, 1951

[Gibson (1983) considers that there is no 'significant reason to distinguish the species' of Ascaridoid nematodes from seals and pygmy sperm whales 'at the generic level'. Therefore,

Pseudoterranova is recognized as the oldest name and *Phocanema* Myers, 1959 becomes a synonym. Listed in Spencer Jones and Gibson (1987).]

Contracaecinea Mozgovoi and Shakhmatova, 1971

[Sprent (1983b) considered this tribe a subfamily of Ascarididae. Fagerholm (1991) considers it a subfamily of Anisakidae and lists *Contracaecum* Railliet and Henry, 1912, *Phocascaris* Host, 1932 and *Galeiceps* Railliet, 1916 in the subfamily.]

Contracaecum Railliet and Henry, 1912

DIAGNOSIS: Ventriculus with posterior appendix. Intestinal caecum present. True interlabia with pulp present or 'interlabial knobs' present or absent. Excretory pore at distal margin of cephalic region. Folded cuticular collar present at distal margin of cephalic region. Parasites of reptiles, aquatic birds and mammals.

Type species: *Contracaecum microcephalum* (Rudolphi, 1809)

[Soleim (1984) emends the diagnosis of *Contracaecum* and includes all species with or without interlabia. Consequently, *Phocascaris* Höst, 1932 is considered a synonym of *Contracaecum*. Hartwich (1974) considers *Phocascaris* a valid genus. Emended diagnosis based on Soleim (1984). Fagerholm (1991) lists this genus in the subfamily Contracaecinae (Mozgovoi and Shakmatova, 1971 tribe).]

Phocascaris Höst, 1932

[Soleim (1984) proposes that *Phocascaris* should be a a synonym of *Contracaecum* Railliet and Henry, 1912. Fagerholm (1991) does not synonymize this genus with *Contracaecum*, including it as a separate genus in the subfamily Contracaecinae (Mozgovoi and Shakmatova, 1971 tribe).]

Goeziinae Travassos, 1919

Goezia Zeder, 1800

DIAGNOSIS: Cuticle with distinct annules. Posteriorly directed spines present or absent on annules. Mouth with flattened lips extending outwards, separating them from body. Oesophagus with slight constriction in middle. Ventriculus with long, single or double appendix. Short intestinal caecum present. Parasites of teleost fish, rarely reptiles.

Type species: *Goezia ascaridoides* (Goeze, 1782) Railliet and Henry, 1915

[The emended diagnosis is based on Kaur and Khera (1991), revised by them to accommodate two species, *G. taunsai* Zaidi and Khan, 1975 and *G. aspinulosa* Arya, 1980, which have annules without spines. Fagerholm (1991) lists this genus in the Raphidascarididae.]

Amphibiogoezia Rao, 1979 (Fig. 222)

DIAGNOSIS: Cuticle with annules, each with small spines on posterior border. Cephalic vesicle present. Mouth with 3 lips; dorsal lip with 2 papillae, subventral lips with single papilla and amphid; each lip with triangular tooth at base. Oesophageal funnel with forwardly directed spines. Distal oesophagus widens into elongate bulb enveloped in fold of intestine, which has a short caecum. *Male*: Spicules subequal. Gubernaculum present. Precloacal sucker present. *Female*: Vulva opens just anterior to mid-body. Parasite in intestine of *Rana cyanophlyctis* in Bangalore, India.

Type species: *Amphibiogoezia spinosoma* Rao, 1979

[Rao (1979) places the genus in the family Goeziidae on the basis of the three lips and a circlet of cuticular spines. This family is considered to be of subfamily rank in Hartwich (1974). Baker (1987) transfers this genus to the subfamily Cruziinae in the Cosmocercoidea. Fagerholm (1991) also indicates that this genus does not belong to the Ascaridoidea.]

Raphidascaridinae Hartwich, 1954

[Considered a subfamily of Ascarididae *sensu lato* by Sprent (1983b). Fagerholm (1991) elevates this subfamily to family rank based on the position of the excretory pore and papillae on the posterior end of the male. He lists *Paranisakis* Baylis, 1923, *Goezia* Zeder, 1800, *Sprentascaris* Petter and Cassone, 1984, *Hysterothylacium* Ward and Magath, 1917, *Raphidascaroides* Yamaguti, 1941, *Raphidascaris* Railliet and Henry, 1912, *Heterotyphlum* Spaul, 1927, *Lappetascaris*, Rasheed, 1965 and *Paraheterotyphlum* Johnston and Mawson, 1948 in the family.]

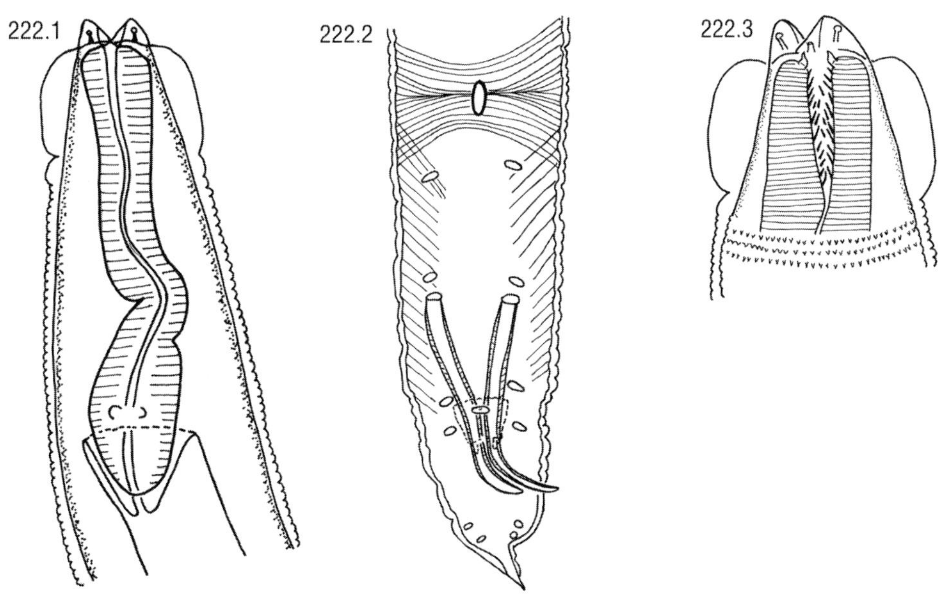

FIG. 222 *Amphibiogoezia* 222.1 Anterior end of female; 222.2 Posterior end of male, ventral view; 222.3 Cephalic region of female. (Redrawn after Rao, 1979.)

Raphidascarididinea Chabaud, 1965

Raphidascaris Railliet and Henry, 1915

DIAGNOSIS: Cuticle with fine annulations. Lateral alae arise between base of dorsal and subventral lips, expand in cervical region, extend to mid-region of tail. Mouth with 3 lips, with lateral flanges, constrict in mid-region, distinct groove at distal margins, pulp bilobed. Dorsal lip with 2 double papillae, subventral lips with 1 double papilla, 1 single papilla and adjacent amphid. Interlabia absent. Ventriculus short, appendage with septum. Intestinal caecum absent. Excretory system posteriorly bilateral, filamentar, reduced right canal, nucleus distinctly posterior to commissure, pore opens posterior to nerve ring. *Male*: Spicules short, alate, subequal. Gubernaculum absent. Median precloacal papilla present. Tail curves ventrally, unornamented, narrow, pointed distal tip. *Female*: Vulva opens in anterior quarter to third of body. Didelphic. Opisthodidelphic. Oviducts not extending anterior to vulva. Parasites of teleost fishes.

Type species: *Raphidascaris acus* (Bloch, 1779) Railliet and Henry, 1915

[Emended diagnosis based on Bruce (1990). Smith (1984) gives a redescription of the type species. Bruce (1990) only lists *Neogoezia* Kreis, 1937 as a synonym of *Raphidascaris*; the other two genera listed by Hartwich (1974), namely *Hysterothylacium* Ward and Magath, 1917 and *Ichthyascaris* Wu, 1949, have been reinstated.]

Subgenus *Raphidascaris* (Railliet and Henry, 1915 genus) Moravec, Kohn and Fernandes, 1990

DIAGNOSIS: Characteristics of the genus. Postlabial cuticular elevations absent or, if present, only poorly developed in ventral interlabial region. *Female*: Mature eggs in uterus contain several blastomeres.

Type species: *Raphidascaris* (*Raphidascaris*) *acus* (Bloch, 1779) Railliet and Henry, 1915

Subgenus *Sprentascaris* (Petter and Cassone, 1984 genus) Moravec, Kohn and Fernandes, 1990 (Fig. 223)

DIAGNOSIS: Characteristics of the genus. Mouth with 3 small lips, membranous horns present or absent. Interlabia and dentigerous ridges absent. Several well-developed postlabial cuticular elevations present. Lateral and cervical alae present or absent. Oesophagus club-shaped. Small ventriculus with short appendix containing nuclei of oesophageal glands. Intestinal caecum absent. Excretory pore opens at level of nerve ring. *Male*: Spicules equal. Gubernaculum absent. Caudal alae present or absent. *Female*: Vulva opens anterior to mid-body. Infundibula unequal in length. Uteri relatively equal, short. Seminal receptacle equal or unequal, may be present between infundibulum and uterus. Mature eggs in uterus embryonated. Parasites in Siluriforme freshwater fish in South America.

Type species: *Raphidascaris* (*Sprentascaris*) *mahnerti* (Petter and Cassone, 1984) Moravec, Kohn and Fernandes, 1990

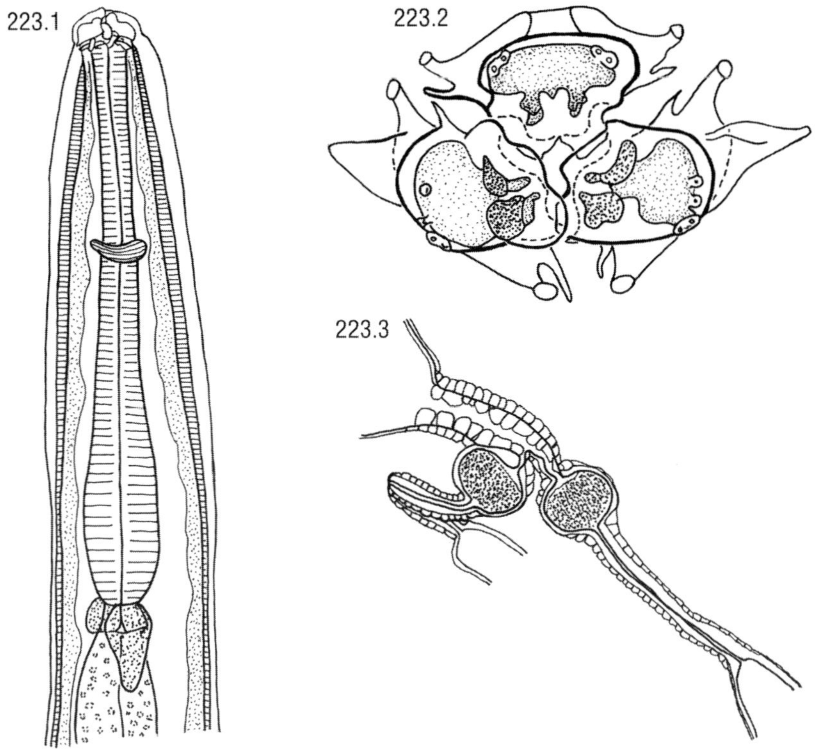

FIG. 223 *Sprentascaris* 223.1 Anterior end of body, ventral view; 223.2 Cephalic region, *en face* view; 223.3 Seminal receptacles at junction of infundibulum and uterus. (Redrawn after Petter and Cassone, 1984.)

[Moravec *et al.* (1990) considered the genus a subgenus of *Raphidascaris* on the basis of the presence of post-labial cuticular ornamentation, short oesophageal appendix and the morphology of the female reproductive system. Fagerholm (1991) lists it as a genus in the new family Raphidascarididae. Listed in Spencer Jones and Gibson (1987) as a genus.]

Hysterothylacium Ward and Magath, 1917

DIAGNOSIS: Lateral alae distinct or indistinct. Three distinct lips, approximately equal in size, cuticular flanges on lateral margins with or without indentations, internal pulp usually pedunculated. Dorsal lip with 2 lateral, double papillae, subventral lips each with medio-lateral double papilla, single lateral papilla and amphid. Dentigerous ridges absent. Interlabia present. Interlabial grooves present or absent. Ventriculus nearly spherical, appendage sack-like or cylindrical with septum dividing appendage into 2 equal longitudinal pouches. Intestinal caecum usually shorter than ventricular appendage. Excretory pore opens near level of nerve ring, duct extends to left lateral chord in which primary excretory canal extends posteriorly. *Male*: Spicules similar, alate, equal or slightly unequal in length. Gubernaculum absent. *Female*: Vulva opens anterior to mid-body. Uterus didelphic, opisthodelphic. Tail conical, tip with or without ornamentation. Medio-ventral preanal organ usually distinct on preanal fold. Parasites of marine, estuarine and freshwater fish.

Type species: *Hysterothylacium brachyurum* Ward and Magath, 1917

[Deardorff and Overstreet (1980, publ. 1981) resurrect the genus, listed as a synonym in Hartwich (1974), and redefine it to include those species previously considered as members of the junior synonyms of *Thynnascaris* Dollfus, 1933 and others described in the genus *Contracaecum* Railliet and Henry, 1912 that are adults in fish. The emended diagnosis is based on Deardorff and Overstreet (1980, publ. 1981) and, for further details, this reference should be consulted. Hartwich (1974) considered this genus a synonym of *Raphidascaris* Railliet and Henry, 1915.]

Heterotyphlum Spaul, 1927

DIAGNOSIS: Anterior body narrower than posterior half. Cuticle with fine annulations. Lateral alae present, extend from subventral lips to end of body. Mouth with 3 lips, half as long as wide, rounded anteriorly. Lateral flanges, constrictions, antero-lateral sockets, dentigerous ridges, distinct distal margin, interlabia absent. Dorsal lip: 2 double papillae. Subventral lips: 1 double papilla; 1 single papilla with adjacent amphid. Ventriculus shorter than wide with long appendage, slender anteriorly, sac-like posteriorly. Intestinal caecum present. Excretory pore opens near nerve ring. *Male*: Fine transverse cuticular crests present anterior to cloaca. Spicules short, alate, approximately 2.2% of body length. Tail broadly rounded, ending in simple mucron. *Female*: Vulva opens in anterior half of body. Didelphic. Opisthodidelphic. Oviducts not extending anterior to vagina.

Type species: *Heterotyphlum himantolophi* Spaul, 1927

[Emended diagnosis based on Bruce (1990).]

Thynnascaris Dollfus, 1933

DIAGNOSIS: Ventriculus with posterior appendix. Intestinal caecum present. 'Semi-interlabia' present. Excretory pore opens at level of nerve ring. Cuticular collar absent. Parasite of teleosts.

Type species: *Thynnascaris cornuta* (Stossich, 1804) Hartwich, 1975 (= *Thynnascaris legendrei* Dollfus, 1933)

[Spencer Jones and Gibson (1987) draw attention to the subgenus *Erschovicaecum* Mozgovoi, 1951 of the genus *Contracaecum*, which Hartwich (1974) includes as a synonym of *Thynnascaris* but spells it *Erschowicaecum*. Soleim (1984) emends the diagnosis of this genus and includes only species from teleosts in the genus. Emended diagnosis based on Soleim (1984) and type species on Hartwich (1975). Sprent (1983b) queries whether this genus is a synonym of *Hysterothylacium*.]

Iheringascaris Pereira, 1935 (Fig. 224)

DIAGNOSIS: Cuticle with overlapping annulations giving serrated appearance without spines. Lateral alae reduced or absent in adult, prominent in L_4 stage. Mouth with 3 equal-sized lips, cuticular flanges on lateral margins. Dentigerous ridges absent. Dorsal lip with 2 double sensory papillae. Subventral lips, each with 1 double papilla, 1 single papilla and 1 amphid. Interlabia present, deep interlabial grooves absent. Posterior border of lip and interlabial peduncles clearly defined. Ventriculus spherical, appendage cylindrical with internal longitudinal septum. Intestinal

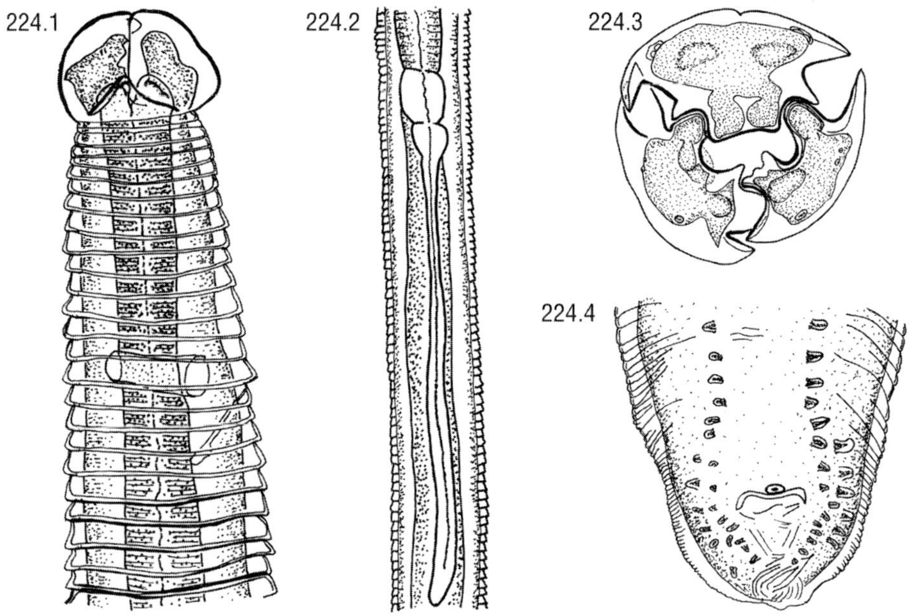

FIG. 224 *Iheringascaris* 224.1 Anterior end, lateral view; 224.2 Intestinal-ventricular junction, lateral view; 224.3 Cephalic region *en face* view; 224.4 Posterior end of male, ventral view. (Redrawn after Deardorff and Overstreet, 1980, publ. 1981.)

caecum present, shorter than ventricular appendage. Excretory system with duct arising from central commissure, bilateral canals extend posteriorly, pore opens near level of nerve ring. *Male*: Spicules approximately equal, similar. Gubernaculum absent. Caudal papillae as for tribe with additional lateral row on each side. Medio-ventral precloacal papilla present. *Female*: vulva opens anterior to mid-body. Didelphic. Opisthodelphic. Tail conical. Parasite of marine fish.

Type species: *Iheringascaris inquies* (Linton, 1901) Deardorff and Overstreet, 1981 (= *Inheringascaris inheringascaris* Pereira, 1935)

[Hartwich (1974) lists *Iheringascaris* Pereira, 1935 as a synonym of *Thynnascaris* Dollfus, 1933. Deardorff and Overstreet (1980, publ. 1981) reinstate the genus on the basis of the cuticle with distinctly overlapping annulations and the additional lateral rows of caudal papillae on the posterior end of the male. Bruce and Cannon (1990) also distinguish the genus from *Hysterothylacium* on the basis of *Iheringascaris* having a posterior bilateral excretory system (unilateral in *Hysterothylacium*) and the presence of distinct distal margins to the interlabial and lip peduncles (absent in *Hysterothylacium*). Deardorff and Overstreet (1980, publ. 1981), following studies of material identified as *Iheringascaris iheringascaris* Pereira, 1935 and specimens collected by Linton in 1899, conclude that this species is a junior synonym of *I. inquies* (Linton, 1901), which now becomes type species of the genus.]

Ichthyascaris Wu, 1949 (Fig. 225)

DIAGNOSIS: Body width greatest in anterior third. Cuticle with fine annulations. Lateral alae extend to near tail tip, join anteriorly to form flange posterior to subventral lips. Mouth with

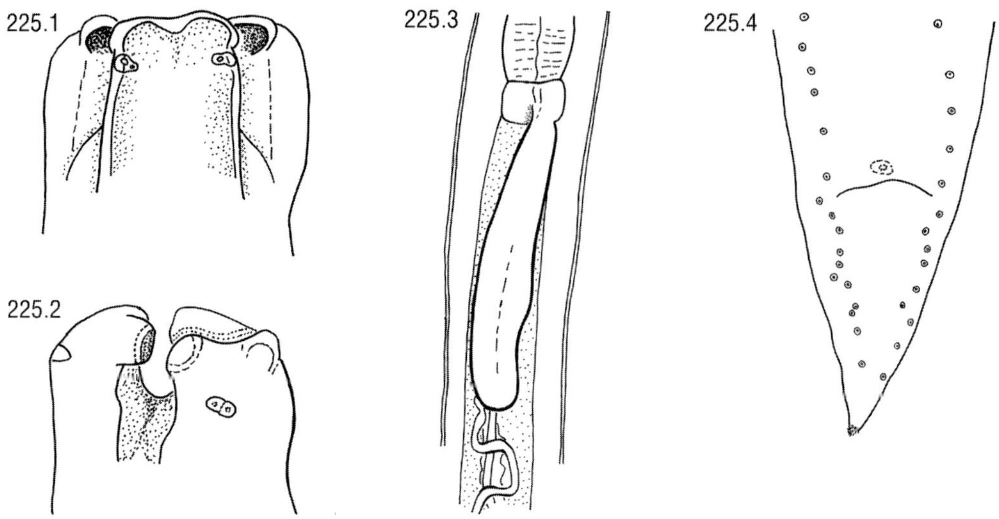

FIG. 225 *Ichthyascaris* 225.1 Female cephalic region, dorsal lip; 225.2 Female cephalic region, between lips; 225.3 Oesophago-ventricular junction; 225.4 Posterior end of male, ventral view. (Redrawn after Bruce, 1990.)

3 lips, long as wide, rounded lateral margins without flanges, teeth, distinct distal margins and lateral constrictions, 'antero-lateral angles with deep sockets', pulp bilobed anteriorly. Dorsal lip: 2 double sensory papillae. Subventral lips: 1 double, 1 single papilla with adjacent amphid. Interlabia absent. Ventriculus shorter than wide with short appendage arising from mid-region with longitudinal septum. Intestinal caecum absent. Excretory system unilateral, filamentary, nucleus placed immediately posterior to excretory commissure, pore opens posterior to nerve ring. *Male*: Spicules short, alate, 2.2–2.9% of body length. Medio-ventral precloacal papilla present. Tail tapers posteriorly, ends in recurved tip with fine 'nodules'. *Female*: Vulva opens in anterior quarter to third of body. Didelphic. Vagina not clearly separated from uterus. Uterus opisthodidelphic, oviducts extend anteriorly but not beyond vagina. Tail tapers posteriorly, ends in recurved tip with fine 'nodules'. Parasites of marine fish (except *I. biwakoensis*).

Type species: *Ichthyascaris lophii* Wu, 1949

[Bruce (1990) reinstates the genus previously considered a synonym of *Raphidascaris* Railliet and Henry, 1915 by Hartwich (1974), and the emended diagnosis is based on his description.]

Paraheterotyphlum Johnston and Mawson, 1948 (Fig. 226)

DIAGNOSIS: Body of even width. Lateral alae not expanded in cervical region, extend from posterior to subventral lips to distal end of body. Mouth with 3 lips: longer than wide; rectangular; without flanges, dentigerous ridges or lateral constrictions; distinct distal margin absent; 'sockets' or 'interlocking facets' present on antero-lateral corners; dorsal lip with 2 double papillae; subventral lips with 1 double, 1 single and an adjacent amphid; interlabia absent. Ventriculus shorter than wide with long appendage. Intestinal caecum present. Excretory system unilateral, filamentar, nucleus near nerve ring, pore opening posterior to nerve ring. *Male*: Spicules alate, short, 2% of body length. Numerous precloacal papillae. Cloaca surrounded with minute 'bosses'. Tail

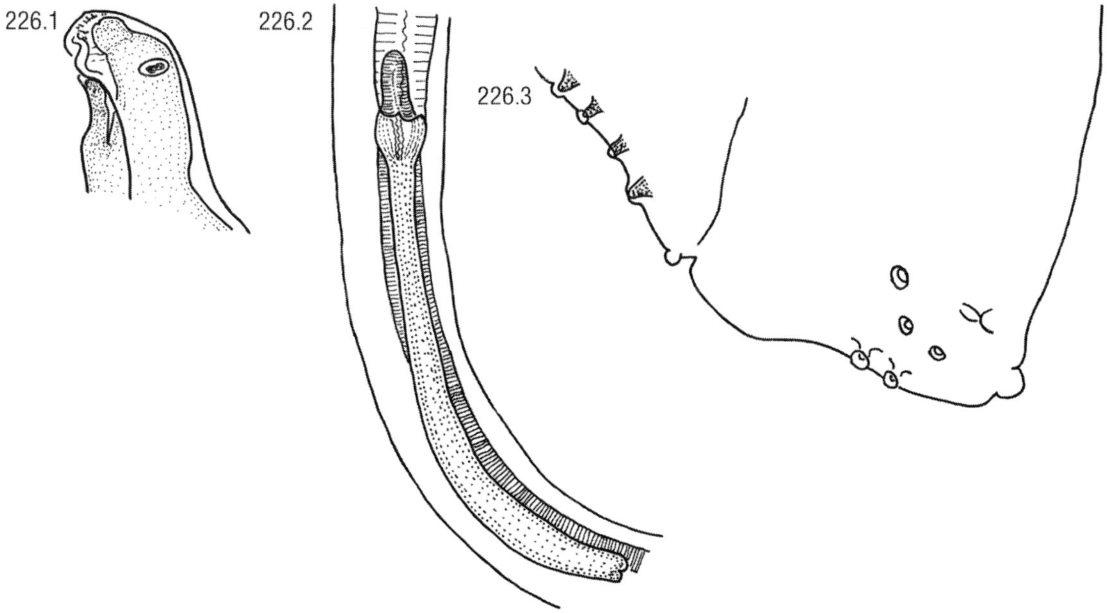

FIG. 226 *Paraheterotyphlum* 226.1 Dorsal lip, lateral view; 226.2 Oesophago-ventricular region; 226.3 Posterior end of male, lateral view. (Redrawn after Sprent, 1978a.)

straight, narrowly rounded with distal tip with minute 'bosses'. *Female*: Vulva opens in anterior third of body. Didelphic. Opisthodelphic. Oviducts not extending anterior to vagina. Tail short, spines on distal tip. Parasite of sea snakes in Australian waters and off Taiwan.

Type species: *Paraheterotyphlum australe* Johnston and Mawson, 1948

[Emended diagnosis based on Sprent (1978a) and Bruce (1990). Hartwich (1974) excluded the genus from his key as it was incompletely described. Listed in Spencer Jones and Gibson (1987).]

Raphidascaroides Yamaguti, 1941

DIAGNOSIS: Cuticle with fine annulations. Mouth with 3 lips longer than wide, each with lateral flanges, dentigerous ridges, constriction in mid-region, bilobed pulp. Dorsal lip with 2 double sensory papillae, each subventral lip with 1 double, 1 single papilla and adjacent amphid. Interlabia present. Ventriculus short with appendage. Intestinal caecum absent. Excretory pore opens posterior to nerve ring. *Male*: Spicules alate, subequal. Gubernaculum absent. Tail slightly curved ventrally, minutely spinose distal tip. *Female*: Vulva opens one-third of body length from anterior. Oviducts not extending anterior to vulva. Parasites of teleost fish.

Type species: *Raphidascaroides nipponensis* Yamaguti, 1941

[Emended diagnosis based on Bruce (1990). *Ryjikovascaris* Mozgovoi, 1951 is listed as a synonym of *Raphidascaroides* by Hartwich (1974), and Bruce (1990) indicates that a redescription of the type species is required to assess the validity of the synonymy. Khalil and Oyetayo (1988) give a key to the species.]

Alibagascaris Kalyankar, 1970 (Fig. 227)

DIAGNOSIS: Mouth with 3 lips, each with dentigerous ridge and tooth. Interlabia absent. Oesophagus simple. Ventriculus with appendix. Intestinal caecum absent. *Male*: Distal end with numerous ventral papillae. Spicules equal, similar. Gubernaculum present. Tail tapers posterior to anus with tuft of spines at distal tip. *Female*: Vulva opens anterior to mid-body. Tail with spines at distal tip. Parasites in intestine of marine fish.

Type species: *Alibagascaris pseudosciaenae* Kalyankar, 1970

[Kalyankar (1970) placed *Alibagascaris* in the family Heterocheilidae, subfamily Filocapsulariinae Yamaguti, 1961. Hartwich (1974) retained only the genus *Heterocheilus* Diesing, 1839 in the family Heterocheilidae, which are parasites of sirenians and not marine fish. Kalyankar (1970) considers *Raphidascaris* and *Raphidascaroides*, both parasites of fish, as the nearest genera; consequently, *Alibagascaris* is reassigned to the tribe Raphidascarididinea. Listed in Spencer Jones and Gibson (1987).]

Sprentascaris Petter and Cassone, 1984

[Moravec *et al.* (1990) consider this genus a subgenus of *Raphidascaris*. See under genus *Raphidascaris* for diagnosis, and Fig. 223.]

Maricostula Bruce and Cannon, 1989 (Fig. 228)

DIAGNOSIS: Body maximum width in anterior third to half of length. Cuticle with distinct annulation. Lateral alae present. Mouth with 3 lips: equal or dorsal lip smaller; pedunculate; deep postlabial grooves present; prominent lateral flanges present; constricted on lateral edges

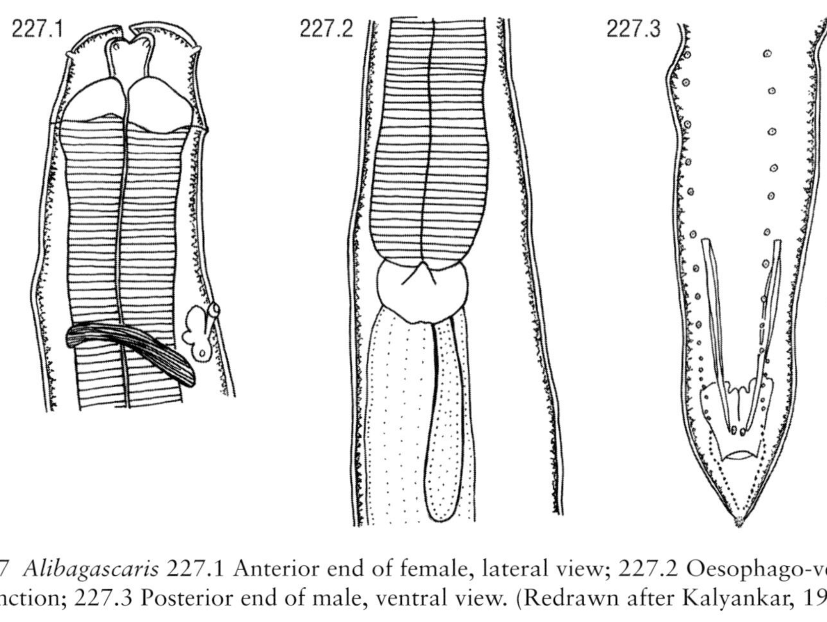

FIG. 227 *Alibagascaris* 227.1 Anterior end of female, lateral view; 227.2 Oesophago-ventricular junction; 227.3 Posterior end of male, ventral view. (Redrawn after Kalyankar, 1970.)

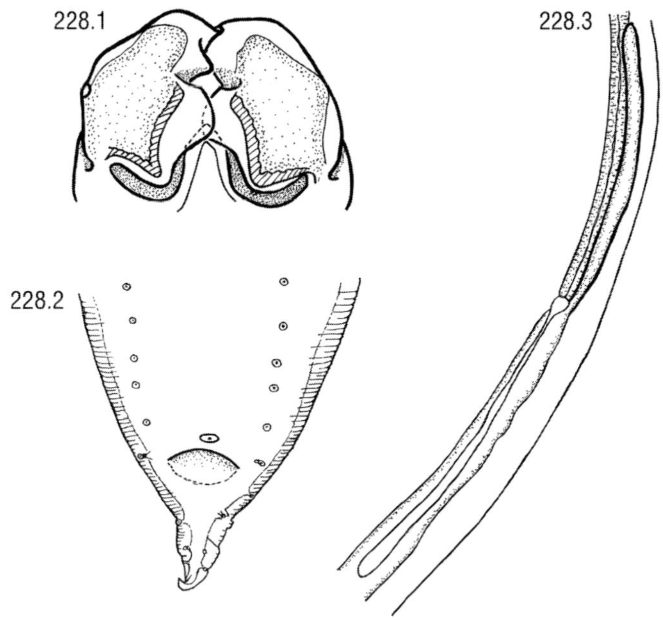

FIG. 228 *Maricostula* 228.1 Cephalic region, ventral view; 228.2 Posterior end of male, ventral view; 228.3 Oesophago-ventricular region showing ventricular appendix and intestinal caecum. (Redrawn after Bruce and Cannon, 1989.)

anterior to flanges; pulp pedunculated, anteriorly bilobed. Dorsal lip with 2 double sensory papillae, each subventral lip with 1 double, 1 single papilla and adjacent amphid. Interlabia present. Ventriculus longer than wide, pyriform, appendage widens posteriorly, anterior portion with septum, posterior portion glandular. Intestinal caecum long, over 50% of oesophagus length. Excretory system generally unilateral, filamentar, right side not extending to ventriculus, left side glandular, nucleus posterior to excretory canal, pore opens posterior to nerve ring. *Male*: Cuticular annulation prominent anterior to cloaca, rows of transverse crests present. Medio-ventral papilla anterior to cloaca. Spicules filiform, alate. Gubernaculum absent. 'Postcloacal pad absent'. Tail sharply pointed without ornamentation. *Female*: Vulva opens at or anterior to mid-body. Vagina clearly defined. Didelphic. Opisthodidelphic. Parasites of marine fish.

Type species: *Maricostula makairi* Bruce and Cannon, 1989

Lappetascaridinea (Rasheed, 1965 subfamily)

Mehdiascaris Kalyankar, 1969 (Fig. 229)

DIAGNOSIS: Body long, stout. Mouth with 3 lips, each with 2 teeth. Interlabia absent. Lappets arise at proximal margins of lip. Cephalic region distinctly separated from body. Oesophagus long. Ventriculus small with long appendix. Intestinal caecum very short. *Male*: Caudal alae absent. Papillae numerous. Spicules equal or subequal. Gubernaculum absent. *Female*: Ophisthodelphic. Vulva opens in posterior half of body. Parasite of marine fish.

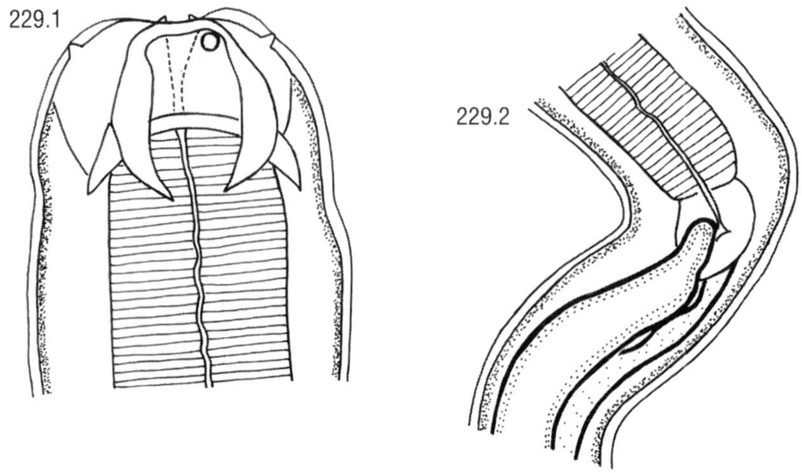

Fig. 229 *Mehdiascaris* 229.1 Anterior end, dorsal view; 229.2 Oesophago-ventricular junction. (Redrawn after Kalyankar, 1969.)

Type species: *Mehdiascaris bombayensis* Kalyankar, 1969

[Kalyankar (1969) places the genus in the subfamily Lappetascaridinae. Hartwich (1974) moves this subfamily to tribe status. Fagerholm (1991) indicates that this genus might be a synonym of *Hysterothylacium*. Listed in Spencer Jones and Gibson (1987).]

Paranisakinea Hartwich, 1974

Ortoanisakis (Mozgovoi, 1951 subgenus) Yamaguti, 1961

DIAGNOSIS: Dentigerous ridges absent. Oval ventriculus. Excretory pore near nerve ring. *Male*: Gubernaculum absent. Parasite of teleost fish.

Type species: *Ortoanisakis lophii* (Yamaguti, 1935)

[Sprent (1990a) proposes that this genus be reinstated. Hartwich (1974) lists this genus as a synonym of *Paranisakis* Baylis, 1923. Sprent (1990a) did not have the type specimens to examine and was unable completely to redefine the genus. The diagnosis given is based on Sprent (1990a). Sprent (1990a) lists three additional species and a subspecies that are included in the genus.]

Mawsonascaris Sprent, 1990 (Fig. 230)

DIAGNOSIS: Body stout. Mouth with 3 lips, each with wide alate margins, dentigerous ridges extending only from anterior to anterior edge of alate margin. Pulp with narrow extensions into alate margins. Excretory pore opens just posterior to nerve ring, nucleus placed in anterior end of larger left lateral filament. Oval ventriculus present. Intestinal caecum present or absent. *Male*: Spicules alate, unequal, long, slender. Gubernaculum present. Fewer than 12 precloacal papillae. *Female*: Vulva

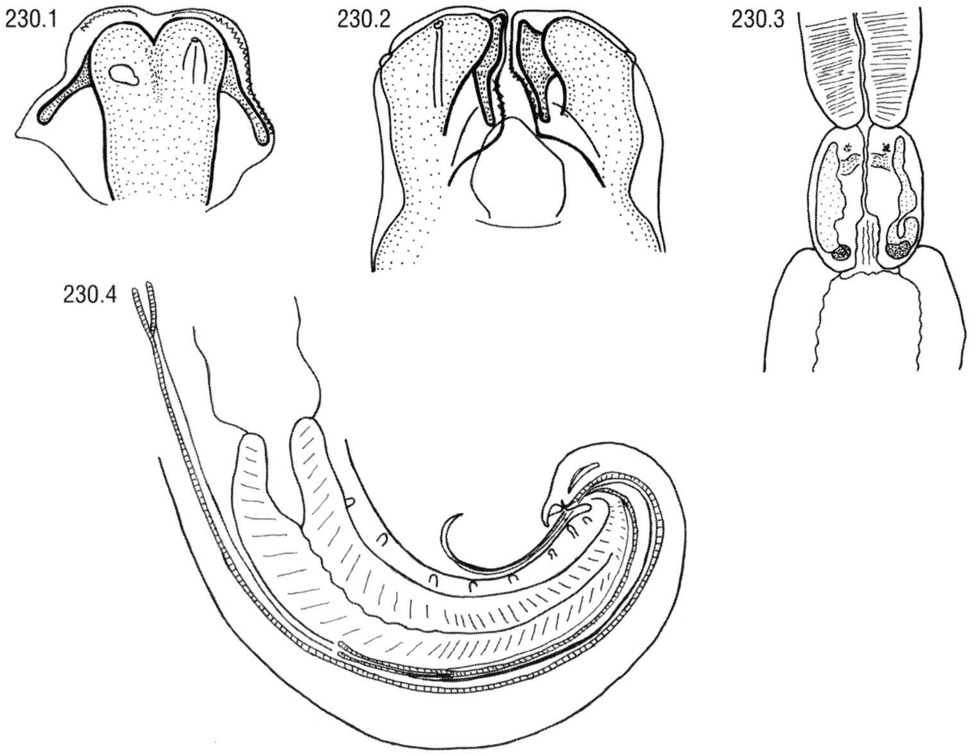

FIG. 230 *Mawsonascaris* 230.1 Subventral lip; 230.2 Cephalic region, lateral view; 230.3 Oesophago-ventricular region; 230.4 Posterior end of male, lateral view. (Redrawn after Sprent, 1990a.)

opens anterior to mid-body. Parasites in stomach, spiral valve and intestine of fish in Australian, former USSR (Peter the Great Bay), West African and Indian marine and estuarine coastal waters.

Type species: *Mawsonascaris australis* (Johnston and Mawson, 1943) Sprent, 1990

[Placed provisionally in Paranisakinea as type species; originally a species of the genus *Paranisakis*. Petter *et al.* (1991) place this in the subfamily Paranisakinae (Hartwich, 1974 tribe).]

Ascarididae Baird, 1853

[Sprent (1983b) divides the family Ascardidae *sensu lato* into five subfamilies, namely Ascaridinae, Heterocheilinae, Raphidascaridinae, Anisakinae and Contracaecinae, mainly on the basis of the excretory system, ventriculus morphology and the shape of the oesophageal gland nuclei. This represents a major change from Hartwich (1974). For the details of the subfamily diagnosis and a key to the subfamilies, Sprent (1983b) should be consulted.

Sprent (1983b) lists *Heterocheilus*, *Paradujardinia*, *Hartwichia*, *Trispiculascaris*, *Dujardinascaris*, *Ortleppascaris*, *Gedoelstascaris*, *Krefftascaris*, *Brevimulticaecum* and *Typhlophorus* in the **Heterocheilinae**; *Ascaris*, *Baylisascaris*, *Toxascaris*, *Parascaris*, *Orneoascaris*, *Ophidascaris*, *Travassosascaris*, *Lagochilascaris*, *Hexametra*, *Polydelphis*, *Angusticaecum*, *Toxocara* and *Porrocaecum* in the **Ascaridinae**; *Raphidascaris*, *Hysterothylacium* (? = *Thynnascaris*),

Iberingascaris, *Raphidascaroides*, *Paraheterotyphlum*, *Heterotyphlum*, *Lappetascaris* and *Goezia* in the **Raphidascaridinae**; *Anisakis*, *Pseudoterranova*, *Pulchrascaris*, *Sulcascaris*, *Terranova* (= *Phocanema*) and *Paranisakiopsis* in the **Anisakinae**; and *Contracaecum*, *Galeiceps* and *Phocascaris* in the **Contracaecinae**.

Fagerholm (1991) includes the Heterocheilidae and Raphidascarididae as separate families, the subfamilies Contracaecinae and Anisakinae in the family Anisakidae and accepts the validity of the subfamilies Toxocarinae and Angusticaecinae in the Ascarididae.]

Ascaridinae (Baird, 1853 family) *sensu lato**

[*Sprent (1982) considers that Ascaridinae *sensu lato* includes Angusticaecinae and Toxocarinae *sensu* Hartwich (1974). Sprent (1982) considers that the morphological features of *Bairdascaris dasypodina* and *Lagochilascaris turgida* support the merger of Ascaridinae *sensu lato* with Angusticaecinae.

Fagerholm (1991) lists *Parascaris* Yorke and Maplestone, 1926, *Ascaris* Linnaeus, 1758, *Baylisascaris* Sprent, 1968 and *Toxascaris* Leiper, 1907 in this subfamily.]

Bairdascaris Sprent, 1982 (Fig. 231)

DIAGNOSIS: Mouth with 3 lips, each with indentation anterior margin, dentigerous ridges, labial pulp divided into 2 lobes (asymmetrical in subventral lips), extensions of pulp with

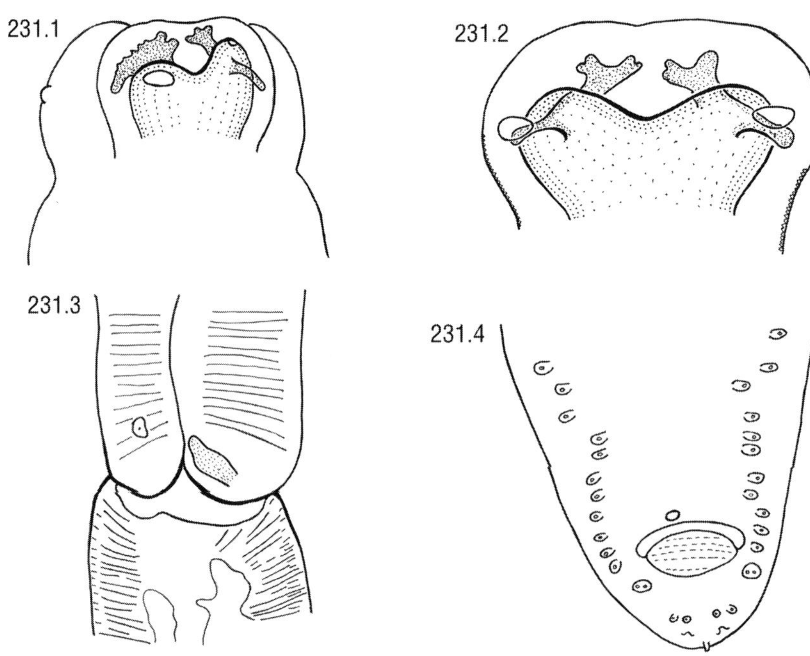

Fig. 231 *Bairdascaris* 231.1 Cephalic region showing subventral lip; 231.2 Dorsal lip; 231.3 Oesophago-intestinal junction, lateral view; 231.4 Posterior end of male, ventral view. (Redrawn after Sprent, 1982.)

sickle-shaped process. Interlabia absent. Excretory pore opens at level of nerve ring, nucleus 'in commissure at junction of posterior bilateral filaments'. Ventriculus absent. Dorsal oesophageal gland nucleus oval, in 'dorsal sector of oesophagus'. Intestinal caecum absent. *Male*: Thirty-plus precloacal papillae. Spicules alate. Gubernaculum absent. Just posterior to margin of cloaca cuticle roughened. *Female*: Vulva opens in anterior half of body. Didelphic. Parasites in small intestine of armadillo (*Cabassous unicinctus*, *Tolypeutes matacos*, *Tatus* sp.) in South America, Paraguay and Brazil.

Type species: *Bairdascaris dasypodina* (Baylis, 1922) Sprent, 1982

[Sprent (1982) considers this genus belongs to Ascaridinae *sensu lato*. Fagerholm (1991) lists this genus in the Angusticaecinae of the Ascarididae. Genus listed in Spencer Jones and Gibson (1987).]

Toxocarinae (Hartwich, 1954 family) Osche, 1958

[Fagerholm (1991) lists only *Toxocara* Stiles, 1905 and *Porrocaecum* Railliet and Henry, 1912 in this subfamily.]

Neoascaris Travassos, 1927

[Hartwich (1974) retains this genus as a synonym of *Toxocara* Stiles, 1905. Paggi *et al.* (1985), using biochemical techniques, suggest that *Neoascaris* and *Toxocara* are distinct genera resulting from an 'ancient process of evolutionary divergence'. The synonymy of the two genera is widely accepted.]

Paradujardinia Travassos, 1933 (Fig. 232)

DIAGNOSIS: Body large. Mouth with 3 smoothly rounded lips, each with internal transverse groove and marginal notches. Dentigerous ridges absent. Interlabia present. Cervical papillae posterior to nerve ring. Excretory pore opens near nerve ring, excretory filament bilateral. Cervical alae absent. Ventriculus present. Intestinal caecum present. *Male*: Body slightly smaller than female. Four precloacal papillae, 2 paracloacal papillae, 2 subventral postcloacal papillae on each side of tail. Spicules alate in mid-region, stop short of distal tips. Gubernaculum absent. *Female*: Vulva opens anterior to mid-body. Didelphic. Parasite in stomach and small intestine of sirenians (*Dugong dugong* in Sprent, 1980a) in Indian Ocean, Red Sea, Bay of Bengal, Philippine Sea, Coral Sea, Sea of Japan.

Type species: *Paradujardinia halicoris* (Owen, 1833)

[Emended diagnosis based on Sprent (1980a), who moves this genus to the subfamily Heterocheilinae in the Ascarididae. Hartwich (1974) includes this genus in the Toxocarinae.]

Viverranisakis Soota and Chaturvedi, 1971 (Fig. 233)

DIAGNOSIS: Mouth with 3 lips. Dentigerous ridges present. Interlabia present. Oesophagus 'club-shaped'. Ventriculus present without appendix. Intestinal caecum absent. *Male*: Spicules

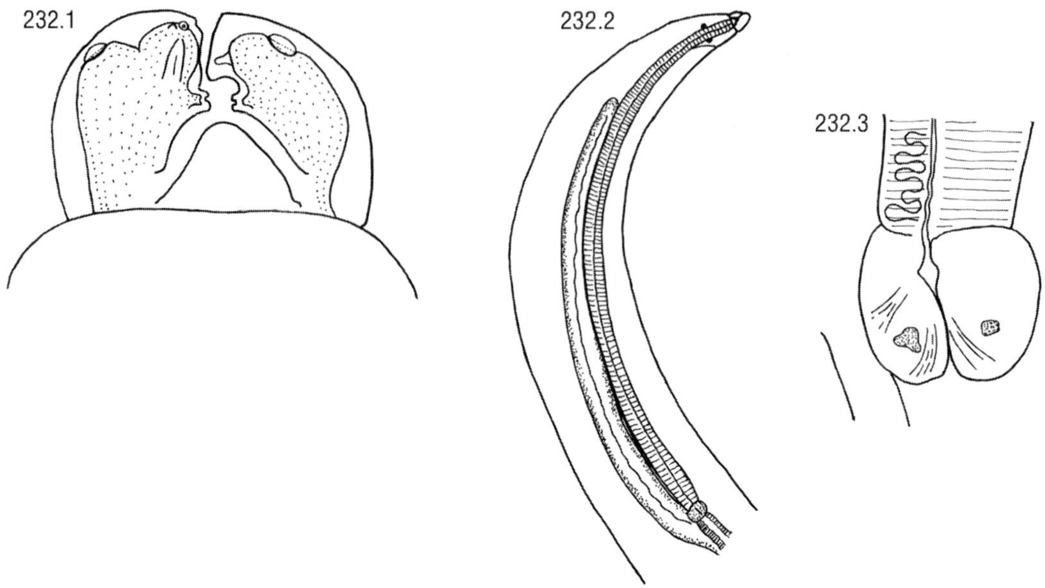

FIG. 232 *Paradujardinia* 232.1 Cephalic region, ventral view; 232.2 Anterior end showing oesophagus, ventriculus and intestinal caecum; 232.3 Oesophago-ventricular junction, lateral view. (Redrawn after Sprent, 1980a.)

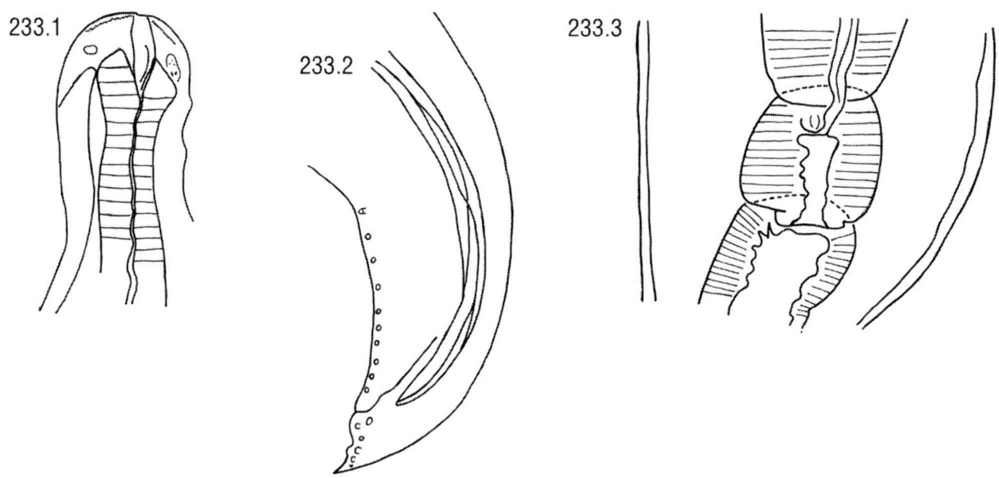

FIG. 233 *Viverranisakis* 233.1 Anterior end of male; 233.2 Posterior end of male, lateral view; 233.3 Oesophago-ventricular junction, female. (Redrawn after Soota and Chaturvedi, 1971.)

equal. Gubernaculum absent. Numerous papillae on posterior end. Tail with 'constriction'. *Female*: Eggs rounded. Parasite in intestine of Civet cat, *Viverricula indica*, and jungle cat, *Felis chaus*, in Goa.

Type species: *Viverranisakis molemensis* Soota and Chaturvedi, 1971

[The features of this genus indicate that it is nearer to members of the subfamily Toxocarinae than to the Heterocheilidae, where it was originally placed by Soota and Chaturvedi (1971). Re examination of thc typc specimens may reveal that the species is a member of the genus *Toxocara*. Listed in Spencer Jones and Gibson (1987).]

Multicaecinae (Hartwich, 1954 family) Campana-Rouget, 1960

[Sprent (1980a) does not include this subfamily, in Hartwich (1974), in his classification.]

Multicaecinea Campana-Rouget, 1960

***Multicaecum* Baylis, 1923 (Fig. 234)**

DIAGNOSIS: Mouth with 3 lips, rounded, smooth, 'non-alate'. Dentigerous ridges present. Interlabia short. Excretory pore opens at or just posterior to level of nerve ring, system bilateral, excretory cell extends anteriorly. Cervical papillae posterior to nerve ring. Cervical alae present. Ventriculus present with long 'finger-like' projections. Intestinal caecum present. *Male*: Caudal alae absent. Spicules alate. Gubernaculum present. *Female*: Vulva opens anterior to mid-body. Vagina short. Uterus with undivided and divided region forming 2 opisthodelphic branches. Parasite in stomach of *Crocodylus* spp. in Africa, India and Australia.

Type species: *Multicaecum agile* (Wedl, 1861) Baylis, 1923

[Emended diagnosis based on Sprent (1979a). Sprent (1980a) moves *Multicaecum* to the Heterocheilinae.]

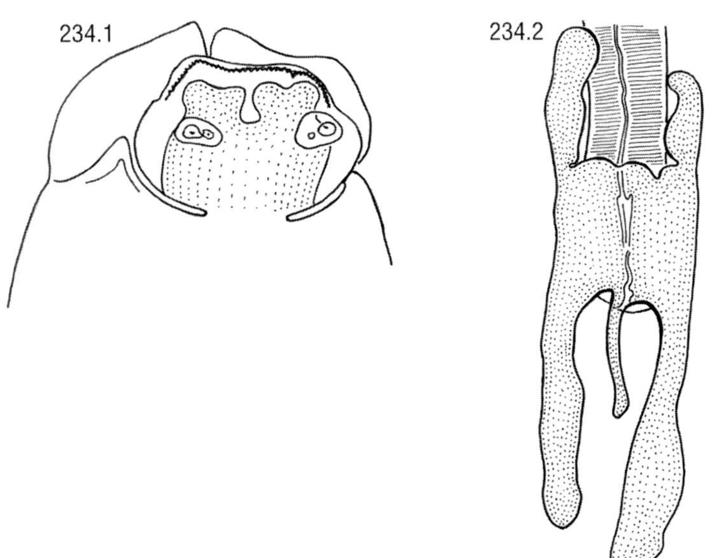

FIG. 234 *Multicaecum* 234.1 Dorsal lip; 234.2 Ventriculus region. (Redrawn after Sprent, 1979a.)

Brevimulticaecum Mozgovoi, in Skrjabin, Shikhobalova and Mozgovoi, 1952 (Fig. 235)

DIAGNOSIS: Mouth with 3 diamond-shaped lips, borders alate, dentigerous ridges absent. Distal margin of lips separated from cephalic region by distinct groove. Interlabia well developed. Excretory pore opens just anterior or at level of nerve ring, system bilateral without anterior extensions of excretory cell. Cervical alae present or absent. Ventriculus present with short appendices. Intestinal caecum present. *Male*: Spicules alate. Gubernaculum slender. Caudal alae absent. *Female*: Uterus with undivided and divided region forming 2 uterine branches opisthodelphic or amphidelphic. Parasites in oesophagus and stomach of New World caimans and alligators.

Type species: *Brevimulticaecum baylisi* (Travassos, 1933) Sprent, (1979)

[Hartwich (1974) includes the subgenus *Brevimulticaecum* as a synonym of the genus *Multicaecum*. Sprent (1979a) reinstates and raises the subgenus to generic rank. The emended diagnosis is based on Sprent (1979a). Sprent (1980a) moves this genus to the Heterocheilinae. Sprent (1990b) considers *Gedoelstascaris* Sprent, 1978 a synonym of *Brevimulticaecum*. Fagerholm (1991) lists both *Gedoelstascaris* and *Brevimulticaecum* in the Heterocheilidae.]

Angusticaecinae Skrjabin and Karokhin, 1945

[Fagerholm (1991) lists *Seuratascaris* Sprent, 1985, *Orneoascaris*, Skrjabin, 1916, *Raillietascaris* Sprent, 1985, *Angusticaecum* Baylis, 1920, *Ophidascaris* Baylis, 1921, *Polydelphis* Dujardin, 1845, *Hexametra* Travassos, 1919, *Travassosascaris* Sprent, 1978, *Bairdascaris* Sprent, 1982, *Freitasascaris* Sprent, 1983 and *Lagochilascaris* Leiper, 1909 in this subfamily.]

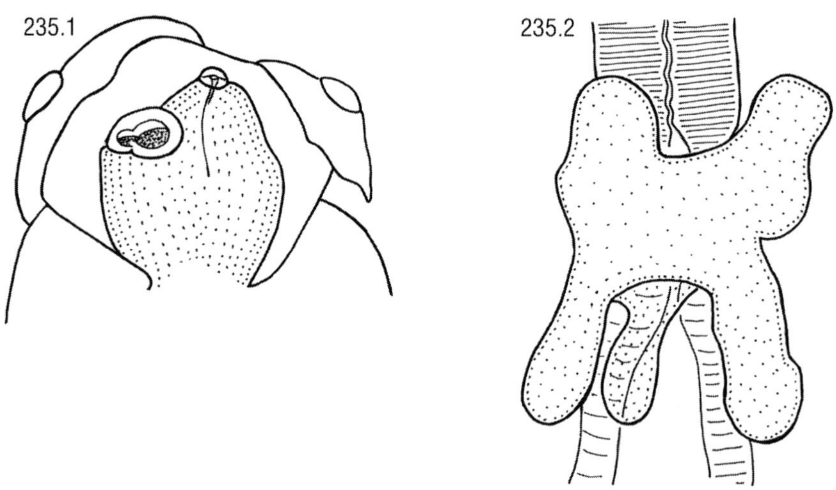

FIG. 235 *Brevimulticaecum* 235.1 Cephalic region showing subventral lip; 235.2 Ventriculus region. (Redrawn after Sprent, 1979a.)

Genera not designated to a tribe

Orneoascaris Skrjabin, 1916 (Fig. 236)

DIAGNOSIS: Body small to medium. Mouth with 3 lips, 'slightly bulbous', isthmus narrow, dentigerous ridge present along whole margin, deep division at anterior margin. Pulp divided anteriorly forming 2 rounded lobes, median lobe absent. Interlabia present. Cervical alae absent. Excretory pore opens just posterior to nerve ring, system bilateral, nucleus on left side of commissural part of cell. Cervical papillae just posterior to nerve ring. Distal end of oesophagus lobed, distinct ventriculus absent. Nuclei of oesophageal glands fairly large. Intestinal caecum extends over half length of oesophagus. *Male*: Spicules slender, shorter than ejaculatory duct. Gubernaculum absent. Ornamentation anterior to cloaca. Lateral cuticle of tail widens. *Female*: Vulva opens anterior to mid-body. Didelphic. Opisthodelphic. Parasite in stomach and intestine of Old World frogs and toads, tortoises, frog-eating snakes, chamaeleons, monitor lizards, crocodiles (occasionally).

Type species: *Orneoascaris chrysanthemoides* Skrjabin, 1916

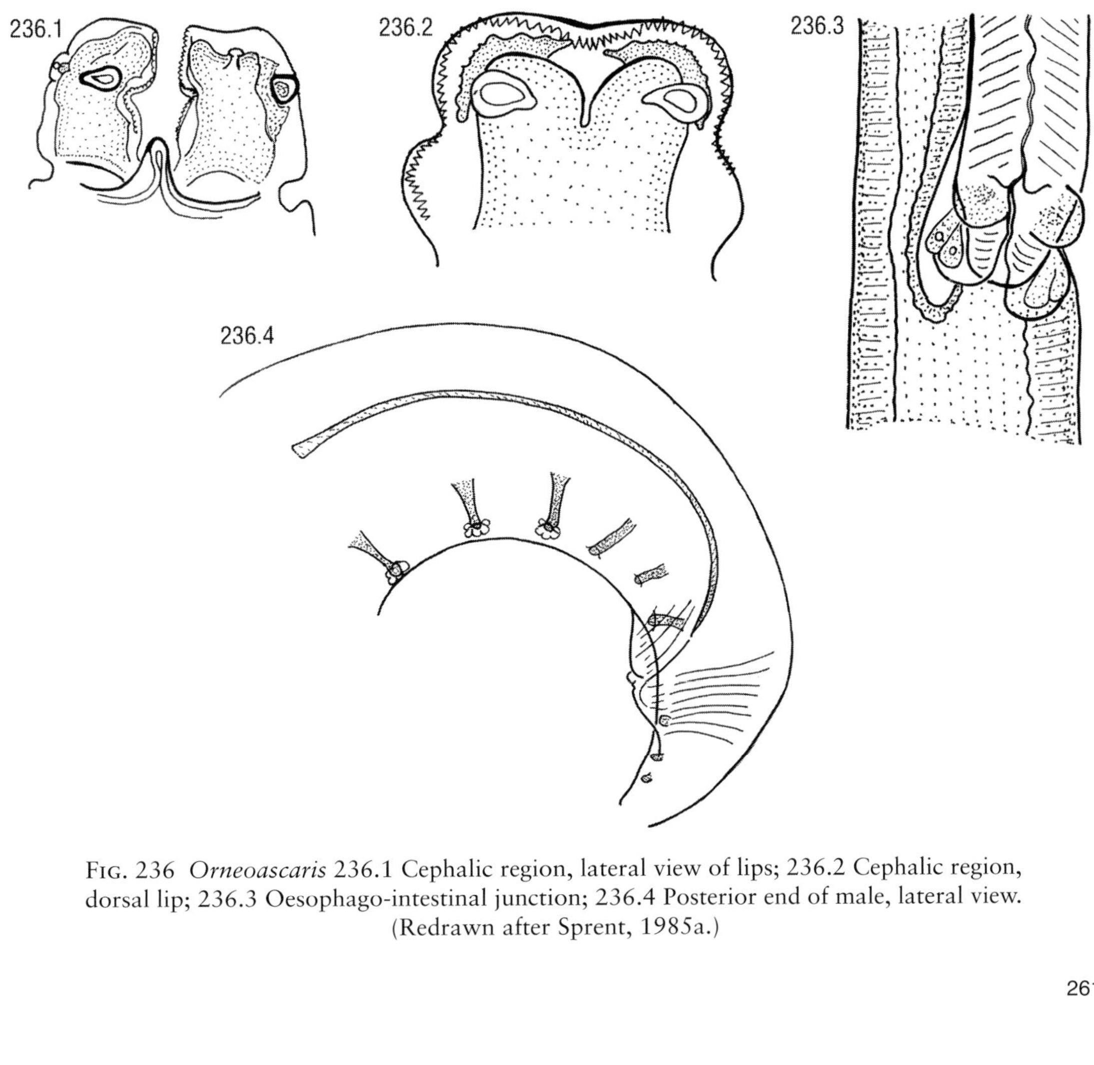

FIG. 236 *Orneoascaris* 236.1 Cephalic region, lateral view of lips; 236.2 Cephalic region, dorsal lip; 236.3 Oesophago-intestinal junction; 236.4 Posterior end of male, lateral view. (Redrawn after Sprent, 1985a.)

[*Orneoascaris* was considered to be a synonym of *Amplicaecum* by Hartwich (1974). Sprent (1985a) revalidates the genus, and the emended diagnosis is based on his description. Sprent (1985a) indicates that some of the characters 'may indicate affinities with [the] Heterocheilinae'. Hartwich (1974) placed *Amplicaecum* in the Angusticaecinae, a subfamily which Sprent (1983b) considers should be merged with the Ascaridinae. Fagerholm (1991) lists this genus in the Angusticaecinae.]

Freitasascaris Sprent, 1983 (Fig. 237)

DIAGNOSIS: Body medium-sized. Mouth with 3 lips, each narrowing anteriorly 'with deep notch continuous with oral groove'. Pulp deeply divided anteriorly, median lobe absent. Posterior margins of lips 'with narrow isthmus', adjacent lip margins forked. Interlabia present linked by postlabial ring. Excretory system bilateral, nucleus in commissure. Cervical alae present. Ventriculus absent. Intestinal caecum present. *Male*: Spicules without alae. Gubernaculum absent. *Female*: Posterior body enlarged when mature. Vulva opens in posterior half of body. Didelphic. Parasite in stomach and intestine of tejid lizards, *Tupinambis nigropunctatus* and *T. teguixin*, in Brazil, Trinidad and Uruguay.

Type species: *Freitasascaris alata* (Baylis, 1947) Sprent, 1983

[The nucleus of the dorsal oesophageal gland is enlarged in the type species of this genus, which is a unique feature in reptiles and shows affinities with ascaridoids from mammals. Sprent (1983c) considers this adds further support to his proposals that Angusticaecinae should be combined within the Ascaridinae (see Sprent, 1982). Sprent (1983b) lists *Angusticaecum* in the Ascaridinae, which would indicate a merger of the two subfamilies. Listed in Spencer Jones and Gibson (1987).]

Raillietascaris Sprent, 1985 (Fig. 238)

DIAGNOSIS: Body small to medium. Mouth with 3 lips each with dentigerous ridge along whole margin, oral groove, posterior margin forming isthmus or 'pillar', pulp deeply divided anteriorly, median lobe absent. Interlabia present linked to isthmus of lips by 'collar'. Cervical alae present. Excretory pore opens near nerve ring, system with bilateral filaments.

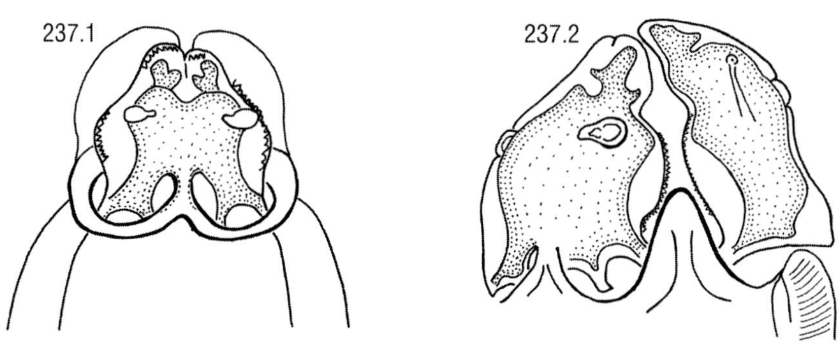

FIG. 237 *Freitasascaris* 237.1 Cephalic region, dorsal view; 237.2 Cephalic region, lateral view. (Redrawn after Sprent, 1983b.)

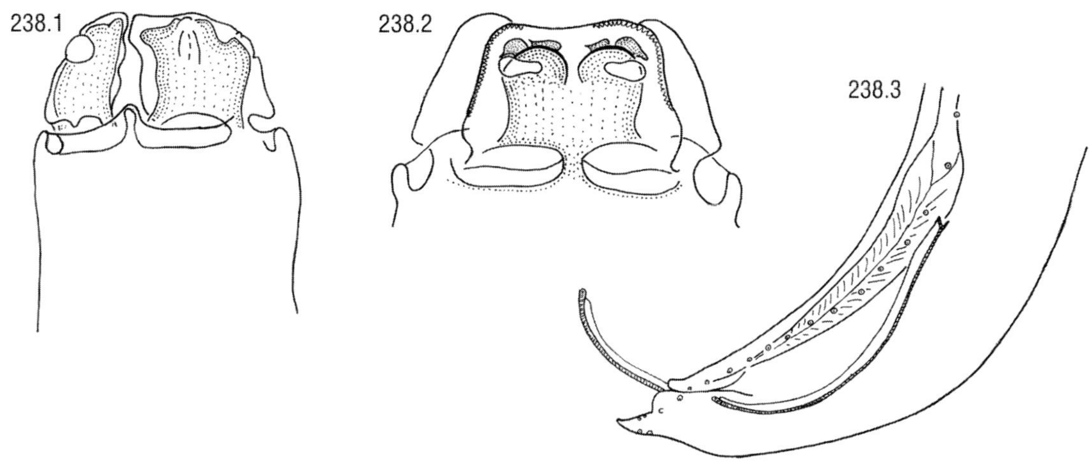

FIG. 238 *Raillietascaris* 238.1 Cephalic region showing lips with membranous collar and pillars, lateral view; 238.2 Cephalic region showing dorsal lip; 238.3 Posterior end of male, latero-ventral view. (Redrawn after Sprent, 1985c.)

Ventriculus absent. Nuclei of oesophageal glands large within oesophagus. Intestinal caecum present. *Male*: Spicules alate. Twenty-plus pairs precloacal, 1 double pair paracloacal, 2 subventral, 2 subdorsal postcloacal papillae. Ornamentation anterior to cloaca absent. *Female*: Vulva opens anterior to mid-body. Two uterine branches. Parasite in stomach and intestine of *Varanus* spp. and *Physignathus cocincinus* in Thailand, Australia, Philippine Islands and Africa.

Type species: *Raillietascaris varani* (Baylis and Daubney, 1922) Sprent, 1985

[The type species is formerly a species of *Amplicaecum*, which was placed in the subfamily Angusticaecinae. Sprent (1982) considers Angusticaecinae should be merged with the Ascaridinae. Listed in Spencer Jones and Gibson (1987).]

Seuratascaris Sprent, 1985 (Fig. 239)

DIAGNOSIS: Body small to medium. Mouth with 3 lips, just narrower than body, anterior margin with cleft, posterior margin with 'wide isthmus', dentigerous ridges present. Pulp without deep anterior division, median lobe absent. Cervical alae absent. Excretory pore opens just posterior to nerve ring, cervical papillae at same level as excretory pore. Excretory system bilateral, nucleus on left-hand side of 'commissural part of excretory cell'. Ventriculus absent. Intestinal caecum present. Rectal glands present. *Male*: Spicules short, stout. Gubernaculum absent. Lateral tail cuticle slightly expanded. *Female*: Vulva opens anterior to mid-body. Didelphic. Opisthodelphic. Parasites in stomach and intestine of Old World anurans, *Rana* spp., *Bufo* spp., *Kaloula* spp., *Polypedetes leucomystax* and *Megophrys nasicornis*.

Type species: *Seuratascaris numidica* (Seurat, 1917) Sprent, 1985

[Listed in Spencer Jones and Gibson (1987).]

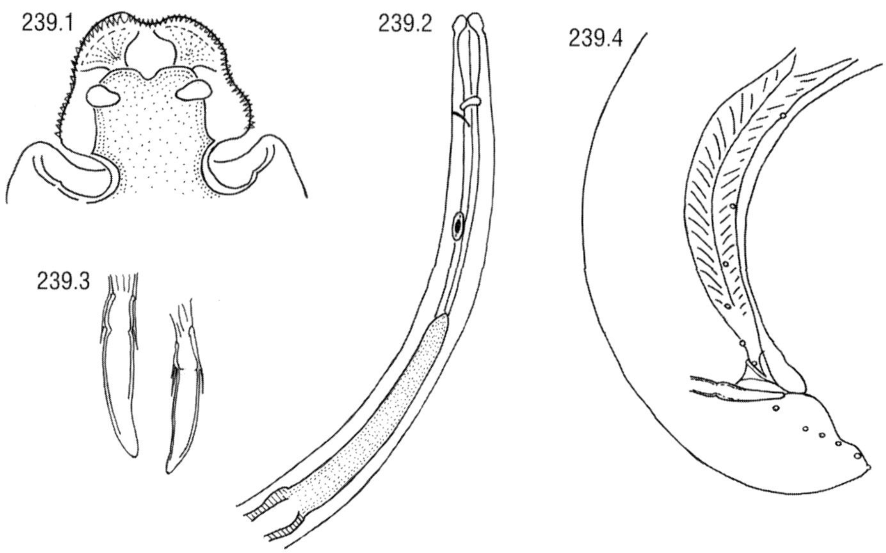

FIG. 239 *Seuratascaris* 239.1 Dorsal lip; 239.2 Anterior end, showing caecum, lateral view; 239.3 Spicules; 239.4 Posterior end of male, lateral view. (Redrawn after Sprent, 1985b.)

Angusticaecinea Chabaud, 1965

Angusticaecum Baylis, 1920

DIAGNOSIS: Body medium to large. Mouth with 3 trapezoidal, rounded lips, pulp with hammer-shaped anterior prolongation, median lobe of pulp absent or poorly developed. In adults interlabial and postlabial grooves absent. Dentigerous ridges present. Excretory pore opens just posterior to nerve ring, excretory filament bilateral. Cervical papillae posterior to excretory pore. 'Cuticular bars' present in lateral field, 'salient alae' absent in adult. Ventriculus absent. Intestinal caecum present, narrow. *Male*: Spicules alate. Gubernaculum absent. Twenty or more pairs of precloacal, 2 pairs of single or 1 pair of double paracloacal, 4 pairs postcloacal papillae. *Female*: Vulva opens anterior to mid-body. Didephic. Opisthodelphic. Adults parasitic in oesophagus, stomach, small intestine, more rarely large intestine; larvae in lungs of Chelonians (*Testudo* spp., *Emys orbicularis*, *Psammobates geometrica*, *Terrapine carolina* and *Kinyxis belliana*) in Europe, Armenia, Azerbaijan, Turkestan, Caspian region, Iran, Morocco, Brazil and Central Africa.

Type species: *Angusticaecum holopterum* (Rudolphi, 1819) Baylis, 1920

[Emended diagnosis based on Sprent (1980b).]

Amplicaecum Baylis, 1920

DIAGNOSIS: Three well-developed lips with dentigerous ridges. Interlabia present or absent. Posterior oesophageal bulb or ventriculus absent. Intestinal caecum present. *Male*: Caudal alae present or absent. Numerous precloacal papillae, few postcloacal papillae. Gubernaculum absent. *Female*: Vulva opens in anterior part of body. Parasites of fish, amphibians, reptiles and birds.

Type species: *Amplicaecum colurum* Baylis, 1920

[Emended diagnosis based on Gupta and Srivastava (1984, publ. 1985).]

Ophidascaridinea (Hartwich, 1954 subfamily) Chabaud, 1965

Travassosascaris Sprent, 1978 (Fig. 240)

DIAGNOSIS: Mouth with 3 flat lips, attachment to body thin, hammer-shaped extensions of pulp, dentigerous ridges present. Transverse groove present at base of lips. Interlabia present. Lateral alae present, narrow with cuticular bars, extend from level of nerve ring until posterior to level of oesophago-intestinal junction. Cervical papillae just posterior to excretory pore. Dorsal sector of oesophagus 'displaced laterally'. Ventriculus absent. Dorsal oesophageal gland extends from posterior oesophagus, opening between nerve ring and dorsal lip. Subventral glands open near nuclei in posterior oesophagus. Intestinal caecum absent. Excretory pore opens just posterior to nerve ring, system bilateral, nucleus in left filament extends into commissure. *Male*: Spicules alate. Gubernaculum absent. *Female*: Vulva opens just anterior to mid-body, 4 uterine branches. Parasites of rattlesnakes (*Crotalus durissus territicus*) in Brazil and Central America.

Type species: *Travassosascaris araujoi* Sprent, 1978 (= *nom. nov.* for *Polydelphis quadrangularis* (Schneider, 1866) of Araujo, 1969)

[This genus is placed in the Ophidascaridinea a tribe included in Hartwich (1974), on the basis that is established for a former species of *Polydelphis*. Hartwich (1974) included *Polydelphis* Dujardin, 1845, *Ophidascaris* Baylis, 1921 and *Hexametra* Travassos, 1919 in this tribe. Listed in Spencer Jones and Gibson (1987).]

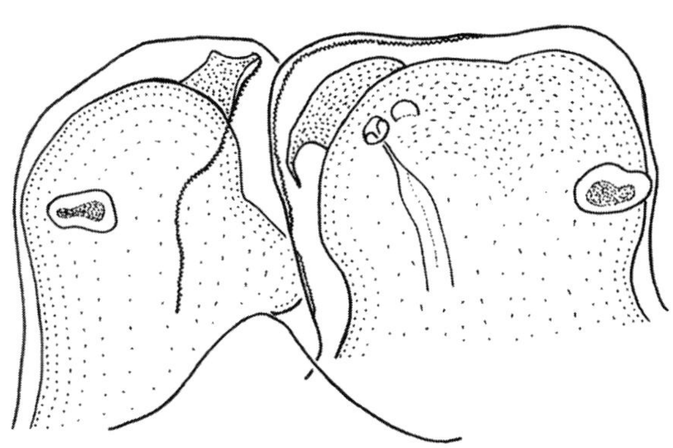

FIG. 240 *Travassosascaris* Cephalic region, lateral view. (Redrawn after Sprent, 1978c.)

SUBULUROIDEA

Subuluridae (Travassos, 1914 subfamily) Yorke and Maplestone, 1926

Subulurinae Travassos, 1914

Travassallodapa López-Neyra, 1945

DIAGNOSIS: Parasite in the intestine of birds and mammals.

Type species: *Travassallodapa allodapa* (Creplin, 1853) López-Neyra, 1945

[López-Neyra (1945) erected the genus on the basis of spicule morphology. Skrjabin *et al.* (1964) list the type species as a synonym of *Subulura allodapa* (Creplin, 1853) Railliet and Henry, 1913. The type species of López-Neyra's genus is currently the type species of *Allodapa* Diesing, 1861, i.e. *Allodapa allodapa* and, consequently, *Travassallodapa* is considered a synonym of *Allodapa*. *Allodapa* is now the type genus of the subfamily Allodapinae Inglis, 1958. Listed in Spencer Jones and Gibson (1987).]

Allodapinae Inglis, 1958

Inglisubulura Devamma, 1975 (Fig. 241)

DIAGNOSIS: Cervical alae generally present. Hexagonal mouth opening with 6 lips. Extensions of buccal cavity lining narrower than lobes of lips. Buccal cavity circular in cross-section. Proximal end of oesophagus straight, not lobed or twisted. Parasites of Indian birds.

Type species: *Inglisubulura turnicis* (Maplestone, 1931) Devamma, 1975

[Listed in Spencer Jones and Gibson (1987).]

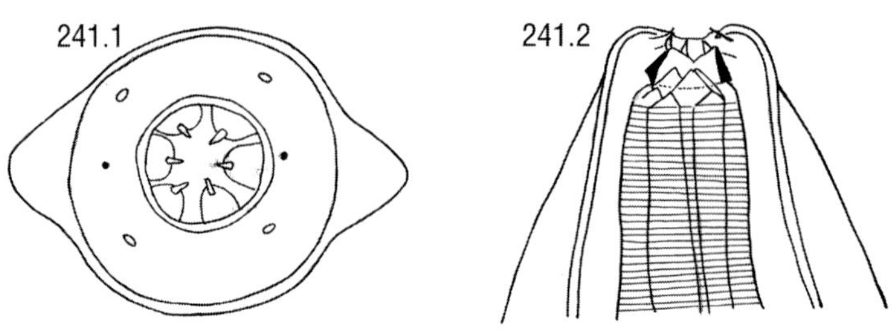

FIG. 241 *Inglisubulura* 241.1 Cephalic region, *en face* view; 241.2 Anterior end, dorsal view. (Redrawn after Devamma, 1975.)

Gallioxynema Shamim and Deshmukh, 1980 (Fig. 242)

DIAGNOSIS: Cervical alae present. Mouth opening with '3 lips'. Buccal capsule shorter than wide without teeth. Oesophagus with corpus, very short isthmus and bulb. *Male*: Precloacal sucker with ornamented cuticular membrane. Caudal alae narrow. Eleven pairs of papillae present on posterior end. Spicules equal, alate. Gubernaculum present. *Female*: Unknown. Parasite in intestine and caecum of *Gallus gallus domesticus*.

Type species: *Gallioxynema thapari* Shamim and Deshmukh, 1980

[Shamim and Deshmukh (1980, publ. 1981) distinguish their genus from *Subulura* and consider it nearest to *Heteroxynema* Hall, 1916. The characters illustrated, however, appear closest to those used by Chabaud (1978) in his key to the Subuluroidea.]

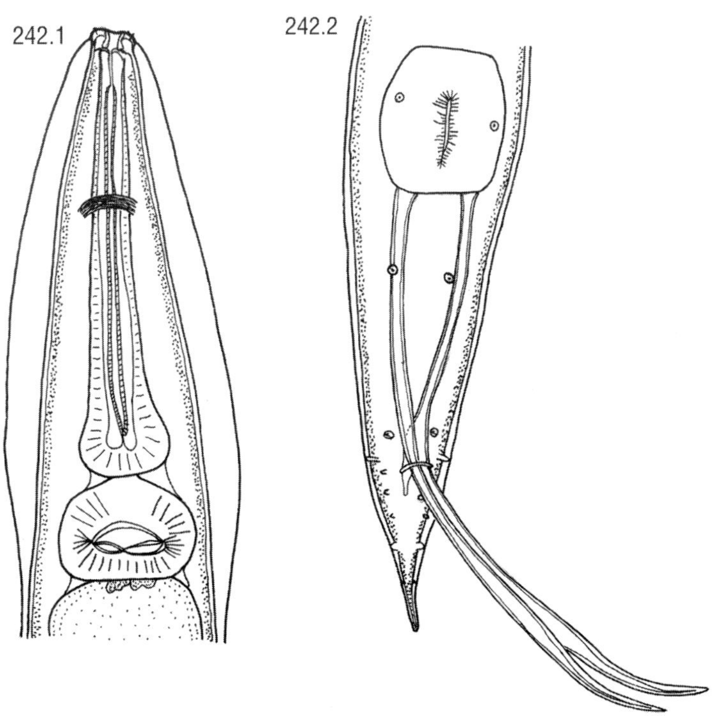

FIG. 242 *Gallioxynema* 242.1 Anterior end of male, lateral view; 242.2 Posterior end of male, ventral view. (Redrawn after Shamim and Deshmukh, 1980, publ. 1981.)

6 SPIRURIDA

Spiroptera Rudolphi, 1819

[The genus erected by Rudolphi (1819) originally contained 30 species. Although Yorke and Maplestone (1926) indicate that the genus is synonymized in part with the genera *Acuaria* Bremser, 1811 and *Spirura* Blanchard, 1819, the many species assigned to this genus are now spread between several genera, including *Acuaria* Bremser, 1811, *Cosmocephalus* Molin, 1858, *Chevreuxia* Seurat, 1918, *Echinuria* Soloviev, 1912, *Habronema* Diesing, 1861, *Schistorophus* Railliet, 1916, *Sciadiocara* Skrjabin, 1916, *Seuratia* Skrjabin, 1916, *Spirura* Blanchard, 1819 and *Synhimantus* Railliet, Henry and Sisoff, 1912. The current position of all the species of this genus is beyond the remit of this volume. The genus listed in Spencer Jones and Gibson (1987) is no longer considered valid.]

CAMALLANINA

[Tikhomirova (1975) proposed a new superfamily Skrjabillanoidea in the suborder Camallanina for the Skrjabillanidae Schigin and Schigina, 1958. Chabaud (1975a) includes the subfamily Skrjabillaninae in the Anguillicolidae of the Dracunculoidea.]

CAMALLANOIDEA

Camallanidae (Railliet and Henry, 1915 subfamily)

Camallaninae Railliet and Henry, 1915

DIAGNOSIS: Buccal capsule divided into 2 lateral valves, posterior division absent or, if present, distinctly smaller, deep lateral groove present or absent, longitudinal bands in 1–2 groups, continuous, rough or beaded. Tridents present or absent or reduced to a single rod.

Type genus: *Camallanus* Railliet and Henry, 1915

[Chabaud (1975a) did not divide the family into subfamilies. However, his key uses the same criteria to separate the two groups of genera as that of Petter (1979c), who reinstates the subfamily. Later, Bilqees and Akram (1982) revised the Camallanidae and reinstated three subfamilies, including the Camallaninae. Petter (1979c) also included *Paracamallanus* Yorke and Maplestone, 1926, *Oncophora* Diesing, 1851, *Serpinema* Yeh, 1960 and *Camallanides* Baylis and Daubney, 1922 in the subfamily. Bilqees and Akram (1982) reinstated Paracamallaninae for the genus *Paracamallanus*. The emended diagnosis is based on Petter (1979c) and Bilqees and Akram (1982).]

Type genus: *Camallanus* Railliet and Henry, 1915

Subgenus *Zeylanema* (Yeh, 1960 genus) Moravec and Scholz, 1991

DIAGNOSIS: Buccal capsule with 10 longitudinal ribs each armed with 2–5 prominent teeth. Parasites in intestine of freshwater fish (Anabantidae, Cyprinidae, Belontiidae, Clariidae).

Type species: *Camallanus* (*Zeylanema*) *anabantis* Pearse, 1933

[Moravec and Scholz (1991) consider *Zeylanema* should be retained as a subgenus of *Camallanus*. Later, Moravec (1994) lists *Zeylanema* as a synonym of *Camallanus*. Chabaud (1975a) lists it as a synonym of *Camallanus* in his key.]

Neoparacamallanus Bilqees and Akram, 1982 (Fig. 243)

DIAGNOSIS: Buccal capsule with paired lateral valves, posterior division greater than one-third of length. Longitudinal bands form 1 group, continuous or discontinuous, smooth or beaded. Tridents present or absent. Parasites of freshwater fish.

Type species: *Neoparacamallanus sweeti* (Moorthy, 1937) Bilqees and Akram, 1982

[Bilqees and Akram (1982) separate this genus from *Camallanus* on the length of the posterior division of the buccal capsule, one-third or less of length in *Camallanus*. The type species was later referred to the subgenus *Dentocamallanus* of the genus *Paracamallanus* by Moravec and Scholz (1991) in the subfamily Paracamallaninae. Listed in Spencer Jones and Gibson (1987).]

Neocamallanus Ali, 1957 (Fig. 244)

DIAGNOSIS: Tridents in buccal capsule absent. Buccal capsule with well-developed distal ring, smooth longitudinal bars present on internal surface of lateral valves.

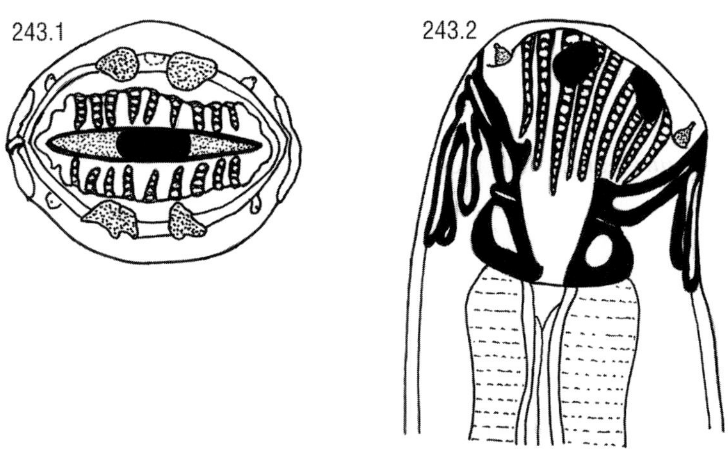

FIG. 243 *Neoparacamallanus* 243.1 Cephalic region, *en face* view; 243.2 Cephalic region, lateral view. (Redrawn after Moorthy, 1937.)

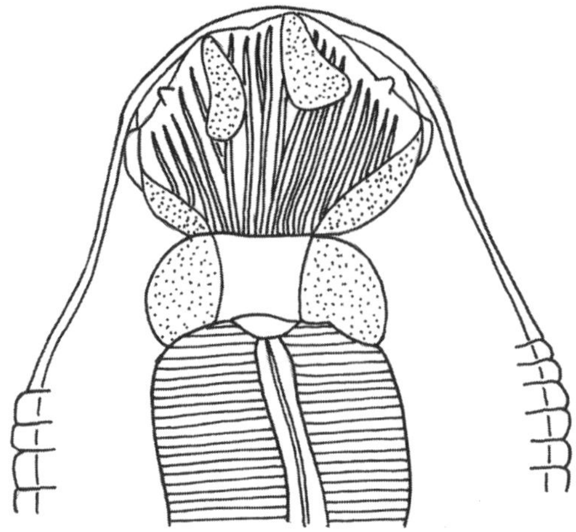

FIG. 244 *Neocamallanus* Cephalic region, lateral view. (Redrawn after Moravec and Sey, 1988a.)

Type species: *Neocamallanus singhi* Ali, 1957

[The validity of this genus has been discussed by many authors; Chabaud (1975a) considered it a synonym of *Camallanus*, while Petter (1979c) considered it in part synonym of *Camallanus* and *Paracamallanus*; Moravec and Sey (1988a) considered it should be retained, and the diagnosis is based on their publication.]

Procamallaninae Yeh, 1960

DIAGNOSIS: Buccal capsule continuous, not separated into 2 valves. Sclerotized bands present or absent or chitinous plates with 'leaf crowns' present. Trident absent.

[Petter (1979a) revised the group and reinstated the subfamily Procamallaninae in the family Camallanidae. Chabaud (1975a) did not use these subfamilies in his key.]

Type genus: *Procamallanus* Baylis, 1923

DIAGNOSIS: Buccal capsule with smooth internal walls, not separated into 2 valves. Parasites in stomach and intestine of fish and amphibians.

Type species: *Procamallanus laeviconchus* (Wedl, 1862) Railliet and Henry, 1915

[Emended diagnosis based on Petter (1979a) and Bilqees and Akram (1982). Chabaud (1975a) did not separate this genus into subgenera.]

Subgenus *Procamallanus* (Baylis, 1923 genus) Ali, 1957

DIAGNOSIS: Inner surface of buccal capsule smooth. *Male*: Two unequal spicules.

Type species: not designated. ?*Procamallanus* (*Procamallanus*) *laeviconchus* (Wedl) Baylis, 1923

[Rodrigues *et al.* (1991) support the subgeneric status of *Procamallanus*.]

Subgenus *Monospiculus* Ali, 1957

DIAGNOSIS: Characters of the genus. *Male*: Single spicule present.

Type species: not designated. ?*Procamallanus* (*Monospiculus*) *parasiluri* Fujita, 1927

[Listed in Spencer Jones and Gibson (1987).]

Subgenus *Isospiculus* Ali, 1957

DIAGNOSIS: Characters of the genus. *Male*: 2 equal spicules.

Type species: not designated. ?*Procamallanus* (*Isospiculus*) *mehrii* Agarwal, 1930

[For the complete list of species assigned to the subgenera, Ali (1957) should be consulted. Listed in Spencer Jones and Gibson (1987).]

Subgenus *Spirocamallanus* (Olsen, 1952 genus) Moravec and Sey, 1988

DIAGNOSIS: Spiral thickening present on internal surface of buccal capsule in both sexes. Parasites of fish and amphibians.

Type species: *Procamallanus* (*Spirocamallanus*) *spiralis* (Baylis, 1923)

[Moravec and Sey (1988a) consider that, as the genus *Spirocamallanus* is separated from *Procamallanus* only by the presence of spiral thickenings in the buccal capsule, it should become a subgenus to accommodate species with spiral thickenings in the buccal capsules of both sexes. Chabaud (1975a) and Petter (1979a) consider *Spirocamallanus* a separate genus. Rodrigues *et al.* (1991) support the subgeneric status of *Spirocamallanus* within the genus *Procamallanus*.]

Subgenus *Spirocamallanoides* Moravec and Sey, 1988

DIAGNOSIS: Spiral thickenings present on internal surface of buccal capsule in female only.

Type species: *Procamallanus* (*Spirocamallanoides*) *siluri* Osmanov, 1964

Subgenus *Punctocamallanus* Moravec and Scholz, 1991

DIAGNOSIS: Distinct punctations present on inner surface of buccal capsule. Parasites in stomach of freshwater fish in Laos.

Type species: *Procamallanus* (*Punctocamallanus*) *punctatus* Moravec and Scholz, 1991

Subgenus *Denticamallanus* Moravec and Thatcher, 1997

DIAGNOSIS: *Male*: Inner surface of buccal capsule with approximately 10 large teeth. *Female*: Inner surface of buccal capsule with 4–5 spiral thickenings. Parasite in intestine of characid fish, *Bryconops alburnoides*, Uburu River, Amazonas State, Brazil.

Type species: *Procamallanus* (*Denticamallanus*) *dentatus* Moravec and Thatcher, 1997

Onchocamallanus Petter, 1979a (Fig. 245)

DIAGNOSIS: Buccal capsule globular, not divided into 2 lateral valves. Discontinuous transverse striations on inner surface of buccal capsule. Three sclerotized, rounded processes protruding from base of buccal capsule.

Type species: *Onchocamallanus bagarii* (Karve and Naik, 1951) Petter, 1979

[Moravec and Sey (1988a) do not consider the features described for separating the genus are sufficient, and consider this genus a synonym of *Procamallanus*. They list the type species in the subgenus *Spirocamallanus* temporarily. Chabaud (1975a) accepts the validity of *Spirocamallanus* as a genus. Listed in Spencer Jones and Gibson (1987).]

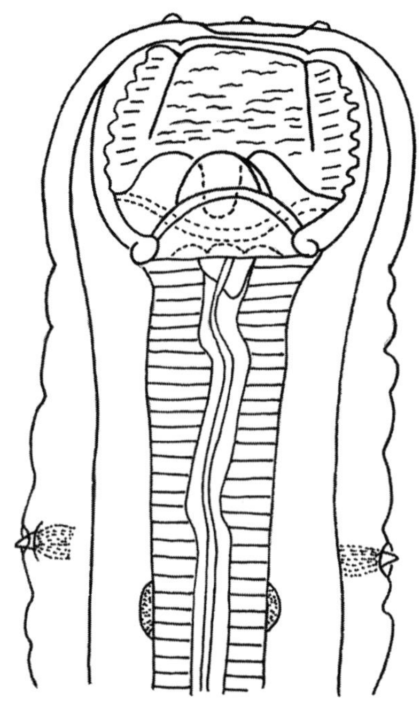

FIG. 245 *Onchocamallanus* Anterior end of female body, dorsal view. (Redrawn after Karve and Naik, 1951 from Ivashkin *et al*., 1971.)

Malayocamallanus Jothy and Fernando, 1970

DIAGNOSIS: Buccal capsule with internal wall ornamented with longitudinal ridges. Parasite in the freshwater fish *Fluta alba*, Symbranchii, in Malaysia.

Type species: *Malayocamallanus intermedius* Jothy and Fernando, 1970

[Placed in this subfamily by Bilqees and Akram (1982).]

Platocamallanus Bilqees and Akram, 1982 (Fig. 246)

DIAGNOSIS: Buccal capsule continuous, smooth. Chitinous plates with a leaf crown present on proximal margin. Parasites of freshwater fish.

Type species: *Platocamallanus mehrii* (Agrawal, 1930) Bilqees and Akram, 1982

[Listed in Spencer Jones and Gibson (1987).]

Batrachocamallanus Jackson and Tinsley, 1995 (Fig. 247)

DIAGNOSIS: Mouth opening circular. Cephalic sensory papillae: 6 labial; 4 submedian; 2 lateral amphids. Buccal capsule with smooth, spiral, transverse or longitudinal thickenings on internal surface. Excretory pore opens near nerve ring. *Male*: Smaller, narrower than female. Testis at level of and posterior to oesophagus. Posterior end ventrally coiled. Eight to 11 precloacal pedunculate papillae, 3 pairs adcloacal, 3 postcloacal pedunculate papillae, 3 pairs postcloacal sessile papillae. Spicules unequal. *Female*: Monodelphic. Ovary near distal end of oesophagus. Vulva opens slightly posterior to middle of body. Tail short with numerous mucrons. Parasites of African amphibians, *Xenopus* spp.

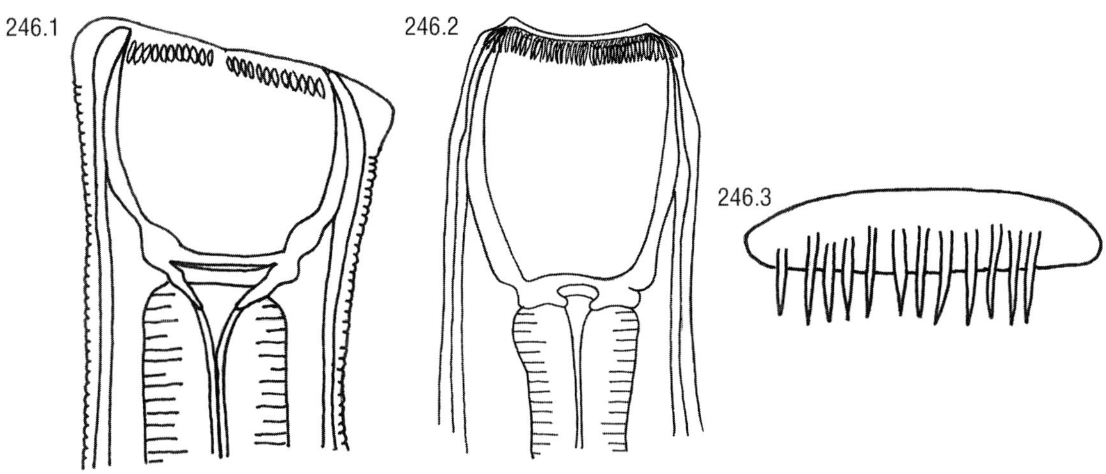

FIG. 246 *Platocamallanus* 246.1 Male cephalic region; 246.2 Female cephalic region; (246.1/2 Redrawn after Agarwal, 1930); 246.3 Chitinous plate from anterior margin of buccal capsule showing leaf-like processes. (Redrawn after Khera, 1955.)

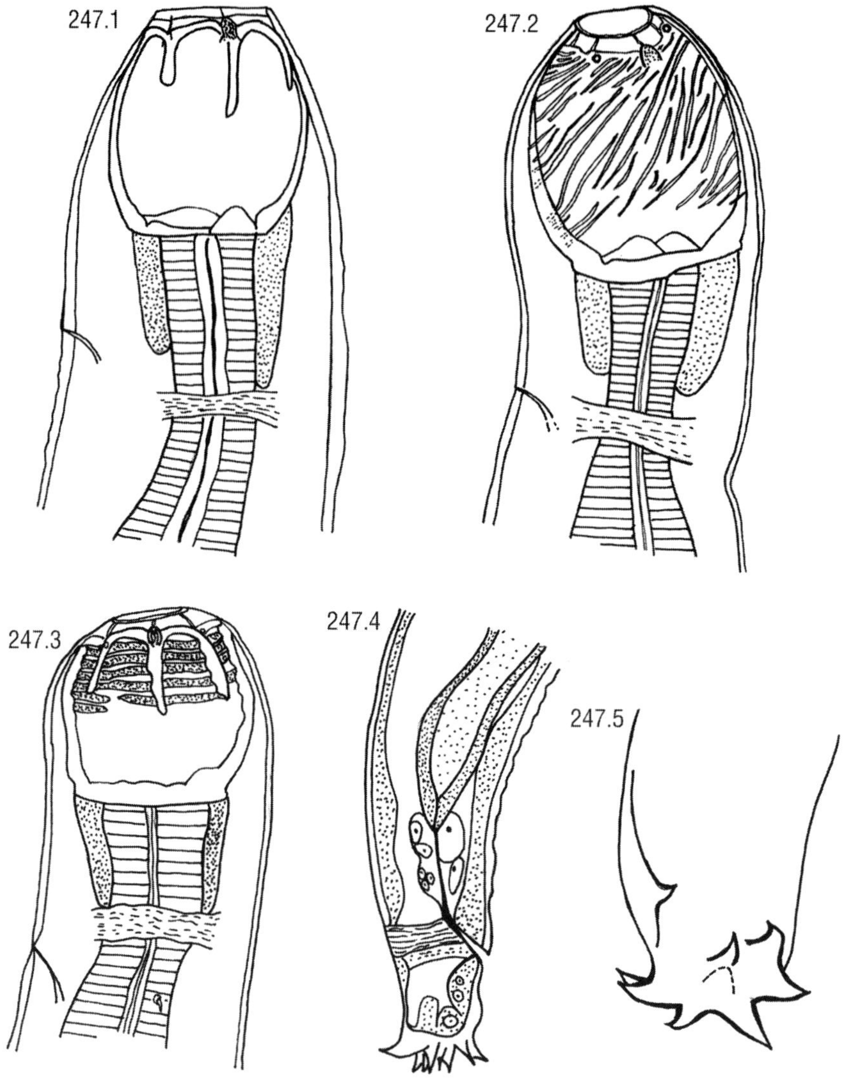

FIG. 247 *Batrachocamallanus* 247.1 Cephalic region, lateral view showing buccal capsule with smooth walls; 247.2 Cephalic region, lateral view showing buccal capsule with spiral thickenings; 247.3 Cephalic region, lateral view showing buccal capsule with transverse thickenings; 247.4 Female tail, lateral view; 247.5 Female tail. (Redrawn after Jackson and Tinsley, 1995.)

Type species: *Batrachocamallanus xenopodis* (Baylis, 1929) Jackson and Tinsley, 1995

Paracamallaninae Stromberg and Crites, 1974

DIAGNOSIS: Buccal capsule separated into paired lateral valves, anterior and posterior divisions nearly equal in depth. Sclerotized bands smooth or denticulate. Chitinous ring present or absent. Tridents present.

Type genus: *Paracamallanus* Yorke and Maplestone, 1926

[Bilqees and Akram (1982) reinstate this subfamily and the emended diagnosis is based on their publication. Petter (1979c) included the genus *Paracamallanus* in the subfamily Camallaninae.]

Type genus: *Paracamallanus* Yorke and Maplestone, 1928

Subgenus *Dentocamallanus* Moravec and Scholz, 1991 (see Fig. 243)

DIAGNOSIS: Presence of teeth on ribs of buccal capsule.

Type species: *Paracamallanus* (*Dentocamallanus*) *sweeti* (Moorthy, 1937)

[Bilqees and Akram (1982) created a new genus, *Neoparacamallanus*, for this type species in the Camallaninae. Moravec and Scholz (1991) moved the species and placed it in this new subgenus of *Paracamallanus*.]

DRACUNCULOIDEA

[Moravec (2004) reviews the taxonomy of species of this superfamily in fish. Later, Moravec (2006) revises the classification of this superfamily and lists the families Dracunculidae, Skrjabillanidae, Guyanemidae, Philometridae, Micropleuridae, Daniconematidae, Lucionematidae and Tetanonematidae.

Moravec (2006) separates the Anguillicolidae into a separate superfamily, Anguillicoloidea Yamaguti, 1935. Moravec (2006) characterizes the two superfamilies as follows: (i) Dracunculoidea, absence or presence of poorly developed buccal capsule; oesophagus divided or undivided occasionally with 'free oesophageal glands' or prominent dorsal oesophageal gland; generally well-defined, sclerotized spicules, gubernaculum, copulatory plate or genital cone; vulva present or absent in gravid female; generally first stage larvae released; (ii) Anguillicoloidea, well-developed buccal capsule; oesophagus undivided, well-defined, sclerotized spicules, gubernaculum, copulatory plate or genital cone absent; distinct vulva opening in posterior half of body; generally sheathed second stage larvae released or released retained within egg.]

Genus not assigned to a family

Lockenloia Adamson and Caira, 1991 (Figs 248.1, 248. 2 and 248.3)

DIAGNOSIS: *Male*: Unknown. *Female*: (based on fragments) Cephalic region bluntly rounded. Body greatly increases in width posterior to cephalic region. Oral opening triangular, 4 pairs of submedian papillae, amphids with wide opening; buccal capsule reduced. Undivided oesophagus gradually widening posteriorly, subspherical distal end. Anal opening present, tail conical with sharp distal tip. Ovoviviparous. Parasite in heart of sharks (*Ginglymostoma cirratum*, Nurse shark).

Type species: *Lockenloia sanguineus* Adamson and Caira, 1991

[The authors were unable to assign this genus to any of the families of the Dracunculoidea owing to both the fragmented nature of the specimens and the absence of males. Moravec (2004, 2006) considers this genus *incertae sedis*.]

Dracunculidae (Stiles, 1907 subfamily) Leiper, 1912

[Chabaud (1975a) and Moravec (2006) list *Dracunculus* Reichard, 1759 and *Avioserpens* Wehr and Chitwood, 1934 in this family.]

Fuellebornius Leiper, 1926

DIAGNOSIS: Parasite of humans.

Type species: *Fuellebornius medinensis* (Linnaeus, 1758) Leiper, 1926

[Leiper (1926) erected this genus on the grounds that the generic name *Dracunculus* was preoccupied, having been assigned by Wiegmann, 1834 to a genus of reptiles. Muller (1971) points out that 'Gallandant in his thesis [of 1773] called the parasite *Dracunculus sive vena medinensis*'. 'Reichard, 1759 called the parasite *Dracunculus* possibly in the vernacular sense but the Gallandant, 1773 citation is still valid' (R. Muller, pers. comm.). The ICZN (opinion 66 of 1915) approved the generic name of *Dracunculus*, and the accepted type species is therefore *Dracunculus medinensis* (Linnaeus, 1758) Gallandant, 1773. The genus *Fuellebornius* was not included as a synonym in Chabaud (1975a), but was referred to in Spencer Jones and Gibson's (1987) list of genera. *Fuellebornius* is clearly a synonym of *Dracunculus* Reichard, 1759. Moravec (2006) also lists this genus as a synonym of *Dracunculus*.]

Anguillicolidae Yamaguti, 1935

[Moravec (2006) moves this family out of the Dracunculoidea, where it is placed in Chabaud (1975a), to the reinstated superfamily Anguillicoloidea Yamaguti, 1935. Moravec (2006) characterizes the superfamily as follows: well-developed buccal capsule, oesophagus undivided; sclerotized spicules, gubernaculum or associated structures absent; functional vulva present; oviparous; parasites of fish; including the single family Anguillicolidae.]

Anguillicola Yamaguti, 1935

Subgenus *Anguillicola* (Yamaguti, 1935 genus) Moravec and Taraschewski, 1988

DIAGNOSIS: Body filiform. Cuticle with fine spines. Oesophagus markedly expanded at proximal end, cylindrical posteriorly.

Type species: *Anguillicola* (*Anguillicola*) *globiceps* Yamaguti, 1935

[Diagnosis based on Moravec and Taraschewski (1988). Moravec (2006) no longer includes this subgenus within *Anguillicola*.]

Genus *Anguillicoloides* Moravec and Taraschewski, 1988 (Moravec and Taraschewski, 1988 subgenus) (Figs 248.4, 248.5 and 248.6)

DIAGNOSIS: Body filiform or fusiform. Cuticle without spines, occasionally random 'excrescenses' on anterior and posterior ends. Cephalic sensory organs: 4 submedian papillae, 2 lateral amphids. Buccal capsule with row of teeth near proximal margin. Oesophagus not expanded at proximal end, posterior end markedly expanded, undivided, well-developed valve. *Male*: Testes originate near distal end. Well-developed seminal vesicle. Large rectal glands. Cloaca opens on ventral caudal projection. *Female*: Vulva opens in posterior half of body on ventral projection. Eggs with sheathed second-stage larvae. Parasites of swimbladder of eels.

Type species: *Anguillicoloides crassus* (Kuwahara, Niimi and Itagaki, 1974) Moravec and Taraschewski, 1988

[Formerly erected as a subgenus of *Anguillicola*, Moravec (2006) raises this subgenus to generic level. Emended diagnosis based on Moravec (2006).]

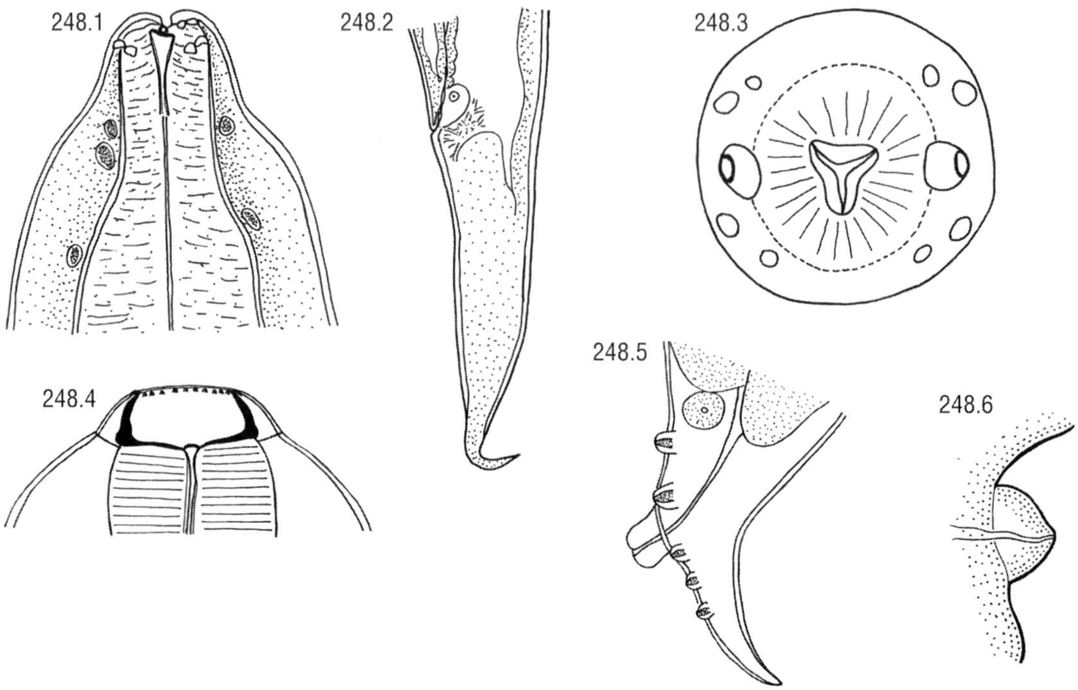

FIGS 248.1, 248.2 and 248.3. *Lockenloia* 248.1 Female anterior end, lateral view; 248.2 Posterior end of female, lateral view; 248.3 Cephalic region of female, *en face* view. (Redrawn after Adamson and Caira, 1991.)
FIGS 248.4, 248.5 and 248.6. *Anguillicoloides* 248.4 Female buccal capsule; 248.5 Posterior end of male; 248.6 Vulva, lateral view. (Redrawn from Moravec, 2006 after Moravec and Taraschewski, 1988.)

Skrjabillanidae Shigin and Shigina, 1958

DIAGNOSIS: Body very slender. Eight cephalic papillae, 2 lateral amphids. Buccal capsule present. Oesophagus undivided with external oesophageal gland or occasionally divided. *Male*: Body much smaller than female. Large, pedunculate papillae incorporated into caudal alae. Spicules absent. Copulatory plate present. *Female*: Vulva opens in anterior half of body. Monodelphic. Viviparous. Distal tip of tail may have 3 processes. Parasites generally in the peritoneal cavity of freshwater fishes.

[Emended diagnosis based on Moravec (2006), who includes two subfamilies, Skrjabillaninae and Esocineminae.]

Skrjabillaninae (Shigin and Shigina, 1958 family) Chabaud, 1965

[Moravec (1977) accepts the family Skrjabillanidae; however, Chabaud (1975a) does not but retains the subfamily Skrjabillaninae. Moravec (2006) lists *Skrjabillanus* Shigin and Shigina, 1958, *Kalmanmolnaria* Sokolov, 2006 (= *Molnaria* Moravec, 1968 preoccupied) and *Sinoichthyonema* Wu, 1965 publ. in Chen, 1973 in the subfamily.]

Kalmanmolnaria Sokolov, 2006 (= *Molnaria* Moravec, 1968 preoccupied)

[Sokolov (2006) replaces the preoccupied generic name *Molnaria*, a fossil Formanifera, with *Kalmanmolnaria nom. nov.*]

Sinoichthyonema Wu, 1965 publ. in Chen, 1973 (Fig. 249)

DIAGNOSIS: Cephalic region bluntly rounded. Cephalic sensory organs: 4 double submedian papillae; 2 lateral amphids. Buccal capsule small. Oesophagus divided, large external oesophageal gland present. *Male*: Spicules, gubernaculum or copulatory plate absent. Genital cone very large. Caudal alae wide with several pairs of pedunculated papillae. *Female*: Body length at least twice that of male. Monodelphic. Vulva opens in anterior half of body. Viviparous. Parasite in abdominal cavity of cyprinid fish.

Type species: *Sinoichthyonema amuri* (Garkavi, 1972) Moravec, 1982

[The generic diagnosis is based on Moravec (1994). Moravec (1982b) indicates that the species *Sinoichthyonema ctenopharyngodoni* Wu, 1965 publ. 1973 from the abdominal cavity of grass carp, *Ctenopharyngodon idella*, in China is identical to *Skrjabillanus amuri* Garkavi, 1972 from grass carp introduced into the Krasnodar region, former USSR. *Garkavillanus* was erected by Lomakin and Chernova, 1980 to accommodate Garkavi's species on the basis that it differed from all other genera in the Skrjabillanidae (this is a subfamily in Chabaud, 1975a) by the presence of a sclerotized genital cone. Moravec (1982b) considers the two species are identical, and therefore Wu's species becomes a synonym of *S. amuri* and the genus *Garkavillanus* becomes a synonym of *Sinoichthyonema*.]

Garkavillanus Lomakin and Chernova, 1980

DIAGNOSIS: Cephalic sensory organs: 4 pairs submedian papillae; 2 large lateral amphids. Oesophagus divided with 1 large oesophageal gland. *Male*: Caudal alae trilobed. Seven pairs of papillae, 2 precloacal, 4 adcloacal, 1 postcloacal. Sclerotized genital cone present.

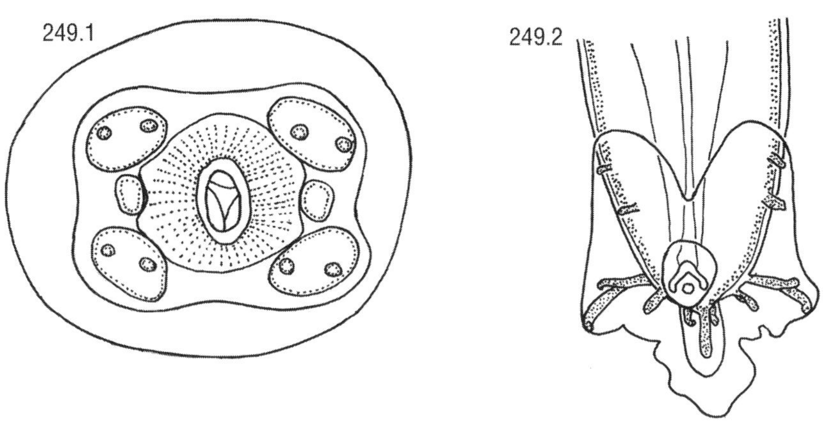

FIG. 249 *Sinoichthyonema* 249.1 Cephalic region, *en face* view; 249.2 Posterior end of male, ventral view. (Redrawn from Moravec, 1994 after Lomakin and Chernova, 1980.)

Type species: *Garkavillanus amuri* (Garkavi, 1972) Lomakin and Chernova, 1980

[Moravec (1982b) indicates that the species *Sinoichthyonema ctenopharyngodoni* Wu, 1965 publ. 1973 is identical to *G. amuri*, the type species of the genus *Sinoichthyonema* Wu, 1965 publ. 1973. *Garkavillanus* is considered a synonym of *Sinoichthyonema*.]

Esocineminae Moravec, 2006

DIAGNOSIS: Anterior end tapers. Buccal capsule long, narrow. Oesophagus divided. Postcloacal papillae restricted to distal half of tail.

Type and only genus *Esocinema* Moravec, 1977

[Diagnosis based on Moravec (2006).]

Esocinema Moravec, 1977 (Fig. 250)

DIAGNOSIS: Body thread-like. Cephalic region bluntly rounded. Buccal capsule cylindrical, thin-walled, deeper than wide. Oesophagus divided. *Male*: Tail conical with wide sublateral alae. Postcloacal papillae in posterior half of tail. Copulatory plate present, weekly sclerotized. *Female*: Approximately twice as long as male. Vulva opens in region of glandular oesophagus. Uterus containing larvae fills most of body. Tail tapers to a point. Parasite under the serosa of the air bladder of pike, *Esox lucius*, in North Bohemia (Elbe River, drainage system).

Type species: *Esocinema bohemicum* Moravec, 1977

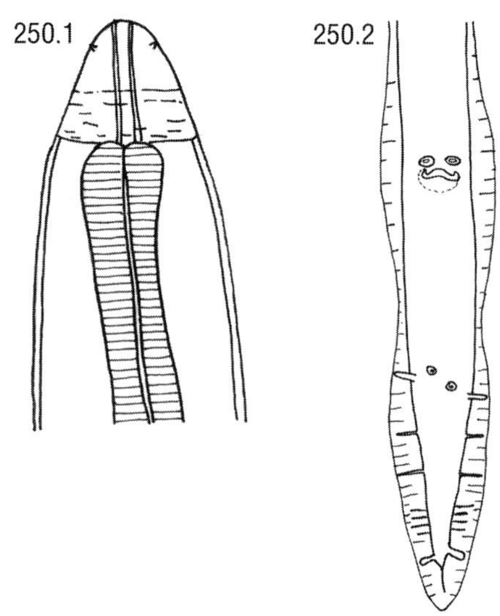

FIG. 250 *Esocinema* 250.1 Anterior end of female; 250.2 Posterior end of male, ventral view. (Redrawn after Moravec, 1977.)

Guyanemidae Petter, 1975

DIAGNOSIS: Cephalic region rounded. Oesophagus divided with or without 'free posterior appendage'. *Male*: Tail elongate, caudal alae present or absent, caudal papillae paired. Two spicules present. *Female*: Vulva generally opens in anterior part of body in gravid female. Monodelphic or didelphic. Parasites in peritoneal cavity and tissues of fish.

[Emended diagnosis based on Moravec (2006). Moravec (2006) includes two subfamilies, Guyaneminae Petter, 1975 (Petter, 1975 family) and Travassosneminae Moravec, 2006. Chabaud (1975a) included a single genus, *Guyanema* Petter, 1975, in the family.]

Guyaneminae Petter, 1975 (Petter, 1975 family)

DIAGNOSIS: Oesophagus with long glandular region without 'free posterior appendage'.

[Diagnosis based on Moravec (2006), who lists *Guyanema* Petter, 1975, *Moravecia* Ribu and Lester, 2003, *Histodytes* Aragort, Álvarez, Iglesias, Leiro and Sanmatín, 2002, and *Pseudodelphis* Adamson and Roth, 1990 in the subfamily.]

Pseudodelphis Adamson and Roth, 1990 (Figs 251.1, 251.2, 251.3 and 251.4)

DIAGNOSIS: Long, slender worms. Cephalic region rounded, mouth opening triangular. Cephalic sensory papillae: 6 inner papillae, 4 pairs of submedian papillae; 2 lateral amphids. Buccal capsule reduced, forms peribuccal ring. Oesophagus long, divided with muscular and glandular portions. *Male*: Caudal alae absent. Distal tip of tail subconical. Single spicule. *Female*: Didelphic. Anterior uterus blind-ending sac without attachment to oviduct and ovary. Vulva opens in oesophageal region. Parasite of peritoneal cavity and mesenteries surrounding intestine of marine fish (*Oligocottus maculosus*) in coastal waters of British Columbia, Canada.

Type species: *Pseudodelphis oligocotti* Adamson and Roth, 1990

Histodytes Aragort, Álvarez, Iglesias, Leiro and Sanmatín, 2002 (Figs 251.5 and 251.6)

DIAGNOSIS: Buccal capsule absent. Oesophagus divided into short, narrow, anterior muscular region, long, wide, posterior glandular region. Cephalic sensory organs: 3 pairs inner circle, 4 pairs outer circle. *Male*: Distal end conical, ventral surface ornamented. Caudal alae absent. Spicules subequal, similar. Copulatory plate absent. *Female*: Amphidelphic. Didelphic. Viviparous. Vulva opens in anterior part of body posterior to oesophagus. Parasites in tissues of fish.

Type species: *Histodytes microocellatus* Aragort, Álvarez, Iglesias, Leiro and Sanmatín, 2002

[Emended diagnosis based on Moravec (2006).]

Moravecia Ribu and Lester, 2004 (Fig. 252)

DIAGNOSIS: Body slender. Cuticle transversely striated. Cephalic region rounded, dorsal and ventral anterior projections present. Oesophagus divided into muscular and glandular regions.

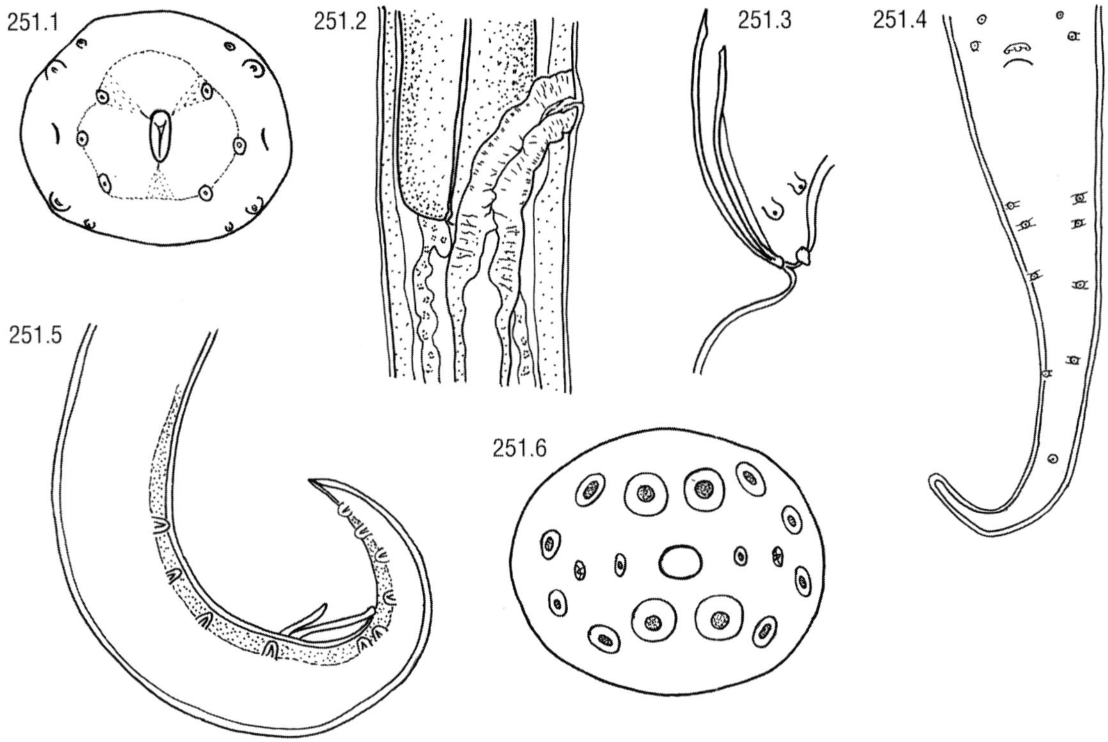

Figs 251.1, 251.2, 251.3 and 251.4 *Pseudodelphis* 251.1 Cephalic region, *en face* view; 251.2 Vulvar region, lateral view; 251.3 Cloacal region showing spicule; 251.4 Posterior end of male, ventral view. (Redrawn after Adamson and Roth, 1990.)
Figs 251.5 and 251.6 *Histodytes* 251.5 Posterior end of male, lateral view; 251.6 Cephalic region, *en face* view. (Redrawn from Moravec, 2006 after figures adapted from Aragort *et al.*, 2002.)

Male: Narrow caudal alae. Spicules equal. Tail conical. *Female*: Body length approximately 1.5 times that of male. Vulva protrudes slightly above body surface, opens at distal end of glandular oesophagus. Monodelphic. Parasite of gill filaments of green porcupine fish, *Tragulichthys jaculiferus*, in Moreton Bay, Queensland, Australia.

Type species: *Moravecia australiensis* Ribu and Lester, 2004

Travassosneminae Moravec, 2006

DIAGNOSIS: Oesophagus with short glandular region with 'free posterior appendage'. Vulva opens in anterior or middle of body or atrophied.

[Diagnosis based on Moravec (2006), who lists *Travassosnema* De Araújo Costa, Mareira and De Oliveira, 1991 and *Ichthyofilaria* Yamaguti, 1935 in the subfamily. Chabaud (1975a) placed the latter genus in the Philometrinae.]

252.1 252.2

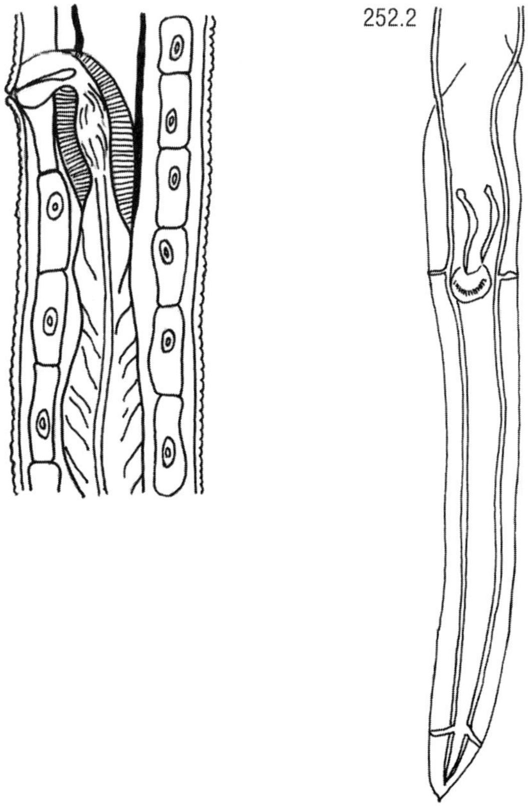

FIG. 252 *Moravecia* 252.1 Vulvar region, lateral view; 252.2 Posterior end of male, ventral view. (Redrawn after Ribu and Lester, 2004.)

Travassosnema De Araújo Costa, Mareira and De Oliveira, 1991 (Figs 253.1 and 253.2)

DIAGNOSIS: Buccal capsule absent. Mouth oval, lips absent. Cephalic sensory organs: 6 papillae of inner ring; 4 submedian papillae; 2 lateral amphids. Oesophagus divided into muscular and glandular regions, glandular part with posterior appendix near junction with intestine. *Male*: Posterior end spirally coiled with well-developed caudal alae. Ten to 11 pairs of pedunculated caudal papillae, 5–6 pairs precloacal, 5 pairs postcloacal. Spicules similar, slender. Gubernaculum present. Tail tip rounded. *Female*: Monodelphic. Vulva opens anteriorly, well developed in mature female. Anus functional. Viviparous. Parasitic in tissues behind eyes of fish, *Acestrorhynchus lacustris*, from Três Marias Reservoir, Mina Gerais State, Brazil.

Type species: *Travassonema travassosi* De Araújo Costa, Moreira and De Oliveira, 1991

Philometridae Baylis and Daubney, 1926

[Moravec (2006) includes three subfamilies in this family: Philometrinae Baylis and Daubney, 1926, Alineminae Moravec, 2006 and Neophilometroidinae Moravec, Salgado-Maldonado and Aquilar-Aguilar, 2002. Chabaud (1975a) included the subfamilies Philometrinae and Philoneminae, the latter for the single genus *Philonema* Kuiten-Ekbaum, 1933. Moravec (2006) moves this genus to the family Micropleuridae.]

Afrophilometra Moravec, Charo-Karisa and Jirků, 2009 (Fig. 253.3)

DIAGNOSIS: *Male*: Unknown. *Female*: Body with ornamented annulations, smooth in lateral field. Cephalic sensory organs: 4 small pairs of papillae in outer circle, 6 papillae in inner circle. Oesophagus short, expanded at proximal end. Mononucleate oesophageal gland present. Ventriculus present, small. Gravid females with vulva and anus atrophied. Parasites in tissues of African freshwater fish.

Type species: *Afrophilometra hydrocyoni* (Fahmy, Mandour and El-Nafar, 1976) Moravec, Charo-Karisa and Jirků, 2009

[Diagnosis based on Moravec *et al.*, 2009.]

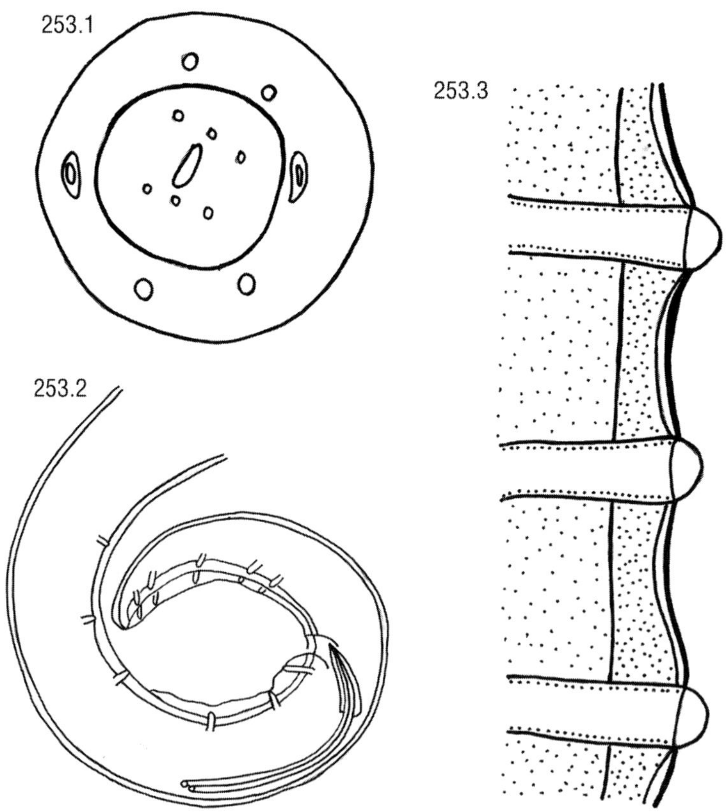

FIGS 253.1 and 253.2 *Travassosnema* 253.1 Cephalic region, *en face* view; 253.2 Posterior end of male, lateral view. (Redrawn after De Araújo Costa *et al.*, 1991.)
FIG. 253.3 *Afrophilometra* Section of cuticle showing inflated annulations. (Redrawn after Moravec *et al.*, 2009.)

Philometrinae (Baylis and Daubney, 1926 family)

DIAGNOSIS: *Male*: Generally lobed distal end. *Female*: Proximal and distal ends round or distal end blunt. Gravid female with small, occasionally large, cephalic sensory organs generally in 2 circles, smooth cuticle, atrophied vulva and uterus.

[Emended diagnosis based on Moravec (2006), who includes *Philometra* Costa, 1845, *Spirophilometra* Parukhin, 1971, *Nilonema* Khalil, 1960, *Buckleyella* Rasheed, 1963, *Paraphilometroides* Moravec and Shaharom-Harrison, 1989, *Philometroides* Yamaguti, 1935, *Rumai* Travassos, 1960, *Clavinema* Yamaguti, 1935 and *Dentiphilometra* Moravec and Wang, 2002. *Dentiphilometra* was added to the subfamily by Moravec and Gui Tang Wang (2002) and Moravec *et al.* (2008a) add the genus *Caranginema*.]

Icthyonema Diesing, 1861

DIAGNOSIS: Body long, filiform. Mouth circular, terminal. *Male*: Caudal extremity with 2 parallel lobes, genital aperture subterminal. *Female*: Viviparous. Parasites of body cavity of marine fish.

Type species: not designated. ?*Icthyonema fuscum* Diesing, 1861

[Yorke and Maplestone (1926) list this genus as a synonym of *Philometra* Costa, 1845. Chabaud (1975a) did not include this genus as a synonym of *Philometra*. Spencer Jones and Gibson (1987) list this genus in the superfamily Dracunculoidea.]

Paraphilometroides Moravec and Shaharom-Harrison, 1989 (Fig. 254)

DIAGNOSIS: *Male*: Unknown. *Female*: Body relatively short when gravid, narrower in posterior part. Cuticle with numerous bosses except in oesophageal region. Head rectangular in lateral view, arch-like in dorso-ventral view. Cephalic sensory organs: 4 small papillae surround mouth; 4 pairs of submedian papillae; 2 lateral amphids. Curved cephalic alae on dorsal and ventral sides, supported by inner muscular layer. Oesophagus expanded at proximal end. Dorsal oesophageal gland prominent with 1 large nucleus. Uteri opposed, ovaries near extremities. Viviparous. Anus atrophied. Tail rounded. Parasite in dorsal fin and operculum of the marine perciform fish, *Nemipterus peronii*, in coastal waters off Kuala Terengganu, Malaysia.

Type species: *Paraphilometroides nemipteri* Moravec and Shaharom-Harrison, 1989

[The authors distinguish their genus from *Philometroides* Yamaguti, 1935, which Chabaud (1975a) places in the subfamily Philometrinae. *Paraphilometroides* is placed in the same subfamily in Moravec (2006).]

Margolisianum Blaylock and Overstreet, 1999 (Fig. 255)

DIAGNOSIS: *Male*: Body small. Cephalic sensory papillae minute, 4 pairs. Oesophagus without additional 'bulbous region' adjoining anterior margin. Dorsal oesophageal gland

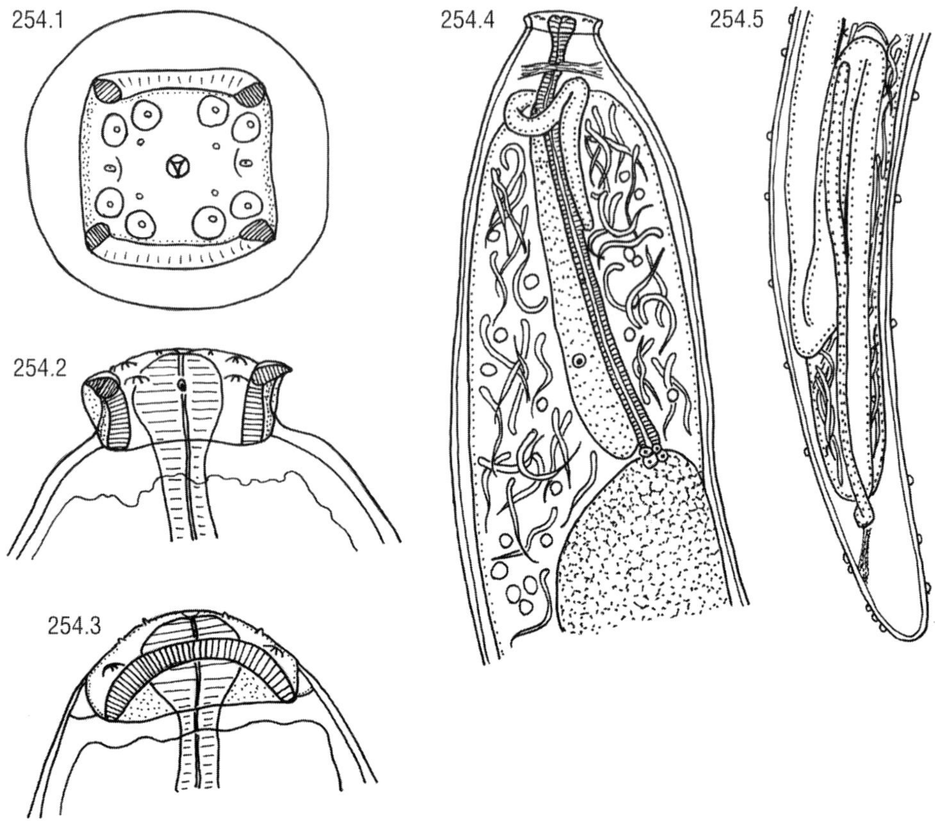

FIG. 254 *Paraphilometroides* 254.1 Cephalic region, *en face* view; 254.2 Cephalic region, lateral view; 254.3 Cephalic region, dorso-ventral view; 254.4 Anterior end of gravid female; 254.5 Posterior end of gravid female. (Redrawn after Moravec and Shaharom-Harrison, 1989.)

large, uninucleate. Spicules subequal. Gubernaculum present. Tail with lobe-like bipartite extension. *Female*: Cuticle with bosses. Cephalic sensory papillae, large, 4 pairs. Mouth trilobed, with 'ring of bead-like teeth'. Oesophagus muscular with 'bulbous region' at anterior margin. Dorsal oesophageal gland large, uninucleate. Rectum vestigial. Didelphic. Amphidelphic. Ovaries reflexed near junction with oviduct. Vulva opens in posterior third, non-functional in gravid females. Tail bilobed, less marked than in males. Parasite in eye (immature females), subcutaneous tissues of mouth and head (mature and gravid females) and muscle adjacent to dorsal fin just posterior to head (males) of southern flounder, *Paralichthys lethostigma*, in Ocean Springs, Mississippi Sound, Mississippi and Galveston Bay, Texas, USA.

Type species: *Margolisianum bulbosum* Blaylock and Overstreet, 1999

[Moravec (2006) indicates that, according to Moravec and de Buron, 2006, *Margolisianum* is a *genus inquirendum* and the type species is a *species inquirenda* and species *incertae sedis*.]

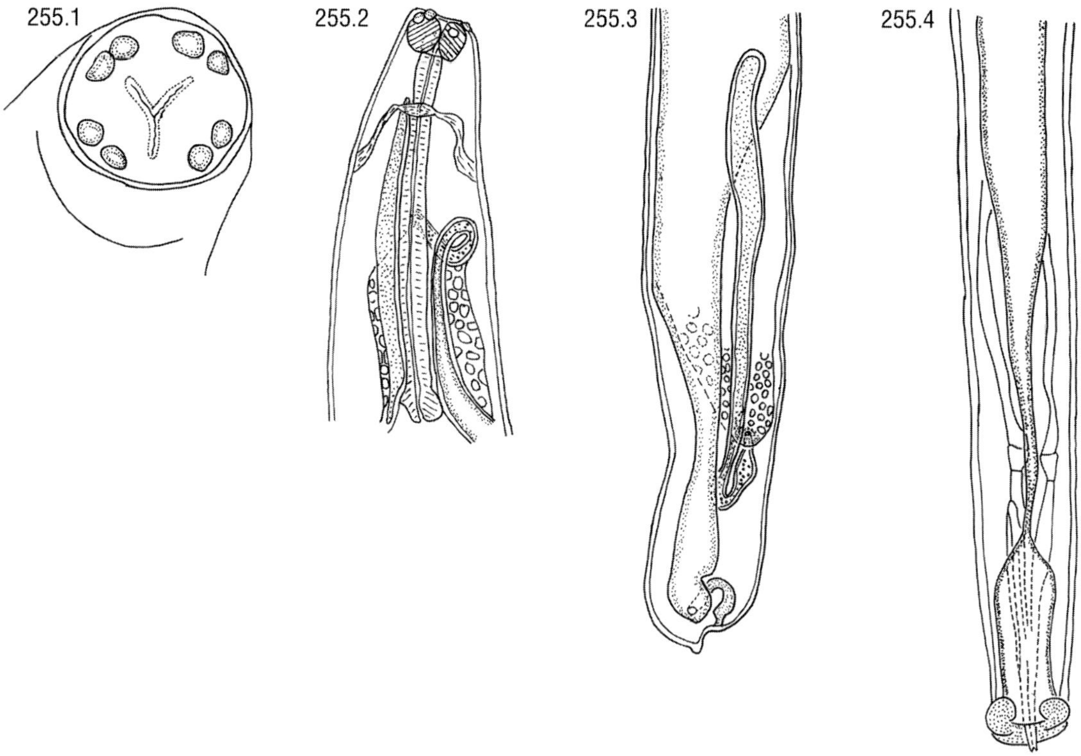

FIG. 255 *Margolisianum* 255.1 Female cephalic region, *en face* view; 255.2 Anterior end of female, lateral view; 255.3 Posterior end of female, lateral view; 255.4 Posterior end of male, ventral view. (Redrawn after Blaylock and Overstreet, 1999.)

Dentiphilometra Moravec and Gui Tang Wang, 2002
(Figs 256.1, 256.2 and 256.3)

DIAGNOSIS: Cuticle smooth. Cephalic sensory organs: 4 pairs small submedian papillae in outer circle; 4 papillae in inner circle. Oesophagus short, expanded at proximal end. Oesophageal gland external, mononucleate. Ventriculus present, small. *Male*: Spicules present. Gubernaculum present. Distal end lobed with few papillae adjacent to cloaca. *Female* (gravid): Oral ring armed internally with many irregular peribuccal teeth. Buccal capsule absent. Vulva, vagina, anus atrophied. Parasites in abdominal cavity of freshwater fish.

Type species: *Dentiphilometra* Moravec and Gui Tang Wang, 2002

[Diagnosis based on Moravec and Gui Tang Wang, 2002.]

Caranginema Moravec, Montoya-Mendoza and Salgado-Maldonado, 2008
(Figs 256.4 and 256.5)

DIAGNOSIS: *Male*: Unknown. *Female*: Body elongate with 2 parallel cordons adjacent to margins of lateral fields, irregular ornamentation present on dorso-ventral surfaces. Mouth with

3 large oesophageal teeth. Cephalic sensory organs with 4 submedian papillae in outer circle, 4 submedian papillae in inner circle. Oesophagus with expanded proximal end, gland mononucleate, not prominent. Small ventriculus present. Anus, vulva, vagina atrophied, intestine extends to near distal end. Parasites in subcutaneous tissue of marine fish.

Type species: *Caranginema americanum* Moravec, Montoya-Mendoza and Salgado-Maldonado, 2008

[Diagnosis based on Moravec *et al.* (2008a).]

Alineminae Moravec, 2006

DIAGNOSIS: *Male*: distal end not lobed. *Female*: Gravid female with small cephalic sensory papillae in 2 circles, smooth cuticle, buccal capsule absent, peribuccal rim supported by sclerotized structures, oesophagus with proximal bulb, vulva opening near distal end, functional.

Type genus: *Alinema* Rasheed, 1963

[Diagnosis based on Moravec (2006), who includes one genus, *Alinema* Rasheed, 1963. Chabaud (1975a) includes this as a subgenus of *Philometra*.]

Neophilometroidinae Moravec, Salgado-Maldonado and Aguilar-Aguilar, 2002

DIAGNOSIS: *Male*: Cloaca opens well anterior to rounded distal end. Distal end not lobed. *Female*: Buccal capsule with proximal row of peribuccal teeth. Cuticle with bosses. Oesophagus without proximal bulb. Amphidelphic. Gravid female with atrophied vagina and vulva.

Type genus: *Neophilometroides* Moravec, Salgado-Maldonado and Aguilar-Aguilar, 2002

[Diagnosis based on Moravec *et al.* (2002c) and Moravec (2006).]

Neophilometroides Moravec, Salgado-Maldonado and Aguilar-Aguilar, 2002 (Figs 256.6 and 256.7)

DIAGNOSIS: *Male*: Distal end not lobed. Cloaca opens some distance from distal end. One pair precloacal papillae, postcloacal papillae absent. Spicules present. Gubernaculum present. *Female*: Cuticle with bosses on anterior body. Cephalic sensory papillae, small outer circle with 4 submedian pairs, inner circle with 4 single papillae. Buccal capsule small with 'numerous small peribuccal teeth'. Oesophagus short without proximal bulb with external oesophageal gland. Ventriculus present, small. Gravid female with atrophied vagina and vulva, functional anus. Parasites in swimbladder of neotropical freshwater catfish.

Type species: *Neophilometroides caudatus* (Moravec, Schulz and Vivas-Rodríguez, 1995) Moravec, Salgado-Maldonado and Aguilar-Aguilar, 2002

[Diagnosis based on Moravec *et al.* (2002c) and Moravec (2006).]

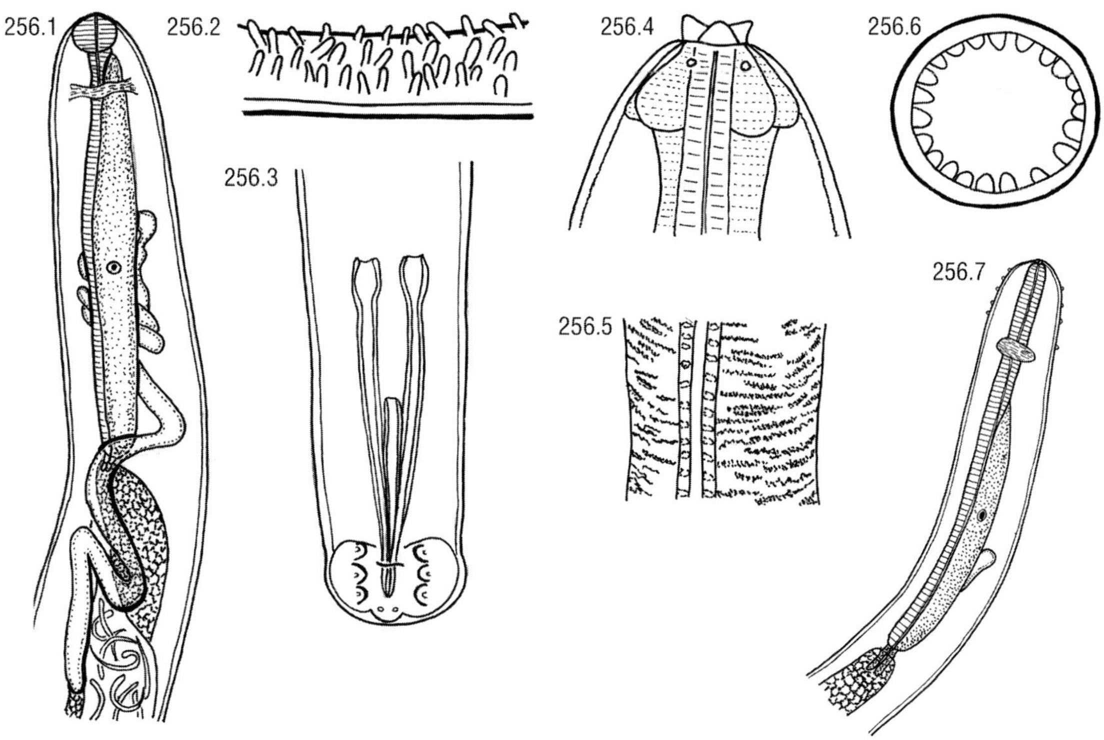

Figs 256.1, 26.2 and 256.3 *Dentiphilometra* 256.1 Anterior end of gravid female; 256.2 Part of peribuccal ring of female, adapted from a scanning electron micrograph; 256.3 Posterior end of young male, ventral view. (Redrawn after Moravec and Gui Tang Wang, 2002.)
Figs 256.4 and 256.5 *Caranginema* 256.4 Gravid female, cephalic region, lateral view; 256.5 Section of body showing cuticular ornamentation, lateral view. (Redrawn after Moravec *et al.*, 2008a.)
Figs 256.6 and 256.7 *Neophilometroides* 256.6 Mouth opening, *en face* view; 256.7 Subgravid female, anterior end. (Redrawn after Moravec *et al.*, 2002.)

Phlyctainophorinae (Roman, 1965 family)

DIAGNOSIS: *Male*: Body filiform. Oesophagus divided. *Female*: Body dorsally coiled, 2 or more subventral rows of swellings present. Oesophagus in gravid female undivided. Parasites in elasmobranchs.

[Emended diagnosis based on Moravec (2006), who moves this subfamily to the Micropleuridae and lists only the single genus, *Phlyctainophora* Steiner, 1921. Chabaud (1975a) includes this subfamily in the Philometridae.]

Phlyctainophora Steiner, 1921

[Adamson *et al.* (1987) describe males of *P. squali* Mudry and Dailey, 1969 for the first time from elasmobranchs. They share the following characters with the Philometridae: rounded head, reduced caudal alae and reduced number of caudal papillae. They differ in the presence of a long, glandular

long posterior glandular region with large dorsal nuclei near distal end. Nerve ring around posterior part of muscular oesophagus. Excretory pore opens just posterior to nerve ring in posterior half of muscular oesophagus, lateral excretory canals extend along most of body incorporate dark corpuscles. *Male*: Cuticle with fine transverse striations. Distal end spirally coiled. Caudal alae poorly developed. Caudal papillae: 2–3 pairs precloacal; 1 pair adcloacal; 6 pairs postcloacal. Spicules slender, pointed distal tips. Gubernaculum present or absent. Tail conical, distal tip pointed. *Female*: Mature females with markedly dense transverse cuticular ornamentation. Juvenile females with non-functional vagina near mid-body. Anus atrophied. Tail pointed. Parasites of bull shark, *Carcharhinus leucas*, in Lake Borgne, Louisiana, USA.

Type species: *Granulinema carcharini* Moravec and Little, 1988

Kamegainema Hasegawa, Doi, Araki and Miyata, 2000 (Fig. 259)

DIAGNOSIS: Cuticle with prominent bosses. Cephalic region with dorsal, ventral 'elevations'. Six internal labial papillae, 4 pairs of submedian cephalic papillae, 1 pair of well-defined amphidial pores. Numerous nuclei in lateral chords. Peribuccal ring present. Buccal capsule absent. *Male*: Unknown. *Gravid female*: Vulva, anus atrophied. Viviparous. First-stage larva long, slender, sharply pointed tail, large phasmids. Parasite in subcutaneous tissue of amphibians.

Type species: *Kamegainema cingulum* (Linstow, 1902) Hasegawa, Doi, Araki and Miyata, 2000.

[Moravec (2006) lists this in the Micropleuridae.]

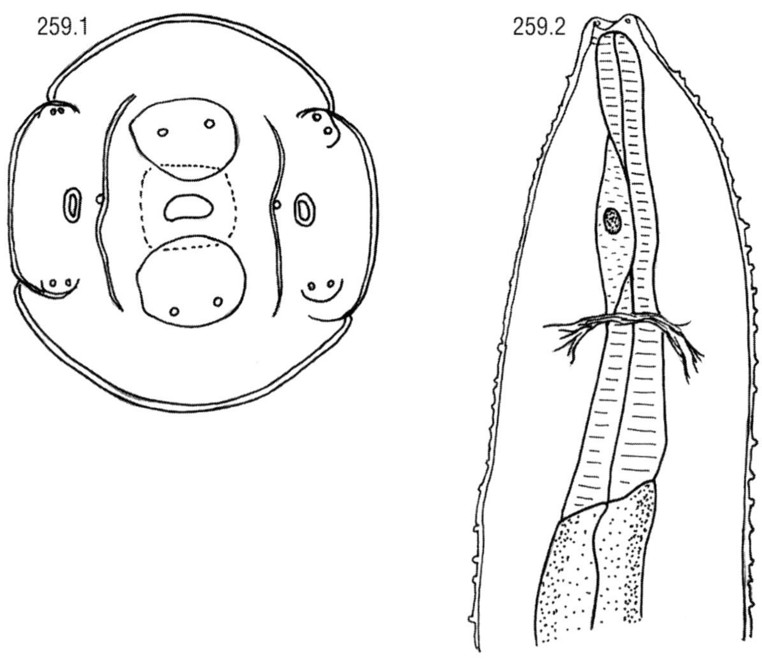

FIG. 259 *Kamegainema* 259.1 Female cephalic region, *en face* view; 259.2 Anterior end of female, lateral view. (Redrawn after Hasegawa *et al*., 2000.)

Daniconematidae Moravec and Køie, 1987

DIAGNOSIS: Body thread-like. Cephalic sensory organs: 1 pair of prominent internal labial papillae; 8 external paired papillae. Buccal capsule absent. Oesophagus long, muscular with long, posteriorly extending oesophageal gland. *Male*: Tail long, caudal alae absent. Several pre- and post-cloacal papillae present. Spicules or sclerotized copulatory plate absent. Cloaca opens on projection. *Female*: Vulva opens in anterior part of body. Monodelphic. Viviparous. Parasites of fish.

Type genus: *Daniconema* Moravec and Køie, 1987

[Moravec (2006) includes *Daniconema* Moravec and Køie, 1987, *Mexiconema* Moravec, Vidal and Salgado-Maldonado, 1992 and *Syngnathinema* Moravec, Spangenberg and Frasca, 2001 in this family.]

Daniconema Moravec and Køie, 1987 (Fig. 260)

DIAGNOSIS: Cephalic region bluntly rounded. Cephalic sensory organs: Pair of large dorso-ventral dome-shaped papillae; 8 external paired papillae; 2 lateral amphids. Nerve ring around anterior part of oesophagus. Excretory pore opens just posterior to nerve ring. *Male*: Caudal papillae: 4 pairs pre-, 5 pairs postcloacal papillae. Spicules, gubernaculum or sclerotized copulatory plate absent. Cloaca opens on genital projection. Tail conical. *Female*: Body twice as long as male. Vulva opens in anterior half of body. Monodelphic. Viviparous. Anus functional. Tail conical. Parasite under serosa of swimbladder and intestine of eels, *Anguilla anguilla*, Lake Esrum, northern Zealand, Denmark.

Type species: *Daniconema anguillae* Moravec and Køie, 1987

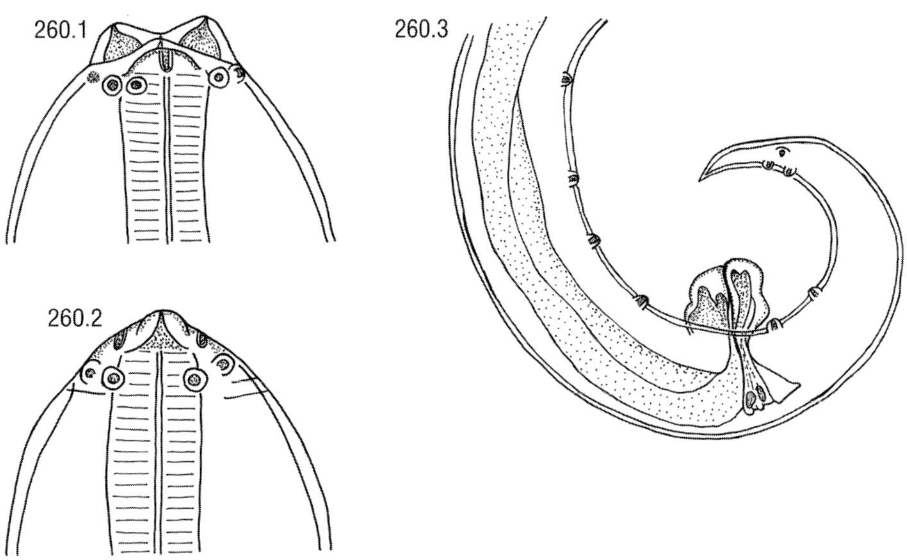

Fig. 260 *Daniconema* 260.1 Female cephalic region, lateral view; 260.2 Female cephalic region, dorsal view; 260.3 Posterior end of male, lateral view. (Redrawn from Moravec, 1994 after Moravec and Køie, 1987.)

Mexiconema Moravec, Vidal and Salgado-Maldonado, 1992 (Figs 261.1, 261.2, 261.3 and 261.4)

DIAGNOSIS: Cephalic region bluntly rounded. Cephalic sensory organs: 4 pairs of submedian papillae; 2 lateral amphids. Mouth oval to triangular. Oesophagus with short anterior muscular region and long posterior glandular region, which incorporates 2 large nuclei. Nerve ring around middle of muscular oesophagus. Excretory pore opens just posterior to level of nerve ring. *Male*: Caudal alae absent. Four pairs sessile postcloacal and adcloacal papillae present. Tail conical, 3 processes at distal tip. *Female*: Vulva opens near posterior end of oesophagus. Vagina short. Uterus with larvae fills most of body. Ovary single, in posterior of body. Viviparous. Tail conical, 3 processes at distal tip. Parasite in abdominal cavity, viscera, rarely skin of cichlids, *Cichlasoma* spp. in coastal lagoon Celestun, North Yucatan, Mexico.

Type species: *Mexiconema cichlasomae* Moravec, Vidal and Salgado-Maldonado, 1992

Syngnathinema Moravec, Spangenberg and Frasca, 2001 (Fig. 261.5)

DIAGNOSIS: Cuticle smooth. Mouth opening large, oval with peribuccal ring surrounded by cephalic papillae, 4 in outer circle, 6 in inner circle, 2 large amphids. Buccal capsule absent.

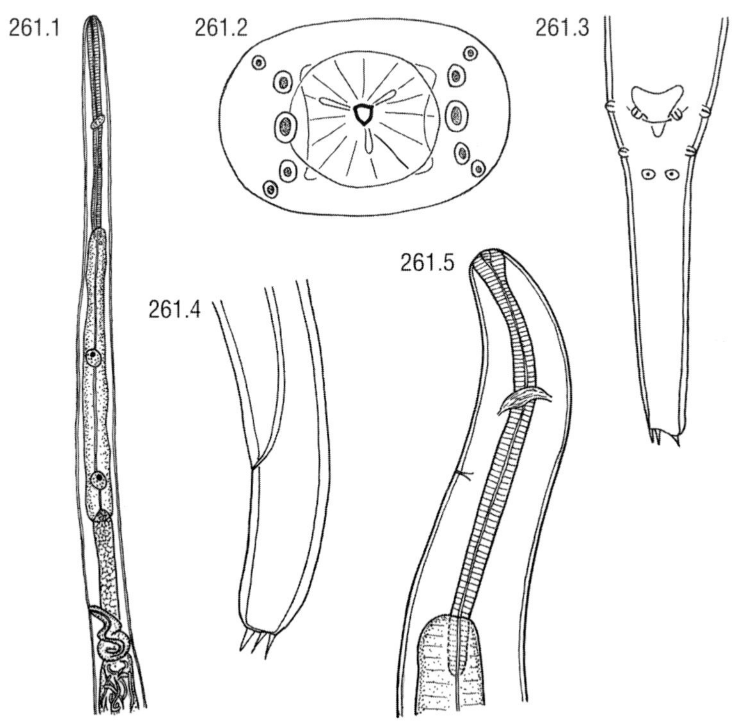

FIGS 261.1, 261.2, 261.3 and 261.4 *Mexiconema* 261.1 Gravid female, anterior end; 261.2 Gravid female, cephalic region, *en face* view; 261.3 Posterior end of male, ventral view; 261.4 Posterior end of female, lateral view. (Redrawn after Moravec *et al.*, 1992c.)
FIG. 261.5 *Syngnathinema* Anterior end of female showing muscular oesophagus, lateral view. (Redrawn after Moravec *et al.*, 2001.)

Oesophagus divided, proximal end expanded. *Male*: Unknown. *Female*: Monodelphic. Vulva, well developed, opens posterior to oesophagus. Anus functional. Parasites in vascular system of the Bay pipefish, *Syngnathus leptorhynchus*, in California, USA.

Type species: *Syngnathinema californiense* Moravec, Spangenberg and Frasca, 2001

[Moravec *et al.* (2001) provisionally assign this genus to the Daniconematidae, later confirmed in Moravec (2006).]

Lucionematidae Moravec, Molnár and Székely, 1998

DIAGNOSIS: Body tapers both ends. Eight cephalic papillae present. Buccal capsule absent. Oesophagus undivided. *Male*: Caudal alae, precloacal papillae and spicules absent. Few pairs of postcloacal papillae present. Sclerotized copulatory plate present. Tail long. *Female*: Vulva opens near mid body. Monodelphic. Viviparous. Parasites of fish.

Type genus: *Lucionema* Moravec, Molnár and Székely, 1998

Lucionema Moravec, Molnár and Székely, 1998 (Fig. 262)

DIAGNOSIS: Cephalic region rounded. Cephalic sensory organs: 8 submedian papillae in 2 circles; 2 lateral amphids. Proximal end of oesophagus swollen. Nerve ring near middle region

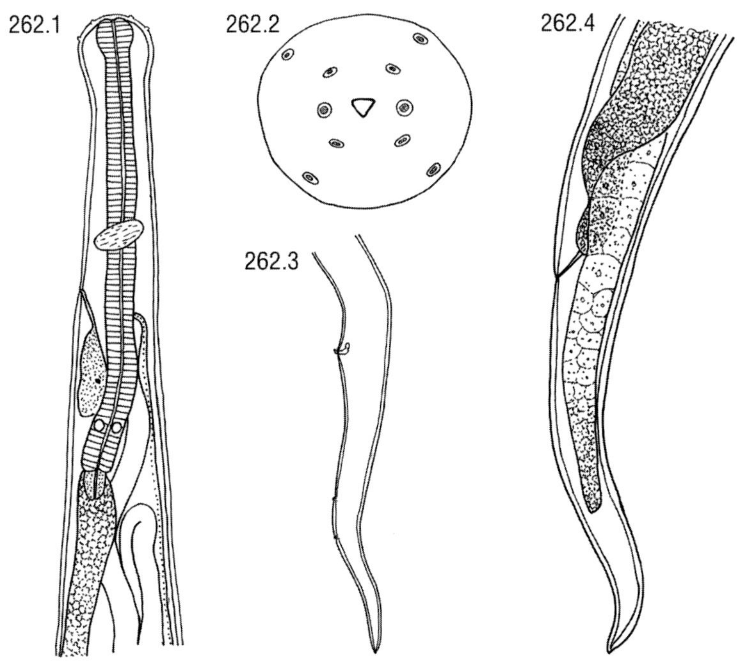

FIG. 262 *Lucionema* 262.1 Anterior end of gravid female, lateral view; 262.2 Female cephalic region, *en face* view; 262.3 Posterior end of male, lateral view; 262.4 Posterior end of female, lateral view. (Redrawn after Moravec *et al.*, 1998.)

of oesophagus. Excretory pore opens just posterior to level of nerve ring. *Male*: Two pairs of postcloacal papillae. Copulatory plate present. Tail conical. *Female*: Vulva opens near mid-body. Uterus filled with larvae, extends anteriorly to distal end of oesophagus. Anus functional. Tail conical with distal point.

Type species: *Lucionema balatonense* Moravec, Molnár and Székely, 1998

[Moravec *et al*. (1998) include a key to the families of the Dracunculoidea]

SPIRURINA

[Moravec (2007) reviews the spirurines of fish belonging to approximately 300 species in four superfamilies, namely Gnathostomatoidea, Habronematoidea, Physalopteroidea and Thelazioidea. Moravec (2007) considers the classification and taxonomy of species of this suborder in fish requires re-evaluation using new techniques, including scanning electron microscopy and molecular biology. He suggests that a wide range of morphological features, as well as host details, should also be studied and that such a major revision will require multidisciplinary international collaboration.]

GNATHOSTOMATOIDEA

[Skrjabin and Ivashkin (1973) review the Gnathostomatoidea and propose that it should be in its own suborder, Gnathostomatata, within the Spirurida. They include Gnathostomatoidea Ivaschkin, 1962, Gnathostomatidae Railliet, 1895 emend. Nicoll, 1927, Gnathostomatinae (Railliet, 1895 family), Ancyracanthinae Yorke and Maplestone, 1926, Echinocephalinae Yamaguti, 1961, Tanquinae Yamaguti, 1961 and Spiroxyinae Baylis and Lane, 1920. Chabaud (1975a), however, maintains the superfamily within the Spirurida and places it in the suborder Spirurina. He does not include the subfamilies Echinocephalinae and Tanquinae, placing *Echinocephalus* Molin, 1858 and *Tanqua* Blanchard, 1904 in the Gnathostomatinae.]

Gnathostomatidae Railliet, 1895

Ancyracanthinae Yorke and Maplestone, 1926

Elaphocephalus Molin, 1860

DIAGNOSIS: Body spinose with rings of 'posteriorly directed spines'. Mouth with pseudolabia, each with median conical process. Head with 4 processes directed posteriorly, 2 laterals shorter and bicuspid at tips, 2 median longer and serrated at tips. *Male*: Unknown. *Female*: Vulva opens near mouth. Parasite in feet of birds, *Psittacus macao*.

Type species: *Elaphocephalus octocornutus* Molin, 1860

[Diagnosis based on Molin (1860) and Yorke and Maplestone (1926). Molin (1860) indicates this genus has affinity with *Ancyracanthus* Diesing, 1858. Yorke and Maplestone (1926) place the genus in the Filariidae of the Filarioidea. Spencer Jones and Gibson (1987) consider the genus *incertae cedis*, and it was not a genus included in the *CIH Keys*. Although the taxonomic position of the genus is in doubt, as Molin considered the genus has some links to *Ancyracanthus* it is placed in the same subfamily in the Gnathostomatoidea until further study.]

PHYSALOPTEROIDEA

Physalopteridae (Railliet, 1893 subfamily) Leiper, 1908

Genus not assigned to a subfamily

Metaleptus Machida, Ogawa and Okiyama, 1982 (Fig. 263)

DIAGNOSIS: Cephalic collarette absent. Pseudolabia each with 2 groups of 2–3 submedian teeth, 1 group of 2–3 median teeth, 2 submedian cephalic papillae and lateral amphid. Oesophagus undivided. Oesophago-intestinal valve well developed. Excretory pore and cervical papillae at same level posterior to nerve ring (at middle of anterior half of oesophagus). *Male*: Posterior body spirally coiled. Caudal alae poorly developed. Caudal papillae pedunculate: 4 pairs precloacal; 4 pairs postcloacal; 2 anteriormost pairs do not extend into caudal alae. Spicules equal, similar. Tail tip blunt. *Female*: Vulva opens in posterior half of body, not close to anus. Amphidelphic. Tail bluntly pointed. Parasite in alimentary canal of selachians (*Chlamydoselachus anguineus*) off Pacific coast of Central Honshu, Japan.

Type species: *Metaleptus rabuka* Machida, Ogawa and Okiyama, 1982

[Listed in Spencer Jones and Gibson (1987).]

263.1

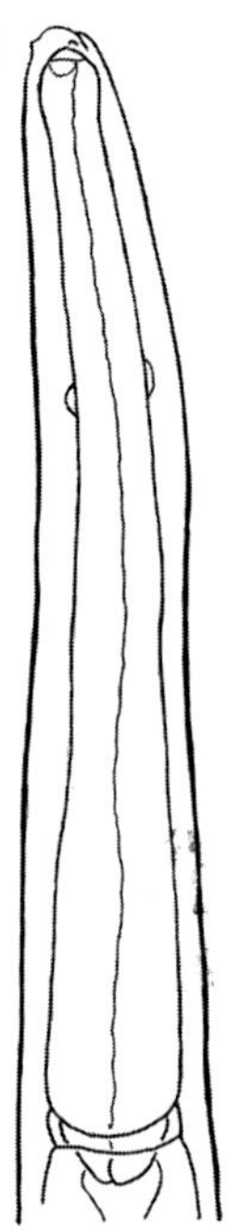

263.2

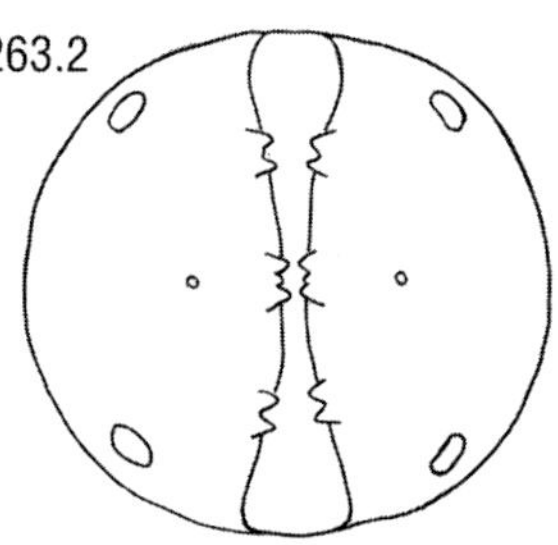

263.3

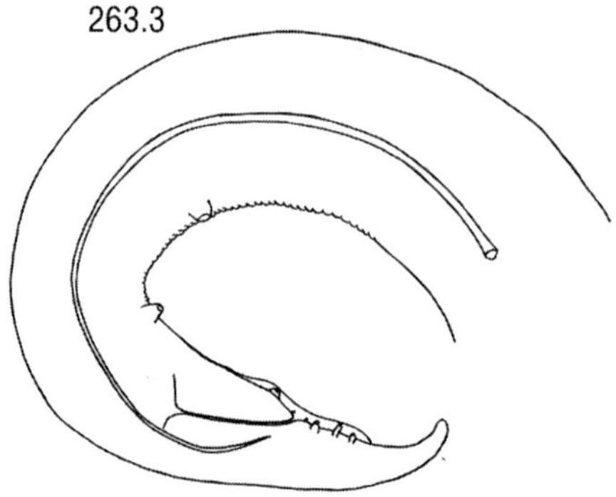

FIG. 263 *Metaleptus* 263.1 Anterior end of male; 263.2 Cephalic region, *en face* view; 263.3 Posterior end of male, lateral view. (Redrawn after Machida *et al.*, 1982.)

Physalopterinae Railliet, 1893

Physaloptera Rudolphi, 1819

Subgenus *Physaloptera* (Rudolphi, 1819 genus) Gupta and Johri, 1987

DIAGNOSIS: *Male*: Presence of closed prepuce-like sheath on posterior end. *Female*: Absence of prepuce-like sheath on posterior end.

Type species: *Physaloptera* (*Physaloptera*) *clausa* Rudolphi, 1819

Subgenus *Chlamydonema* (Hegt, 1910 genus) Gupta and Johri, 1987

DIAGNOSIS: Presence of an open prepuce-like sheath at distal end. *Male*: Sheath only extends slightly beyond posterior end of tail. *Female*: Sheath greatly extends beyond posterior end of tail.

Type species: *Physaloptera* (*Chlamydonema*) *praeputiale* (Linstow, 1889) Travassos, 1917

[*Chlamydonema* was erected for the species *C. felineum* Hegt, 1910. Yorke and Maplestone (1926) list the type species as a synonym of *C. praeputiale* (Linstow, 1889). Skrjabin and Sobolev (1964) list *C. praeputiale* as a species of the genus *Physaloptera*. *Chlamydonema* is not included in Chabaud (1975a). It is listed by Spencer Jones and Gibson (1987).]

Leptosoma Travassos, 1920

DIAGNOSIS: Male: ?unknown. Female: Four uteri [according to Travassos (1920)].

Type species: ?*Leptosoma leptosoma* (Gervais, 1848)

[Yorke and Maplestone (1926) suggest the type species is *L. leptosoma* (Gervais, of Seurat, 1917) and record that Travassos (1920) divided the genus *Physaloptera* Rudolphi, 1819 into several genera. They note that the name '*Leptosoma*' is preoccupied and that 'according to Seurat, 1917' the 'type species, *P. leptosoma*, has only two uteri'. Skrjabin and Sobolov (1964) place the type species in the genus *Abbreviata* as *A.* (*Didelphyoptera*) *leptosoma* (Gervais, 1848). The genus is listed in Spencer Jones and Gibson (1987).]

Skrjabinoptera Shulz, 1927

[Schulz (1927) originally divided his genus into two subgenera, namely *Skrjabinoptera* and *Didelphysoma*. Chabaud (1975a) did not accept these subgenera. *Didelphysoma* is listed in Spencer Jones and Gibson (1987).]

Paraphysaloptera Gupta and Kazim, 1979 (Fig. 264)

DIAGNOSIS: Mouth with 2 large pseudolabia, each with 4 pairs of teeth. Oesophagus divided with anterior muscular and posterior glandular regions. Lateral alae present along length of body. *Male*: Caudal alae well developed, longitudinal striations present. Caudal papillae

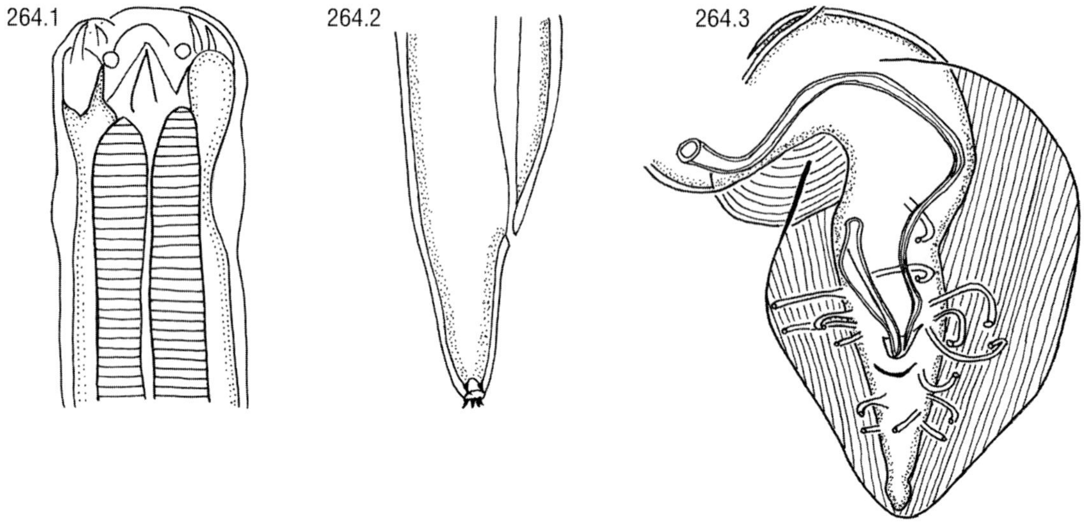

FIG. 264 *Paraphysaloptera* 264.1 Anterior end of male; 264.2 Female tail, lateral view; 264.3 Posterior end of male, ventral view. (Redrawn after Gupta and Kazim, 1978.)

pedunculated, 5 pairs precloacal, 3 pairs postcloacal. Spicules dissimilar: right long, slender and pointed; left short, broader and curved. Gubernaculum present. *Female*: Vulva opens in posterior part of body. Uteri parallel. Oviparous. Tail tapers to point with 6 spines. Parasite of gizzard lining in Hoopoe, *Upupa epops*.

Type species: *Paraphysaloptera alii* (Gupta and Kazim, 1978) Gupta and Kazim, 1979

[Diagnosis based on Gupta and Kazim (1978). This genus was first named in the abstracts of the 58th Indian Science Congress by Gupta and Jehan (1971) with *Paraphysaloptera thapari* Gupta and Jehan, 1971 as type species. Under the rules of the International Commission of Zoological Nomenclature this name is not valid. Later the genus was referred to as *Parpahysaloptera* Gupta and Kazim, 1978, *Parapaysaloptera* Gupta and Kazim, 1978 and *Paraphyaloptera* Gupta and Kazim, 1978 with *P. alii* as type species and described again under the name *Paraphysaloptera* by Gupta and Kazim (1979) with the same type species. The name *Paraphysaloptera* is accepted here as the intended name for the genus and the other names are considered synonyms. Listed in Spencer Jones and Gibson (1987).]

Kreisiella Jones, 1985 (Fig. 265)

DIAGNOSIS: Pseudolabia each with 2 submedian cephalic papillae, 1 lateral amphid; 25–30 fine, even-sized denticles on inner margin; no additional teeth. *Male*: Caudal alae narrow, not joined antero-ventrally, stop before distal end of tail, unsupported by papillae. Ventral ornamentation extends between anterior and posterior margins of caudal alae, contained laterally within area of caudal papillae. Ten pairs of sessile caudal papillae: 7 pairs ventro-lateral; 1 pair precloacal; 2 pairs immediately postcloacal; single median precloacal papilla. Spicules unequal, dissimilar. *Female*: Vulva opens just posterior to level of origin of glandular oesophagus. Tail very short, broadly rounded. Parasite in stomach of Australian lizards (*Egernia inornata*).

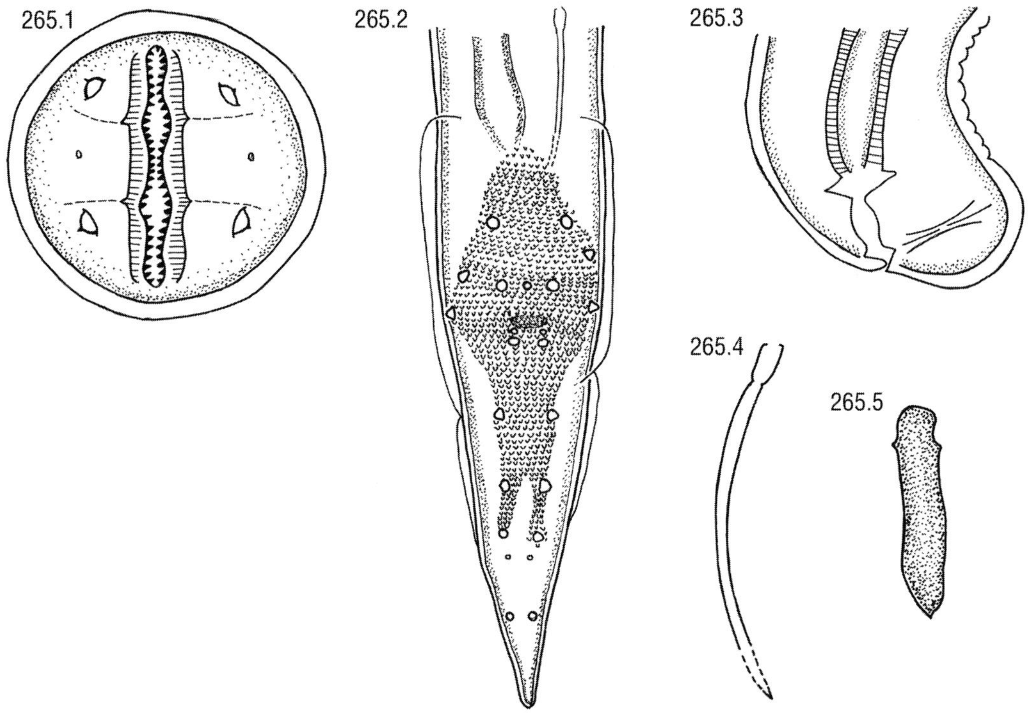

FIG. 265 *Kreisiella* 265.1 Cephalic region *en face* view; 265.2 Posterior end of male, ventral view; 265.3 Posterior end of female, lateral view; 265.4 Left spicule; 265.5 Right spicule. (Redrawn after Jones, 1985.)

Type species: *Kreisiella chrysocampa* Jones, 1985

[Listed in Spencer Jones and Gibson (1987).]

Proleptinae (Schulz, 1927)

Neoleptus Specian, Ubelaker and Dailey, 1975 (Fig. 266)

DIAGNOSIS: Mouth with 2 lateral pseudolabia, each with single interno-lateral tooth, a single row of small denticles on inner anterior margin and 2 submedian cephalic papillae. Cephalic region inflated forming collarette. Buccal capsule absent. Oesophagus divided. *Male*: Caudal alae 'poorly developed'. Nine pairs of sessile caudal papillae: 4 precloacal; 5 postcloacal. Spicules markedly unequal, dissimilar. Gubernaculum absent. *Female*: Vulva opens in posterior half of body. Tail tapers to blunt point. Parasite in fish, *Heterodontus philippi* and *Mustelus antarticus*.

Type species: *Neoleptus australis* (Johnston and Mawson, 1943) Specian, Ubelaker and Dailey, 1975

[Specian, Ubelaker and Dailey (1975) created this genus for a species of *Paraleptus* and place their genus in the subfamily Physalopterinae. They give a key to the genera from fish, namely *Proleptus*

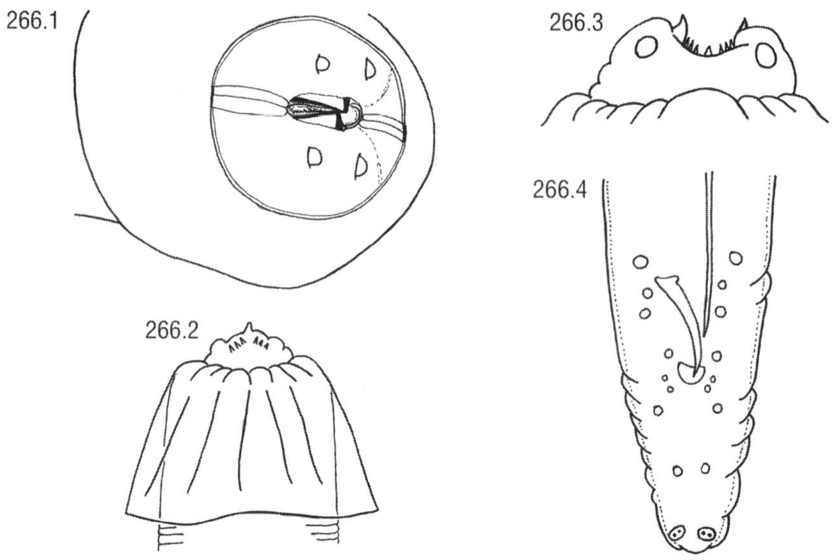

FIG. 266 *Neoleptus* 266.1 Cephalic region (traced from scanning electron micrograph after Specian *et al.*, 1975); 266.2 Cephalic region; 266.3 Cephalic region, higher magnification; 266.4 Posterior end of male, ventral view. (266.2, 266.3 and 266.4 redrawn after Johnston and Mawson, 1943 from Skrjabin and Sobolev, 1964.)

Dujardin, 1845, *Paraleptis* Wu, 1927, *Heliconema* Travassos, 1919, *Dogielina* Sobolev, 1949 (Chabaud, 1975a considers this one of the genera which appears to belong to the Physalopteroidea but is not well enough known to be classified), *Pseudoproleptus* Khera, 1955 (classified by Chabaud, 1975b in the Cystidicolidae of the Habronematoidea) and *Neoleptus*. Chabaud (1975a) places *Proleptus* in a separate subfamily, Proleptinae, which he establishes for genera from fish and includes some of the genera Specian *et al.* (1975) place in the Physalopterinae. Following Chabaud (1975a), *Neoleptus* is placed in the Proleptinae. Moravec (2007) considers *Neoleptus* as a synonym of *Paraleptus*. *Neoleptus* is listed in Spencer Jones and Gibson (1987).]

Mirzalopterinae Wason and Johnson, 1977

DIAGNOSIS: Caudal alae absent in males.

Type genus *Mirzaloptera* Wason and Johnson, 1977 (Fig. 267)

DIAGNOSIS: Mouth opening terminal. Two large lateral pseudolabia present, each with large median tooth and pair of smaller teeth at each lateral margin plus a pair of submedian papillae. Poorly developed cephalic collarette present. Oesophagus divided. *Male*: Caudal papillae sessile. Spicules delicate, similar, subequal. Gubernaculum absent. *Female*: Unknown. Parasite in the stomach of a bat, *Rhinopoma microphyllum*, in Jodhpur, India.

Type species: *Mirzaloptera barbari* Wason and Johnson, 1977

[Listed in Spencer Jones and Gibson (1987).]

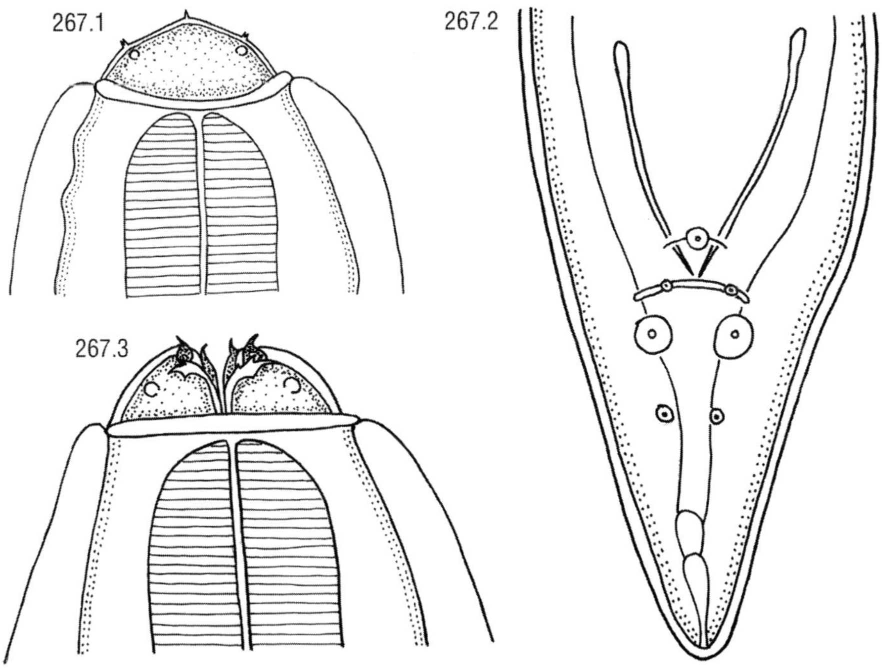

FIG. 267 *Mirzaloptera* 267.1 Cephalic region of male, lateral view; 267.2 Posterior end of male, ventral view; 267.3 Cephalic region of male, dorsal view. (Redrawn after Wason and Johnson, 1977.)

RICTULARIOIDEA

Rictulariidae (Hall, 1915 subfamily) Railliet, 1916

Quentius Chabaud and Bain, 1981 (Fig. 268)

DIAGNOSIS: Mouth opening apical. Buccal capsule ventral wall less high than dorsal. Three oesophageal teeth, each equipped with 2 thin, denticulate vertical plates. Spines numerous, uniform in structure. Large dorsal cuticular ala on the front of the body. Posterior part of body swollen. *Male*: unknown. *Female*: Vulva opens posterior to oesophagus. Parasite in duodenum and small intestine of neotropical marsupials (*Marmosa* sp.) in Cali, Colombia.

Type species: *Quentius kozeki* Chabaud and Bain, 1981

[Listed in Spencer Jones and Gibson (1987).]

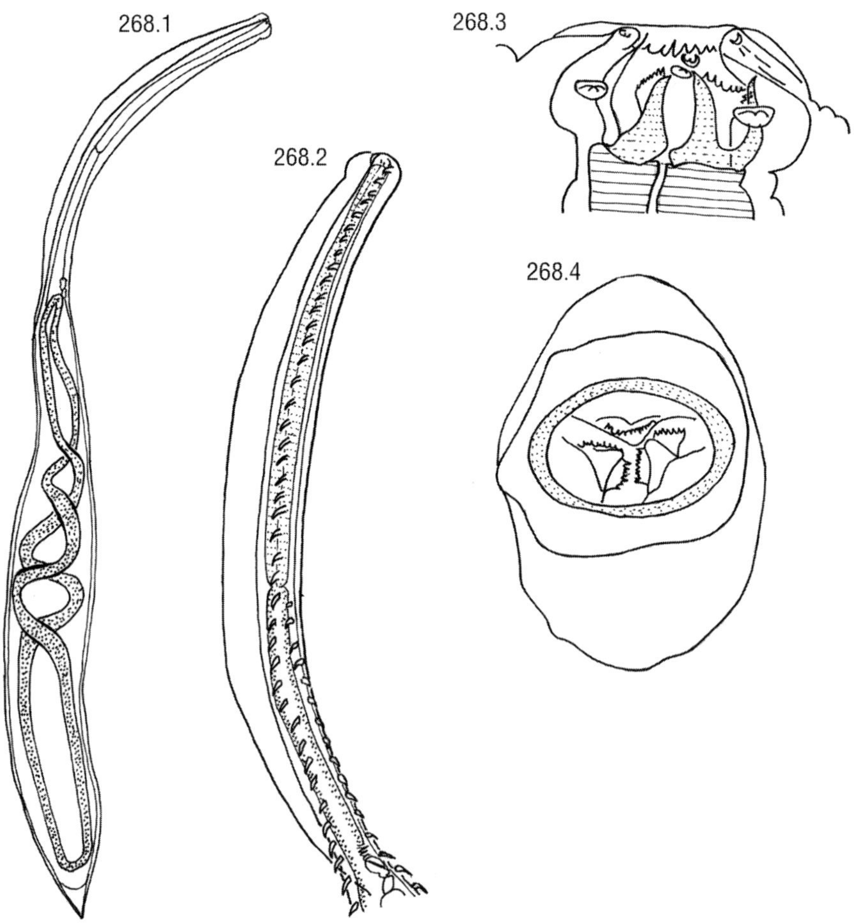

FIG. 268 *Quentius* 268.1 Whole body; 268.2 Anterior end of body; 268.3 Cephalic region, right lateral view; 268.4 Cephalic region, optical section showing oesophageal teeth. (Redrawn after Chabaud and Bain, 1981.)

Shamimana Gupta and Masoodi, 1990 (Fig. 269)

DIAGNOSIS: Cuticle with distinct annulations, longitudinal and transverse rows of spines present. Mouth terminal, 2 well developed lips, circle of denticles present. Teeth present on anterior rim of buccal capsule. *Male*: Unknown. *Female*: Didelphic. Vulva opens in posterior half of body. Tail conical. Eggs smooth shelled. Parasites of intestine of marine fish (*Plotosus arab*) off Trivandrum coast, Kerala, India.

Type species: *Shamimana durdanae* Gupta and Masoodi, 1990

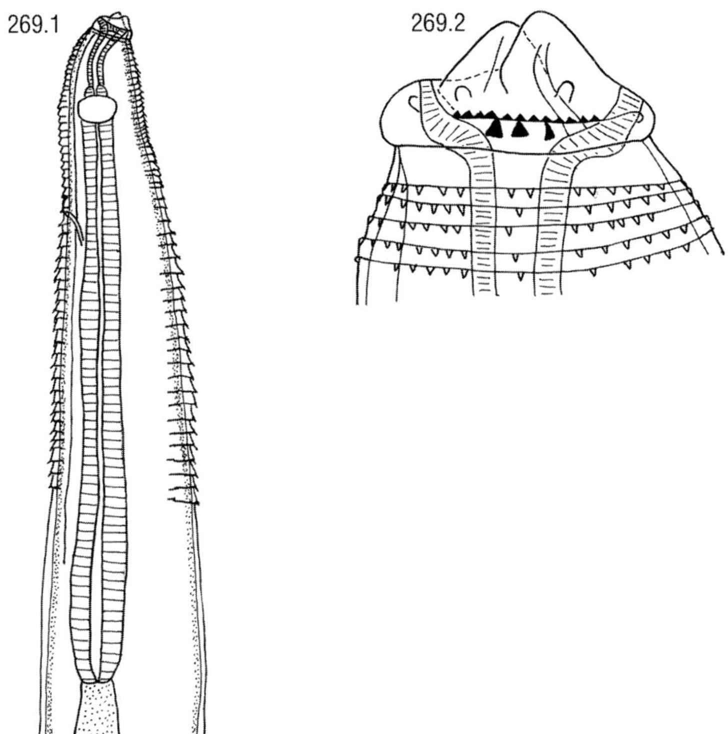

FIG. 269 *Shamimana* 269.1 Anterior end of female; 269.2 Cephalic region of female. (Redrawn after Gupta and Masoodi, 1990.)

THELAZIOIDEA

Thelaziidae Skrjabin, 1915

Thelazo Pearse, 1933

DIAGNOSIS: Mouth without lips. Buccal capsule elongated with curved lateral walls, valves absent. *Male*: Posterior end coiled. Caudal alae absent. Six sessile precloacal papillae present. Spicules unequal. Gubernaculum present. *Female*: Vulva opens near anterior end. Tail tapers with 2 divergent spines at distal tip. Parasite in intestine of *Glossogobius giurus*.

Type species: *Thelazo glossogobii* Pearse, 1933

[Diagnosis based on Pearse (1933), who placed the genus in the Thelaziidae. Ivashkin *et al.* (1971) consider this type species should be included in the genus *Procamallanus* Baylis, 1923 and the genus a synonym of *Procamallanus*. Chabaud (1975a) places *Procamallanus* in the family Camallanidae of the superfamily Camallanoidea. Listed in Spencer Jones and Gibson (1987) in the Camallanoidea.]

Thelaziinae (Skrjabin, 1915 family) Baylis and Daubney, 1926

Thelazia Bosc, 1819

Subgenus *Thelazia* (Bosc, 1819 genus)

DIAGNOSIS: *Male*: Posterior end with single precloacal papilla. *Female*: Viviparous. Parasites in mammals.

Type species: *Thelazia* (*Thelazia*) *rhodesi* (Desmarest, 1827) Railliet and Henry, 1910

Subgenus *Pericyema* Railliet, 1925

DIAGNOSIS: *Female*: Ovoviviparous. Parasites in mammals.

Type species: *Thelazia* (*Pericyema*) *callipaeda* Railliet and Henry, 1910

Subgenus *Isothela* Railliet, 1925

DIAGNOSIS: *Male*: Posterior end with genital papillae paired. *Female*: Viviparous. Parasites in birds.

Type species: *Thelazia* (*Isothela*) *lutzi* Travassos, 1918

[Diagnosis of subgenera based on Railliet (1925) and Skrjabin *et al.* (1967, transl. 1971). Chabaud (1975a) splits the genus into two subgenera, namely *Thelazia* and *Thelaziella*, based on the length of the female tail, the presence or absence of a gubernaculum, thickness of the egg shell and host. Skrjabin *et al.* (1967, transl. 1971) place the type species of subgenera *Thelazia* and *Pericyema* in the genus *Thelazia*, and the type species of *Isothela* is considered a synonym of *Thelaziella anolabiata* (Molin, 1860). Following the classification of Chabaud (1975a) and Skrjabin *et al.* (1967), the subgenus *Pericyema* is considered a synonym of the subgenus *Thelazia* and the subgenus *Isothela* a synonym of the subgenus *Thelaziella*. The subgenera *Pericyema* and *Isothela* are listed in Spencer Jones and Gibson (1987).]

Rhabdochonidae (Travassos, Artigas and Pereira, 1928 subfamily) Skrjabin, 1946

Rhabdochona Railliet, 1916

Subgenus *Globochonoides* Moravec, 1975 (Fig. 270)

DIAGNOSIS: Buccal capsule with 6–8 teeth on anterior rim, basal teeth present. Lateral alae wide, extend from just posterior to cervical papillae to distal end of body. Cervical papillae simple, poorly developed. *Male*: Tail rounded. *Female*: Mature eggs smooth. Tail tip bluntly conical to rounded, crown of minute, tooth-like processes present or absent. Parasites in intestine of freshwater fishes.

Type species: *Rhabdochona* (*Globochonoides*) *coronacauda* Belouss, 1965

[Emended diagnosis based on Moravec and Sey (1988b).]

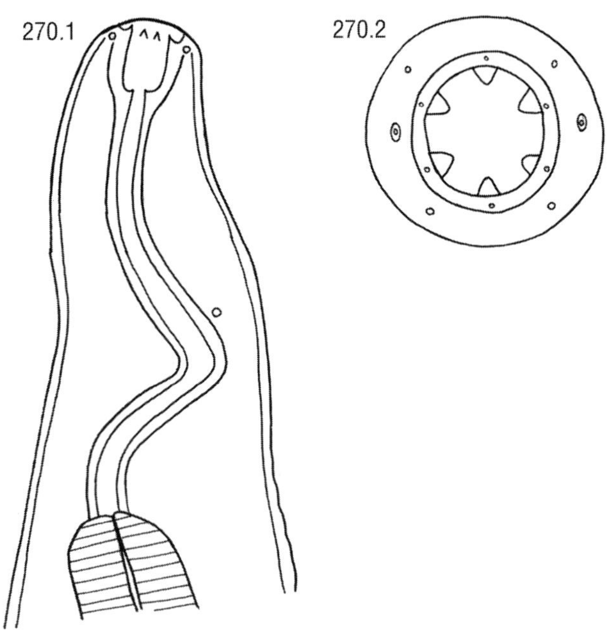

FIG. 270 *Globochonoides* 270.1 Anterior end of female, lateral view; 270.2 Cephalic region, *en face* view. (Redrawn after Moravec and Sey, 1988b.)

Afrochona Puylaert, 1973 (Fig. 271)

DIAGNOSIS: Proximal end of buccal capsule with teeth. *Male*: Spicules equal and similar. *Female*: Unknown. Parasite in intestine of fish, *Aphyosemion cameronensis*, in Olounou, Cameroon.

Type species: *Afrochona camerounensis* Puylaert, 1973

[Puylaert (1973) considers his species close to *Rhabdochona moraveci* but is distinguished by having equal and similar spicules. He placed his genus in the subfamily Rhabdochininae Travassos, Artigas and Pereira, 1928. Chabaud (1975a) did not divide the Rhabdochonidae into subfamilies. Listed in Spencer Jones and Gibson (1987).]

Johnstonmawsonoides Machida, 1975 (Fig. 272)

DIAGNOSIS: Cuticle with transverse striations posterior to cephalic region. Cephalic region slightly swollen. Mouth opening dorso-ventrally, elongate. Cephalic sensory organs: 4 submedian papillae, 2 amphids. Buccal capsule narrow, chitinous walls with fine transverse

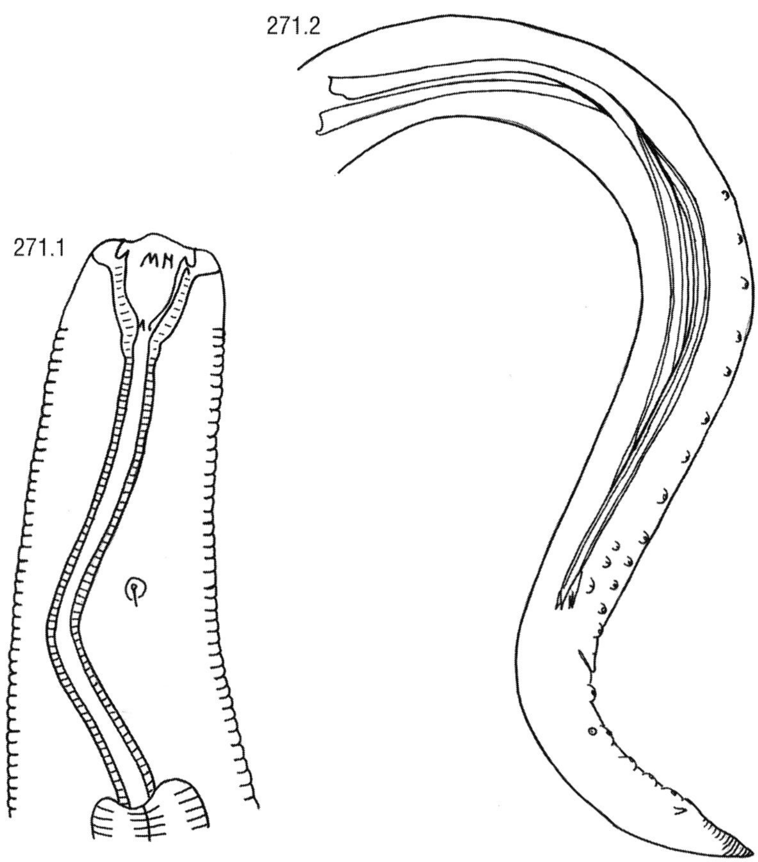

FIG. 271 *Afrochona* 271.1 Anterior end of male, lateral view; 271.2 Posterior end of male, lateral view. (Redrawn after Puylaert, 1973.)

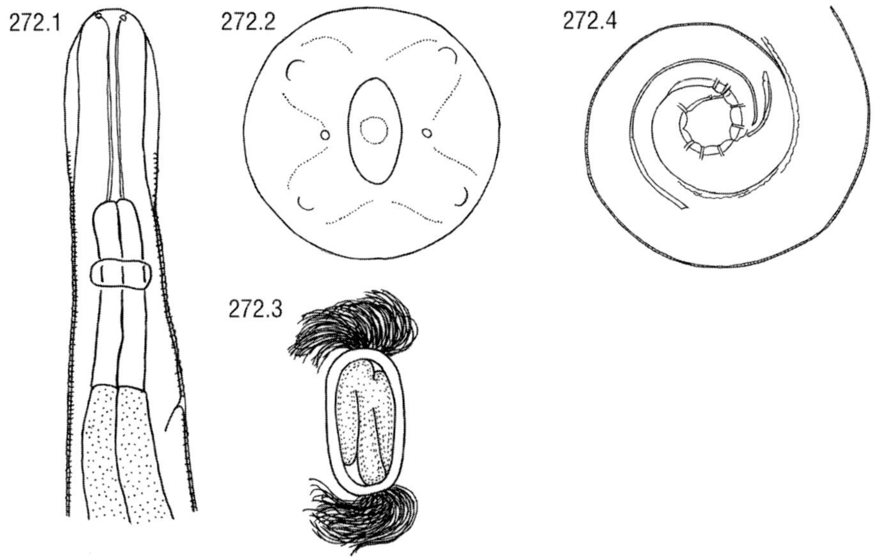

FIG. 272 *Johnstonmawsonoides* 272.1 Female anterior end, lateral view; 272.2 Cephalic region, *en face* view, originally traced from a scanning electron micrograph; 272.3 Egg; 272.4 Posterior end of male, lateral view. (Redrawn after Machida, 1975.)

striations. Oesophagus with anterior short, muscular and posterior long, glandular region. Cervical papillae at level of buccal capsule–oesophageal junction. *Male*: Posterior body spirally coiled. Caudal alae narrow. *Area rugosa* with longitudinal striae. Caudal papillae pedunculate: 4 pairs precloacal; 5 pairs postcloacal. Spicules unequal. Gubernaculum absent. Tail bluntly pointed. *Female*: Vulva opens near mid-body. Uteri opposed. Egg with thick tufts of filaments at each pole. Tail bluntly pointed. Parasite in intestine of marine teleosts, *Nemichthys scolopaceus*, in Suruga Bay, Japan.

Type species: *Johnstonmawsonoides nemichthyos* Machida, 1975

[Machida (1975) placed this genus in the family Rhabdochonidae and the subfamily Ascarophidinae. The subfamily was not accepted by Chabaud (1975a). Listed in the Thelazioidea in Spencer Jones and Gibson (1987). Moravec (2007) lists this genus in the family Cystidicolidae of the Habronematoidea, but considers it 'inadequately known' in his review of the suborder.]

Neoascarophis Machida, 1976 (Fig. 273)

DIAGNOSIS: Cuticle with transverse striations. Mouth opening dorso-ventrally elongate, with 3 pairs of lips. Lateral lips prominent with projecting median process, submedian lips triangular. Cephalic sensory organs: 4 submedian papillae; 2 lateral amphids. Cephalic collarette present or absent. Buccal capsule short. Oesophagus with anterior, short muscular region and posterior, long glandular region. Cervical papillae at level of nerve ring. *Male*: Posterior body spirally coiled. Caudal alae narrow. *Area rugosa* with longitudinal striae anterior to caudal alae. Caudal papillae pedunculate: 4 pairs precloacal; 5 pairs postcloacal. Spicules unequal. Gubernaculum absent. Tail bluntly pointed. *Female*: Vulva opens near mid-body. Uteri opposed. Eggs without

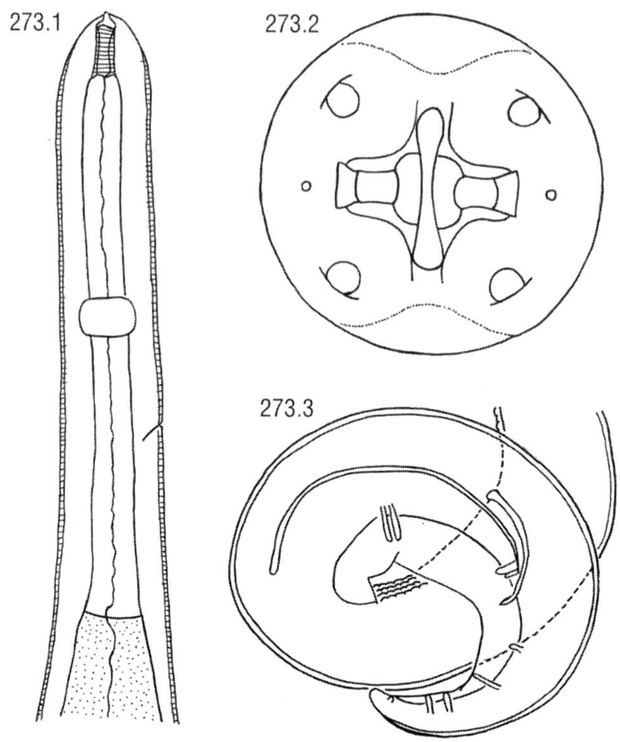

FIG. 273 *Neoascarophis* 273.1 Anterior end of male, lateral view; 273.2 Cephalic region, *en face* view (originally traced from a scanning electron micrograph); 273.3 Posterior end of male, lateral view. (Redrawn after Machida, 1976.)

filaments. Tail bluntly pointed. Parasites in intestine of marine teleosts, *Coelorhynchus multispinulosus* and *Bathygadus garretti*, in Suruga Bay, Japan.

Type species: *Neoascarophis yarihige* Machida, 1976

[Machida (1976) placed this genus in the family Rhabdochonidae and the subfamily Ascarophidinae. The subfamily Ascarophidinae is not accepted by Chabaud (1975a). Listed in the Thelazioidea in Spencer Jones and Gibson (1987). Moravec (2007) lists this genus in the Cystidicolidae of the Habronematoidea.]

Fellicola Petter and Køie, 1993 (Fig. 274)

DIAGNOSIS: Mouth oval. Cephalic sensory organs: inner ring of 6 papillae; 4 submedian cephalic papillae; 2 amphids. Buccal capsule: elongate, narrow, wider at anterior end; proximal walls with 6 internal longitudinal thickenings. Oesophagus divided. Nerve ring surrounds buccal capsule. Excretory pore opens at level of proximal end of oesophagus. *Male*: Spicules unequal, dissimilar. Gubernaculum present. Caudal alae rudimentary, only present at level of last pair of papillae nearest tail tip. Caudal papillae: 1 pair precloacal; 3 pairs postcloacal. *Female*: Vulva opens anterior to mid-body. Didephic. Amphidelphic. Eggs without filaments.

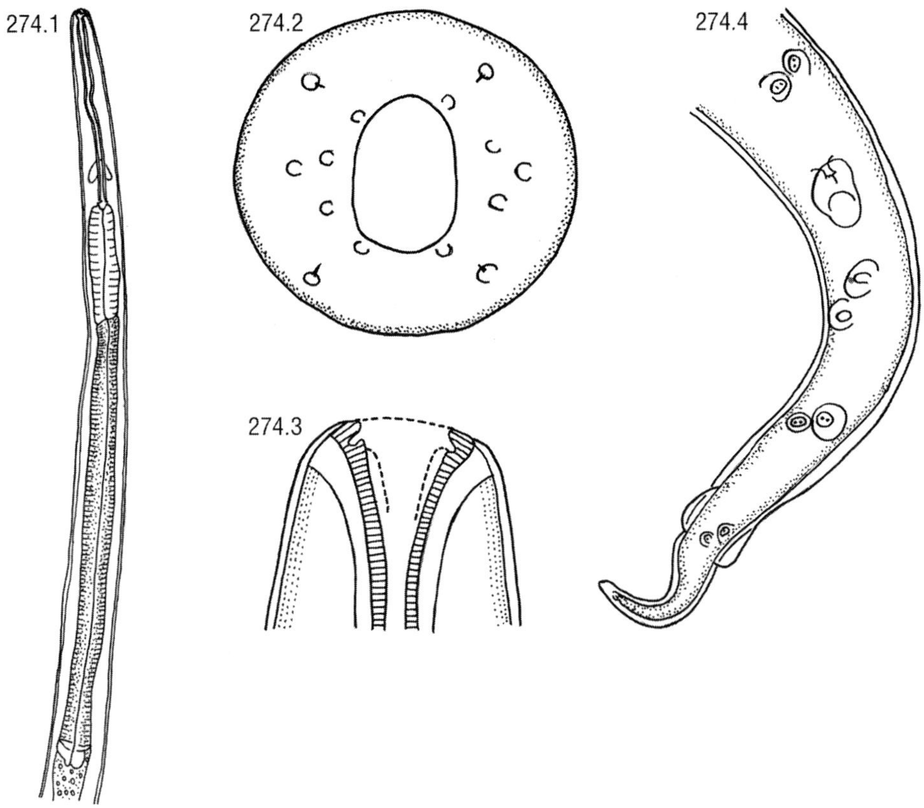

FIG. 274 *Fellicola* 274.1 Anterior end of male; 274.2 Cephalic region of male, *en face* view; 274.3 Cephalic region of female; 274.4 Posterior end of male. (Redrawn after Petter and Køie, 1993.)

Parasite in gall bladder of marine fish, *Coryphaenoides rupestris*, collected in the North Atlantic off the Faroes.

Type species: *Fellicola longispiculus* Petter and Køie, 1993

Beaninema Caspeta-Mandujano, Moravec and Salgado-Maldonado, 2001 (Fig. 275)

DIAGNOSIS: Mouth hexagonal. Cephalic sensory organs: inner circle of 4 submedian papillae, outer circle of 4 submedian papillae, 2 lateral amphids. Buccal capsule elongate, narrow, internal longitudinal thickenings extend into wider proximal end as teeth. *Male*: Pairs of precloacal and postcloacal papillae few in number. Caudal alae, *area rugosa* absent. Spicules unequal, dissimilar. Gubernaculum absent. Tail conical. *Female*: Amphidelphic. Vulva opens near midbody. Tail conical. Parasites in the gallbladder of freshwater fish (*Cichlasoma beani*) from the Santiago River, Tepic, Nayarit, Mexico.

Type species: *Beaninema nayaritense* Caspeta-Mandujano, Moravec and Salgado-Maldonado, 2001

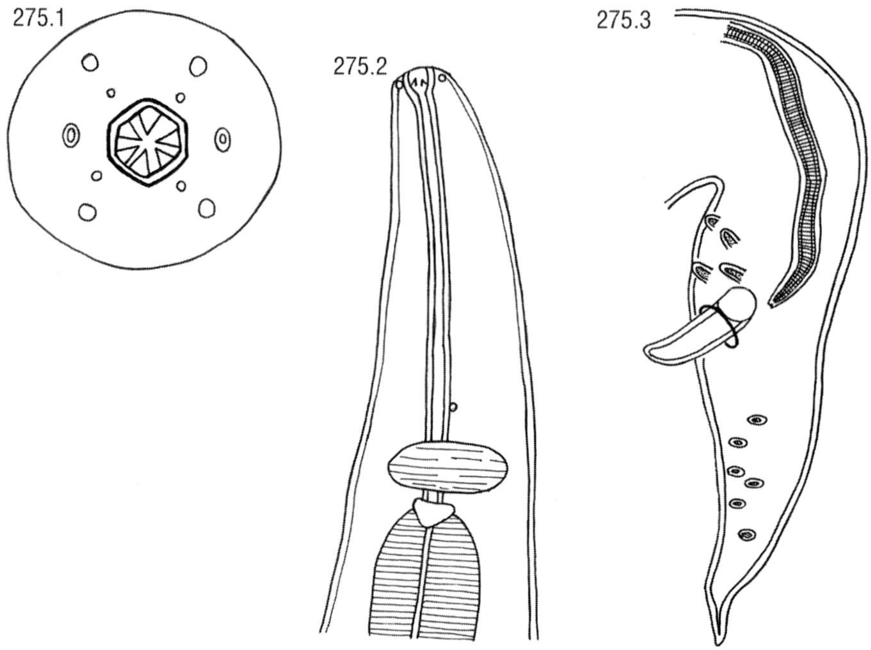

FIG. 275 *Beaninema* 275.1 Cephalic region, *en face* view; 275.2 Anterior end of male, lateral view; 275.3 Posterior end of male, ventro-lateral view. (Redrawn after Caspeta-Mandujano *et al.*, 2001.)

Megachona Mejía-Madrid and Pérez-Ponce de León, 2007 (Fig. 276)

DIAGNOSIS: Mouth opening oval without pseudolabia. Thick circumoral ring present with outer circle of 4 submedian and an inner circle of 4 cephalic papillae. Buccal capsule long, narrow, proximal end funnel-shaped with longitudinal ridges forming large tooth-like structures at their proximal end anterior to randomly placed smaller teeth. Nerve ring encircles proximal part of muscular oesophagus. Cervical papillae large. *Male*: Pairs of precloacal and postcloacal papillae few in number. Spicules unequal, dissimilar. Lateral caudal alae, *area rugosa*, gubernaculum absent. Tail long, broadly rounded distal tip. *Female*: Amphidelphic. Vulva opens near mid-body. Tail long, broadly rounded distal tip. Parasite in the intestinal cecae of blue-striped chubb, *Sectator ocyurus*, from Chamela Bay, Mexico.

Type species: *Megachona chamelensis* Mejía-Madrid and Pérez-Ponce de León, 2007

Prosungulonematinae Skrjabin, Sobolev and Ivashkin, 1967

[Chabaud (1975a) did not include this subfamily, accepting the synonymy of the type genus *Prosungulonema* Roitman, 1963 with *Johnstonmawsonia* Campana-Rouget, 1955 as proposed by Moravec and Puylaert (1970). These two genera are included as synonyms in the Rhabdochonidae by Chabaud (1975a). Later, Caspeta-Mandujano *et al.* (2001) do not include *Prosungulonema* as one of the valid genera in the family Rhabdochonidae. McVicar and Gibson (1975), however, supported the validity of *Prosungulonema* and the subfamily Prosungulonematinae. They included three genera listed below in the subfamily.]

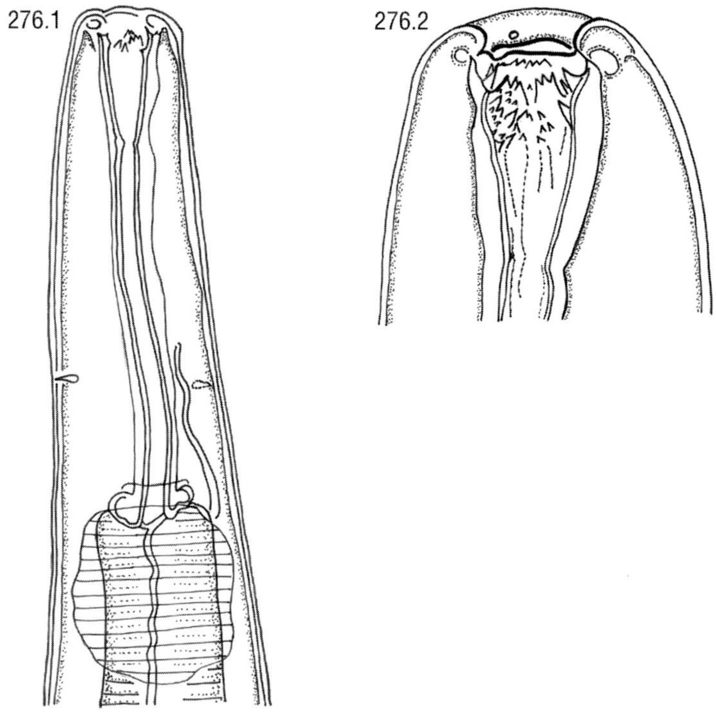

FIG. 276 *Megachona* 276.1 Anterior end of male, dorsal view; 276.2 Cephalic region of female, ventral view. (Redrawn after Mejía-Madrid and Peréz-Ponce de León, 2007.)

Type genus: *Prosungulonema* Roitman, 1963

DIAGNOSIS: Parasites of freshwater teleosts.

Type species: *Prosungulonema siniperca* (Dogiel and Akhmerov, 1959)

Johnstonmawsonia Campana-Rouget, 1955

DIAGNOSIS: Parasites of marine teleosts.

Type species: *Johnstonmawsonia coelorhynchi* (Johnston and Mawson, 1945) Campana-Rouget, 1955

Pancreatonema McVicar and Gibson, 1975 (Fig. 277)

DIAGNOSIS: Body long, thin, forms conical helix. Cuticle with faint transverse striations. Mouth with 2 unarmed pseudolabia. Cephalic papillae: ring of 6 internal papillae; 4 double submedian papillae; 2 lateral amphids. Buccal capsule long, slender, widens at distal end. Oesophagus long, not clearly divided. Cervical papillae near level of distal end of buccal capsule. *Male*: Spicules unequal, dissimilar. Gubernaculum present. Caudal alae absent. Cuticular collar surrounds cloaca. Caudal papillae: 1 pair and single, median precloacal papillae; 4–5 pairs postcloacal

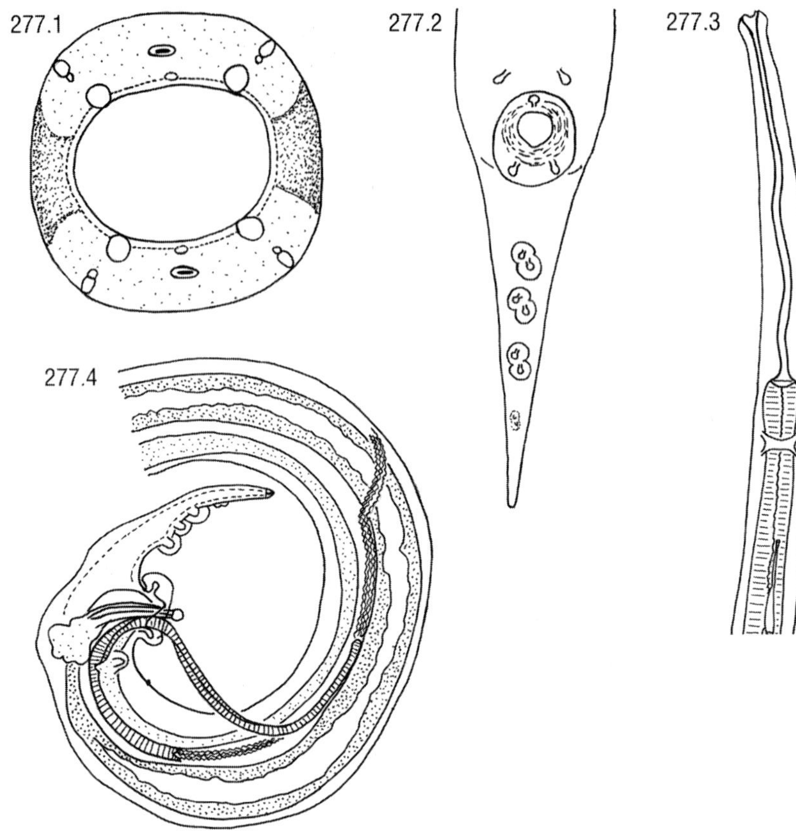

FIG. 277 *Pancreatonema* 277.1 Cephalic region, *en face* view; 277.2 Posterior end of male, hypothetical ventral view based on lateral view observations by McVicar and Gibson, 1975; 277.3 Anterior region of body; 277.4 Posterior end of male, lateral view. (Redrawn after McVicar and Gibson, 1975.)

papillae. Tail tapers to a point. *Female*: Vulva opens near mid-body. Amphidelphic. Eggs without filaments. Tail tapers to a point. Parasite in pancreatic duct of the cuckoo ray, *Raja naevus*, from the North Sea off Aberdeen, Scotland, UK.

Type species: *Pancreatonema torriensis* McVicar and Gibson, 1975

[Listed in Spencer Jones and Gibson (1987).]

Pneumospiruridae Wu and Hu, 1938

***Pneumospirura* Wu and Hu, 1938**

[Wertheim and Chabaud (1977) examined species of the family Pneumospiruridae with the aid of the scanning electron microscope. They concluded that *Pneumospirura* should become a subjective synonym of *Metathelazia* Skinker, 1931 and that the family name should be retained according to Article 40 of the ICZN, London, 1961. *Pneumospirura* was included as a valid genus in the key (Chabaud, 1975a).]

SPIRUROIDEA

Spiruridae Oerley, 1885

Gastronodus Singh, 1934

DIAGNOSIS: Circle of 6 cephalic sensory organs. Buccal capsule with 6 tooth-like processes protruding into buccal cavity. Oesophagus with short, narrow muscular region and long, wide glandular region. *Male*: Caudal papillae: 7 pairs precloacal; 3 pairs postcloacal. Pair of spine-like processes on distal tip of tail. *Female*: Vulva opens at level of distal region of oesophagus. Parasite in nodules on wall of stomach of muskrat, *Crocidura coerulea*, in Hyderabad State, India.

Type species: *Gastronodus strasseni* Singh, 1934

[Singh (1934) placed the genus in the subfamily Spirurinae. The genus is listed in Spencer Jones and Gibson (1987) but not included in the key of Chabaud (1975b).]

Dollfusnema Caballero, 1974 (Fig. 278)

DIAGNOSIS: Body terminates in long caudal appendix with rounded point. Cephalic region wide, lateral pseudolabia present. Four pairs submedian cephalic papillae on pseudolabia. Buccal capsule absent. Oesophagus long with anterior muscular region with funnel-shape proximal end, wider posterior glandular region. *Male*: Caudal extremity curled ventrally. Four pairs of pedunculate, precloacal papillae, 1 median papilla on anterior lip of cloaca, 5 pairs of postcloacal papillae. Spicules subequal, arcuate, finely pointed. Gubernaculum absent. Caudal

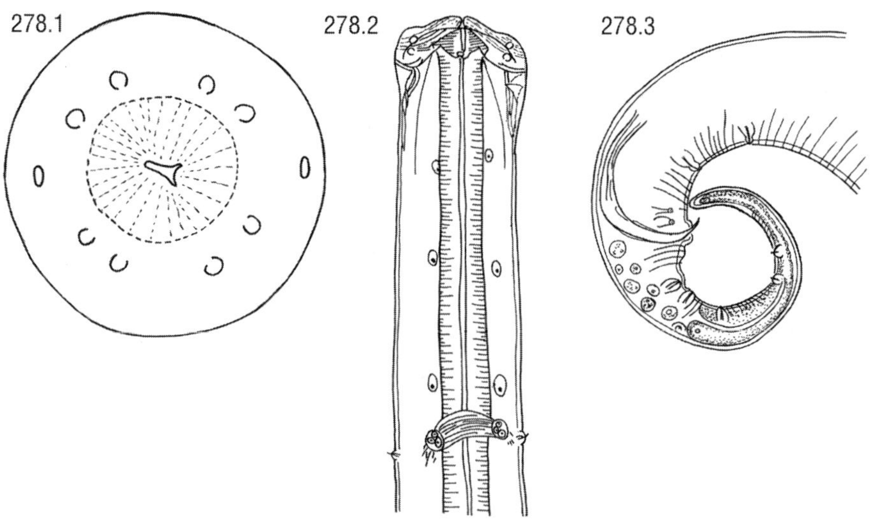

FIG. 278 *Dollfusnema* 278.1 Cephalic region of male, *en face* view; 278.2 Anterior end of male, ventral view; 278.3 Posterior end of male, lateral view. (Redrawn after Caballero, 1974.)

alae absent. *Female*: unknown. Parasites of intestine of marine fish (*Paralabrax clathratus*), Mexico.

Type species: *Dollfusnema piscicola* Caballero, 1974

Paracyrnea Gupta and Jaiswal, 1987 (Fig. 279)

DIAGNOSIS: Body spirally coiled. Cervical alae absent. Two conical pseudolabia each with a single 'papilla'. Interlabia absent. Cylindrical buccal capsule. Oesophagus divided. *Male*: Caudal alae present. Caudal papillae pedunculate: 4 precloacal papillae; 6 postcloacal papillae. Spicules equal. Gubernaculum absent. *Female*: Vulva opens posterior to mid-body. Uteri opposed. Oviparous. Eggs thin-shelled. Parasite of intestine of birds, *Anser indicus*, in the Prince of Wales Zoological Gardens, Lucknow, India.

Type species: *Paracyrnea yamagutii* Gupta and Jaiswal, 1987

Isospirura Sood and Parshad, 1972

DIAGNOSIS: Cuticular ornamentation including ventral bosses absent. Mouth dorso-ventrally elongate, lips absent. Buccal capsule funnel-shaped, walls without ring-like or spiral thickening, teeth absent. Oesophagus divided. *Male*: Spicules dissimilar. *Female*: Vulva opens in mid-region

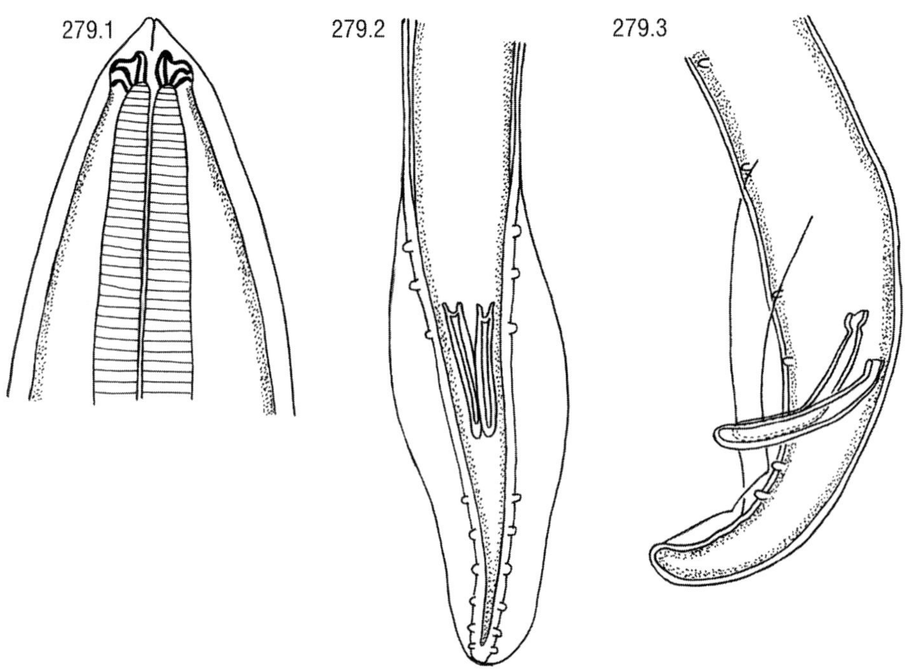

FIG. 279 *Paracyrnea* 279.1 Anterior end of male, lateral view; 279.2 Posterior end of male, ventral view; 279.3 Posterior end of male, lateral view. (Redrawn after Gupta and Jaiswal, 1987.)

of body. Uterine branches opposed. Parasite in stomach of *Millardia meltada*, *Mus musculus bactrianes* and *Mus booduga* in Ludhiana, India.

Type species: *Isospirura meltadi* Sood and Parshad, 1972

[The authors placed their genus in the subfamily Spririrurinae (Sood and Parshad, 1972). Chabaud (1975b) did not include subfamilies in his key. Listed in Spencer Jones and Gibson (1987).]

Gongylonematidae (Hall, 1916 subfamily) Sobolev, 1949

Gongylonema Molin, 1857

Subgenus *Progongylonema* Hernandez-Rodriguez and Gutiérrez-Palomino, 1992

DIAGNOSIS: Absence of cuticular verruciform thickenings. Parasites from mucosa under the tongue of *Pica pica*, *Garrulus glandarius*, *Cyanopica cyanus* and *Corvus monedula* (Passeriformes, Corvidae) in Córdoba Province, southern Spain.

Type species: *Gongylonema* (*Progongylonema*) *pacoi* Hernandez-Rodriguez and Gutiérrez-Palomino, 1992

[Hernandez-Rodriguez and Gutiérrez-Palomino (1992) add this third subgenus to *Gongylonema* on the basis of the morphology of the cuticle.]

Spirocercidae (Chitwood and Wehr, 1932 subfamily)

Spirocercinae Chitwood and Wehr, 1932

Mazzia Khalil and Vogelsang, 1932

DIAGNOSIS: Parasite in dasypodid mammals in Argentina.

Type species: *Mazzia mazzia* Khalil and Vogelsang, 1932

[This genus could not be classified in Chabaud (1975b) as the cephalic structures had not been studied, but it was thought to belong to the Spirocercinae. Chabaud, Navone and Bain (1983) confirm that *Mazzia* belongs to the Spirocercidae and the subfamily Spirocercinae. Chabaud *et al.* (1983) suggest similarities between *Physocephalus* of the subfamily Ascaropsinae and *Mazzia* and propose that the subfamily Ascaropsinae should be a synonym of Spirocercinae.]

Chlamydoprocta Chandler, 1954

DIAGNOSIS: Parasite of skunk, *Mephitis mephitis* in USA.

Type species: *Chlamydoprocta itascensis* Chandler, 1954

[Wong, Watson and Anderson (1980) re-examined the type species and considered it close to *Vigisospirura whitei* (Monnig, 1931) Chabaud, 1959. They transferred the type species as *Vigisospirura itascensis* (Chandler, 1954) Wong, Watson and Anderson, 1980 and considered *Chlamydoprocta* a synonym of *Vigisospirura* Petrow and Potekhina, 1953. Chabaud (1975b) suggested that the two genera may be synonyms.]

Paraspiralatus Gibbons, Nicholls, Bailey and Samour, 2004 (Fig. 280)

DIAGNOSIS: Cuticle with transverse striations. Cephalic vesicle absent. Pseudolabia present. Median teeth or lobes absent. Buccal capsule laterally compressed, funnel-shaped in dorso-ventral view, 3 teeth on each lateral anterior margin. Lateral alae present, wide, on anterior half of body. Cervical papillae large towards anterior end. Oesophagus divided. Oesophago-intestinal valve projects into intestine. *Male*: Caudal papillae: 4 pairs precloacal; 2 pairs post-cloacal; group of sessile papillae near distal tip. Spicules unequal, dissimilar. Gubernaculum present. *Female*: Vulva opens in middle of body. Didelphic. Parasites in stomach of *Falco* spp.

Type species: *Paraspiralatus sakeri* Gibbons, Nicholls, Bailey and Samour, 2004

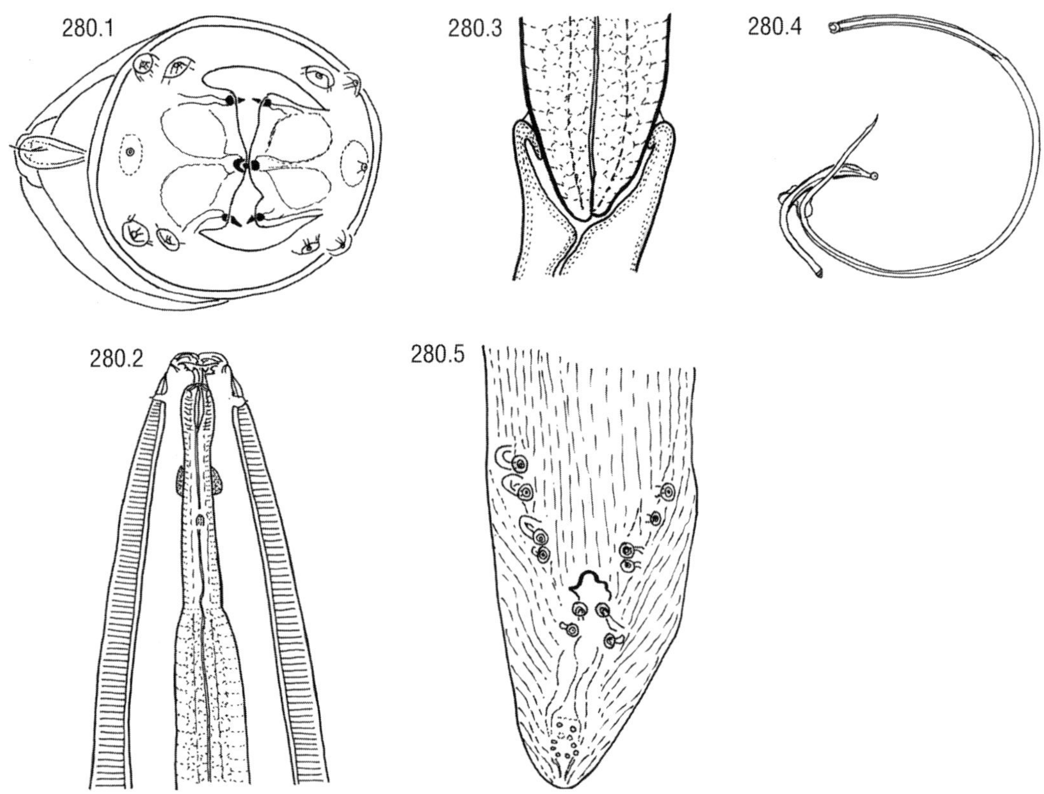

FIG. 280 *Paraspiralatus* 280.1 Female cephalic region, *en face* view; 280.2 Female anterior end, dorso-ventral view; 280.3 Oesophago-intestinal junction; 280.4 Spicules and gubernaculum; 280.5 Posterior end of male, ventral view. (Redrawn after Gibbons *et al.*, 2004.)

Ascaropsinae Alicata and McIntosh, 1933

[Chabaud, Navonc and Bain (1983) suggest similarities between *Physocephalus* of the subfamily Ascaropsinae and *Mazzia*, and propose that the subfamily Ascaropsinae should be a synonym of Spirocercinae.]

Mastigonema Dailey and Perrin, 1973 (Fig. 281)

DIAGNOSIS: Cuticle transversely striated. Mouth with 2 lips, not lobed. Cervical cuticle without inflations or alae. Buccal capsule cylindrical with transverse reticular thickenings. Oesophagus long, undivided with spirurid-type valve. *Male*: Unknown. *Female*: Body thin anteriorly, thick posteriorly. Vulva opens in middle third of body, ovaries posterior to oesophagus, prodelphic. Oviparous. Parasites of forestomach of Cetacea, *Stenella graffmani*, *S. longirostris*, in eastern tropical Pacific Ocean.

Type species: *Mastigonema stellae* Dailey and Perrin, 1973

[Listed in Spencer Jones and Gibson (1987).]

Spirobakerus Chabaud and Bain, 1981 (Fig. 282)

DIAGNOSIS: Pseudolabia present, mouth elongate, narrowest margins arranged laterally with 2 large median teeth. Buccal capsule long ornamented with wrinkles. Cervical papillae asymmetrical. Lateral alae wide and simple. *Male*: Spicules unequal. Gubernaculum small, transparent. Four pairs of precloacal papillae, single median precloacal papilla, 1 pair adcloacal papillae, 1 pair postcloacal papillae, cluster of small, sessile papillae near distal tip. *Female*: Vulva opens just anterior to middle of body. Ovejector short directed towards anterior. Eggs with 2 small rounded projections at 1 pole. Tail short, with tuft of 7–8 spines at distal tip. Parasites of Cricetidae, *Zygodontomys brevicauda*, of Colombia.

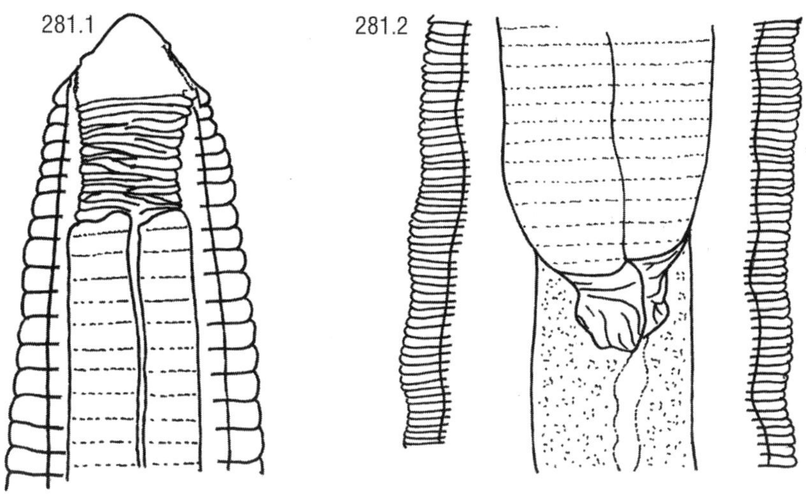

FIG. 281 *Mastigonema* 281.1 Cephalic region showing cuticular thickenings in buccal capsule; 281.2 Oesophago-intestinal valve. (Redrawn after Dailey and Perrin, 1973.)

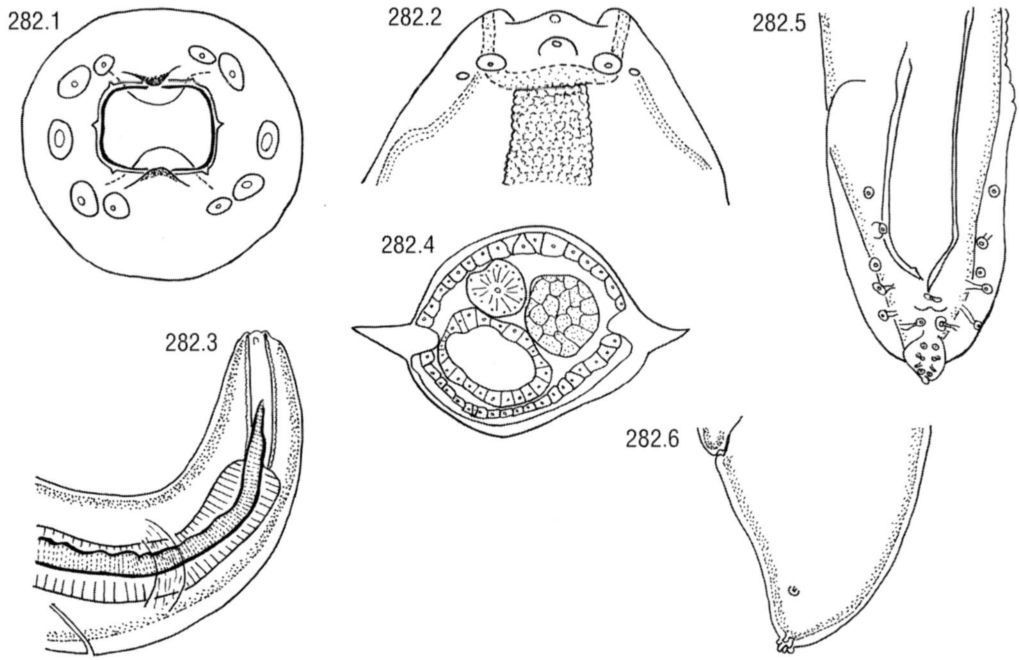

FIG. 282 *Spirobakerus* 282.1 Female cephalic region, *en face* view; 282.2 Male cephalic region, lateral view; 282.3 Anterior end of male, lateral view; 282.4 Female mid-body, transverse section; 282.5 Posterior end of male, ventral view; 282.6 Posterior end of female, lateral view. (Redrawn after Chabaud and Bain, 1981.)

Type species: *Spirobakerus weitzeli* Chabaud and Bain, 1981

[Listed in Spencer Jones and Gibson (1987).]

Spirosprattus Smales, 2004 (Fig. 283)

DIAGNOSIS: Cuticle transversely striated. Mouth oval, with reduced lips. Cephalic sensory papillae: 4 submedian papillae; 2 lateral amphids. Buccal capsule globular anteriorly with internal 'plates and struts', cylindrical in posterior half with annular thickened walls. Oesophagus with short anterior muscular region, long posterior glandular portion. *Male*: Spicules unequal, dissimilar. Caudal alae narrow. *Female*: Body widens posteriorly. Eggs with 'polar ornamentation'. Parasites in cysts in stomach wall of Australian rodents (Cape York rat, *Rattus leucopus*, Muridae).

Type species: *Spirosprattus scyphiformis* Smales, 2004

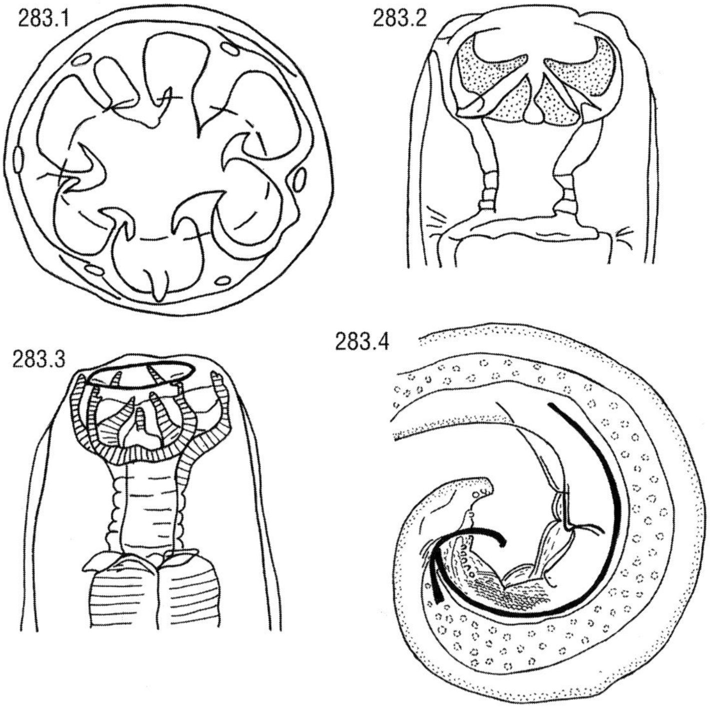

FIG. 283 *Spirosprattus* 283.1 Cephalic region, optical transverse section through the middle of buccal capsule; 283.2 Anterior end, dorso-ventral longitudinal optical section through buccal capsule; 283.3 Anterior end dorso-ventral view; 283.4 Posterior end of male, lateral view. (Redrawn after Smales, 2004.)

HABRONEMATOIDEA

Habronematidae (Chitwood and Wehr, 1932 subfamily) Ivaschkin, 1961

Habronematinae Chitwood and Wehr, 1932

Dermofilaria Rivolta, 1884

DIAGNOSIS: Parasite of equids and bovines.

Type species: *Dermofilaria irritans* Rivolta, 1884

[Yorke and Maplestone (1926) list the type species *Dermofilaria irritans* as a synonym of *Habronema megastoma* (Rudolphi, 1819) and the genus as a synonym of *Habronema* Diesing, 1861. Spencer Jones and Gibson (1987) list the genus in the superfamily Spiruroidea.]

Histiocephalinae Gendre, 1922

Sobolevicephalus Parukhin, 1964

Type species: *Sobolevicephalus chalcyonis* Parukhin, 1964

[A synonym of *Hadjelia* Seurat, 1916 in Chabaud (1975b). Parukhin (1978) considers this genus belongs in the Schistorophidae (subfamily status in the Acuarioidea in Chabaud, 1975) and that *Smetaleksenema* Schmidt and Kuntz, 1972 is a synonym. Chabaud (1975b) considers *Smetaleksenema* a synonym of *Ancyracanthopsis* Diesing, 1861, which he includes in the Schistorophinae in the Acuarioidea. Wong and Lankester (1985a,b) revise and validate the genus and include the genus in the superfamily Acuarioidea (see Acuarioidea for generic diagnosis).]

Tetrameridae Travassos, 1914

Tetramerinae (Travassos, 1914 subfamily)

Tetrameres Creplin, 1846

[ICZN validated in 1969 the generic name *Tetrameres* Creplin, 1846 using plenary powers in response to Dr Chitwood's request (Chitwood, 1967) and placed *Tropisurus* Diesing, 1835 on the list of rejected names. The generic name *Tetrameres* should continue to be used to avoid any further confusion. Case discussed in Ali (1973) and cited in Pilar (2002).]

Echinurioides Thwaite, 1926

DIAGNOSIS: Mouth with 4 lips: 2 lateral; 1 dorsal; 1 ventral. Buccal capsule longer than wide. Oesophagus divided. Four 'cordons' do not anastomose. Cuticle with 4 longitudinal rows of spines. *Male*: Caudal alae absent. Spicules unequal. *Female*: Unknown. Parasite of spurwinged goose, *Plectropterus* sp., in northern Nigeria.

Type species: *Echinurioides plectropteri* Thwaite, 1926

[Diagnosis based on Thwaite (1926). Skrjabin and Sobolev (1963) list *Echinurioides* as a synonym of *Tetrameres* Creplin, 1846. The genus was listed by Spencer Jones and Gibson (1987). Chabaud (1975b) did not include this genus as a synonym in his key.]

Acanthophorus von Linstow, 1876

[The genus *Acanthophorus* is considered in part a synonym of the genus *Tropidocerca* Diesing, 1835 and in part a synonym of *Tropisurus* Diesing, 1835 in Stiles and Hassall (1920). Yorke and Maplestone (1926) list both genera as synonyms of *Tetrameres* Creplin, 1846. Von Linstow (1876) did not designate a type species, but *A. tenuis* is the first species described after the generic diagnosis. *A. tenuis* is listed as a synonym of *Tetrameres fissispina* (Diesing, 1861), a parasite of *Anas* spp., and *Gallus* spp. in Yorke and Maplestone (1926). The genus *Acanthophorus* was not listed in Chabaud (1975b) but is included in the Spencer Jones and Gibson (1987) list of genera. The genus is an accepted synonym of *Tetrameres*.]

Gubernaculomeres Oshmarin and Parukhin, 1963 (Fig. 284)

DIAGNOSIS: *Male*: Gubernaculum present. Fairly long tube formed by the cloacal lips. Left spicule with small, disc-shaped ala at distal tip. Parasite in the proventriculus of *Astur gentilis* and *Aquila clanga*.

Type species: *Gubernaculomeres tubocloacis* (Oshmarin, 1956) Oshmarin and Parukhin, 1963

[Listed in Spencer Jones and Gibson (1987).]

Cystidicolidae (Skrjabin, 1946 subfamily)

[Moravec (2007) lists 23 genera in this family, including *Johnstonmawsonoides* Machida, 1975 and *Neoascarophis* Machida, 1976 – see Thelazioidea for descriptions of these genera, where they were originally placed by Machida (1975) and Machida (1976), respectively, and are listed in Spencer Jones and Gibson, 1987.]

Ascarophis van Beneden, 1871

DIAGNOSIS: Cuticle annulated. Mouth dorso-ventrally elongate. Pseudolabia, each with conical process on proximal margin, medially continuous with lateral walls of buccal capsule. Four submedian labia each with flap-like or bilobed sublabium on inner margin. Cephalic sensory organs: 4 single stalked or sessile submedian papillae; 2 lateral amphids. Buccal capsule long, narrow, widens at proximal end, teeth absent. Oesophagus with short muscular, longer glandular, regions. Cervical papillae anterior to nerve ring. *Male*: *Area rugosa* present. Spicules unequal, dissimilar. Cuticular bosses on ventral surface of caudal alae and tail tip present or absent. Four pairs precloacal papillae, 6 pairs postcloacal papillae, phasmids posterior to last pair of papillae. *Female*: Amphidelphic. Vulva opens from mid-body to posterior half of body. Tail with blunt or knob-like distal tip. Eggs with or

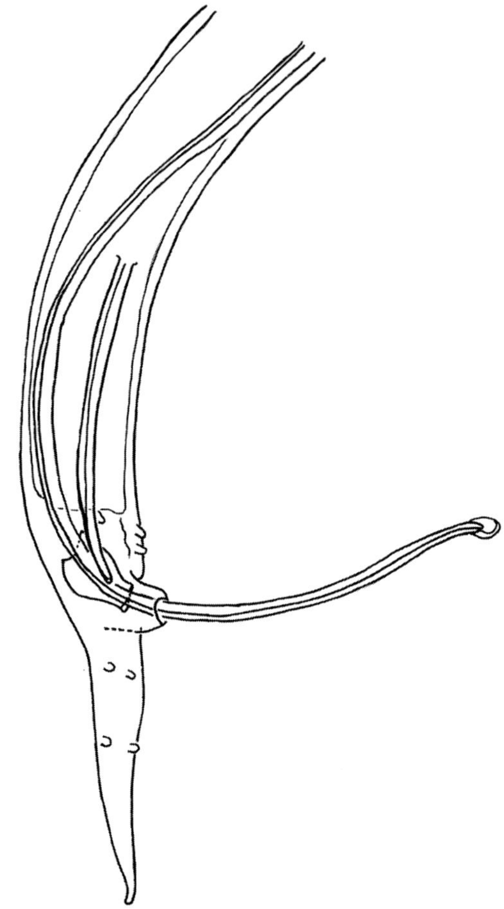

FIG. 284 *Gubernaculomeres* Posterior end of male, ventro-lateral view. (Redrawn after Oshmarin and Parukhin, 1963.)

without filaments arising from knob at one or both poles. Parasites of gastrointestinal tract of fish.

Type species: *Ascarophis morrhuae* Beneden, 1871

[Emended diagnosis based on Ko (1985, 1986), which includes a key to species. Ko placed the family and the genus in the superfamily Spiruroidea; however, Chabaud (1975b) places the family Cystidicolidae and the genus in the superfamily Habronematoidea.]

Similascarophis Muñoz, González and George-Nascimento, 2004 (Fig. 285)

DIAGNOSIS: Cuticle with well-defined transverse striations. Four submedian cephalic papillae, 2 amphids lateral to pseudolabia. Mouth dorso-ventrally elongate. Pseudolabia very reduced: submedial labia absent, sublabia present 'weakly developed'; pseudolabia with 'conical anterior protrusion', medial portion without 'projections towards central oral opening'. Oesophagus

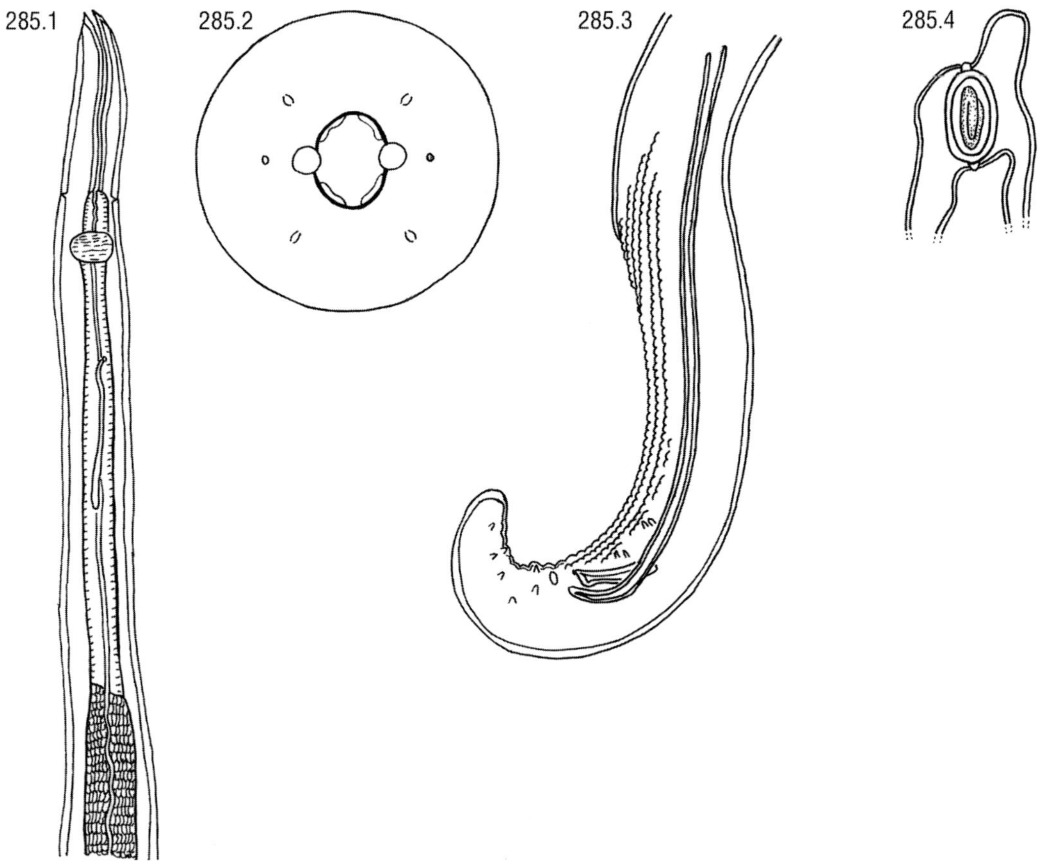

FIG. 285 *Similascarophis* 285.1 Anterior end of body; 285.2 Cephalic region, diagrammatic *en face* view; 285.3 Posterior end of male, lateral view; 285.4 Egg. (Redrawn after Muñoz *et al.*, 2004.)

divided with shorter anterior muscular region, longer posterior glandular region. *Male*: Shorter and narrower than female. Spicules unequal. Monorchic. Four pairs precloacal, 5 pairs postcloacal papillae. Caudal alae narrow. *Female*: Longer and wider than male. Didelphic. Amphidelphic. Vulva opens in posterior half of body. Embryonated eggs with filaments. Parasite of digestive tract of marine fish off the Chilean coast.

Type species: *Similascarophis maulensis* Muñoz, González and George-Nascimento, 2004

[Moravec and González-Solís (2007) consider this genus a junior synonym of *Ascarophis*. Later, Moravec and Justine (2007) suggest that this should be a subgenus of *Ascarophis* for the group of species with very reduced pseudolabia.]

Cystidicoloides Skinker, 1931

DIAGNOSIS: Body wide posteriorly, thinner towards anterior. Mouth with 2 well-developed pseudolabia. On both sides of pseudolabia cuticular cordons border anterior margin of mouth.

Four submedian cephalic papillae, 2 lateral amphids posterior to pseudolabia. Buccal capsule widens anteriorly, dorso-ventrally elongated in cross-section. *Male*: Posterior end spirally coiled. Caudal alae present. *Area rugosa* absent. Four pairs precloacal papillae, 1 single precloacal papilla, 5–6 pairs postcloacal papillae and phasmids present. Spicules unequal, dissimilar. Gubernaculum absent. *Female*: Vulva opens in mid-body or just posterior to it. Tail short. Parasites of South American freshwater fish.

Type species: *Cystidicoloides fischeri* (Travassos, Artigas and Pereira, 1928) Skinker, 1931

[Emended diagnosis based on Petter (1984).]

Sterliadochona Skrjabin, 1948

[Chabaud (1975b) considers this genus a synonym of *Cystidicoloides* Skinker, 1931. Moravec (2007) revalidates the genus. Moravec *et al.* (2008c) consider the type species *S. ssavini* a *species inquirenda* and the genus *Sterliadochona* a *genus inquirendum*.]

Salmonema Moravec, Santos, Brasil-Sato, 2008 (Fig. 286)

DIAGNOSIS: Mouth opening oval, orientated dorso-ventrally, surrounded by wide lateral pseudolabia with a median protuberance projecting from surface and 4 submedian labia. Buccal capsule elongate, narrow, expands at proximal end. *Male*: *Area rugosa* present. Caudal alae present. Four pairs precloacal papillae present, postcloacal papillae present. Spicules unequal, dissimilar. *Female*: Amphidelphic. Vulva opens in posterior part of body. Eggs without filaments. Parasites in digestive tract of freshwater fish.

Type species: *Salmonema ephemeridarum* (Linstow, 1872) Moravec, Santos, Brasil-Sato, 2008

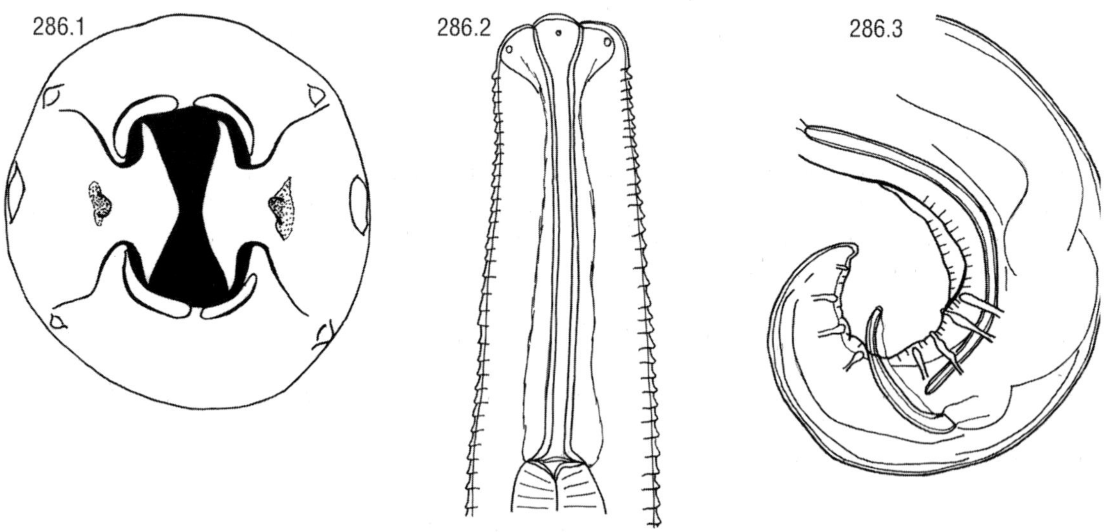

Fig. 286 *Salmonema* 286.1 Cephalic region, *en face* view (adapted from a scanning electron micrograph in Moravec *et al.*, 2008c); 286.2 Anterior end of body, lateral view; 286.3 Posterior end of male, lateral view. (286.2 and 286.3 redrawn after Moravec, 1967 from Moravec, 1994.)

Comephoronema Layman, 1933

DIAGNOSIS: Mouth with well-developed pseudolabia 'without any conical structure'. Denticles absent on proximal margin of stoma. Four submedian cephalic papillae. *Male*: *Area rugosa* present. Caudal alae present. Six to ten pedunculate precloacal papillae. Spicules unequal, dissimilar. *Female*: Vulva opens posterior to mid-body. Eggs with filaments. Parasites in alimentary canal of freshwater fish.

Type species: *Comephoronema werestschagini* Layman, 1933

[Chabaud (1975b) placed this genus as a synonym of *Cystidicola* Fischer, 1798 in his key. Moravec (1994) and Moravec and Nagasawa (1999) regard it as a valid genus. Diagnosis based on Moravec (1994). Moravec *et al.* (2007) provide a key to the valid species of the genus.]

Pseudoproleptus Khera, 1953

[The species of this genus were originally assigned to the subfamily Physalopterinae in the Physalopteroidea. Sahay and Prasad (1993) question the inclusion of this genus in the Cystidicolidae (as in the key by Chabaud, 1975b) and consider the species should be retained in the Physalopteridae until further studies are undertaken. Chabaud (1975b) indicates in his key that the genus is characterized by the presence of a collarette in the cephalic region similar to that seen in physalopterids. The genus is retained as in Chabaud (1975b).]

Pseudascarophis Ko, Margolis and Machida, 1985 (Fig. 287)

DIAGNOSIS: Cuticle striated. Mouth dorso-ventrally elongate, submedian labia absent. Pseudolabia extend into T-shaped process on median margin. Cephalic sensory organs: papillae indistinct; amphids prominent. Four submedian digitiform processes arise at proximal margin of buccal capsule, extend beyond mouth margin. Cervical papillae at level of mid-region of buccal capsule. Buccal capsule long, narrow, widens at proximal end. Oesophagus with short anterior muscular, longer posterior glandular regions. *Male*: *Area rugosa* present. Spicules simple, unequal. Caudal alae with bosses. Three pairs precloacal, 7 pairs postcloacal papillae. *Female*: Amphidelphic. Mature egg with filaments at both poles. Parasites in stomach of fish, *Kyphosus cinerascens*, off south-eastern coast of Japan.

Type species: *Pseudascarophis kyphosi* Ko, Margolis and Machida, 1985

[Listed in Spencer Jones and Gibson (1987).]

Capillospirura Skrjabin, 1924

DIAGNOSIS: Four cephalic papillae present. Mouth dorso-ventrally elongate. Pseudolabia narrow, with ridge-like apex. Submedian labia present. Mouth opening with opposing plates projecting from antero-lateral wall of buccal cavity. Sublabium in each subdorsal and subventral quadrant of mouth opening. Buccal cavity long, slender, dilated dorso-ventrally behind mouth opening. *Male*: Ten pairs of caudal papillae, 4 precloacal and 6 postcloacal. Left spicule longer than right. *Female*: Vulva opens near mid-body, Parasite of digestive tract of sturgeons.

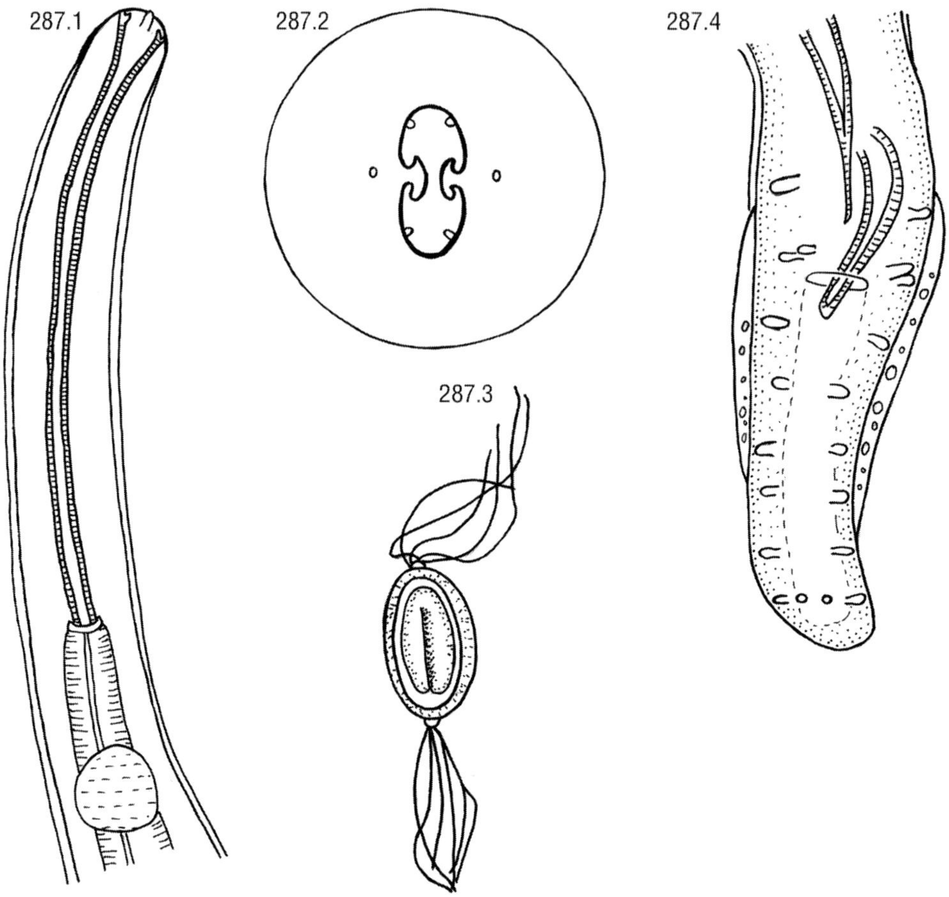

FIG. 287 *Pseudascarophis* 287.1 Anterior end of male, lateral view; 287.2 Cephalic region, diagrammatic *en face* view (Redrawn after Muñoz *et al.*, 2004); 287.3 Egg; 287.4 Posterior end of male, ventral view (287.1, 287.3 and 287.4 redrawn after Ko *et al.*, 1985.)

Type species: *Capillospirura ovotrichuria* Skrjabin, 1924

[Emended diagnosis based on Appy and Anderson (1982). Poljansky (1952) considered this genus a synonym of *Ascarophis* van Beneden, 1871, and Chabaud (1975b) also included it as a synonym of this genus. Appy and Anderson (1982) re-described the type species and validated the genus *Capillospirura*.]

Caballeronema Margolis, 1977 (Fig. 288)

DIAGNOSIS: Cuticle with transverse striations. Mouth opening: dorso-ventrally elongate; 4 submedian labia each with bilobed sublabium; 2 narrow, lateral pseudolabia present; extensions of lateral wall of buccal capsule form a prong aligned either side of pseudolabia. Cephalic sensory papillae: 4 submedian papillae; 2 lateral amphids. Buccal capsule wall thick, sigmoid, deeper than wide. Oesophagus with anterior, short muscular region and posterior glandular

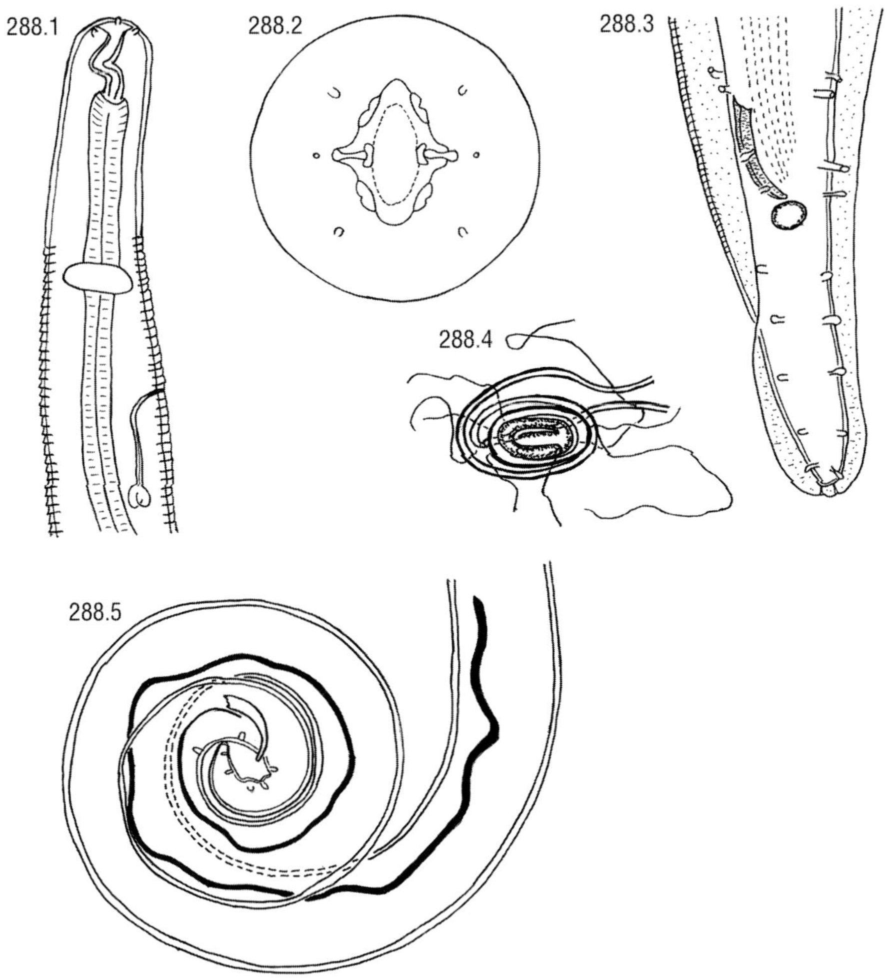

FIG. 288 *Caballeronema* 288.1 Anterior end of female, lateral view; 288.2 Cephalic region, diagrammatic *en face* view (redrawn after Muñoz *et al.*, 2004); 288.3 Posterior end of male, ventral view; 288.4 Egg with long filaments only partially shown; 288.5 Posterior end of male, lateral view. (288.1, 288.3, 288.4 and 288.5 redrawn after Margolis, 1977.)

region. *Male*: Posterior end spirally coiled, distal tip rounded. Caudal alae present. Longitudinal rows of cuticular ridges present anterior to cloaca. Caudal papillae pedunculate: 4 pairs precloacal; 6 pairs postcloacal. Spicules unequal, markedly dissimilar. Gubernaculum absent. *Female*: Vulva opens in mid-body. Amphidelphic. Eggs with polar plugs and filaments on both poles. Parasite in the alimentary canal of the marine fish, *Scorpaenichthys marmoratus*, from Pacific coast of Canada.

Type species: *Caballeronema wardlei* (Smedley, 1934) Margolis, 1977

[The author assigns this genus to the superfamily Spiruroidea and indicates that the new genus is close to *Ascarophis* van Benenden, 1871 on the basis of the cephalic region. Chabaud (1975b) places *Ascarophis* in the family Cystidicolidae in the superfamily Habronematoidea. For these reasons the genus *Caballeronema* is also placed in the same superfamily.]

Prospinitectus Petter, 1979 (Fig. 289)

DIAGNOSIS: Cuticle with spines arranged in transverse annules. Mouth with pseudolabia reduced to 2 lobes each with an external apical tooth. Buccal capsule elongated dorso-ventrally in cross-section. Pharynx absent. *Male*: Posterior end with 4 pairs precloacal and 7 pairs post-cloacal papillae. Spicules unequal. Parasite in intestine of fish, *Euthynnus affinis*, off Kuala Lumpur, Malaysia and in China Sea.

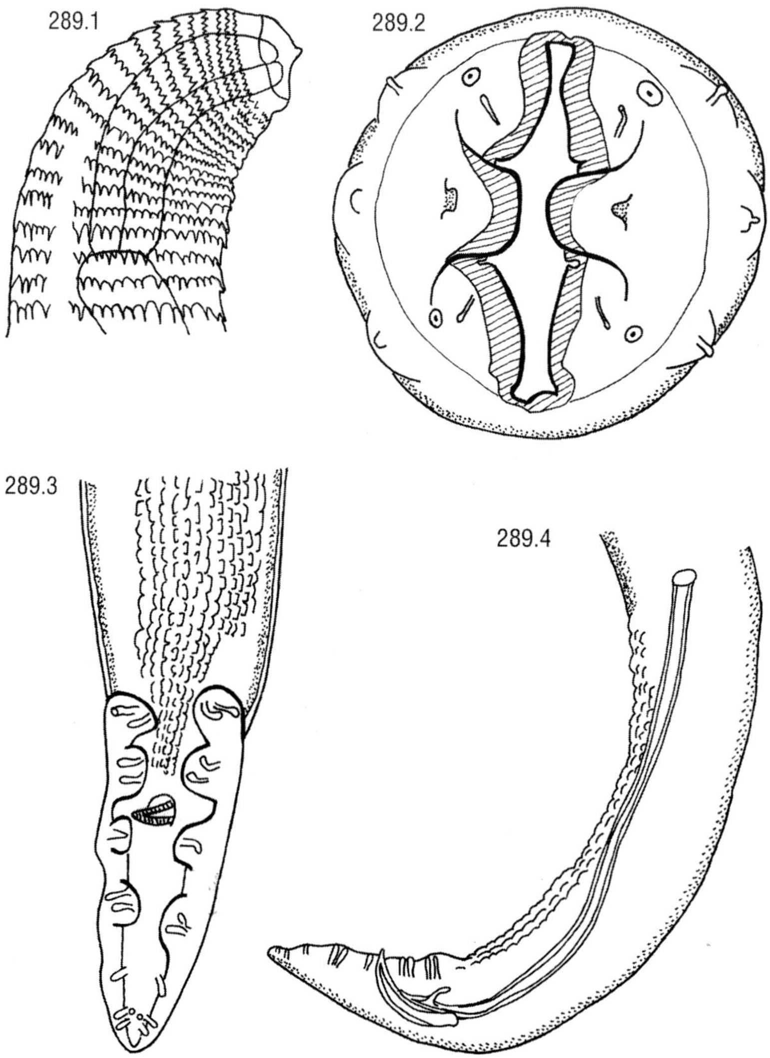

FIG. 289 *Prospinitectus* 289.1 Male anterior end, lateral view; 289.2 Male cephalic region, *en face* view; 289.3 Posterior end of male, ventral view; 289.4 Posterior end of male, lateral view. (Redrawn after Petter, 1979b.)

Type species: *Prospinitectus mollis* (Mameav, 1968) Petter, 1979

[Moravec (2007) refers to Moravec *et al.* (2004), in which this genus was not considered valid. Listed in Spencer Jones and Gibson (1987).]

Crenatobronema Solov'eva, 1987 (Fig. 290)

DIAGNOSIS: *Male*: *Area rugosa* composed of numerous ventral cuticular crests. Spicules unequal, dissimilar. *Female*: Tiny ventral 'spines' at level of junction of muscular and glandular oesophagus. 'Spines' arranged as sharp projections on distal margins of annules. Parasite in fish from Pacific Ocean.

Type species: *Crenatobronema guentheri* (Baylis, 1929) Solov'eva, 1987

[Solov'eva (1987) places the genus in the family Spinitectidae. The type species of the genus is a former member of the genus *Spinitectus*. Chabaud (1975b) places *Spinitectus* in the

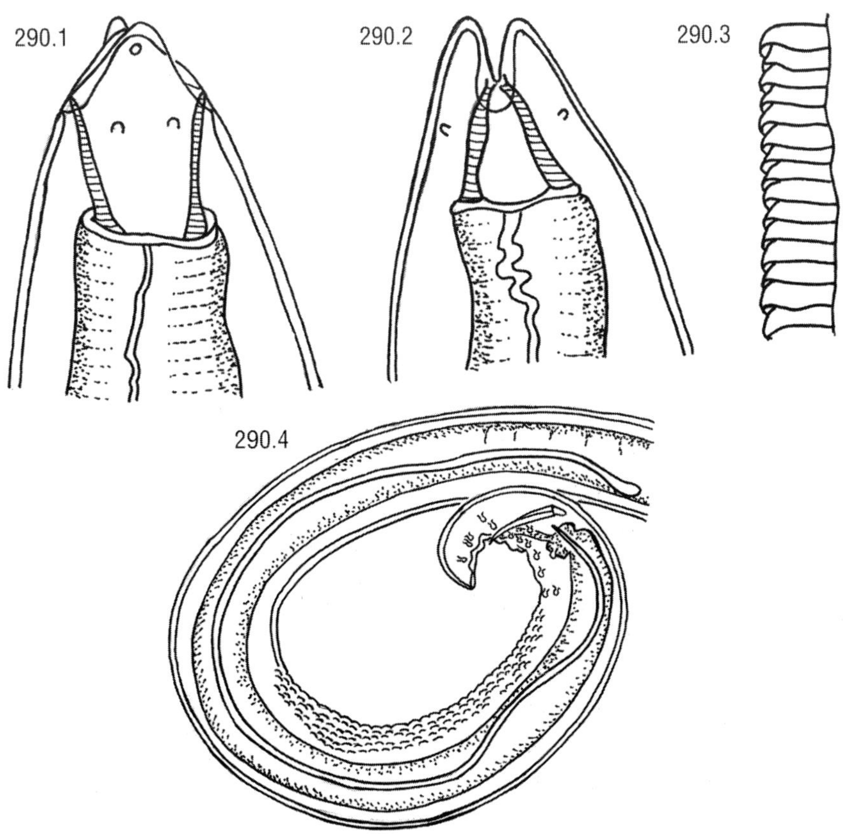

FIG. 290 *Crenatobronema* 290.1 Cephalic region, lateral view; 290.2 Cephalic region dorso-ventral view; 290.3 'Spines' on cuticular annules; 290.4 Posterior end of male, lateral view. (Redrawn after Solov'eva, 1987.)

Cystidicolidae and so *Crenatobronema* is also included in the same family. Moravec (2007) considers this genus 'inadequately known' in his review of the suborder.]

Moravecnema Justine, Cassone and Petter, 2002 (Fig. 291)

DIAGNIOSIS: Mouth opening hexagonal, dorso-ventrally elongate with rudimentary, rounded pseudolabia. Cephalic sensory organs: 4 submedian papillae; 2 lateral amphids. Buccal capsule long, narrow, expands at proximal end with 4 submedian processes near mouth opening. Oesophagus divided. Cervical papillae spinous near distal end of buccal capsule. *Male*: Caudal alae present. *Area rugosa* absent. Spicules unequal, dissimilar. Posterior end with 10 pairs papillae, 4 pairs precloacal (in 2 groups), 6 pairs postcloacal. *Female*: Amphidelphic. Vulva opens at junction of middle and posterior thirds of body. Released eggs with filaments. Parasites of deep-sea hydrothermal fish from Mid-Atlantic Ridge.

Type species: *Moravecnema segonzaci* Justine, Cassone and Petter, 2002

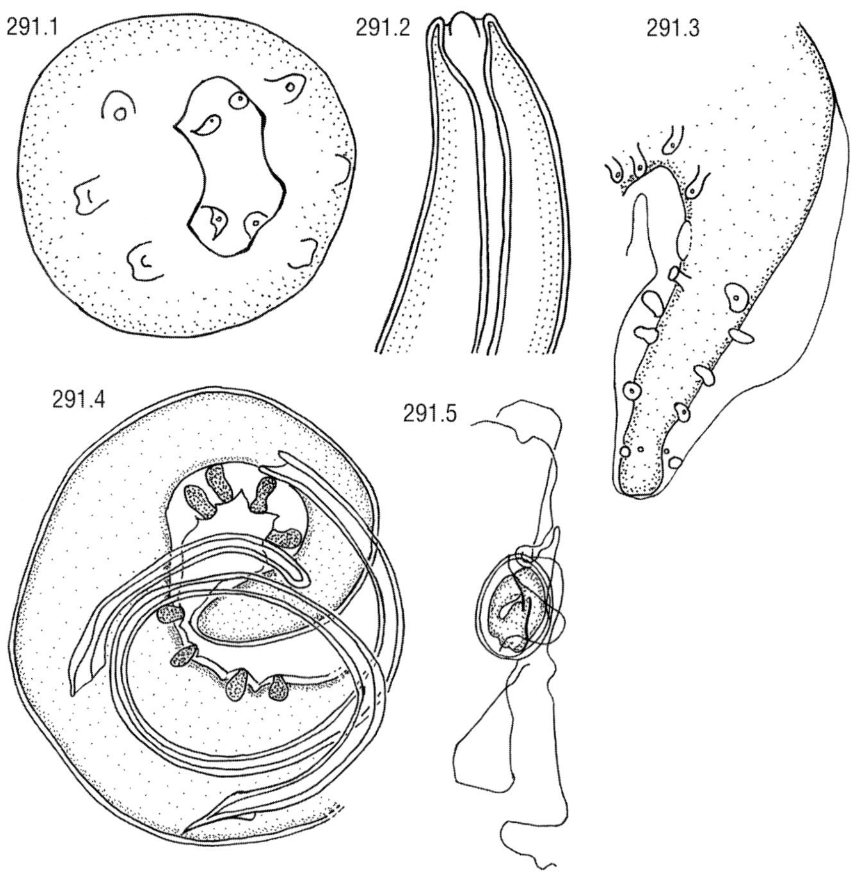

FIG. 291 *Moravecnema* 291.1 Female cephalic region, *en face* view; 291.2 Female anterior region, lateral view; 291.3 Posterior end of male, ventral view (not showing spicules); 291.4 Posterior end of male, lateral view (showing spicules); 291.5 Mature egg free from uterus. (Redrawn after Justine *et al*., 2002.)

ACUARIOIDEA

Acuariidae (Railliet, Henry and Sisoff, 1912 subfamily)

Paracuaria Rao, 1951

DIAGNOSIS: Pseudolabia well developed, each with median protruberance, 2 submedian cephalic papillae and 1 amphid near base. Two interlabia present in dorsal and ventral fields. Cervical papillae tricuspid. Lateral alae present or absent. Buccal capsule long, cylindrical, wall transversely striated. Oesophagus divided. *Male*: Caudal alae present. Caudal papillae pedunculate: 4 precloacal pairs; 5–6 postcloacal pairs. Spicules unequal. *Female*: Vulva opens posterior to mid-body. Oviparous. Amphidelphic. Parasite in submucosa of crop in seabirds or stomach of insectivorous mammals.

Type species: *Paracuaria adunca* (Creplin, 1946)

[Emended diagnosis based on Alvarez *et al.* (1994), who redefined the genus to accommodate their new species, *Paracuaria hispanica*, which has lateral alae. They considered the presence of lateral alae insufficient to erect a new genus.]

Pseudoaviculariella Gupta and Kazim, 1978 (Fig. 292)

DIAGNOSIS: Body slender, coiled. Triangular cordons restricted to cephalic region, anastomose laterally curving slightly towards anterior end. Cuticle not inflated posterior to cordons. Two small pseudolabia present. Long, cylindrical buccal capsule present. Oesophagus divided. *Male*: Tail coiled, caudal alae well developed. Caudal papillae pedunculated: 4 pairs precloacal; 4 pairs postcloacal; fifth pair postcloacal, sessile. Spicules dissimilar: right long, slender; left short, broad. Gubernaculum present. *Female*: Vulva opens near anus. Eggs embryonated. Anus terminal. Parasite of gizzard of cattle egret, *Egretta garzetta*, from Lucknow, India.

Type species: *Pseudoaviculariella srivastavai* Gupta and Kazim, 1978

[The genus was previously named in the abstracts of the 58th and 59th Indian Science Congress by Gupta and Jehan (1971, 1972), with *Pseudoaviculariella shuklai* Gupta and Jehan, 1971 as type species and *P. srivastavai* Gupta and Jehan, 1972 as type species, respectively. Under the rules of the International Commission for Zoological Nomenclature these are not valid. The genus and species were described again as a new genus and species by Gupta and Kazim (1979). Listed in Spencer Jones and Gibson (1987).]

Deliria Vicente, Pinto and Noronha, 1980 (Fig. 293)

DIAGNOSIS: Mouth with 2 conical pseudolabia. Cordons delicate, sinuous, anastomose laterally, restricted to cephalic region. Buccal capsule long, walls striated. *Male*: Striated caudal alae present. Caudal papillae pedunculated: 4 pairs precloacal; 5 pairs postcloacal. Spicules unequal, dissimilar. *Female*: Didelphic. Amphidelphic. Vulva opens in mid-body. Parasite in stomach of birds, *Pitangus sulphuratus*, in Rio de Janeiro State, Brazil.

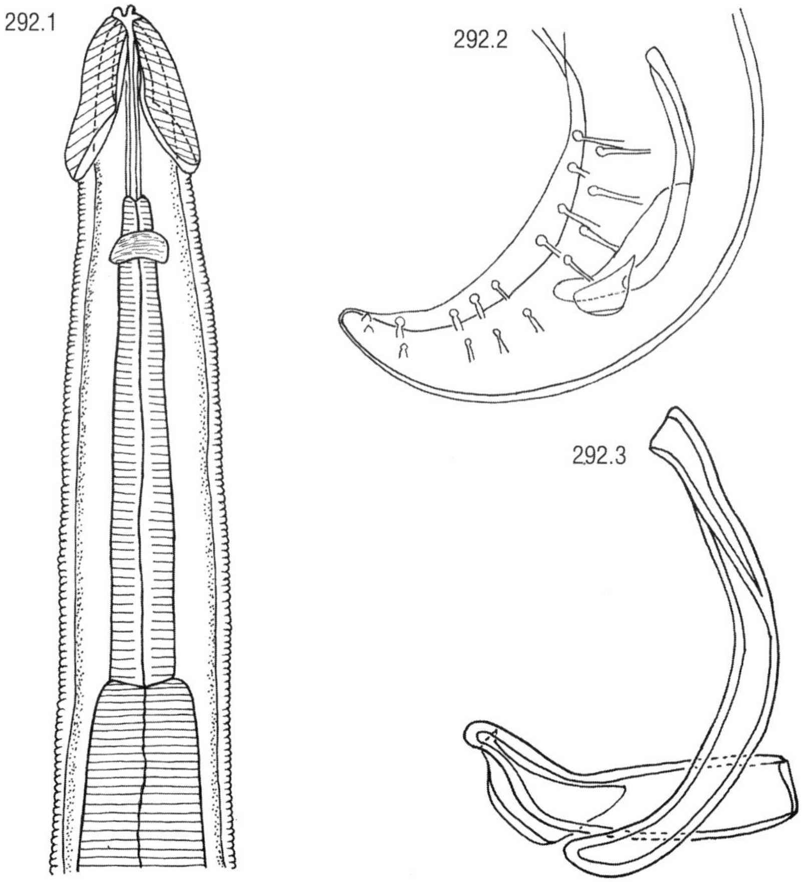

FIG. 292 *Pseudoaviculariella* 292.1 Anterior end of female, lateral view; 292.2 Posterior end of male, lateral view; 292.3 Spicules and gubernaculum. (Redrawn after Gupta and Kazim, 1978.)

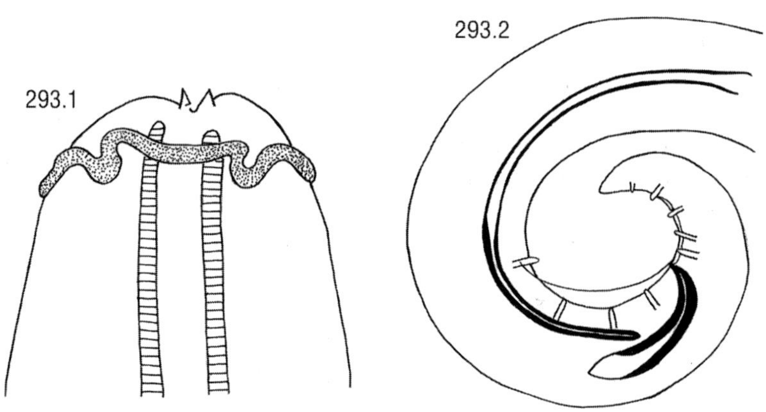

FIG. 293 *Deliria* 293.1 Cephalic region; 293.2 Posterior end of male, lateral view. (Redrawn after Vicente *et al.*, 1980.)

Type species: *Deliria gomesae* Vincente, Pinto and Noronha, 1980

[Listed in Spencer Jones and Gibson (1987).]

Acuariinae Railliet, Henry and Sisoff, 1912

Syncuaria Gilbert, 1927

DIAGNOSIS: Mouth opening elongate, orientated dorso-ventrally. Buccal capsule cylindrical, walls usually transversely striated. Pseudolabia each with median projection continuous with buccal capsule wall. Cephalic sensory organs: 1 pair of submedian cephalic papillae and 1 amphid at base of each pseudolabium. Cordons extend posteriorly from dorsal and ventral sides of mouth, anastomose laterally. Cervical papillae just posterior to cordons. Lateral alae present or absent. *Male*: Spicules unequal. Caudal alae present. Caudal papillae pedunculate: 4 pairs precloacal; 5–6 pairs postcloacal. *Female*: Monodelphic. Vulva opens near anus, cuticular inflation present. Parasites in gizzard of birds (grebe, storks, cormorants).

Type species: *Syncuaria ciconiae* Gilbert, 1927

[Emended diagnosis based on Wong, Anderson and Bartlett (1986). The authors include *Skrjabinocara* Kurashvili, 1940, *Chordocephalus* Alegret, 1941 and *Decorataria* Sobolev, 1949 as synonyms. Chabaud (1975b) maintained *Chordocephalus* as a valid genus, with *Skrjabinocara* as a synonym and *Decorataria* as a synonym of *Synacuaria*.]

Cordonema Schmidt and Kuntz, 1972

DIAGNOSIS: Cuticle at proximal end inflated. Buccal capsule with walls distinctly transversely striated. Pseudolabia each with pair of submedian cephalic papillae and amphid. Two rows of body spines present on lateral sides. Cervical papillae small. Cordons triangular, not sinuous anteriorly, anastomose in lateral field. Parasites of birds.

Type species: *Cordonema venusta* Schmidt and Kuntz, 1972

[Wong and Anderson (1983) revised the genera *Cordonema* Schmidt and Kuntz, 1972 and *Skrjabinoclava* Sobolev, 1943. Emended diagnosis based on Wong and Anderson (1983). They concluded that *Cordonema* was a valid genus. Chabaud (1975b) considered *Cordonema* a synonym of *Skrjabinoclava*.]

Chandleronema Little and Ali, 1980 (Fig. 294)

DIAGNOSIS: Cephalic cordons short, recurrent, anastomosing. Cervical papillae tricuspid, posterior to nerve ring. Lateral alae double, arising immediately posterior to cervical papillae. Buccal capsule cylindrical, long. Parasite in the stomach of raccoons, *Procyon lotor*, and muskrats, USA.

Type species: *Chandleronema longigutturata* (Chandler, 1942) Little and Ali, 1980

[Listed in Spencer Jones and Gibson (1987).]

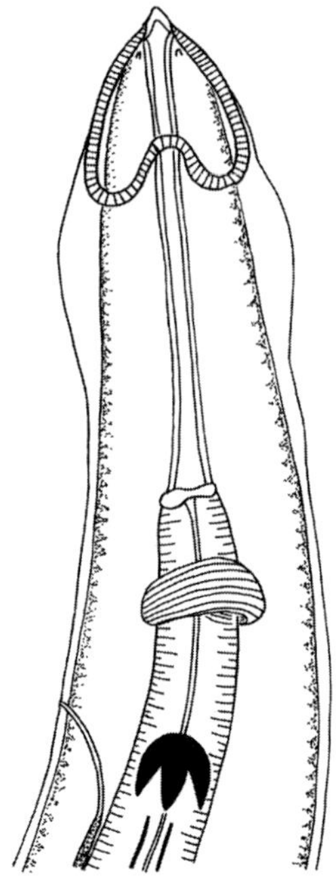

FIG. 294 *Chandleronema* Anterior end of male, lateral view. (Redrawn after Little and Ali, 1980.)

Xenocordon Mawson, 1982 (Fig. 295)

DIAGNOSIS: Cordons straight. Scales over inner section of each cordon. Cervical papillae posterior to nerve ring. *Male*: Caudal alae present. Caudal papillae: 4 pairs precloacal; 5 pairs postcloacal. Spicules unequal, dissimilar. *Female*: Vulva opens near junction of third and fourth quarters of body. Tail tapers to blunt point. Parasite in gizzard of Australian birds, *Phylidonyris novaehollandiae* and *Gymnorhina tibicen*.

Type species: *Xenocordon patonae* Mawson, 1982

[Listed in Spencer Jones and Gibson (1987).]

Willmottia Mawson, 1982 (Fig. 296)

DIAGNOSIS: Cordons symmetrical, recurrent but not anastomosing. Two rows of sublateral hooks on each side of body, extend from cervical papillae to tail. *Male*: Unknown. *Female*: Vulva opens near junction of middle and posterior thirds of body. Tail tapers to rounded point. Parasite of birds, *Malurus cyaneus*, Tasmania.

Type species: *Willmottia australis* Mawson, 1982

[Listed in Spencer Jones and Gibson (1987).]

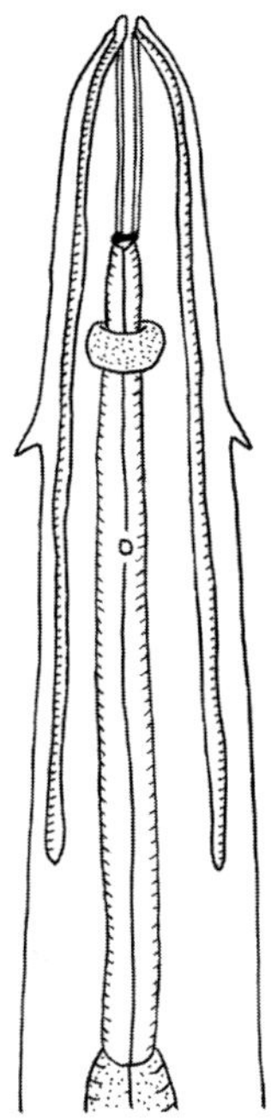

FIG. 295 *Xenocordon* Anterior end of male, ventral view. (Redrawn after Mawson, 1982.)

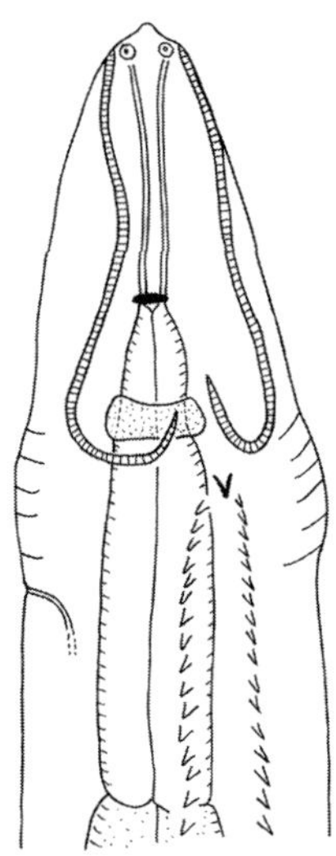

FIG. 296 *Willmottia* Anterior end of female, lateral view. (Redrawn after Mawson, 1982.)

Voguracuaria Wong and Anderson, 1993 (Fig. 297)

DIAGNOSIS: Cephalic vesicle restricted to antero-dorsal region. Buccal capsule 'laterally compressed' at anterior margin, walls transversely striated. Pseudolabia with median conical processes continuous with walls of buccal capsule. Each pseudolabia with pair of submedian papillae and amphid at base. Cordons extend posteriorly from dorsal and ventral sides of

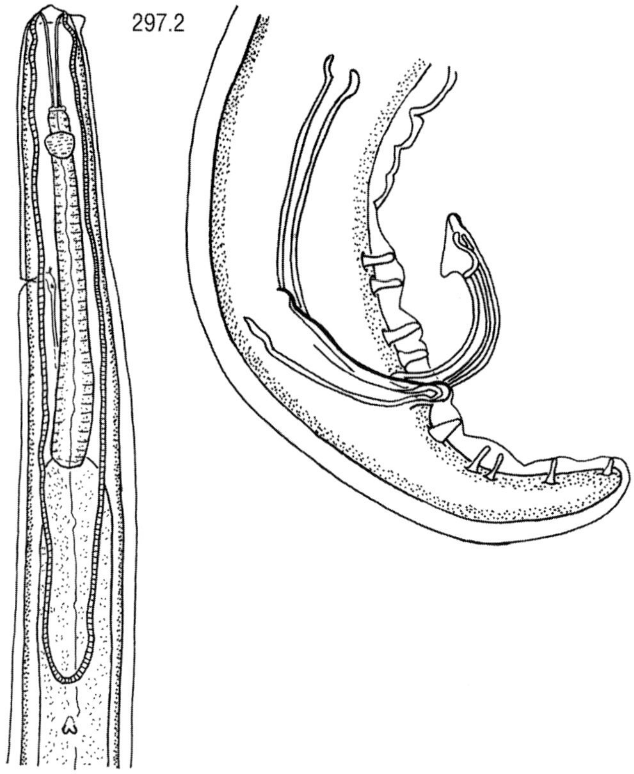

FIG. 297 *Voguracuaria* 297.1 Anterior end of female showing inflated cuticle on dorsal side of cephalic region and cordons, lateral view; 297.2 Posterior end of male, lateral view. (Redrawn after Wong and Anderson, 1993.)

mouth, anastomose laterally. Cervical papillae just posterior to cordons, mainly bifid. Lateral alae absent. Oesophagus divided. *Male*: Spicules unequal. Caudal alae narrow. Caudal papillae: 4 pairs precloacal, elongate; 5 pairs postcloacal, pedunculate; 1 sessile pair near tail tip. Parasite in oesophagus of whimbrel, *Numenius phaeopus phaeopus*, in Vogur, Iceland.

Type species: *Voguracuaria lankesteri* Wong and Anderson, 1993

Seuratiinae Chitwood and Wehr, 1932

Molinacuaria Wong and Lankester, 1985 (Fig. 298)

DIAGNOSIS: Pseudolabia with median tips continuous with buccal capsule walls. Each pseudolabia with pair of submedian papillae and amphid at base. Sublabia absent. Two pairs of grooves extend from dorso-ventral surfaces laterally, ending at distal margin of pseudolabia. Ptilina with divided margins, form 'four shields' posterior to groove. Cervical papillae pointed just anterior to nerve ring. Buccal capsule cylindrical, narrow, longer than wide, walls with transverse striations. Oesophagus

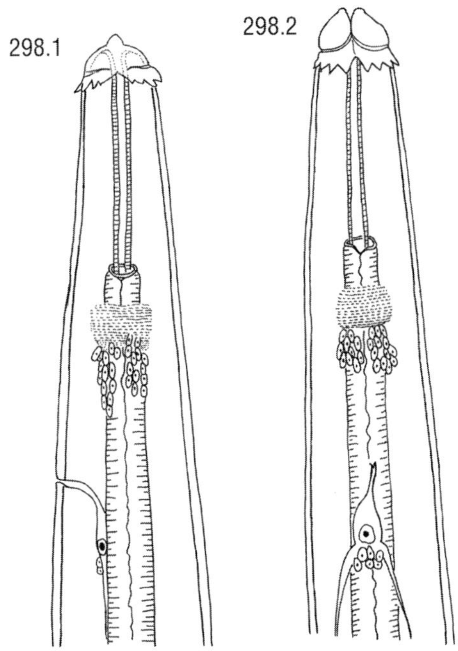

FIG. 298 *Molinacuaria* 298.1 Anterior end of female, lateral view; 298.2 Anterior end of female, ventral view. (Redrawn after Wong and Lankester, 1985.)

divided. Parasite under gizzard lining of birds, *Dendragapus obscurus fuliginosus*, *Gallinula chloropus indica*, *Alcippe brunnea brunnea* in Vancouver Island (Canada), China and Taiwan, respectively.

Type species: *Molinacuaria bendelli* (Adams and Gibson, 1969) Wong and Lankester, 1985

[Listed in Spencer Jones and Gibson (1987).]

Antechiniella Quentin and Beveridge, 1986 (Fig. 299)

DIAGNOSIS: Cordons: salient; transversely striated with median furrow; form triangles on lateral surfaces; curve to lateral line where slightly recurrent, anastomosing. Cuticular collarette and spines absent. Parasite in Australian marsupials.

Type species: *Antechiniella suffodiax* (Beveridge and Barker, 1975) Quentin and Beveridge, 1986

[Listed in Spencer Jones and Gibson (1987).]

Tikusnema Hasegawa, Shiraishi and Rochman, 1992 (Fig. 300)

DIAGNOSIS: Cephalic region inflated followed by distinct constriction. Mouth laterally compressed, 2 pseudolabia triangular in lateral view. Each pseudolabium with apical process, right larger than left; 2 submedian cephalic papillae; amphid. Ptilina posterior

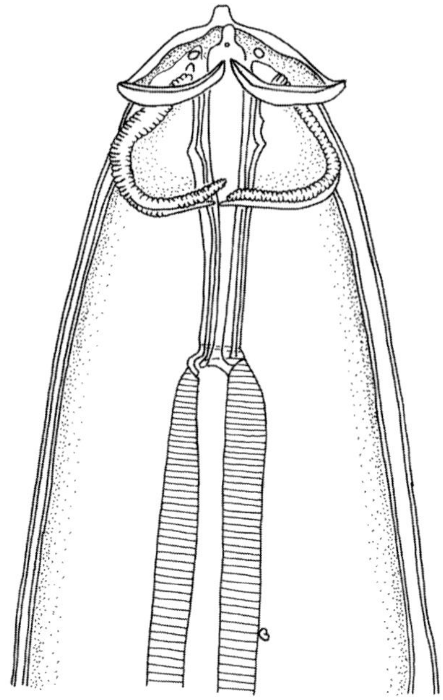

FIG. 299 *Antechiniella* Anterior end of fourth-stage larva, lateral view. (Redrawn after Quentin and Beveridge, 1986.)

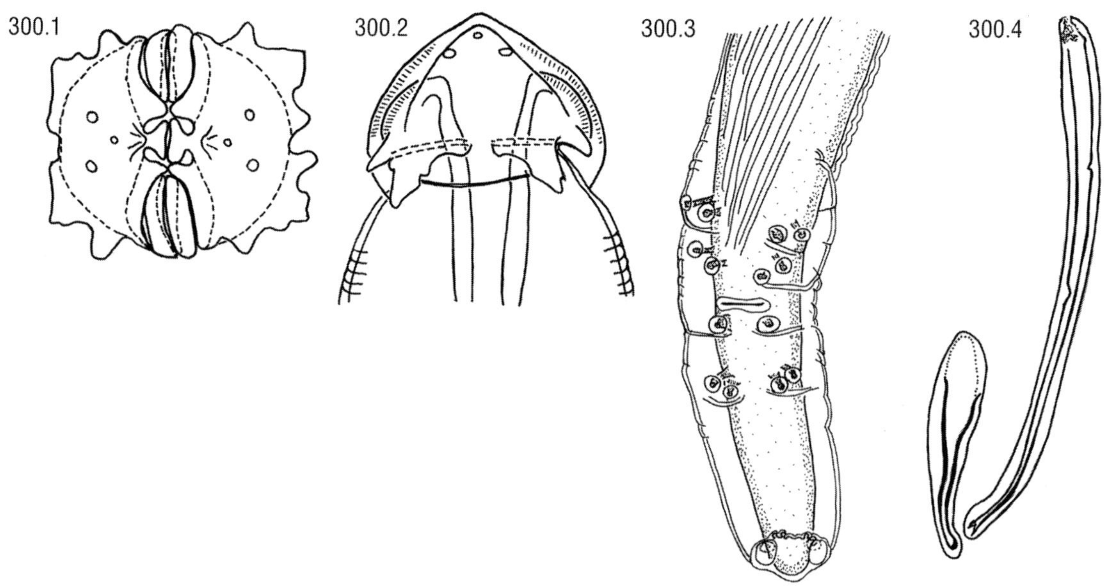

FIG. 300 *Tikusnema* 300.1 Cephalic region of female, *en face* view; 300.2 Cephalic region of female, left lateral view; 300.3 Posterior end of male, ventral view; 300.4 Spicules, ventral view. (Redrawn after Hasegawa *et al.*, 1992.)

to pseudolabia form 4 tripartite shields. Cordons ('two dorsal and two ventral labia' in Gibbons *et al.*, 1992) separate pseudolabia, do not extend laterally or posteriorly. Buccal capsule long, thick-walled. Oesophagus divided. Cervical papillae or deirids bifid, in region of buccal capsule. Bicuspid post-deirids in region of glandular oesophagus. *Male*: Caudal alae, 4 pairs precloacal, 6 pairs postcloacal papillae present. *Area rugosa* with longitudinal cuticular ridges. Spicules dissimilar, right short, left long and bifid at distal tip. *Female*: Vulva opens in middle third of body. Didelphic. Amphidelphic. Tail conical with terminal knob. Parasite in stomach and small intestine of the ricefield rat, *Rattus argentiventer*, in Indonesia.

Type species: *Tikusnema javaense* Hasegawa, Shiraishi and Rochman, 1992

[Gibbons, Crawshaw and Rumpus (1992) described a new species, *Molinacuaria indonesiensis*, from the same host and locality. Smales (1995b) revised the genus *Tikusnema* and regarded this species as a synonym of *T. javaense*. The generic diagnosis is based on Gibbons *et al.* (1992), Hasegawa *et al.* (1992) and Smales (1995b).]

Schistorophinae Travassos, 1918

Schistogendra Chabaud and Rousselot, 1956

[Wong and Lankester (1984) consider *Schistogendra* a synonym of *Quasithelazia* Maplestone, 1932. For a full discussion of this synonymy, Wong and Lankester (1984) should be consulted. Chabaud (1975b) considers *Schistogendra* a valid genus in his key, with *Quasithelazia* a synonym of *Schistorophus* Railliet, 1916.]

Quasithelazia Maplestone, 1932

Type species: *Quasithelazia tenuis* Maplestone, 1932

[Wong and Lankester (1984) consider *Quasithelazia* valid. Chabaud (1975b) considered this genus a synonym of *Schistorophus*. See comments on *Schistogendra* above.]

Sobolevicephalus Parukhin, 1964 (Fig. 301)

DIAGNOSIS: Cuticle transversely striated. Mouth opening with 2 pairs of teeth. Pseudolabia with median conical projections 'continuous with antero-lateral walls of buccal capsule'. Cephalic sensory organs arranged with 1 pair of submedian papillae and 1 amphid associated with each pseudolabium. Sublabia present, 'extend posteriorly beyond pseudolabia'. Ptilina consist of 4 'leaves' irregularly divided. Buccal capsule with transversely striated walls. Oesophagus divided. Parasites under gizzard of birds.

Type species: *Sobolevicephalus chalycyonis* Parukhin, 1964

[Emended diagnosis based on Wong and Lankester (1985b). Chabaud (1975b) listed this genus as a synonym of *Hadjelia* Seurat, 1916 in the superfamily Habronematoidea.]

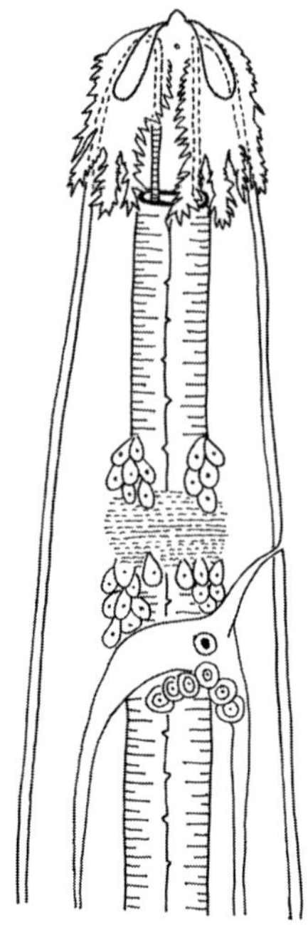

FIG. 301 *Sobolevicephalus* Female anterior end, lateral view. (Redrawn after Wong and Lankester, 1985b.)

FILARIOIDEA

Genus not assigned to a family

Avifilaris Saunders, 1955

DIAGNOSIS: *Male*: Unknown. *Female*: Unknown. Microfilaria: Colourless, transparent. Sheath may or may not be present. Nuclei arranged in 5 groups by nerve ring; excretory pore; genital cells; anus. Nuclei may or may not extend to anterior and posterior extremities. Extremities range from round to tapered. Parasite in blood of *Rhodothraupis*, *Passerina*, *Pitangus* and *Empidonax*.

Type species: *Avifilaris fringillidarum* Saunders, 1955

[Saunders (1955) proposes this name for microfilaria recovered from birds and it is here regarded as a *nomen dubium*. Spencer Jones and Gibson (1987) listed the genus in the superfamily Filarioidea.]

Filariidae (Weinland, 1858) Cobbold, 1879

'*Agamofilaria*' Stiles, 1907

DIAGNOSIS: No description. This represents a collective group name for agamic forms, i.e. forms reproducing asexually or parthenogenetically, e.g. parasite in superficial skin sores of human in Georgia, USA.

Type species: Not designated.

[*Agamofilaria georgiana* Stiles, 1907 is the only species fully described by Stiles (1907). Stiles and Hassall (1920) also list *A. labialis* (Pace, 1864), *A. oculi* (Sieb., 1839), *A. oculi humani* (Nordm., 1832) and *A. palbebralis* (Pace, 1866). For descriptions of these the original references should be consulted.

This is not a valid genus but a grouping of organisms, described by Stiles (1907), 'which have not yet reached a stage in development' that enables their systematic position to be assessed. The 'genus' is listed in the Filarioidea by Spencer Jones and Gibson (1987).]

Filariinac Wcinland, 1858

Cystofilaria Skrjabin and Schikhobalova, 1948

DIAGNOSIS: Mouth opening trapezoid. Well-defined vestibule between mouth opening and oesophagus. Oesophagus undivided. *Male*: Caudal papillae large. Caudal alae small. Spicules dissimilar: 1 long, slender, even width; 1 short, wider at proximal end, tapering at distal end. Tail short. *Female*: Vulva opens in anterior half of oesophageal region. Ovaries in posterior

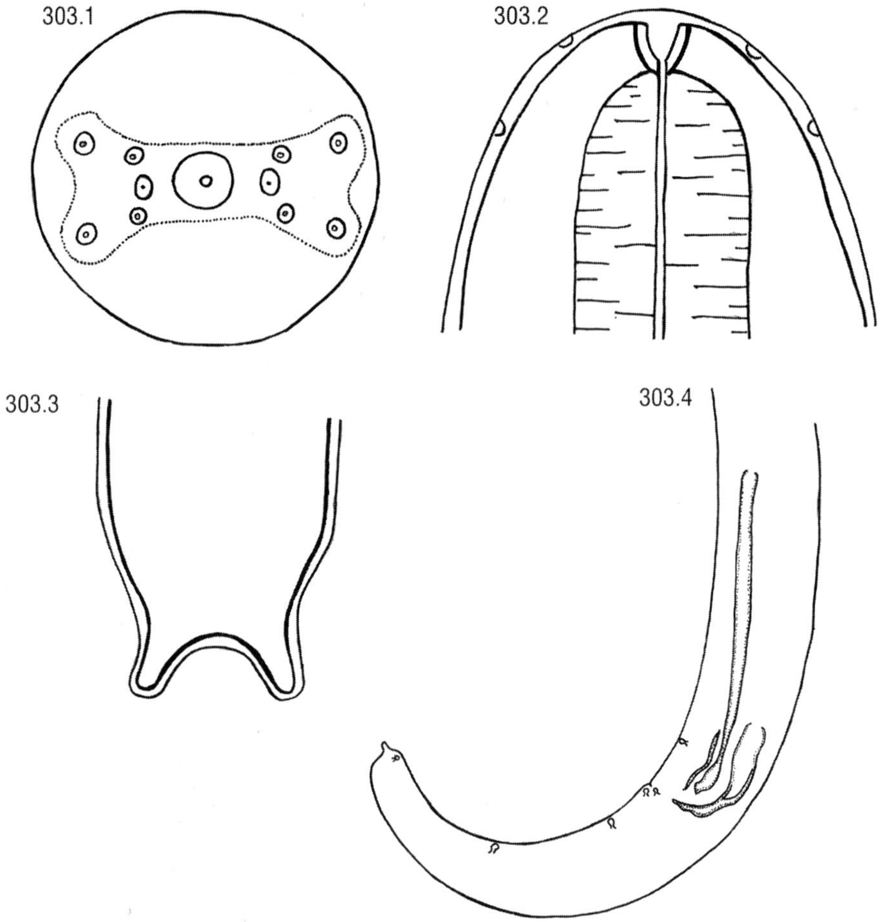

FIG. 303 *Dasypafilaria* 303.1 Female cephalic region, *en face* view; 303.2 Female cephalic region, dorso-ventral view; 303.3 Distal end of female tail, ventral view; 303.4 Posterior end of male, lateral view. (Redrawn after Eberhard, 1982.)

equal in size, few in number. Gubernaculum absent. Distal extremity of tail with 2 large lappets. *Female*: Distal end of tail with 2 large lappets. Microfilaria sheathed. Parasites in the omentum of Dasypodidae (nine-banded armadillo, *Dasypus novemcinctus*), in southern Louisiana, USA.

Type species: *Dasypafilaria averyi* Eberhard, 1982

[Bain *et al.* (1982, publ. 1983) elevated the subgenus of Eberhard (1982) to generic rank. Listed as subgenus in Spencer Jones and Gibson (1987).]

Yatesia Bain, Baker and Chabaud, 1982 publ. 1983 (= *Alafilaria* Yates and Jorgenson, 1983 subgenus) (Figs 304.1, 304.2, 304.3 and 304.4)

DIAGNOSIS: Cuticle thick, prominent transverse striations. Narrow, bluntly rounded lateral alae, extend from level of nerve ring to distal end in adults and microfilariae. Cephalic shield rectangular, 'laterally elongated', with 4 pairs submedian cephalic papillae, 1 pair lateral amphids. Buccal capsule present, well cuticularized. Oesophagus very long, distinctly divided. *Male*: Lateral alae stop after cloaca, absent on tail. Precloacal papillae numerous. Gubernaculum absent. Small caudal projections present. Lateral alae stop after cloaca, absent on tail. *Female*: Approximately twice length of male. Pair of subterminal protruberances present. Small terminal caudal papilla. Microfilaria in skin. Parasite in skeletal muscle fascia of capybara, *Hydrochoerus hydrochaeris*, in Colombia, South America.

Type species: *Yatesia hydrochoerus* (Yates, 1980) Bain, Baker and Chabaud, 1982

[Diagnosis based on Bain, Baker and Chabaud (1982, publ. 1983) and Yates and Jorgenson (1983). Bain, Baker and Chabaud (1982, publ. 1983) created the genus *Yatesia* for the type species originally described by Yates (1980) and therefore the subgenus *Alafilaria* is a synonym of *Yatesia*, which has seniority. Genus *Yatesia* and subgenus *Alafilaria* listed in Spencer Jones and Gibson (1987).]

Loxodontofilaria Berghe and Gillain, 1939 (Figs 304.5 and 304.6)

DIAGNOSIS: Body large. Oesophagus well developed. Buccal cavity shallow. *Male*: Spicules dissimilar. Posterior end with 10 pairs of pre- and postcloacal papillae. Tail with 3 terminal

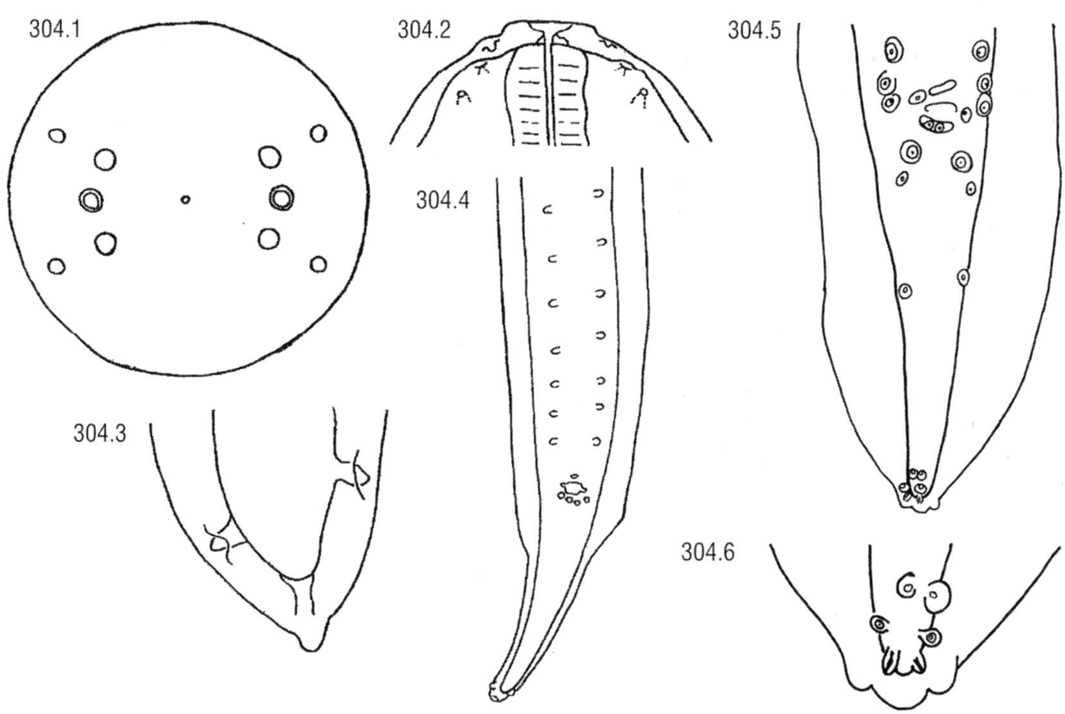

FIGS 304.1, 304.2, 304.3 and 304.4 *Yatesia* (= *Alafilaria*) 304.1 Female cephalic region, *en face* view; 304.2 Female cephalic region, ventral view; 304.3 Distal end of female tail, ventral view; 304.4 Posterior end of male, ventral view. (Redrawn after Yates and Jorgenson, 1983.)
FIGS 304.5 and 304.6 *Loxodontofilaria* 304.5 Posterior end of male, ventral view; 304.6 Distal tip of male, ventral view. (Redrawn after Uni *et al.*, 2006.)

lappets. *Female*: Vagina complex. Tail with 3 terminal lappets. Parasites of elephants in Africa and Burma, serows (Caprinae, Bovidae) in Japan and hippopotamus in Africa.

Type species: *Loxodontofilaria loxodontis* Berghe and Gillain, 1939

[Emended diagnosis based on Uni *et al.* (2006), who described the first male assigned to this genus. *Loxodontofilaria* is a subgenus in Chabaud and Bain (1976). Bain *et al.* (1982, publ. 1983) restored the subgenus to generic rank.]

Mansonella Faust, 1929

DIAGNOSIS: Small, slender worms. Body uniform width for most of length. Annular swellings present on anterior end. Pre-oesophageal ring absent. Oesophagus poorly developed, narrow, not divided. *Male*: Precloacal papillae absent or mixed with papillae surrounding cloaca. Gubernaculum absent. Spicules unequal, dissimilar, long and slender. *Female*: Four terminal papillae on tail. Microfilaria without sheath, in peripheral blood. Parasites in subcutaneous tissues of hosts occurring in Caribbean, Central and South America and Africa.

Type species: *Mansonella ozzardi* (Manson, 1897) Faust, 1929

[Emended diagnosis based on Orihel and Eberhard (1982), who provided the first description of complete adults of the species, and *Tetrapetalonema* is considered a synonym of *Mansonella*. Anderson and Bain (1976) did not consider the genus *Mansonella*, regarded *Tetrapetalonema* as valid and divided it into three subgenera (see below). Eberhard and Orihel (1984b) divided the genus *Mansonella* into five subgenera, including *Tetrapetalonema*. Bain (2002) includes these subgenera, and a sixth (see below) is added by Uni *et al.* (2004). Listed in Spencer Jones and Gibson (1987).]

Subgenus *Mansonella* (Faust, 1929 genus) Eberhard and Orihel, 1984 (Fig. 305.1)

DIAGNOSIS: Cephalic region with axis dorso-ventral, not compressed or narrower than body. *Male*: Tail flattened dorso-ventrally near distal tip. Spicules medium length, ratio 2.6:3.0. *Female*: Vulva opens at level of middle of oesophagus. Microfilaria with nuclei not extending to distal tip of tail. Parasites of man, rodents and carnivores.

Type species: *Mansonella* (*Mansonella*) *ozzardi* (Manson, 1897) Faust, 1929

Subgenus *Tupainema* Eberhard and Orihel, 1984 (Fig. 305.2)

DIAGNOSIS: Cephalic region with major axis lateral, compressed, narrower than body. *Male*: Distal tip of tail not dorso-ventrally flattened. Spicules medium length. *Female*: Vulva opens at or posterior to distal end of oesophagus. Microfilaria with nuclei not extending to distal tip of tail. Parasites of tree shrews in South-east Asia.

Type species: *Mansonella* (*Tupainema*) *dunni* (Mullin and Orihel, 1972) Eberhard and Orihel, 1984

Subgenus *Esslingeria* (Chabaud and Bain, 1976) Eberhard and Orihel, 1984 (Fig. 305.3)

DIAGNOSIS: Cephalic region with major axis lateral, not compressed or narrower than body although constricted, neck-like region may be present in dorso-ventral view. *Male*: Tail flattened dorso-ventrally or not. Spicules long, ratio 2.3:4.7. *Female*: Vulva opens at level of middle of oesophagus. Microfilaria with nuclei extending to distal tip of tail. Parasites of man, African anthropoid apes and South American rodents.

Type species: *Mansonella* (*Esslingeria*) *perstans* (Manson, 1891) Eberhard and Orihel, 1984

[Anderson and Bain (1976) and Chabaud and Bain (1976) include this subgenus in the genus *Tetrapetalonema*.]

Subgenus *Tetrapetalonema* (Faust, 1935 genus) Eberhard and Orihel, 1984 (Fig. 305.4)

DIAGNOSIS: Cephalic region with major axis dorso-ventral or compressed, narrower than body. *Male*: Tail not flattened dorso-ventrally. Spicules short to long, ratio 1.6:4.8. *Female*: Vulva generally opens in mid-oesophageal region. Microfilaria with nuclei extending to distal tip of tail. Parasites of platyrrhine (neotropical) primates.

Type species: *Mansonella* (*Tetrapetalonema*) *marmosetae* (Faust, 1935) Eberhard and Orihel, 1984

[Anderson and Bain (1976) and Chabaud and Bain (1976) include this subgenus in the genus *Tetrapetalonema*.]

Subgenus *Sandnema* (Chabaud and Bain, 1976) Eberhard and Orihel, 1984 (Fig. 305.5)

DIAGNOSIS: Cephalic region rounded, papillae arranged in a square. *Male*: Tail not dorso-ventrally flattened. Caudal papillae distributed along tail. Spicules moderate length, ratio 2.3:2.6. *Female*: Vulva opens at level of posterior oesophagus. Microfilariae with nuclei extending to distal tip of tail. Parasites of Asian primates and insectivores.

Type species: *Mansonella* (*Sandnema*) *digitata* (Chandler, 1929) Eberhard and Orihel, 1984

[Anderson and Bain (1976) and Chabaud and Bain (1976) include this subgenus in *Tetrapetalonema*.]

Subgenus *Cutifilaria* (Bain and Schulz-Key, 1974 genus) Uni, Bain and Takaoka, 2004 (Figs 305.6 and 305.7)

DIAGNOSIS: Cephalic region rounded, papillae arranged in rectangle. Buccal capsule absent. Oesophagus reduced to thin, fibrous tube, anteriorly fused to body muscles. *Male*: Caudal extremity with 4 lappets, almost completely fused to form cone. *Area rugosa* formed of transverse bands of small, pointed projections. Right spicule with stout, spoon-like distal part. Papillae on posterior end not greatly reduced, grouped closely around cloaca, precloacal

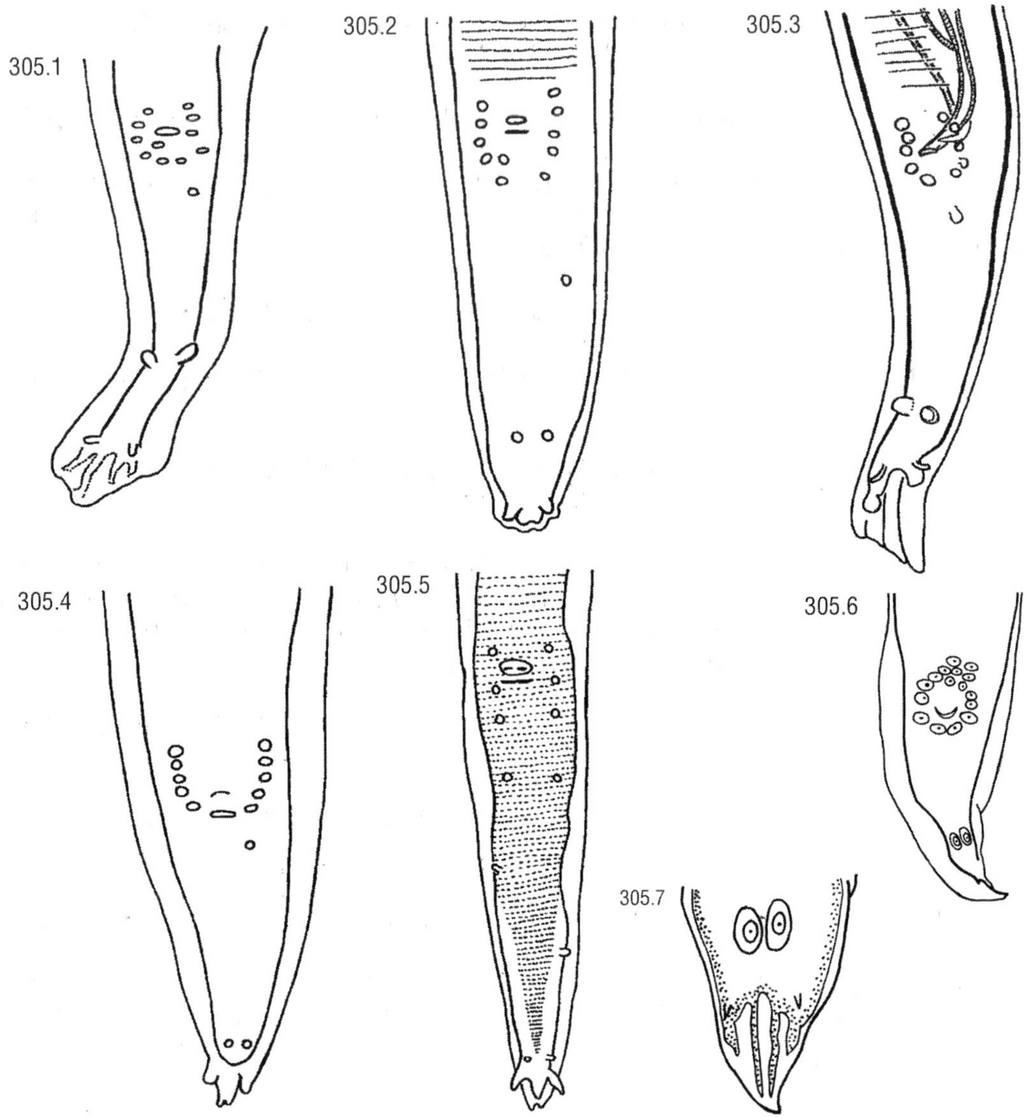

FIG. 305 *Mansonella*, ventral view of male tails of subgenera 305.1 *Mansonella* subgenus; 305.2 *Tupainema* subgenus; 305.3 *Esslingeria* subgenus; 305.4 *Tetrapetalonema* subgenus; 305.5 *Sandnema* subgenus. (305.1, 305.2, 305.3, 305.4 and 305.5 redrawn after Eberhard and Orihel, 1984b); 305.6 *Cutifilaria* subgenus; 305.7 *Cutifilaria* subgenus distal tip of tail. (305.6 and 305.7 redrawn after Uni *et al*., 2004.)

papillae formed roughly into a semicircle. *Female*: Vulva opens posterior to oesophagus. Caudal extremity with 4 lappets almost fused to form cone. Parasite in Cervidae in Europe and Japan.

Type species: *Mansonella* (*Cutifilaria*) *wenki* (Bain and Schulz-Key, 1974)

[Emended diagnosis based on Uni *et al.* (2004). *Cutifilaria* is a genus in Anderson and Bain (1976) and Chabaud and Bain (1976).]

Cruorifilaria Eberhard, Morales and Orihel, 1976 (Fig. 306)

DIAGNOSIS: Body 2–5 cm long, tapers anteriorly for one-fifth of length. Cephalic region rounded without ornamentation. Four pairs of submedian cephalic papillae, conspicuous amphids. Mouth simple, buccal capsule cuticularized, shallow, flask-shaped. Oesophagus long, slender, short anterior muscular region, long posterior glandular region. *Male*: Few caudal papillae. Caudal alae absent. Tail tapers gradually, rounded without lappets. Spicules unequal, dissimilar. Gubernaculum absent. *Female*: Body approximately twice as long as male. Tail sharply pointed without lappets. Didelphic, opisthodelphic. Vulva slit-like, opens posterior to level of oesophago-intestinal junction. Ovejector pyriform, muscular. Vagina long, muscular layer thinning distally. Ovaries extend to level of anus. Microfilaria unsheathed in peripheral blood. Parasites in renal and pulmonary blood vessels, rarely coronary vessels of Capybara, *Hydrochoerus hydrochaeris*, in Colombia.

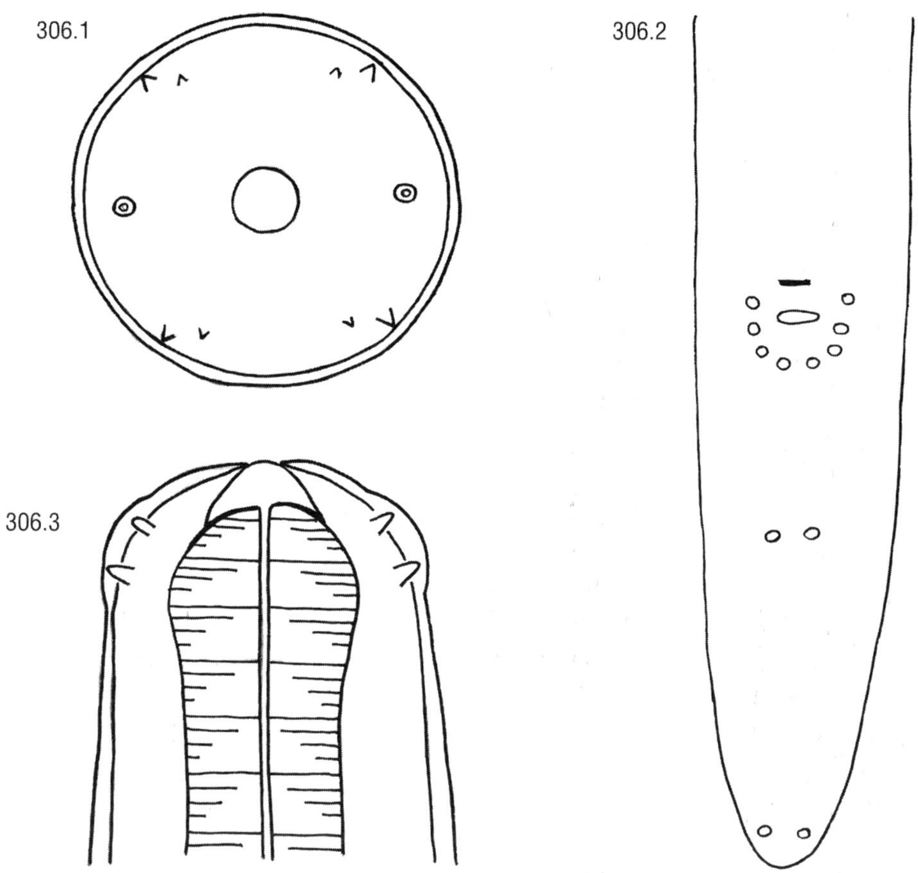

FIG. 306 *Cruorifilaria* 306.1 Female cephalic region, *en face* view; 306.2 Posterior end of male, ventral view; 306.3 Female cephalic region, lateral view. (Redrawn after Eberhard *et al.*, 1976.)

Type species: *Cruorifilaria tuberocauda* Eberhard, Morales and Orihel, 1976

[Listed in Spencer Jones and Gibson (1987).]

Struthiofilaria Noda and Nagata, 1976 (Fig. 307)

DIAGNOSIS: Cuticle with delicate transverse striations and minute bosses in lateral fields. Head dorso-ventrally flattened, separated by 'neck region' from body. Mouth oval, flattened laterally. Cephalic sensory organs: 4 pairs of submedian papillae; 2 lateral amphids. Buccal capsule rudimentary, forms small sclerotized ring. Oesophagus not distinctly divided. Cervical papillae and excretory pore small. *Male*: Caudal alae narrow. *Area rugosa* present anterior to precloacal papillae. Four pairs and single, median precloacal papillae, 6 pairs postcloacal papillae. Spicules unequal, dissimilar. Gubernaculum not clearly seen. *Female*: Approximately twice as long as male. Didelphic, opisthodelphic. Vulva opens in mid-oesophageal region. Microfilaria with sheath, tail folded along body, in blood. Parasite in body cavity of ostrich, *Sruthio camelus*, in Misaki Park Zoo, Osaka Prefecture, Japan (originally from South Africa).

Type species: *Struthiofilaria megalocephala* Noda and Nagata, 1976

[Listed in Spencer Jones and Gibson (1987).]

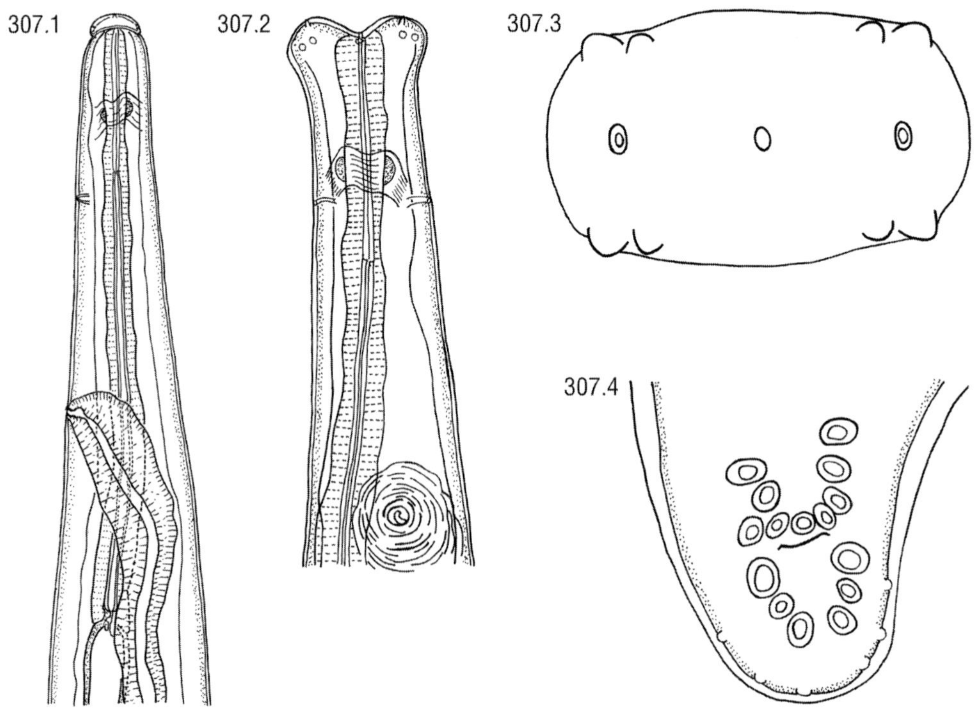

FIG. 307 *Struthiofilaria* 307.1 Female anterior end, lateral view; 307.2 Female anterior end, ventral view; 307.3 Female cephalic region, *en face* view; 307.4 Posterior end of male, ventral view. (Redrawn after Noda and Nagata, 1976.)

Josefilaria Moorhouse, Bain and Wolf, 1979 (Fig. 308)

DIAGNOSIS: Buccal capsule reduced to preoesophageal ring. Oesophagus long, divided. *Male*: Posterior end with lateral cuticular swellings. *Area rugosa* present. Caudal papillae few, grouped close to cloaca. *Female*: Vulva opens in oesophageal region. Tail long with many cuticular projections at tip. Microfilaria small, folded up in sheath. Parasite of the ghost bat, *Macroderma gigas*, in Australia.

Type species: *Josefilaria mackerrasae* Moorhouse, Bain and Wolf, 1979

[Listed in Spencer Jones and Gibson (1987).]

Molossinema Georgi, Georgi, Jiang and Frongillo, 1987 (Fig. 309)

DIAGNOSIS: Mouth small, dorso-ventral, oval in shape. Buccal cavity poorly developed. Cephalic sensory organs: 1 pair submedian ventral papillae; 4 inner submedian papillae; 2 lateral amphids. Oesophagus undivided. *Male*: Spicules dissimilar, unequal and arcuate, left spicule with 'gouge-like' distal end. Caudal alae poorly developed, restricted to cloacal region. Two pairs small, sessile adcloacal papillae present. *Female*: Vulva opens posterior to oesophago-intestinal junction. Lateral field with approximately 60 small, papillae-like structures. Tail without spines. Parasites of cerebral ventricular system of bats, *Molossus ater*, in Trinidad, West Indies.

Type species: *Molossinema wimsatti* Georgi, Georgi, Jiang and Frongillo, 1987

Chabfilaria Bain, Purnomo and Dedet, 1983 (Fig. 310)

DIAGNOSIS: External labial papillae large. Two small lateral cephalic bosses present. Oesophagus very long, clearly divided. *Male*: Spicules dissimilar, left spicule blade surrounded with a large cuticular vesicle. Gubernaculum present. Nine to ten pairs of caudal papillae: 4

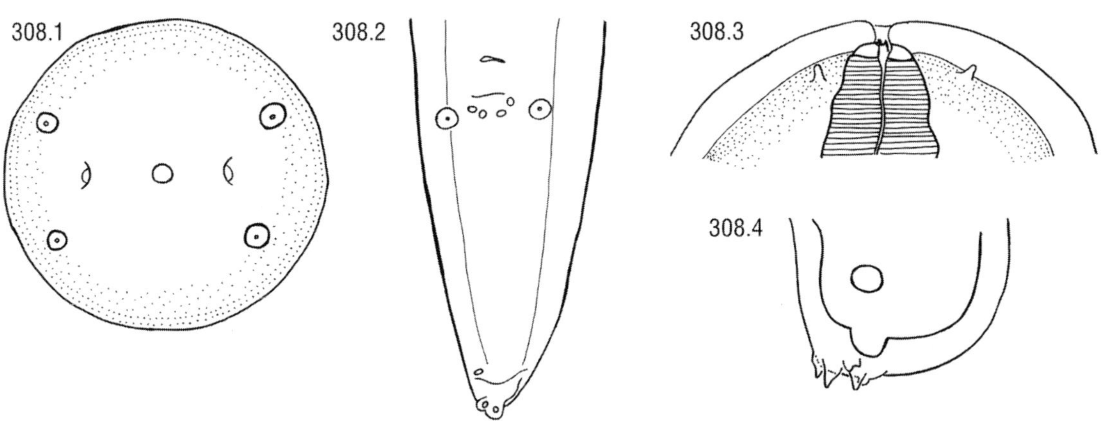

FIG. 308 *Josefilaria* 308.1 Female cephalic region, *en face* view; 308.2 Posterior end of male, ventral view; 308.3 Male cephalic region, lateral view; 308.4 Distal tip of female tail, lateral view. (Redrawn after Moorhouse *et al.*, 1979.)

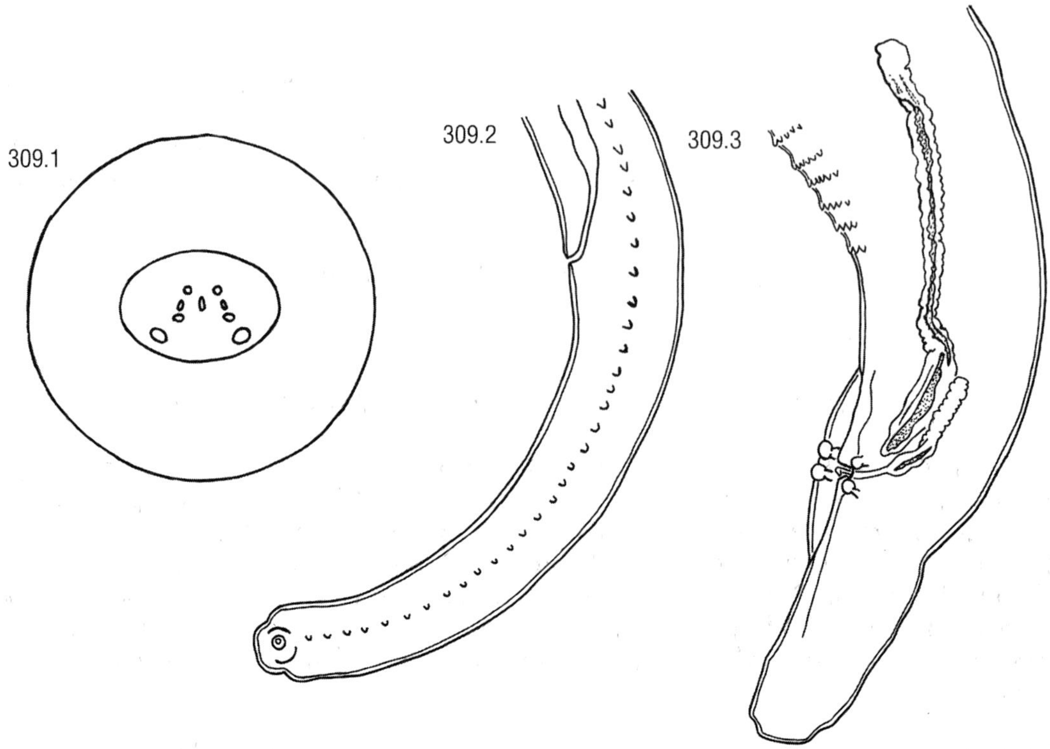

Fig. 309 *Molossinema* 309.1 Cephalic region, *en face* view; 309.2 Posterior end of female, lateral view; 309.3 Posterior end of male, lateral view. (Redrawn after Georgi *et al.*, 1987.)

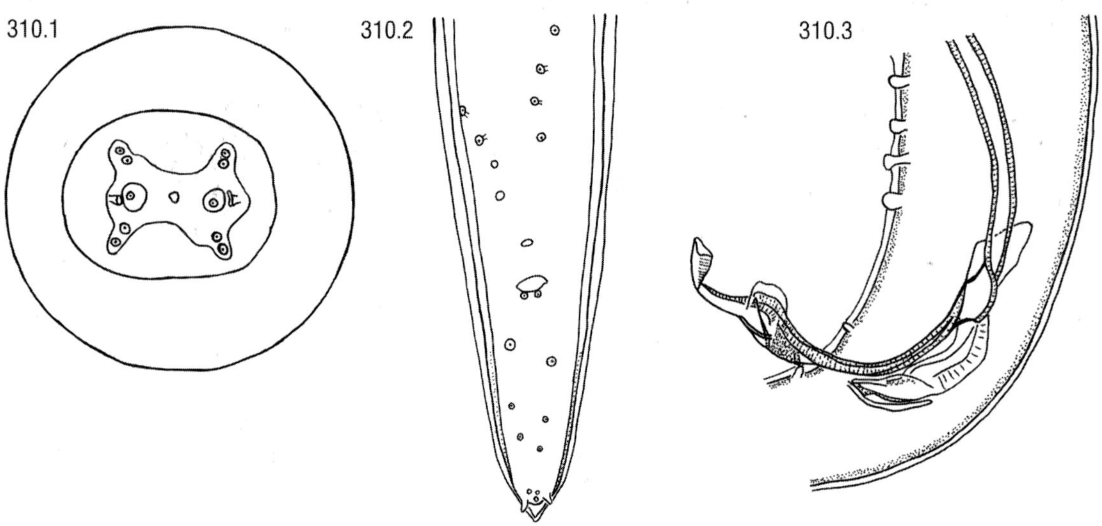

Fig. 310 *Chabfilaria* 310.1 Female cephalic region, *en face* view; 310.2 Posterior end of male, ventral view; 310.3 Spicules and gubernaculum, lateral view. (Redrawn after Bain *et al.*, 1983.)

pairs precloacal, 6 pairs arranged along tail. Petaloid appendages on tail present or absent. *Female*: Vulva opens at level of muscular oesophagus. Vagina complex. Opisthodelphic. Tail curves dorsally, slightly spiral. Parasite of Xenarthra in French Guiana and Guyana.

Type species: *Chabfilaria jonathani* Bain, Purnomo and Dedet, 1983

[Listed in Spencer Jones and Gibson (1987).]

Cherylia Bain, Petit, Jacquet-Viallet and Houin, 1985 (Fig. 311)

DIAGNOSIS: Head rounded, followed by narrower neck region. Lateral alae absent. Buccal capsule with anterior, thin-walled region, posterior, thick-walled region. Oesophagus with glandular region poorly distinguished from muscular region. *Male*: Spicules: dissimilar; left complex, three times as long as right, blade narrow, long with anvil-shaped distal tip. Gubernaculum present. Caudal papillae on tail and surround cloaca, absence of petaloid appendages on tail. *Female*: Vulva opens in oesophageal region. Tail long without petaloid appendages. Vagina complex with large, pleated cells. Microfilariae occur in skin. Infective stage larvae without caudal petaloid appendages. Experimental intermediate host *Ixodes ricinus*. Parasite of the ventral

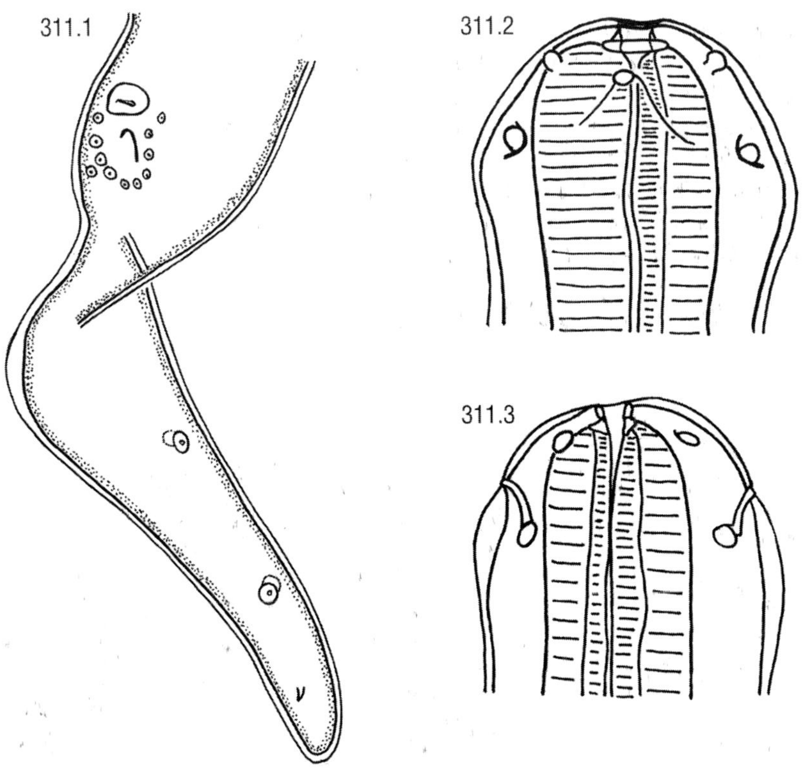

FIG. 311 *Cherylia* 311.1 Posterior end of male, ventral and distally lateral view; 311.2 Male cephalic region, lateral view; 311.3 Male cephalic region, dorso-ventral view. (Redrawn after Bain *et al.*, 1985.)

subcutaneous and perimuscular tissues of the South American marsupial, *Metachirops opossum*, in French Guiana.

Type species: *Cherylia guyanensis* Bain, Petit, Jacques-Viallet and Houin, 1985

[Listed in Spencer Jones and Gibson (1987).]

Strianema Eberhard, Orihel and Campo-Aasen, 1993 (Fig. 312)

DIAGNOSIS: Cephalic plate not greatly extended along lateral or median axis. Cuticular striations prominent, extend from anterior to nerve ring to tail. Buccal cavity and pre-oesophageal cuticular ring absent. Oesophagus muscular, not divided. *Male*: Caudal alae present. 3–4 pairs precloacal, 3–4 pairs adcloacal, 5 pairs postcloacal pedunculated caudal papillae. Small, button-like tail tip present. Spicules unequal, dissimilar. Gubernaculum, *area rugosa* absent. *Female*: Vulva opens in oesophageal region. Tail with 3 similar-sized lobes at distal tip. Microfilaria without sheath, with filamentous distal end, occurs in skin and sometimes blood. Parasites in subcutaneous tissues of Venezuelan armadillos, *Dasypus* spp.

Type species: *Strianema venezuelensis* Eberhard, Orihel and Campo-Aasen, 1993

Bisbalia Bain and Guerrero, 2003 (Fig. 313)

DIAGNOSIS: Cephalic papillae reduced. Buccal capsule only a small ring. Oesophagus <400 μm, no obvious division into muscular or glandular regions. Caudal lappets absent. *Male*: Two to three pairs adcloacal papillae, grouped together. *Area rugosa* present. Spicules dissimilar; left with membraneous blade; right with lateral margins extended, rounded at distal tips. *Female*: Opisthodelphic. Muscle cells flattened, 3–4 per quadrant. Vagina folded, sphincter present. Tail with 2 pairs of terminal bosses. Microfilaria: sheath absent; minute cephalic hook present; tail nucleated. Parasitic in membraneous 'pocket' in pleural cavity of *Heteromys anomalus* (Rodentia: Geomyoidea) in northern Venezuela.

Type species: *Bisbalia vossi* Bain and Guerrero, 2003

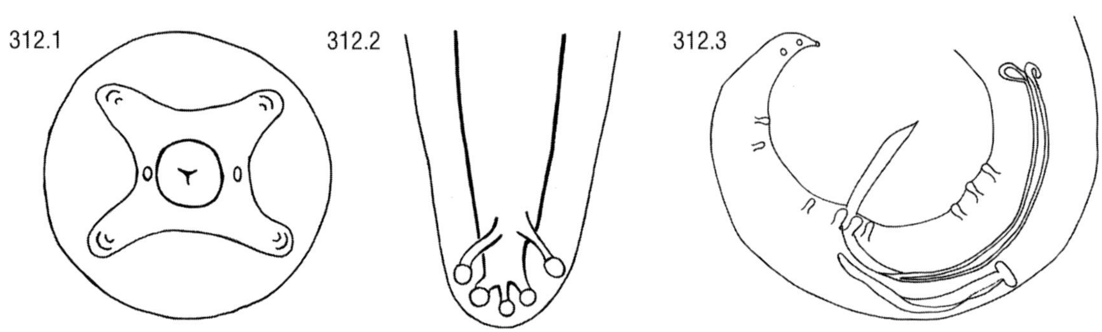

FIG. 312 *Strianema* 312.1 Cephalic region, *en face* view; 312.2 Distal end of male tail, ventral view; 312.3 Posterior end of male, lateral view. (Redrawn after Eberhard *et al*., 1993.)

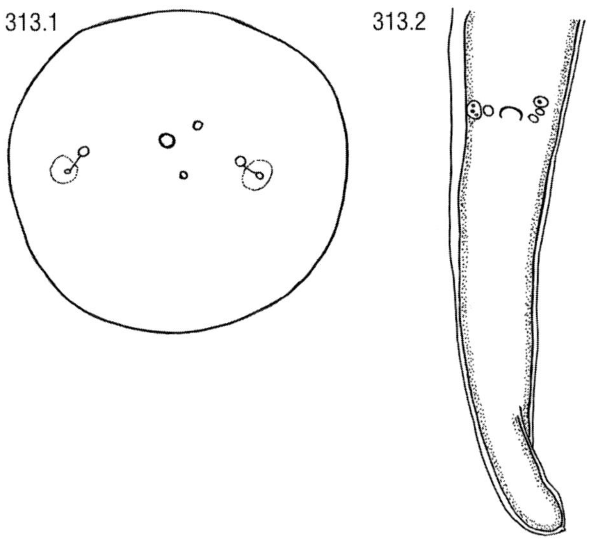

FIG. 313 *Bisbalia* 313.1 Female cephalic region, *en face* view; 313.2 Posterior end of male, ventral view. (Redrawn after Bain and Guerrero, 2003.)

Waltonellinae Bain and Prod'hon, 1974

Foleyellides Caballero, 1935

DIAGNOSIS: Cephalic region with pair of flap-like structures adjacent to mouth. Four pairs submedian cephalic sensory papillae, outer papillae broad-based with distal projection. Distinct buccal capsule absent. Lateral alae present. Annular bands of minute, longitudinally orientated bosses in mid-body absent. Oesophagus divided, shorter anterior muscular region, wider and longer posterior glandular region. *Male*: Caudal alae present. *Female*: Vulva opens at level of oesophago-intestinal junction. Microfilaria with sheath. Parasites of anuran amphibians, mainly Ranidae.

Type species: *Foleyellides striatus* (Ochoterena and Caballero, 1932) Caballero, 1935

[Emended diagnosis based on Esslinger (1986b). Esslinger (1986b) considered the subgenus *Waltonella* Schacher, 1975 (= *Waltonia* Schacher and Crans, 1973 preoccupied in Schacher, 1975) and the genus *Waltonella* (Schacher, 1975) Bain and Prod'hon, 1974 to be synonyms of *Foleyellides*, but retained the subfamily Waltonellinae. Schacher and Crans (1973) considered *Waltonella* a subgenus of *Foleyella* in the Dirofilariinae. Anderson and Bain (1976) considered *Foleyellides* a synonym of *Foleyella* Seurat, 1917 in the Dirofilariinae and *Waltonella*, a valid genus.]

Ochoterenella Caballero, 1944

DIAGNOSIS: Cuticle, apart from extremities, with annular bands of minute, longitudinally orientated bosses. Pair of cuticular, flap-like structures present adjacent to mouth. Cephalic plate lateral axis longer than dorso-ventral axis. Four pairs of submedian cephalic papillae, broad-based with slender distal projection. Lateral alae absent. *Male*: Caudal alae absent. *Female*:

Vulva opens in glandular region of oesophagus. Microfilaria with sheath. Parasites of body cavity of anuran amphibians, mainly neotropical Bufonidae.

Type species: *Ochoterenella digiticauda* Caballero, 1944

[The emended diagnosis is based on Esslinger (1986a). The study by Esslinger (1986a) led to the erection of the new genus *Paramadochotera*.]

Paramadochotera Esslinger, 1986

DIAGNOSIS: Body abruptly tapering at both ends. Cuticular, flap-like structures adjacent to mouth absent from cephalic region. Distinct buccal capsule present. Cephalic papillae broad-based with conspicuous distal tip. Lateral alae absent. *Male*: Features as described for both sexes. *Female*: Cuticle with dorsal and ventral transverse ridges or bosses. Parasite of *Mantidactylus redimitus*, a racophorid frog in Madagascar.

Type species: *Paramadochotera guibei* (Bain and Prod'hon, 1974) Esslinger, 1986

Dirofilariinae Sandground, 1921

[Bartlett (1986) considered that *Edesonfilaria* Yeh, 1960 and *Macacanema* Schad and Anderson, 1963 probably belonged to the subfamily Lemdaninae López-Neyra, 1956 rather than the Dirofilariinae, as proposed by Anderson and Bain (1976) in their key.]

Pelecitus Railliet and Henry, 1910

DIAGNOSIS: Body rounded at both ends. Cuticle with transverse striations. Bosses and longitudinal striations absent. Lateral alae present from cervical region to distal tip of body. One to two posterior deirids within alae in posterior half of body. Four pairs of submedian cephalic sensory organs. Pre-oesophageal cuticular ring present or absent. Oesophagus divided into 2 regions or not. *Male*: Spicules equal or not, dissimilar. *Female*: Vulva opens in pre- to post-oesophageal region. Ovejector simple. Microfilaria bluntly rounded at both ends, body tapering from one-third to one-fifth of body length to distal tip, sheath present. Parasites associated with tendons and muscles near leg joints and feet, rarely wings, of birds and mammals.

Type species: *Pelecitus helicinus* (Molin, 1860)

[Emended diagnosis based on Bartlett and Greiner (1986). Anderson and Bain (1976) regarded this genus as only from birds. Bartlett and Greiner (1986) add two new species and transfer two species from mammals to the genus.]

Foleyella Seurat, 1917

DIAGNOSIS: Head and tail bluntly rounded. Cuticle smooth or with transverse striations, without longitudinal striations or bosses. Lateral alae present or absent. Cephalic region with 4 pairs of papillae. Pre-oesophageal cuticular ring present or absent. Oesophagus well developed, divided or undivided. *Male*: Post-deirids present. Caudal alae present, wide. Spicules unequal. *Female*: Vulva opens in pre- to post-oesophageal region. Ovejector simple. Microfilaria anterior and posterior ends bluntly rounded, body width tapers in tail region, loose oval to elongate sheath present. Parasites in subcutaneous and intermuscular connective tissues and body cavities in chameleonid reptiles.

Type species: *Foleyella candezei* (Fraipont, 1882) Seurat, 1917

[Emended diagnosis based on Bartlett (1986). This genus was considered a subgenus by Schacher and Crans (1973), who also included *Waltonella* as a second subgenus. See comments after *Foleyellides*, above.]

Edesonfilaria Yeh, 1960

[Bain *et al.* (1983a) considered *Makifilaria* Krishnasamy, Singh and Iyamperumal, 1981 a synonym of *Edesonfilaria*. See *Makifilaria*, Lemdaninae.]

Loaina Eberhard and Orihel, 1984 (Fig. 314)

DIAGNOSIS: Anterior and posterior ends rounded. Cuticular bosses absent. Lateral alae present or absent. Oesophagus not divided. *Male*: Body one-third to one-half length of female. Tail short, not spirally coiled. Caudal papillae few, large, paired, close together at posterior end. *Female*: Vulva opens posterior to oesophagus. Tail short. Microfilaria sheathed, nuclei to tip of tail, circulate in blood. Parasites of North American rabbits.

Type species: *Loaina uniformis* (Price, 1957) Eberhard and Orihel, 1984

Splendidofilariinae Chabaud and Choquet, 1953

Splendidofilaria Skrjabin, 1923

[Lopez Caballero and Jimenez Millan (1979) propose four subgenera based on the morphology of adults and microfilariae.]

Subgenus *Splendidofilaria* (Skrjabin, 1923 genus) Lopez Caballero and Jimenez Millan, 1979

DIAGNOSIS: Adult cuticle with protuberances. Microfilaria 2–4 μm wide, up to 150 μm long.

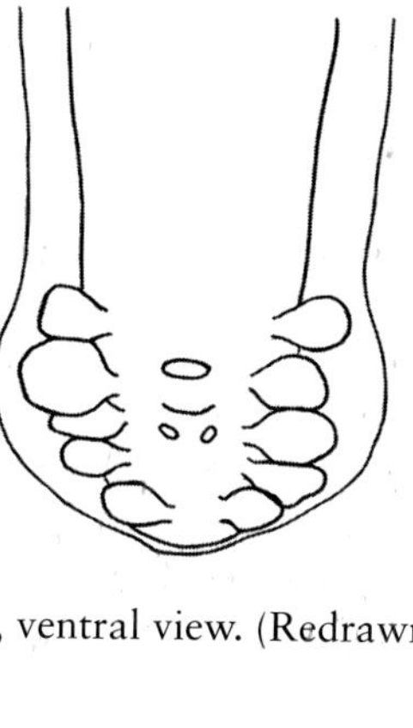

FIG. 314 *Loaina* Posterior end of male, ventral view. (Redrawn after Eberhard and Orihel, 1984a.)

Type species: *Lemdana marthae* Seurat, 1917

[Emended diagnosis based on Bartlett and Anderson (1987). The authors list *Singhfilaria* Rasheed, 1960, *Ameeria* Ali, 1961, *Lemdanella* Sonin, 1963 and *Singhnema* Ali, 1969 as synonyms of *Lemdana*. Anderson and Bain (1976) consider *Singhfilaria* (= *Lemdanella*) and *Ameeria* as valid genera. Bartlett and Anderson (1987) included a key to avian filarioid genera and, for full details, this reference should be consulted.]

Eulimdana Founikoff, 1934

DIAGNOSIS: Characteristics of the subfamily Lemdaninae *sensu* Anderson and Prestwood (Anderson and Prestwood, 1969). Cuticle smooth or with fine transverse striations, sometimes irregular small longitudinal thickenings present. Lateral alae absent. Cephalic region slightly swollen or bluntly rounded. Four pairs of submedian cephalic sensory papillae. Supra- and sub-oral cuticular inflations present or absent. Cuticular oral ring present or absent. Oesophagus muscular, well developed, clear oesophago-intestinal junction. *Male*: Spicules equal or unequal. Caudal alae poorly developed or absent. Variable number of caudal papillae present. *Female*: Vulva opens well posterior to oesophagus. Tail rounded. Microfilaria with sheath. Parasites of birds.

Type species: *Eulimdana clava* (Wedl, 1856)

[The genus was validated by Bartlett *et al.* (1985), and the generic diagnosis is based on their description. Bartlett *et al.* (1985) consider *Heimnema* Chabaud, Brygoo and Richard, 1964 a synonym of *Eulimdana*. Anderson and Bain (1976) previously considered *Eulimdana* a synonym of *Pelecitus* Railliet and Henry, 1910 in the subfamily Dirofilariinae, and *Heimnema* a valid genus in the subfamily Lemdaninae.]

Makifilaria Krishnasamy, Singh and Iyamperumal, 1981 (Fig. 320)

DIAGNOSIS: Cuticle with fine transverse striations. Head rounded, cephalic papillae small, indistinct. Mouth opening simple. Buccal cavity absent. Oesophagus divided with short,

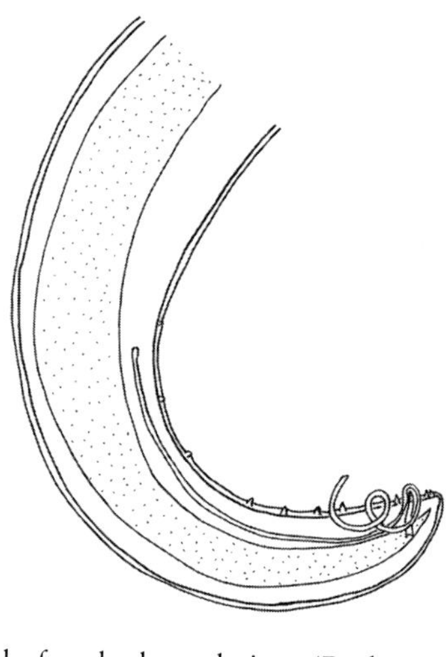

FIG. 320 *Makifilaria* Posterior end of male, lateral view. (Redrawn after Krishnasamy *et al.*, 1981.)

narrow, muscular and long, wide glandular regions. *Male*: Tail short, curved. Caudal alae absent. Sixteen pairs of papillae on posterior end, 2 small papillae at tail tip. Spicules unequal, dissimilar. Gubernaculum absent. *Female*: Vulva opens near level of junction of muscular and glandular regions of oesophagus. Ovaries extend to near anus. Anus subterminal. Tail with 2 small papillae at tip. Microfilariae short, unsheathed with rounded anterior tip, narrower tail. Parasite in the peritoneal cavity of island flying fox, *Pteropus hypomelanus*, in Pulau Langkawi, Malaysia.

Type species: *Makifilaria inderi* Krishnasamy, Singh and Iyamperumal, 1981

[Bain *et al.* (1983) considered this genus a synonym of *Edesonfilaria* Yeh, 1960, placed in the subfamily Dirofilariinae in Anderson and Bain (1976). Listed in Spencer Jones and Gibson (1987).]

APROCTOIDEA

Aproctidae (Yorke and Maplestone, 1926 subfamily) Skrjabin and Schikhobalova, 1945

Genus not assigned to a subfamily

Hovorkonema Jurášek, 1977 (Fig. 321)

DIAGNOSIS: Cuticle smooth, rudimentary somatic papillae present. Anterior and posterior ends of body rounded. Mouth without lips. Cordons or epaulettes absent. Oesophagus not divided, distal end bulb-like. *Male*: Caudal alae absent. Six pairs of lobe-like papillae. Spicules short, similar, equal, needle-like, slightly wider in mid-region. *Female*: Unknown. Parasite of stomach of Carpathian wild boar, *Sus scrofa atilla*, in Lučenec, Slovak Socialist Republic.

Type species: *Hovorkonema gastrofilaria* Jurášek, 1977

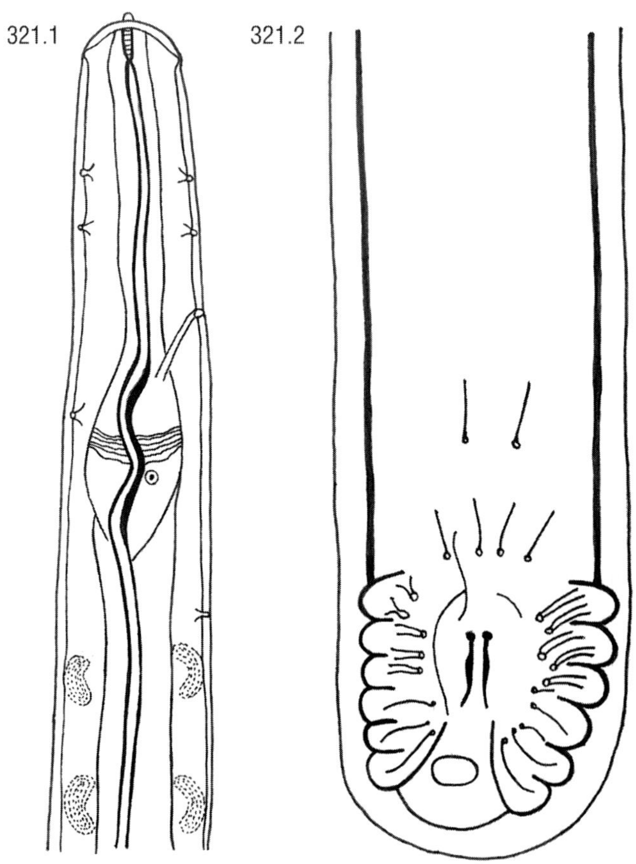

FIG. 321 *Hovorkonema* 321.1 Male anterior end, lateral view; 321.2 Posterior end of male, ventral view. (Redrawn after Jurášek, 1977.)

[Listed in Spencer Jones and Gibson (1987) as *Hovorkonema* Jurášek, 1977 *nec.* Turemuratov, 1963.]

Aproctinae Yorke and Maplestone, 1926

Squamofilaria Schmerling, 1925

[Bain and Mawson (1981) reinstate *Lissonema* Linstow, 1903 (listed as a synonym of *Aprocta* in Anderson and Bain, 1976). They consider the species of the genus *Squamofilaria* fall into the same two morphological groups as occur in the genera *Aprocta* Linstow, 1883 and *Lissonema* Linstow, 1903. They transfer the type species *Squamofilaria coraciae* (Gmelin, 1870) to *Lissonema* and consequently synonymize *Squamofilaria* with *Lissonema*. For full details the paper by Bain and Mawson (1981) should be consulted.]

Lissonema Linstow, 1903

DIAGNOSIS: Cervical papillae present. *Male*: Single large precloacal papillae present, 0–4 papillae associated with cloaca. Spicules usually equal. *Female*: Ovejector long. Larvae in excess of 400 µm in length.

Type species: *Lissonema rotunda* Linstow 1903

[Reinstated by Bain and Mawson (1981) – see *Squamofilaria* above. Emended diagnosis based on Bain and Mawson (1981).]

Anomofilaria Wehr, 1936

[Cited in Spencer Jones and Gibson (1987) as *nomen nudum* and placed in the Filarioidea. The details are published from a doctoral thesis and Wehr (1936) does not include a generic diagnosis.]

Pseudodiomedenema Gupta and Johri, 1988 (publ. 1989) (Fig. 322)

DIAGNOSIS: Cuticle smooth, body rounded both ends. Mouth simple. Lips and buccal capsule absent. Cephalic sensory organs 4 submedian papillae, 2 lateral amphids. Oesophagus divided, regions similar in width. *Male*: Tail curves ventrally, 1 pair precloacal, 2 pairs postcloacal sessile papillae. Caudal alae absent. Spicules similar, equal or unequal. Gubernaculum absent. *Female*: Vulva opens in anterior part of body, post-oesophageal. Ovoviviparous. Tail short, anus atrophied. Parasites of the pleural cavity of Hoopoe, *Upupa epops* in Lucknow, India.

Type species: *Pseudodiomedenema cameroni* Gupta and Johri, 1988 (publ. 1989)

Parasaurositus Gupta and Johri, 1989 (Fig. 323)

DIAGNOSIS: Cuticle transversely striated. Anterior end pointed. Mouth simple. Lips and buccal capsule absent. Cephalic sensory organs 4 submedian cephalic papillae, 2 lateral amphids. Oesophagus divided, anterior muscular region short and narrow, posterior glandular region

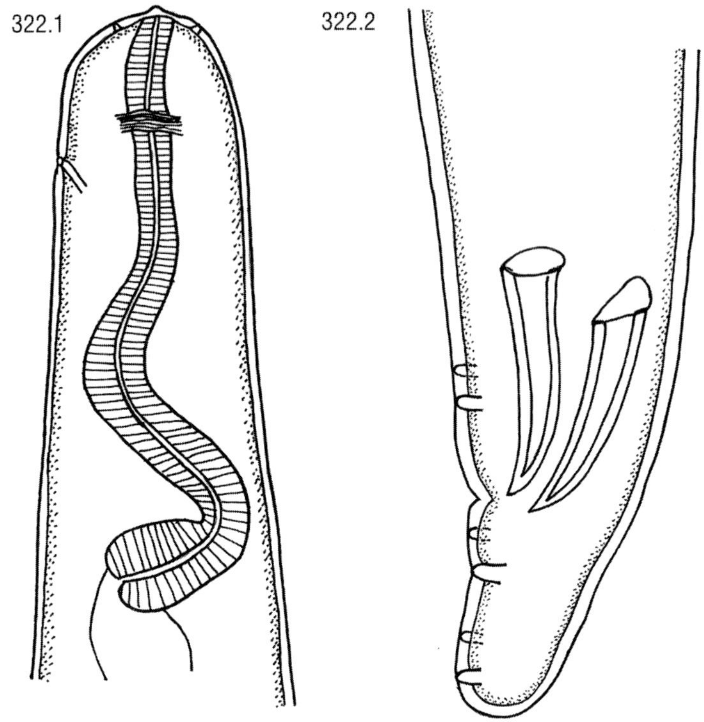

FIG. 322 *Pseudodiomedenema* 322.1 Male anterior end, lateral view; 322.2 Male posterior end, lateral view. (Redrawn after Gupta and Johri, 1988 (publ. 1989).)

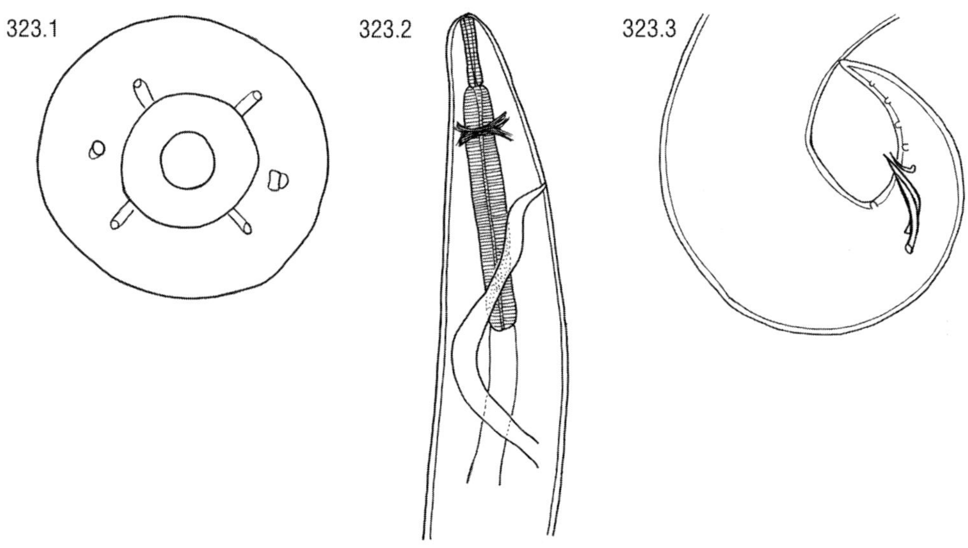

FIG. 323 *Parasaurositus* 323.1 Cephalic region, *en face* view; 323.2 Female anterior end, lateral view; 323.3 Posterior end of male, lateral view. (Redrawn after Gupta and Johri, 1989.)

wide. *Male*: Tail coiled, pointed with 1 pair precloacal, 4 pairs postcloacal papillae. Caudal alae absent. Spicules similar, subequal. Gubernaculum present. *Female*: Vulva opens adjacent to oesophagus or immediately posterior to it. Opisthodelphic. Tail short, approximately 2 lobes on distal tip, anus functional. Parasite of intra hepatic spaces of the Indian soft shell turtle, *Aspideretes gangeticus* (= *Trionyx gangeticus*) in India.

Type species: *Parasaurositus yamagutii* Gupta and Johri, 1989

Desmidocercidae Cram, 1927

Desmidocercella Yorke and Maplestone, 1926

Subgenus *Desmidocercella* (Yorke and Maplestone, 1926 genus) Gushanskaya, 1953

DIAGNOSIS: *Female*: Vulva opens in anterior part of body.

Type species: *Desmidocercella* (*Desmidocercella*) *numidica* (Seurat, 1920)

[Not included by Anderson and Bain (1976). Listed in Spencer Jones and Gibson (1987).]

Subgenus *Skrjabinocercella* Gushanskaya, 1953

DIAGNOSIS: *Female*: Vulva opens in posterior part of body.

Type species: *Desmidocercella* (*Skrjabinocercella*) *incognita* Ssolonitzin, 1932

[Not included by Anderson and Bain (1976). Listed in Spencer Jones and Gibson (1987).]

DIPLOTRIAENOIDEA

Diplotriaenidae (Skjrabin, 1916 subfamily) Anderson, 1958

Diplotriaeninae Skrjabin, 1916

Spinodiplotriaena Kalyankar and Pallawadar, 1989 (Fig. 324)

DIAGNOSIS: Cuticle covered in small spines. Chitinous trident on each side of cephalic region. Rows of sessile papillae along body margins. Mouth simple, lips absent. Oseophagus long, not divided. *Male*: Caudal alae present, 5 pairs sessile papillae. Spicules unequal, dissimilar. *Female*: Vulva opens at anterior extremity. Oviparous. Parasite in body cavity of the common mynah bird, *Acridotheres tristis*, in India.

Type species: *Spinodiplotriaena urmilii* Kalyankar and Pallawadar, 1989

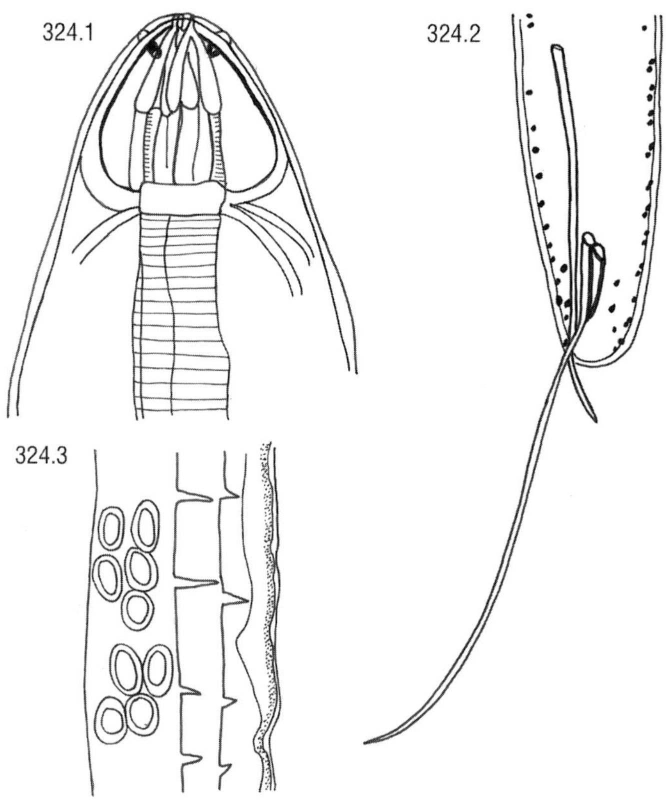

FIG. 324 *Spinodiplotriaena* 324.1 Male cephalic region; 324.2 Posterior end of male, ventral view; 324.3 Anterior end of female showing spines in oesophageal region. (Redrawn after Kalyankar and Pallawadar, 1989.)

Dicheilonematinae Wehr, 1935

Versternema Bain, Chabaud and Burger, 1992 (Fig. 325)

DIAGNOSIS: Cephalic region oval, expanded laterally, narrower dorso-ventrally. Mouth opening very small. Cephalic sensory organs: internal labial papillae absent; 4 external labial papillae, 4 submedian cephalic papillae, 1 pair of lateral amphids. Oesophagus short, not clearly separated into muscular and glandular regions. *Male*: Spicules dissimilar, right short, left long with handle and blade. Twenty-one large caudal papillae. *Female*: Vulva opens in oesophageal region. Probably oviparous. Tail rounded. Parasite of the body cavity of ostrich, *Struthio camelus*, in Botswana.

Type species: *Vesternema struthionis* Bain, Chabaud and Burger, 1992

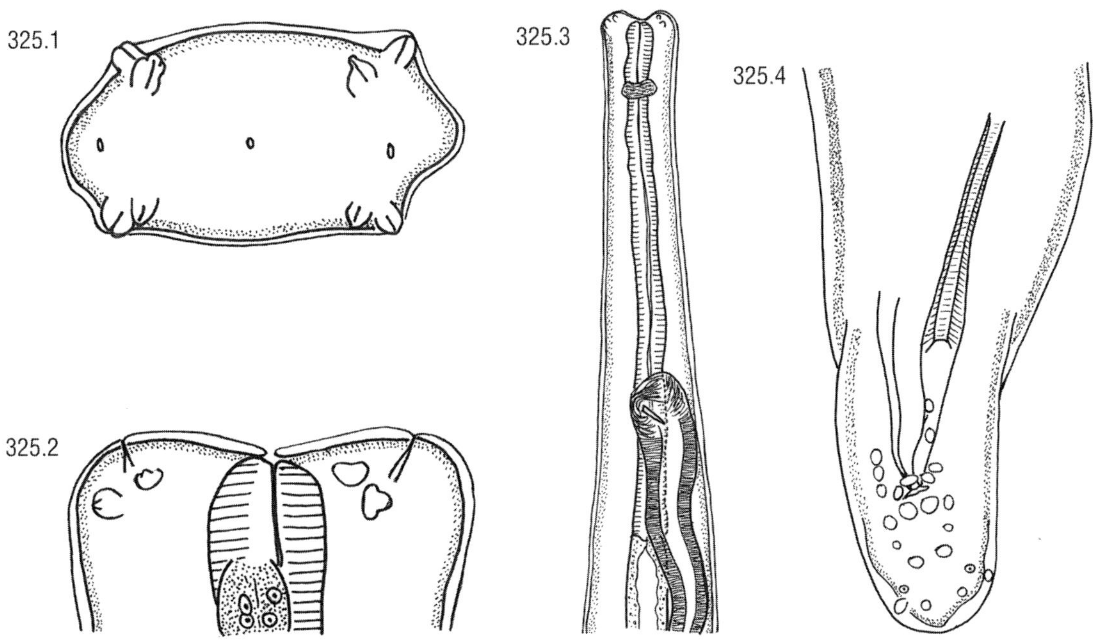

FIG. 325 *Vesternema* 325.1 Male cephalic region, *en face* view; 325.2 Male cephalic region, dorso-ventral view; 325.3 Female anterior end, ventral view; 325.4 Posterior end of male, ventral view. (Redrawn after Bain *et al.*, 1992.)

BIBLIOGRAPHY

ADAMSON, M.L. & BACCAM, D. 1988. Systematic revision of the Atractidae *sensu* Chabaud (1978) (Nematoda: Cosmocercoidea): *Maracaya belemensis.* n.sp. and *Aplectana albae* n.sp. from *Amphisbaena alba* in Brazil. *Canadian Journal of Zoology*, **66** (**8**), 1857–1864.

ADAMSON, M.L. & CAIRA, J.N. 1991. *Lockenloia sanguinis* n.gen., n.sp. (Nematoda: Dracunculoidea) from the heart of a nurse shark, *Ginglymostoma cirratum*, in Florida. *Journal of Parasitology*, **77**, 663–665.

ADAMSON, M.L. & ROTH, M. 1990. Prevalence and intensity of *Pseudodelphis oligocotti* n.gen., n.sp. (Dracunculoidea; Guyanemidae) in the tidepool sculpin, *Oligocottus maculosus* (Scorpaeniformes; Cottidae). *Journal of Parasitology*, **76**, 509–514.

ADAMSON, M.L., DEETS, G.B. & BENZ, G.W. 1987. Description of male and redescription of female *Phlyctainophora squali*. Mudry and Dailey, 1969 (Nematoda: Dracunculoidea) from elasmobranchs. *Canadian Journal of Zoology*, **65** (**12**), 3006–3019.

AGARWAL, M.P. 1930. A new nematode *Procamallanus mehrii* n.sp. from a local siluroid fish, *Wallago attu. Allahabad University Studies*, **6**, 59–64.

AGRAWAL, V. 1965. Two new spirurid nematode parasites from freshwater fishes of India. *Proceedings of the Helminthological Society of Washington*, **32**, 246–249.

AKHTAR, S.A. 1955. On nematode parasites of rats and mice of Lahore with some remarks on the genus *Aspicularis* Schulz, 1924 and two new species of the genus. *Pakistan Journal of Scientific Research*, 7, 104–111.

ALI, M.M. 1973. Notes and observations on a nomenclatorial crisis: *Tetrameres* vs. *Tropisurus. Rivista di Parassitologia*, **34** (**3**), 233–235.

ALI, S.M. 1957. Studies on the nematode parasites of fishes and birds found in Hyderabad State. *Indian Journal of Helminthology*, **8**, 1–83.

ALI, S.M. & FAROOQUI, N. 1970. A new ascarid from the grey pelican, *Pelecanus philippensis* from India. In: *H.D. Srivastava Commemoration Volume*, (edited by K.S. Singh & B.K. Tandan), pp.139–141.

ALLGÉN, C. 1921. Über die Natur und die Bedeutung der Fasersyteme im Oesophagus einiger Nematoden. *Zoologischer Anzeiger*, **53**, 76–84.

ALLGÉN, C. 1925. *Odontogeton phacochoeri* n.g., n.sp. eine neue oxyuride aus Natal. *Göteborgs Kungl. Vetenskaps–och Vitterhets–Samhälles Handlingar*, **29**, 1–37.

ALVAREZ, F., GIJÓN-BOTELLA, H., QUINTEIRO, P., REY, J., LÓPEZ-ROMÁN, F & SANMARTIN, M.L. 1994. *Paracuaria hispanica* n.sp. (Nematoda: Acuariidae), a stomach parasite of the pyrenean desman *Galemys pyrenaicus* Geoffr. (Insectivora: Talpidae), with a redefinition of the genus *Paracuaria* Rao, 1951. *Systematic Parasitology* **29** (**2**), 105–112.

ANDERSON, R.C. 1978. Keys to genera of the superfamily Metastrongyloidea. In: *CIH Keys to the Nematode Parasites of Vertebrates*. No. 5. (edited by R.C. Anderson, A.G. Chabaud & S. Willmott). Commonwealth Agricultural Bureaux, Farnham Royal, UK, 40 pp.

ANDERSON, R.C. & BAIN, O. 1976. Keys to the order Spirurida. No.3, Part 3. Diplotriaenoidea, Aproctoidea and Filarioidea. In: *CIH Keys to the Nematode Parasites of Vertebrates*. No. 3. (edited by R.C. Anderson, A.G. Chabaud & S. Willmott). Commonwealth Agricultural Bureaux, Farnham Royal, UK, pp. 59–116.

ANDERSON, R.C. & BAIN, O. 1982. Keys to the genera of the superfamilies Rhabditoidea, Dioctophymatoidea, Trichinelloidea and Muspiceoidea. In: *CIH Keys to the Nematode Parasites of Vertebrates*. No. 9. (edited by R.C. Anderson, A.G. Chabaud & S. Willmott). Commonwealth Agricultural Bureaux, Farnham Royal, UK, 26 pp.

ANDERSON, R.C. & LIM, L.H.S. 1996. *Synodontisia moraveci* n.sp. (Oxyuroidea: Pharyngodonidae) from *Osteochilus melanopleurus* (Cyprinidae) of Malaysia, with a review of pinworms in fish and a key to species. *Systematic Parasitology* **34**, 157–162.

ANDERSON, R.C. & PRESTWOOD, A.K. 1969. *Singhfilaria hayesi*, n.sp. from *Meleagris gallopavo* and a revision of Lemdaniinae (= Eufilariinae). *Canadian Journal of Zoology*, **47**, 1015–1019.

ANDERSON, R.C., CHABAUD, A.G. & WILLMOTT, S (eds). 1974. No. 1. General introduction; glossary S. Willmott; keys to subclasses, orders and superfamilies A.G. Chabaud. In: *CIH Keys to the Nematode Parasites of Vertebrates*. Commonwealth Agricultural Bureaux, Farnham Royal, UK, 17 pp.

ANDERSON, R.C., PRESTWOOD, A.K. & STRELIVE, U.R. 1979. *Procyonostrongylus muelleri* gen. et sp.n. (Metastrongyloidea: Angiostrongylidae) from the raccoon (*Procyon I. Iotor*). *Journal of Parasitology*, **65**, 811–813.

ANTONUCCI, A. 1929. Alcune considerazioni sistematiche sul genere *Cucullanus* Mueller 1777 e notizie biologiche ed etologiche su *Cucullanus (Stelmioides) ammocoetis* Linstow, 1897. *Rendiconti dell' Accademia delle Scienze Fisiche e Matematiche, Napoli*, **35**, 105–114.

APPY, R.G. & ANDERSON, R.C. 1982. The genus *Capillospirura Skrjabin*, 1924 (Nematoda: Cystidicolidae) of sturgeons. *Canadian Journal of Zoology* **60**, 194–202.

ARANTES, I.G. & ARTIGAS, P. de T. 1983. *Hydrochoerisnema anomalobursata* Arantes & Artigas, 1980 (Nematoda; Trichostrongyloidea) parasito de capivara (*Hydrochoerus hydrochaeris hydrochaeris*) Linnaeus, 1766. Estabelecimento de nova subfamilia: Hydrochoerisnematinae. *Arquivos do Instituto Biologico*, **50**, 39–45.

ARAUJO, P. 1977. Um novo oxiurideo *Gynaecometra bahiensis* n.gen., n.sp. (Nematoda: Oxyuroidea: Oxyuridae), encontrado em Lacertilio. *Memorias do Instituto Butantan*, **40/41**, 251–257.

ARTIGAS, P. & DE, T. 1936. Estudios helminthologicos I. *Paraoxyuronema brachytelesi* g.n., sp. n. parasita de *Brachyteles arachnoides* (Geoff., 1806); Oxyuronemoides, fam. n. *Memorias de Instituto de Butantan*, **10**, 77–85.

ARYA, S.N. & JOHNSON, S. 1977. A new genus of the Cylicostrongylid nematodes from the marine fish *Cybium guttatum*. *Revista Brasileira de Biologia*, **37**, 567–568.

ASAKAWA, M. & OHBAYASHI, M. 1986. A new nematode, *Yatinema japonicum* gen. et sp.n. (Heligmonellidae: Nematoda), from voles, *Eothenomys smithi* Thomas and *E. niigatae* (Anderson). *Japanese Journal of Veterinary Research*, **34**, 7–23.

ASAKAWA, M. & SATOH, R. 1987. Discovery of the genus *Heligmosomum* Railliet et Henry, 1909 (Heligmosomidae: Nematoda) from the Japanese *Clethrionomys* and establishment of *Paraheligmosomum* n.subgen. *Journal of the College of Dairying*, **12**, 111–129.

ASAKAWA, M., OHBAYASHI, M. & OW YANG, C.K. 1986. Studies on the parasite fauna of Malaysia. I. A redescription of *Strongylus brauni* Linstow, 1897, and the establishment of a new genus, *Viverrostrongylus*. *Japanese Journal of Veterinary Research*, **34**, 195–201.

BAIN, O. 2002. Evolutionary relationships among filarial nematodes. In: Klei, T.R. and Rajan, T.V. (eds) *The Filaria*. Kluwer Academic Publishers, Boston, Massachusetts. World Class Parasites **5**, 21–29.

BAIN, O. & GUERREO, R. 2003. *Bisbalia vossi* n.g., n.sp. (Nematoda: Onchocercidae), a filarial worm from a geomyoid rodent, *Heteromys anomalus*, in Venezuela. *Systematic Parasitology*, **54**, 145–151.

BAIN, O. & MAWSON, P. 1981. On some oviparous filarial nematodes mainly from Australian birds. *Records of the South Australian Museum* **18** (13), 265–284.

BAIN, O. & NIKANDER, S. 1983. Un nématode aphasmidian dans les capillaires de l'oreille du renne, *Lappnema auris* n.g., n.sp. (Robertdollfusidae). *Annales de Parasitologie Humaine et Comparée*, **58**, 383–390.

BAIN, O., CHABAUD, A.G., DUKE BOL, KOUZNETZOV, R., MULLER, R. & WENK, P. 1978. *Litomosa wite* Krepkogorskaya, 1933 (Nematoda); proposed correction to *Litomosa viteae*. Z.N. (S.) 2203. *Bulletin of Zoological Nomenclature* **35**, Part 12, 51.

BAIN, O., BAKER, M.R & CHABAUD, A.G. 1982 (publ. 1983). Nouvelles données sur la lignée *Dipetalonema* (Filarioidea, Nematoda). *Annales de Parasitologie, Humaine et Comparée* **57**, 593–620.

BAIN, O., KOUYATE, B. & CASSONE, J. 1983a. *Edesonfilaria cynocephali* n.sp., filaire parasite de Dermoptère en Malaisie. *Annales de Parasitologie Humaine et Comparée*, **58**, 185–191.

BAIN, O., PURNOMO & DEDET, J.P. 1983b. Une nouvelle filariae, *Chabfilaria jonathani* n.gen., n.sp., Onchocercidae parasite de Xénarthra. *Annales de Parasitologie Humaine et Comparée*, **58**, 583–591.

BAIN, O., PETIT, G., JACQUET-VIALLET, P. & HOUIN, R. 1985. *Cherylia guyanensis* n.gen., n.sp., filaire d'un marsupial sub-américain, transmise par tique. *Annales de Parasitologie Humaine et Comparée*, **60**, 727–737.

BAIN, O., CHABAUD, A.G. & BURGER, W.P. 1992. *Versternema struthionis* n.gen., n.sp., filaire archaique à morphologie peu spécialisée. *Annales de Parasitologie Humaine et Comparée*, **67**, 141–143.

BAKER, M.R. 1980a. *Bufonerakis andersoni* n.gen. (Nematoda: Heterakoidea) from *Bufo arenarum* of South America. *Journal of Helminthology*, **54**, 49–53.

BAKER, M.R. 1980b. Reclassification of *Oxysomatium inglisi* Anderson, 1964 and *Aplectana gigantica* Olsen, 1938 (Nematoda: Cosmocercoidea) from North American frogs. *Systematic Parasitology* **1** **(3/4)**, 245–253.

BAKER, M.R. 1980c. A revision of the genus *Oxysomatium* Railiet & Henry, 1916 (Nematoda, Cosmocercidae). *Bulletin du Muséum National d'Histoire Naturelle, Paris, 4e série* 2 **(Sect. A.3)**, 707–718.

BAKER, M.R. 1980d. Revision of Old World species of the genus *Aplectana* Railliet & Henry, 1916 (Nematoda, Cosmocercidae). *Bulletin du Muséum National d'Histoire Naturelle, Paris 4e serie* 2 **(Sect. A.4)**, 955–998.

BAKER, M.R. 1981a. *Cordonema* n.g. (Cosmocercoidea: Kathlaniinae) from the salamander *Cryptobranchus alleghenienis* (Cryptobranchidae) of North America. *Systematic Parasitology*, **3**, 59–63.

BAKER, M.R. 1981b. *Dollfusnema amphisbaenia* n.gen., n.sp. (Nematoda: Cosmocercinae) from the lizard *Leposternon phocaena* (Amphisbaenia) of Brazil. *Canadian Journal of Zoology*, **59**, 138–140.

BAKER, M.R. 1981c. Corrigendum. *Systematic Parasitology*, **3**, 264.

BAKER, M.R. 1981d. *Austraplectana* n.gen. (Cosmocercidae, Austraplectaninae n. sous–fam.), Nematode parasite d'Amphibiens australiens. *Bulletin du Muséum National d'Histoire Naturelle, Paris*, Ser 4, **(Sect A)**, 111–116.

BAKER, M.R. 1982a. On two new nematode parasites (Trichostrongyloidea: Molineidae) from amphibians and reptiles. *Proceedings of the Helminthological Society of Washington*, **49**, 252–257.

BAKER, M.R. 1982b. Systematic relationships of the Atractidae and Cosmocercidae (Nematoda: Cosmocercoidea): two new atractids parasitic in amphibians and fish. *Canadian Journal of Zoology*, **60**, 2395–2402.

BAKER, M.R. 1983. Two trichostrongyle genera (Nematoda) parasitic in Malaysian amphibians: *Batrachostrongylus* Yuen, 1963 (Amphibiophilidae) and *Batrachonema* Yuen, 1965 (Nicollinidae). *Bulletin du Muséum National d'Histoire Naturelle,* 4e Sér, 5 section A. No. 1, 181–185.

BAKER, M.R. 1984. The systematic position of the Inglisonematinae Mawson, 1968 (Nematoda). *Proceedings of the Helminthological Society of Washington*, **51** **(1)**, 69–72.

BAKER, M.R. 1987. Synopsis of the nematoda parasitic in amphibians and reptiles. Occasional papers in biology No.11. Memorial University of Newfoundland, St Johns, Newfoundland, Canada, 325pp.

BAKER, M.R. & VAUCHER, C. 1983. Parasitic helminths from Paraguay IV: cosmocercoid nematodes from *Phyllomedusa hypochondrialis* (Daudin) (Amphibia: Hylidae). *Revue Suisse de Zoologie*, **90**, 325–334.

BAKER, M.R. & VAUCHER, C. 1985. Parasitic helminths from Paraguay VII: systematic position of *Oxyascaris* Travassos, 1920 (Nematoda: Cosmocercoidea). *Revue Suisse de Zoologie*, **92** **(2)** 303–310.

BAKER, M.R. & VAUCHER, C. 1988. Parasitic helminths from Paraguay XV: Atractidae (Nematoda: Cosmocercoidea) from frogs. *Revue Suisse de Zoologie*, **95** **(2)**, 421–431.

BAKER, M.R., GOATER, T.M. & ESCH, G.W. 1987. Descriptions of three nematode parasites of salamanders (Plethodontidae: Desmognathinae) from the southeastern United States. *Proceedings of the Helminthological Society of Washington*, **54**, 15–23.

BARTLETT, C.M. 1986. The reptilian filarioid genus *Foleyella* Seurat, 1917 (Onchocercidae: Dirofilariinae) and its relationship to other dirofilariinae genera. *Systematic Parasitology*, **9** (**1**), 43–56.

BARTLETT, C.M. & ANDERSON, R.C. 1987. *Lemdana wernaarti* n.sp. and other filarioid nematodes from *Bubo virginianus* and *Asio otus* (Strigiformes) in Ontario, Canada, with a revision of *Lemdana* and a key to avian filarioid genera. *Canadian Journal of Zoology*, **65** (**5**), 1100–1109.

BARTLETT, C.M. & BAIN, O. 1987. New avian filarioids (Nematoda: Splendidofilariinae): *Dessetfilaria guianensis* gen.n., sp.n., *Andersonfilaria africanus* gen.n., sp.n. and *Splendidofilaria chandenieri* sp.n. *Proceedings of the Helminthological Society of Washington*, **54**, 1–14.

BARTLETT, C.M. & GREINER, E.C. 1986. A revision of *Pelecitus* Railliet & Henry, 1910 (Filarioidea, Dirofilariinae) and evidence for the 'capture' by mammals of filaroids from birds. *Bulletin du Muséum National d'Histoire Naturelle, A (Zoologie, Biologie et Ecologie Animales)*, **8** (**1**), 47–99.

BARTLETT, C.M., WONG, P.L. & ANDERSON, R.C. 1985. *Eulimdana lari* (Yamaguti, 1935) n.comb. (Nematoda: Filaroidea) from Phalaropus spp. (Charadriiformis) in Canada and a review of the genus *Eulimdana* Founikoff, 1934. *Canadian Journal of Zoology*, **63** (**3**), 666–672.

BARUŠ, V & SERGEEVA, T.P. 1990a. A new genus of capillariids from birds, *Tridentocapillaria* gen., n. (Nematoda: Capillariidae). *Folia Parasitologica*, **67**, 67–75.

BARUŠ, V. & SERGEEVA, T.P. 1990b. A new genus of capillariids from birds, *Ornithocapillaria* gen.n. (Nematoda: Capillariidae). *Folia Parasitologica*, **37**, 237–248.

BARUŠ, V. & SERGEEVA, T.P. 1990c. Capillariids parasitic in birds in the Palaearctic region (4). Genera *Pterothominx* and *Aonchotheca*. *Přirodĕvdné práce ústavú Československé akademie vĕd v Brnĕ*, **24** (**12**), 48 pp.

BEN SLIMANE, B., CHABAUD, A.G. & DURETTE-DESSET, M.C. 1996. Les nématodes Trichostrongylina parasites d'amphibiens et de reptiles: problèmes taxonomiques, phylétiques et biogeographiques. *Systematic Parasitology*, **35** (**3**), 179–206.

BERNARD, J. 1987. Faune des nématodes parasites des mammifères de Tunisie et des contrées voisines. *Archives Institute Pasteur Tunis*, **64**, 265–319.

BEVERIDGE, I. 1981a. *Trigonostonema* gen.n. (Nematoda: Strongyloidea) from the pademelon *Thylogale stigmatica* (Marsupialia) in Australia, with two new species, *T. trigonostoma* sp.n. and *T. longibursata* sp.n. *Journal of Parasitology*, **67**, 94–100.

BEVERIDGE, I. 1981b. *Thylonema* gen.n. (Nematoda: Strongyloidea) from the pademelon *Thylogale stigmatica* (Marsupialia) in Australia, with three new species. *Journal of Parasitology*, **67**, 101–107.

BEVERIDGE, I. 1982. A taxonomic revision of the Pharyngostrongylinea Popova (Nematoda: Strongyloidea) from Macropodid marsupials. *Australian Journal of Zoology, Supplementary Series*, **83**, 45–131.

BEVERIDGE, I. 1983. Taxonomic revision of the Zoniolaiminae (Popova) (Nematoda: Strongyloidea) from macropodid marsupials. *Australian Journal of Zoology, Supplementary Series*), **91**, 1–88.

BEVERIDGE, I. 1986. A review of the nematode tribe *Macropostrongylinea* Lichtenfels, 1980 (Strongyloidea, Cloacininae) from Australian marsupials with the erection of a new tribe, Coronostrongylinea. *Bulletin de Muséum National d'Histoire Naturelle, A. (Zoologie, Biologie et Ecologie Animales)* **8** (**4**), 777–788.

BEVERIDGE, I. 1987. The systematic status of Australian Strongyloidea (Nematoda). *Bulletin du Muséum National d'Histoire Naturelle, A. (Zoologie, Biologie et Ecologie Animals)* **9** (**1**), 107–126.

BEVERIDGE, I. 1997. *Macropostrongyloides dendrolagi* n. sp. and *Mbaisonema coronatum* n.g., n. sp., two new species of nematodes (Strongyloidea: Cloacinidae) from tree-kangaroos, *Dendrolagus* spp. (Marsupialia: Macropodidae) from Irian Jaya, Indonesia. *Systematic Parasitology*, **38**, 25–71.

BEVERIDGE, I. 1998a. *Woodwardostrongylus petrogale* sp. nov. (Nematoda: Cloacinindae), from the stomachs of Rock Wallabies (*Petrogale* spp.) from Arnhem Land. *Transactions of the Royal Society of Southern Australia*, **122** (**2**), 73–78.

BEVERIDGE, I. 1998b. Taxonomic revision of the genus *Cloacina* von Linstow (Nematoda: Strongyloidea) from macropodid marsupials. *Invertebrate Taxonomy*, **12**, 273–508.

BEVERIDGE, I. 1999. *Rugonema labiatum* n.g., n.sp. (Nematoda: Strongyloidea) from the stomach of *Macropus irma* (Marsupialia: Macropodidae) from Westerna Australia. *Systematic Parasitology*, **44**, 229–234.

BEVERIDGE, I. 2001. *Cervonemella reardoni* gen. et sp. nov. (Nematoda: Cloacinidae) from the stomachs of scrub wallabies, *Dorcopsis* spp., in Papua New Guinea. *Transactions of the Royal Society of South Australia*, **125** (**2**), 141–145.

BEVERIDGE, I. & CHILTON, N.B. 1998. *Corollonema thylogale* n.g., n.sp. (Nematoda: Strongyloidea), parasitic in the pademelons *Thylogale thetis* and *T. stigmatica* (Marsupialia: Macropodidae) from Queensland, Australia. *Systematic Parasitology*, **39**, 159–164.

BEVERIDGE, I. & DURETTE-DESSET, M.-C. 1986. New species of *Austrongylus* Chandler, 1924 (Nematoda, Trichostrongyloidea), from Australian marsupials, with a redescription of *A. minutus* Johnston & Mawson, 1938, and description of a new genus, *Sutarostrongylus*. *Bulletin du Muséum National d'Histoire Naturelle*, **8**, 145–170.

BEVERIDGE, I., & JOHNSON, P.M. 1981. Three new genera of nematodes (Strongyloidea: Cloacininae) from the red-legged pademelon, *Thylogale stigmatica* Gould (Marsupialia: Macropodidae) from eastern Australia. *Systematic Parasitology*, **3**, 77–89.

BILQEES, F.M. & AKRAM, M. 1982. Revision of the family Camallanidae Railliet & Henry, 1915 (Nematoda, Camallanoidea). *Biologia*, **28**, 45–60.

BISERKOV, V.Y., DURETTE-DESSET, M.-C. & GENOV, T. 1995. *Spalacina yanchevi* n.g., n.sp. (Nematoda: Heligmonellidae) from the lesser mole rat *Spalax leucodon* (Rodentia: Spalacidae). *Systematic Parasitology*, **32**, 27–32.

BLAYLOCK, R.B. & OVERSTREET, R.M. 1999. *Margolisianum bulbosum* n.gen., n.sp. (Nematoda: Philometridae) from the Southern flounder, *Paralichthys lethostigma* (Pisces: Bothidae), in Mississippi Sound, USA. *Journal of Parasitology*, **85** (**2**), 306–312.

BOEV, S.N. 1975. Protostrongylids. *Fundamentals of Nematodology*. Helminthological Laboratory, Academy of Sciences of the U.S.S.R., Moscow. [English translation by US Department of Agriculture, Washington, D.C., Amerind Publishing Company, New Delhi, 1984.]

BOOMKER, J., HORAK, I.G. & DE VOS, V. 1981. *Paracooperioides peleae* gen. et sp.n. (Nematoda: Trichostrongylidae) from the Vaal Ribbok, *Pelea capreolus* (Forster, 1790). *Onderstepoort Journal of Veterinary Research*, **48**, 169–174.

BOWIE, J.Y. 1985. *Cyathostoma cacatua* n.sp. (Nematoda: Syngamidae) from a sulphur-crested cockatoo *Cacatua galerita* (Aves: Psittaciformes), with a key to the species of *Cyathostoma*. *New Zealand Journal of Zoology*, **12** (**3**), 425–429.

BRUCE, N.L. 1990. *Hysterothylacium* Ward and Magath, 1917, and *Icthyascaris* Wu, 1949, ascaridoid nematodes from Australian demersal fishes. *Memoires of the Queensland Museum*, **28** (**2**), 389–426.

BRUCE, N.L. & CANNON, L.R.G. 1989. *Hysterothylacium*, *Iheringascaris* and *Maricostula* new genus, nematodes (Ascaridoidea) from Australian pelagic marine fishes. *Journal of Natural History*, **23** (**6**), 1397–1441.

BRUCE, N.L. & CANNON, L.R.G. 1990. Ascaridoid nematodes from sharks from Australia and the Solomon Islands, South Western Pacific Ocean. *Invertebrate Taxonomy*, **4** (**4**), 763–783.

BRUCE, N.L., ADLARD, R.D. and CANNON, L.R.G. 1994. Synoptic checklist of ascardioid parasites (Nematoda) from fish hosts. *Taxonomy*, **8**, 583–674.

BULIGINSKAYA, M.A., VLADIMIROV, V.L. & MARKOV, G.S. 1959. [*Paracanthocheilonema* Vladimirov nov.gen.] *Trudy Gelminthologicheskoi Laboratorii. Akademiya Nauk SSSR*, **9**, 54–58.

CABALLERO, R.G. 1974. Contribucion al conocimiento de los nematodos de peces de los litorales de Mexico. III. Dos neuvas formas. *Publicaciones Biologicas Instituto de Investigaciones Cientificas*, **1**, 33–40.

CALVENTE, I.G. 1948. Revisión del género pharyngodon y descripión de especies neuvas. *Revista Ibérica de Parasitología*, **8**, 367–410.

CAMERON, T.W.M. 1935. Studies on the endoparasitic fauna of Trinidad mammals 1. Some parasites of Trinidad deer. *Canadian Journal of Research*, Sect. D, **13**, 89–96.

CAMERON, T.W.M. & PARNELL, I.W. 1933. The internal parasites of land mammals in Scotland. *Proceedings of the Royal Physical Society*, **22**, 133–154.

CARTER, J.H. 1881. Supplementary report on specimens dredged up from the Gulf of Manaar together with others from the sea in the vicinity of the Basse Rocks and from Basse's Straits, respectively.

Presented to the Liverpool Free Museum by Captain H. Cawrie Warren. *Annals and Magazine of the Natural History Museum of Great Britain*, 7, 361–385.

CASPETA-MANDUJANO, J.M., MORAVEC, F. & SALGADO-MALDONADO, G. 2001. Two new species of rhabdochonids (Nematoda: Rhabdochonidae) from freshwater fishes in Mexico, with a description of a new genus. *Journal of Parasitology*, **87**, 139–143.

CASSONE, J. & BACCAM, D. 1985. Le genre *Filarinema* Mönnig, 1929 (Nematoda, Trichostrongyloidea), parasite de marsupiaux australiens. *Bulletin du Muséum National d'Histoire Naturelle, A (Zoologie, Biologie et Ecologie Animales)*, 7 (2), 349–382.

CASSONE, J. & DURETTE-DESSET, M.C. 1991. Cinq espéce (dont trois nouvelles) de nématodes trichostrongyloides coparasites de *Dasyprocta axarae* au Paraquay. *Revue Suisse de Zoologie*, **98** (**1**), 229–242.

CASTELLI, A. 1926. Una nova specie di nematode (*Micronematoden* ovis, n.sp.) patogena per gli ovini. *La Nuova Veterinaria*, **4**, 64–68.

CHABAUD, A.G. 1975a. Keys to genera of the order Spirurida. Camallanoidea, Dracunculoidea, Gnathostomatoidea, Physalopteroidea, Rictularioidea and Thelazioidea. *CIH Keys to the Nematode Parasites of Vertebrates*. No. 3, Part 1. (edited by R.C. Anderson, A.G. Chabaud and S. Willmott). Commonwealth Agricultural Bureaux, Farnham Royal, UK, pp. 1–27.

CHABAUD, A.G. 1975b. Keys to genera of the order Spirurida. Spiruroidea, Habronematoidea and Acuarioidea. *CIH Keys to the Nematode Parasites of Vertebrates*. No. 3, Part 2. (edited by R.C. Anderson, A.G. Chabaud and S. Willmott). Commonwealth Agricultural Bureaux, Farnham Royal, UK, pp. 29–59.

CHABAUD, A.G. 1978. Keys to genera of the superfamilies Cosmocercoidea, Seuratoidea, Heterakoidea and Subuluroidea. *CIH Keys to the Nematode Parasites of Vertebrates*. No. 6. (edited by R.C. Anderson, A.G. Chabaud and S. Willmott). Commonwealth Agricultural Bureaux, Farnham Royal, UK, 71 pp.

CHABAUD, A.G. & ANDERSON, R.C. 1959. Nouvel essai de classification des filaires (superfamilies des Filarioidea) II. *Annales de Parasitologie Humaine et Comparée*, **34**, 64–87.

CHABAUD, A.G. & BAIN, O. 1976. La lignée *Dipetalonema*. Nouvel essai de classification. *Annales de Parasitologie Humaine et Comparée*, **51**, 365–397.

CHABAUD, A.G. & BAIN, O. 1981a. Description de *Spirobakerus weitzeli* n.g., n.sp. et remarques sur les nématodes Spirocercidae. *Annales de Parasitologie Humaine et Comparée*, **56**, 73–78.

CHABAUD, A.G. & BAIN, O. 1981b. *Quentius kozeki* n.g., n.sp., nématode rictulaire parasite d'un marsupial Américain. *Annales de Parasitologie Humaine et Comparée*, **56**, 173–178.

CHABAUD, A.G. & DURETTE-DESSET, M.-C. 1973. [Description of a new oesophagostome parasitic in *Hyemoschus* in Gabon and considerations on the genus *Oesophagostomum*.] *Bulletin du Muséum Nationale d'Histoire Naturelle 3e série. No.184, Zoologie* **123**, 1415–1424.

CHABAUD, A.G. & GOLVAN, Y.J. 1957. *Megalobatrachonema campanae* n.sp. (Nematoda Kathalaniinae) parasite de tritons de la region parisienne. *Annales de Parasitologie Humaine et Comparée*, **32**, 243–263.

CHABAUD, A.G., SEUREAU, C., BEVERIDGE, I., BAIN, O. & DURETTE-DESSET, M.-C. 1980. Sur les nématodes Echinonematinae. *Annales de Parasitologie Humaine et Comparée*, **55**, 427–443.

CHABAUD, A.G., NAVONE, G.T. & BAIN, O. 1983. Description de *Mazzia bialata* n.sp. Parasite de Dasypodidés. Attribution de genre aux Nématodes Spirocercidae. *Bulletin de Muséum National d'Histoire Naturelle A*. **5** (**1**), 175–179.

CHEN, C.-L. (editor) 1973. [An illustrated guide to the fish diseases and causative pathogenic fauna and flora in the Hubei Province.] *Academia Sinica Press, Peking,* 456 pp.

CHILTON, N.B., HUBY-CHILTON, F., GASSER, R.B. & BEVERIDGE, I. 2006. The evolutionary origins of nematodes within the order Strongylida are related to predilection sites within hosts. *Molecular Phylogenetics and Evolution*, **40**, 118–128.

CHITWOOD, M.B. 1967. *Tetrameres* Creplin, 1846 (Nematoda: Spiruroidea). Proposed validation under plenary powers. Z.N. 1783. *Bulletin of Zoological Nomenclature*, **24**, 57–59.

CHU, H.J. 1931. Nematodes from flying lemurs in the Philippine Islands and from birds in China. *Journal of Parasitology*, **17**, 155–164.

CRAM, E.B. 1925. *Cylicostomias* n.g. (*C. aegyptiacum* type sp.). *Journal of Parasitology*, **11**, 229.

DAILEY, M.D. & PERRIN, W.F. 1973. Helminth parasites of porpoises of the genus *Stenella* in the eastern tropical Pacific, with descriptions of two new species: *Mastigonema stenellae* gen. et sp.n. (Nematoda: Spiruroidea) and *Zalphotrema pacificum* sp.n. (Trematoda: Digenea). *Fisheries Bulletin*, **71**, 455–471.

DE, N.C. 1989. Morphology of *Cucullanus ritai* Karve, 1952 and remarks on the validity of the genus *Paracucullanellus* Agrawal, 1965 and its type species. *Folia Parasitologica*, **36 (2)**, 153–160.

DE, N.C. & MAITY, R.N. 1995. *Indocapillaria gibsoni* n.g., n.sp. (Nematoda: Trichuroidea) from the estuarine fish *Sillaginopsis panijus* (Hamilton) of West Bengal, India. *Systematic Parasitology*, **30**, 153–157.

DE, N.C. & MAITY, R.N. 1996. *Pseudocapillaria (Discocapillaria) margolisi* n.subg., n.sp. (Nematoda: Trichuroidea) from Freshwater Fishes of West Bengal, India. *Systematic Parasitology*, **34**, 49–52.

DE ARAÚJO COSTA, H.M., MOREIRA, N.I.B. & DE OLIVEIRA, C.L. 1991. *Travassosnema* gen.n. with the description of *T. travassosi* sp.n. (Dracunculoidea: Guyanemidae) parasite of *Acestrorhynchus lacustris* Rienhardt, 1874 (Characidae) from Três Marias Reservoir, MG, Brazil. *Memórias do Instituto Oswaldo Cruz*, **86**, 437–439.

DEARDORFF, T.L. 1987. Redescription of *Pulchrascaris chiloscylli* (Johnston and Mawson, 1951) (Nematoda: Anisakidae), with comments on species in *Pulchrascaris* and *Terranova*. *Proceedings of the Helminthological Society of Washington*, **54 (1)**, 28–39.

DEARDORFF, T.L. & OVERSTREET, R.M. 1980 (publ. 1981). Review of *Hysterothylacium* and *Iheringascaris* (both previously = *Thynnascaris*) (Nematoda: Anisakidae) from the northern Gulf of Mexico. *Proceedings of the Biological Society of Washington*, **93**, 1035–1079.

DE LEY, P. & BLAXTER, M. 2002. Systematic position and phylogeny. *The Biology of Nematodes* (edited by D.L. Lee). Taylor and Francis, London and New York, pp. 1–30.

DE OLIVEIRA RODRIGUES, H. & SODRE RODRIGUES, S. 1971. Sôbre um nôvo gênero e nova espécie da subfamilia Oxyascaridinae Freitas, 1958 (Nematoda, Subuluroidea). *Atas da Sociedade de Biologia do Rio de Janeiro*, **15**, 15–17.

DE OLIVEIRA RODRIGUES, H., CARVALHO VARELA, M., SODRÉ RODRIGUES, S. & CRISTÓFARO, R. 1973. Alguns nematódeos de peixes do Oceano Atlântico – costa continental Portuguesa e costa do Norte da África. *Memórias do Instituto Oswaldo Cruz*, **71**, 247–259.

DEVAMMA, R. 1975. Studies on the subulurid nematodes from birds in Hyderabad (A.P.) India. *Indian Journal of Helminthology*, **27**, 139–173.

DIESING, K.M. 1861. Familia XXI. Trichinidea. XXII. Filaridea. LXXI Ichthyonema Diesing. *Revision der Nematoden*. K.K. Hof und Staatsdruckerei, Vienna, pp. 620, 698–699.

DIGIANI, M.C. & DURETTE-DESSET, M.-C. 2003a. *Suttonema delta* n.g., n.sp. (Nematoda: Trichostrongylina: Heligmosomoidea) from *Oxymycterus rufus* (Rodentia: Sigmodontinae) in Argentina. *Systematic Parasitology*, **55**, 33–38.

DIGIANI, M.C., SUTTON, C.A. & DURETTE-DESSET, M.-C. 2003b. A new genus of Nippostrongylinae (Nematoda: Heligmonellidae) from the water rat *Scapteromys aquaticus* (Sigmodontinae) in Argentina. *Journal of Parasitology*, **89**, 124–132.

DIKOV, G.I. 1961. [*Trichostrongyella schulzi* gen.n., sp.nov. (Nematoda: Trichostrongylidae) – a new species from sheep.] *Helminthologia*, **3**, 81–84.

DINNIK, J.A. & BOEV, S.N. 1982. A new species of lung nematodes – *Dukerostrongylus kenyae* gen.n. et sp.n. from the African antelope. *Helminthologia*, **19**, 115–119.

DRÓŻDŻ, J. 1995. Polymorphism in the Ostertagiinae Lopez-Neyra, 1947 and comments on the systematics of these nematodes. *Systematic Parasitology*, **32 (2)**, 91–99.

DUJARDIN, F. 1843. Mémoire sur les helminthes des musaraignes et en particulier sur les trichosomes, les distomes et les *Taenias* sur leurs métamorphoses et leurs transmigrations. *Annales des Sciences Naturelle. Zoologie et Biologie Animale*, **20**, 329–349.

DURETTE-DESSET, M.-C. 1970a. Les systèmes d'arêtes cuticulaires chez les nématodes héligmosomes. IV. Description de six espèces nouvelles parasites de Scuiridés africaines; redescription d'*Heligmonina magna* Baylis, 1928 et d'*Heligmonella streptocara* Baylis, 1928. *Cahiers de la Maboké*, **8 (1)**, 41–62.

DURETTE-DESSET, M.-C. 1970b. Nématodes héligmosomes d'Amérique du Sud VII. Étude de trois espèces nouvelles, parasites de *Proechimys semispinosus* (Ronguers Echimyidés). *Bulletin du Muséum d'Histoire Naturelle 2e Série*, **42**, 601–608.

DURETTE-DESSET, M.-C. 1971. Essai de classification des nématodes héligmososmes. Corrélations avec la paléobiogéographie des hôtes. *Mémoires du Muséum National d'Histoire Naturelle Nouvelle Série. Série A*, **LXIX**, 126 pp.

DURETTE-DESSET, M.-C. 1983. Keys to genera of the superfamily Trichostrongyloidea. *CIH Keys to the Nematode Parasites of Vertebrates*. No. 10. (edited by R.C. Anderson and A.G. Chabaud) Commonwealth Agricultural Bureaux, Farnham Royal, UK, 85 pp.

DURETTE-DESSET, M.-C. 1985. Trichostrongyloid nematodes and their vertebrate hosts: reconstruction of the phylogeny of a parasitic group. (edited by J.R. Baker & R. Muller) *Advances in Parasitology*, **24**, 239–306.

DURETTE-DESSET, M.-C. 1989. Nomenclature proposée pour les espéces décrites dans la sous-famille des Ostertagiinae Lopez-Neyra, 1947. *Annales de Parasitologie Humaine et Comparée*, **64** (5), 356–373.

DURETTE-DESSET, M.-C. & CHABAUD, A.G. 1992. *Laurostrongylus hoineffae* n.gen., n.sp., a relict trichostrongylid from an endemic Madagascan rodent. *Memórias do Instituto Oswaldo Cruz*, **87** **(Suppl. l)**, 91–94.

DURETTE-DESSET, M.-C. & CHABAUD, A.G. 1993. Nomenclature des Strongylida au-dessus de groupe-famille. *Annals de Parasitologie Humaine et Comparée*, **68**, 111–112.

DURETTE-DESSET, M.-C. & JUSTINE, J.-L. 1991. A cladistic analysis of the genera in the subfamily Pudicinae (Nematoda, Trichostrongyloidea, Heligmonellidae). *International Journal for Parasitology*, **21**, 579–587.

DURETTE-DESSET, M.-C. & VAUCHER, C. 1988. Trichostrongyloidea (Nematoda) parasites de Chiroptéres néotropicaux. II. Nouvelles donnés sur le genre *Cheiropteronema* Sandground, 1929. *Revue Suisse de Zoologie*, **95** (3), 889–899.

DURETTE-DESSET, M.-C. & VAUCHER, C. 1989. Trichostrongyloidea (Nematoda) parasites de Chiroptères néotropicaux. III. *Carostrongylus touzeti* gen.n., sp.n. chez *Carollia* spp. (Phyllostomatidae). *Revue Suisse de Zoologie*, **96**, 697–706.

DURETTE-DESSET, M.-C. & VAUCHER, C. 1996. *Molostrongylus acanthocolpos* gen.n., sp.n. (Nematoda, Trichostrongylina, Molineoidea) parasite de *Molossops temminki* (Chiroptera, Molossidae) au Paraguay. *Revue Suisse de Zoologie*, **103** (4), 905–913.

DURETTE-DESSET, M.-C., BAKER, M.R. & VAUCHER, C. 1985. Parasitic helminths of Paraguay IX. Rearrangement and redefinition of the genus *Schulzia* Travassos, 1937. *Revue Suisse de Zoologie*, **92**, 811–821.

DURETTE-DESSET, M.-C., CHABAUD, A.G., ASHFORD, R.W., BUTYNSKI, T. & REID, G.D.F. 1992. Two new species of the Trichostrongylidae (Nematoda: Trichostrongyloidea) parasitic in *Gorilla gorilla beringei* in Uganda. *Systematic Parasitology*, **23**, 159–166.

DURETTE-DESSET, M.-C., BEN SLIMANE, B., CASSONE, J., BARTON, D.P. & CHABAUD, A.G. 1994. *Johnpearsonia* gen.nov. and Johnpearsoniinae subf.nov. (Molineoidea: Nematoda) from *Bufo marinus*, with comments on the primitive trichostrongyle parasites of amphibians and reptiles. *Parasite*, **1**, 153–160.

DURETTE-DESSET, M.-C., CHABAUD, A.G. & SUTTON, C.A. 1997. *Tapironema coronatum* n.gen., n.sp. (Trichostrongyloidea, Cooperiinae, Obeliscoidinae), a parasite of *Holochilus brasiliensis* and *Tapirus terrestris*. *Parasite*, **4** (3), 227–232.

DURETTE-DESSET, M.-C, HUGOT, S.P., DARLU, P. & CHABAUD, A.G. 1999. A cladistic analysis of the Trichostrongyloidea (Nematoda). *International Journal for Parasitology*, **29**, 1065–1086.

DURETTE-DESSET, M.-C., LETHONEN, J.T. & HAUKISALMI, V. 2002. Trichostongylina (Nematoda) from Malagasy Muridae. I. Description of two new species of Heligmonellidae in *Nesomys* spp. *Parasite*, **9**, 29–35.

DURETTE-DESSET, M.-C., GONÇALVES, A.Q. & PINTO, R.M. 2006. Trichostrongylina (Nematoda, Heligmosomoidea) coparasites in *Dasyprocta fuliginosa* Wagler (Rodentia, Dasyproctidae) from

Brazil, with re-establishment of the genus *Avellaria* Freitas and Lent and the description of two new species. *Revista Brasileira de Zoologia*, **23**, 509–519.

DVOĬNOS, G.M. 1982. [Systematics and phylogeny of nematodes of the superfamily Strongyloidea Weinland, 1958, parasitic in horses.] In: Parazity I parazitozy chelovekai zhivotnykh. (Sbornik nauchnykh trudov). Naukova Dumka, Kiev, USSR, pp.106–114.

DVOĬNOS, G.M. & KHARCHENKO, V.A. 1986. [A new species of nematodes (Strongyloidea) – a parasite of horses.] *Vestnik Zoologii* **4**, 13–18.

DVOĬNOS, G.M. & KHARCHENKO, V.A. 1988. [Helminth species (Nematoda, Strongyloidea) from horses, new for the USSR Fauna.] *Vestnik Zoologii* **6**, 22–26.

DVOĬNOS, G.M. & KHARCHENKO, V.A. 1994. [Strongylids of domestic and wild horses.] Naukova dumka, Kiev, Ukraine, 234 pp. [ISBN 5–12–003570–1 in Russian.]

DVOĬNOS, G.M., KHARCHENKO, V.A. & LICHTENFELS, J.R. 1994. *Coronocyclus ulambajari* n.sp. (Nematoda: Strongyloidea) from horses of Mongolia. *Journal of Parasitology*, **80**, 312–316.

EBERHARD, M.L. 1980. *Dipetalonema* (*Cercopithifilaria*) *kenyensis* subgen. et sp.n. (Nematoda: Filarioidea) from African baboons, *Papio anubis*. *Journal of Parasitology*, **66**, 551–554.

EBERHARD, M.L. 1982. *Dipetalonema (Dasypafilaria) averyi* subgen. et sp.n. (Nematoda: Filarioidea) from the nine-banded armadillo, *Dasypus novemcinctus* in Louisiana. *Journal of Parasitology*, **68**, 325–328.

EBERHARD, M.L. & ORIHEL, T.C. 1984a. *Loaina* gen.n. (Filarioidea: Onchocercidae) for the Filariae parasitic in rabbits in North America. *Proceedings of the Helminthological Society of Washington*, **51**, 49–53.

EBERHARD, M.L. & ORIHEL, T.C. 1984b. The genus *Mansonella* (syn. *Tetrapetalonema)*: a new classification. *Annales de Parasitologie Humaine et Comparée* **59** (**5**), 483–496.

EBERHARD, M.L., MORALES, G.A. & ORIHEL, T.C. 1976. *Cruorifilaria tubercauda* gen. et sp.n. (Nematoda: Filarioidea) from the capybara, *Hydrochoerus hydrochaeris* in Colombia. *Journal of Parasitology*, **62**, 604–607.

EBERHARD, M.L., ORIHEL, T.C. & CAMPO-AASEN, I. 1993. *Strianema venezuelensis* gen. et sp.n. (Filarioidea: Onchocercidae) from Venezuelan armadillos (Dasypus *spp.*). *Annales de Parasitologie Humaine et Comparée*, **68**, 234–238.

ELIAS, E. & DURETTE-DESSET, M.-C. 2003. *Euzetoda ludovici* n. gen., n. sp. (Nematoda, Trichostrongylina, Heligmosomoidea) a parasite of *Microtus limnophilus* (Rodentia, Arvicolinae) from China (Gansu and Huang He). *Taxonomie, Écologie et Evolution des Métazoaires Parasites* [Taxonomy, ecology and evolution of metazoan parasites]. (Livre-homage à Louis Euzet; Tome I edited by C. Combes & J. Jourdane). Presses Universitaires de Perpignan, Perpignan, France, pp. 291–301.

ELIAS, E., BAO, G. & DURETTE-DESSET, M.-C. 2002. Two new species of *Heligmoptera* Nadtochiy, 1977 (Nematoda: Trichostrongylina: Heligmosomoidea) from myospalacine rodents in China (Gansu), with a redefinition of the genus. *Systematic Parasitology*, **51**, 73–80.

ERKULOV, E. & MOLDOPIYAZOVA, P. 1975 [Syphaciidae in rodents in Southern Kirgizia, USSR]. [*Helminthological Studies in Kirgizia*] (edited by M.M. Tokobaev).] Frunze, USSR, Izdatel'stvo 'ILIM', pp. 36–49.

ERKULOV, K. 1975. [A new genus and species of nematodes from *Alticola argentatus* in southern Kirgizia, USSR] [*Helminthological Studies in Kirgizia* (edited by M.M. Tokobaev).] Frunze, USSR, Izdatel'stvo 'ILIM', pp. 50–53.

ESSLINGER, J.H. 1986a. Redescription of *Ochoterenella digiticauda* Caballero, 1944 (Nematoda: Filarioidea) from the toad, *Bufo marinus,* with a redefinition of the genus *Ochoterenella* Caballero, 1944. *Proceedings of the Helminthological Society of Washington* **53** (**2**), 210–217.

ESSLINGER, J.H. 1986b. Redescription of *Foleyellides striatus* (Ochoterena and Caballero, 1932) (Nematoda: Filarioidea) from a Mexican frog, *Rana montezumae,* with reinstatement of the genus *Foleyellides* Caballero, 1935. *Proceedings of the Helminthological Society of Washington* **53** (**2**), 218–223.

FAGERHOLM, H.-P. 1991. Systematic implications of male caudal morphology in ascaridopid nematode parasites. *Systematic Parasitology*, **19**, 215–228.

FALCÓN-ORDAZ, J., GUZMÁN-CORNEJO, C., GARCIA-PRIETO, L. & GARDENER, S.L. 2006. *Tadaridanema delicates* (Schwartz, 1927) n. gen., n. comb. (Trichostrongylina: Molineidae) parasite of Molossidae bats. *Journal of Parasitology*, **92**, 1035–1042.

FAROOQ, M. & KHATOON, N. 1995. *Lepusius nigricollisi* n.gen., n.sp. (Nematoda: Oxyuridae) from *Lepus nigricollis dayarus* (Blandford) in Karachi, Pakistan. *Proceedings of Parasitology, Karachi*, **19**, 63–68.

FAUST, E.C. 1929. *Human Helminthology* (3rd edn., 1949). Henry Kimpton, London, 744 pp.

FERRAZ, E. & THATCHER, V.E. 1988. *Bacudacnitis grandistomis* gen. et sp. nov. (Nematoda: Cucullanidae) parasita intestinal de um bagre, *Pseudodoras niger* (Valenciennes) da Amazonia brasileira. *Amazonia* **10** (**3**), 249–253.

FOTEDAR, D.N. & BAMBROO, N. 1971. New trichostrongylid nematode *Neocooperia magnivota* gen. et sp.nov. from sheep in Kashmir. *Proceedings of the 58th Session of the Indian Science Congress*, Bangalore, January 1971. Part III (Abstracts), pp. 554–555.

FREITAS, T.J.F. & LENT, H. 1935. Capillariinae de animais de sangue frio (Nematoda: Trichuroidea). *Memórias do Instituto Oswaldo Cruz*, **30**, 241–284.

FREITAS, T.J.F. & MENDONÇA, J.M. 1960. Novo nematódeo parasito de *Procyon cancrivorus* Cuv.: *Pearsonema pearsoni* gen. nov., sp. nov. (Trichuroidea, Capillariidae). *Atas da Sociedade de Biologia do Rio de Janeiro*, **4**, 63–66.

FUKUMOTO, S., KAMIYA, M. & OHBAYASHI, M. 1986. A new genus of the intestinal nematode, *Lagostrongylus* (Heligmonellidae) from lagomorphs: a review of related genera of subfamilies Heligmonellinae and Brevistriatinae. *Japanese Journal of Veterinary Research*, **34**, 214–250.

GAIROLA, D. & MALHOTRA, S.K. 1988. Nematodes of a sub-humid riverine ecosystem. I. *Furconema viviparous* gen. et sp.n. from *Clarias batrachus* and revised key to genera of Cosmocercinae. *Indian Journal of Helminthology*, **40**, 58–63.

GARDNER, S.L. & HUGOT, J.P. 1995. A new pinworm, *Didelphoxyuris thylamisis* n. gen., n. sp. (Nematoda: Oxyurida) from *Thylamys elegans* Waterhouse, 1839) (Marsupialia: Didelphidae) in Bolivia. *Research and Reviews in Parasitology, Barcelona*, **55** (**3**), 139–147.

GARGARIN, V.G. & NAZAROVA, N.S. 1966. [*Armillocapillaria moschiferi* gen., nov. sp. (Nematoda: Capillaridae).] *Materialy Nauchnoi Konferentsii Vsesoyuznogo Obshchestva Gel'mintologov*, **3**, 67–71.

GEORGI, J.R., GEORGI, M.E., JIANG, J. & FRONGILLO, M. 1987. *Molossinema wimsatti* gen. et sp.n. (Nematoda: Onchocercinae) from the brain of *Molossus ater* (Chiroptera: Molossidae). *Proceedings of the Helminthological Society of Washington*, **54**, 225–230.

GIBBONS, L.M. & KAN, S.P. 1987. *Neorhabditostomum yeni* n.g., n.sp. (Nematoda, Strongyloidea) from *Muntiacus muntjak* in Sabah, Malaysia. *Systematic Parasitology* **9** (**3**), 173–178.

GIBBONS, L.M. & KHALIL, L.F. 1982a. A key for the identification of the nematode family Trichostrongyloidea Leiper, 1912. *Journal of Helminthology*, **56**, 185–233.

GIBBONS, L.M. & KHALIL, L.F. 1982b. *Cervicaprastongylus*, a new genus proposed for the nematode species *Ostertagia skrjabini* Singh & Pande, 1963 (Trichostrongyloidea, Trichostrongylidae). *Systematic Parasitology*, **4**, 93–98.

GIBBONS, L.M. & KHALIL, L.F. 1988. A revision of the genus *Dictyocaulus* Railliet & Henry, 1907 (Nematoda: Trichostrongyloidea) with the description of *D. africanus* n. sp. from African artiodactylids. *Journal of African Zoology*, **102**, 151–175.

GIBBONS, L.M. & KRISHNASAMY, M. 1986. *Malayometastrongylus diardinematus* n.g., n.sp. (Metastrongyloidea: Angiostrongylidae) from *Rattus rattus diardii* in Malaysia and a redescription of *Thaistrongylus harinasutai* Ohbayashi, Kamiya & Bhaibulaya, 1979. *Systematic Parasitology*, **8**, 107–115.

GIBBONS, L.M. & LICHTENFELS, J.R. 1999. Case 3075. *Strongylus tetracanthus* Mehlis, 1831 (currently *Cyathostomum tetracanthum* and *C. catinatum*) Looss, 1900 (Nematoda): proposed conservation of usage by the designation of a neotype for *C. tetracanthum*. *Bulletin of Zoological Nomenclature*, **56**, 230–234.

GIBBONS, L.M. & PLATT, T.R. 2006. *Rhinoclemmysnema* n.g. and three new species of nematodes of the family Atractidae (Cosmocercoidea), with notes on the helminth fauna of *Rhinoclemmys pulcherrima* (Testudines: Bataguridae) in Costa Rica. *Journal of Helminthology*, **80**, 333–340.

GIBBONS, L.M., CRAWSHAW, M.T. & RUMPUS, A.E. 1992. *Molinacuaria indonesiensis* n.sp. (Nematoda, Acuarioidea) from *Rattus argentiventer* in *Indonesia. Systematic Parasitology*, **23**, 175–181.

GIBBONS, L.M., KHALIL, L.F. & MARINKELLE, C.J. 1995. A new nematode genus, *Podocnematractis*, for *Atractis ortlepi* Thapar, 1925 (Cosmocercoidea: Atractidae) and the description of another new species, *P. colombiaensis* from turtles *Podocnemis* spp. in Colombia. *Systematic Parasitology*, **30**, 47–56.

GIBBONS, L.M., KNAPP, S.E. & KRECEK, R.C. 1996. *Diceronema versterae* gen.n.; sp.n. (Atractidae: Cosmocercoidea) from the Black Rhinoceros, *Diceros bicornis bicornis* in South Africa. *Journal of the Helminthology Society of Washington*, **63 (1)**, 98–104.

GIBBONS, L.M., KHALIL, L.F. & MARINKELLE, C.J. 1997. *Paraorientatractis semiannulata* n.g., n.sp. (Cosmocercoidea: Atractidae) from the large intestine of the side-necked turtle, *Podocnemis unifilis* Troschel, 1848 (Testudines: Pelomedusidae) in Brazil. *Memórias do Instituto Oswaldo Cruz,* **92 (3)**, 359–364.

GIBBONS, L.M., NICHOLLS, P.K., BAILEY, T. & SAMOUR, J. 2004. *Paraspiralatus sakeri* n. g., n. sp. (Nematoda: Spiruroidea, Spirocercidae) from saker falcons, *Falco cherrug* in Saudi Arabia and the first report of larvae from the subcutaneous tissues of houbara bustards, *Chlamydotis undulata macqueeni* in Pakistan. *Journal of Helminthology*, **78**, 33–40.

GIBSON, D.I. 1973. The genus *Pseudanisakis* Layman and Borokova, 1926 (Nematoda: Ascaridoidea). *Journal of Natural History*, **7**, 319–340.

GIBSON, D.I. 1983. The systematics of ascaridoid nematodes – a current assessment. *Concepts in Nematode Systematics*. Systematics Association Special Volume No.22 (edited by A.R. Stone, H.M. Platt & L.F. Khalil), pp. 321–338.

GIBSON, D.I. 1992. Helminth parasites of aquatic organisms: taxonomic problems. *Bulletin of the Scandinavian Society for Parasitology* **2 (2)**, 58–67.

GIBSON, D.I. 1998. Nature and classification of parasitic helminths. *Topley and Wilson's Microbiology and Microbial Infections, 9th edn., Vol. 5 Parasitology* (edited by L. Collier, A. Balows and M. Sussman). Arnold, London, pp. 453–477.

GIBSON, D.I. & COLIN, J.A. 1982. The *Terranova* enigma. (British Society for Parasitology: Proceedings) *Parasitology*, **85**, xxxvi–xxxvii.

GIBSON, D.I. & TAYLOR, A.L. 1976. A comment on the taxonomic significance and function of the 'excretory system' of ascaridoid nematodes. [Proceedings of the Spring Meeting, British Society for Parasitology, Swansea, Wales, UK, 7–9 April 1976, Abstract]. *Parasitology*, **73 (2)**, v.

GONÇALVES, A.Q., PINTO, R.M. & DURETTE-DESSET, M.-C. 2007. Parasitism of two zoonotic reservoirs *Dasyprocta leporina* and *D. fuliginosa* (Rodentia) from Amazonas, with Trichostrongylina nematodes (Heligmonellidae): description of a new genus and new species. *Memórias do Instituto Oswaldo Cruz*, **102**, 763–768.

GONÇALVES DA COSTA, S.C. 1960. Sobre um novo gênero da família Cobboldinidae Skrjabin, 1948. *Atas da Sociedade de Biologia do Rio de Janeiro,* **4**, 62–63.

GONÇALVES DA COSTA, S.C. 1961. Sobre a sistemática da família Cobboldinidae Skrjabin, 1948 (Nematoda). *Atas da Sociedade de Biologia do Rio de Janeiro,* **5**, 30–32.

GONZALES-ORTEGA, M. 1984. Estudio taxonomico de algunos nematodos parasites de roedores y lagomorfos de Mexico. Thesis, Fucultad de Ciencias, Universidad Nacional Autonoma de Mexico, Mexico D.F., 56 pp.

GRABDA-KAZUBSKA, B. 1985. The identity of *Neyraplectana schneideri* (Travassos, 1931) and *Neoraillietnema praeputiale* (Skrjabin, 1916) *sensu* Semenow, 1929. Emendation of the genus *Neoraillietnema* Ballasteros Marques, 1945 (Nematoda: Cosmocercidae). *Systematic Parasitology* **7 (2)**, 81–90.

GRASSÉ, P.P. 1965. *Traité de Zoologie, Nemathelminthes*, **4 (2)**, 731 pp. Masson et Cie, Paris.

GROVE, D.I. 1990. *A History of Human Helminthology* CAB International, Wallingford, UK, viii + 848 pp.

GUERRERO, C.A. & ROJAS, J.E. 1969. *Graphinema* aucheniae n.gen., n.sp. (Nematoda) en Auquén idos. *Boletin Chileno de Parasitologica*, **24**, 134–136.

GUERRERO, R. 1982. Trichostrongyloidea (Nematoda) parasitos de mamiferos silvestres de Venezuela. Los generos *Bradypostrongylus* Price, 1928; *Longistriata* Schulz, 1926 y *Durettestrongylus* n.gen. *Acta Biologica Venezuelica*, **11**, 111–131.

GUERRERO, R. & HUGOT, J.-P. 2003. *Monodelphoxyuris dollmeiri* n. g., n. sp. (Nematoda: Oxyurida) from *Monodelphis emiliae* in Peru. *Systematic Parasitology*, **55**, 73–79.

GUPTA, P.C. & MASOODI, B.A. 1982. Three new and one known piscine nematodes from Kanpur. *Kanpur University Research Journal (Science)*, **3**, 57–70.

GUPTA, P.C. & MASOODI, B.A. 1990. *Shamimana durdanae* n.gen., n.sp. (Nematoda: Rictulariidae) from *Plotosus arab* (Ble.) from Trivandrum, Kerala, India. *Proceedings of Parasitology*, **9**, 45–48.

GUPTA, S.P. & JEHAN, M. 1971. Studies on some nematode parasites of birds from Lucknow. *Proceedings of the Indian Science Congress*, **58, No. 39**, 537–538. [Abstract].

GUPTA, S.P. & JEHAN, M. 1972. On a new genus *Pseudoaviculariella srivastavai* n.g., n.sp. from the tunic of gizzard of a bird Cattle Egret, *Egretta garzetta* (Linnaeus). *Proceedings of the Indian Science Congress*, **59, No.78**, 432. [Abstract].

GUPTA, S.P. & KAZIM, M. 1978. Spirurid nematodes. *Indian Journal of Helminthology*, **30 (1)**, 68–89.

GUPTA, S.P. & KAZIM, M. 1979. Two new nematode genera, *Paraphysaloptera* and *Pseudoaviculariella*, from avian hosts. *Indian Journal of Parasitology*, **3**, 145–148.

GUPTA, S.P. & SRIVASTAVA, A.B. 1984 (publ. 1985). On some nematode parasites of marine fishes from Puri, Orissa. *Indian Journal of Helminthology* **36 (2)**, 97–110.

GUPTA, V. & JAISWAL, R.K. 1987. *Paracyrnea yamagutii* gen.nov. et sp.nov. (Nematoda: Spirurinae) from a bird *Anser indicus* from Lucknow. *Indian Journal of Helminthology*, **39**, 96–100.

GUPTA, V. & JOHRI, S. 1987a. On a new genus *Pseudocyrostomum* (Atractidae: Nematoda) from *Kachuga dhongoka* Gray from Lucknow, U.P. *Indian Journal of Helminthology*, **39**, 83–87.

GUPTA, V. & JOHRI, S. 1987b (publ. 1988). Nematode parasites of vertebrates. III. A new nematode, *Physaloptera (Chlamydonema) guptai*, new species (Physalopteridae) from the stomach of a domestic cat *Felis catus domesticus* from Lucknow, India. *Pakistan Journal of Zoology* **19 (2)**, 141–145.

GUPTA, V. & JOHRI, S. 1988 (publ. 1989). On some filariid nematode parasites of birds from Lucknow. *Indian Journal of Helminthology*, **40**, 109–120.

GUPTA, V. & JOHRI, S. 1989. On a new genus *Parasaurositus* (Filariidae: Nematoda) from *Trionyx gangeticus* from Lucknow, U.P. *Indian Journal of Helminthology,* **41 (Suppl.)**, *Fotedar Commemoration Volume*, 132–136.

GUPTA, V. & NAIYER, N. 1989 (publ. 1990). On a new nematode *Paratractis indica* gen. et sp.nov from a reptilian host from Lucknow (Nematoda: Atractidae). *Indian Journal of Helminthology*, **41**, 80–85.

GUSHANSKAYA, L.K. 1953. [Desmidocercidae Cram, 1927. *Papers on Helminthology Presented to Academician K.I. Skrjabin on his 75th Birthday*.] Akademii Nauk, Moscow, SSSR, pp.188–204 [In Russian].

HARRIS, E.A. 1985. Some helminths of the volcano rabbit *Romerolagus diazi*, including a description of the nematode *Teporingonema cerropeladoensis* gen.nov., sp.nov. (Trichostrongylidae: Libyostrongylidae). *Journal of Natural History*, **19**, 1239–1248.

HARTWICH, G. 1974. Keys to genera of the Ascaridoidea. *CIH Keys to the Nematode Parasites of Vertebrates*. No. 2. (edited by R.C. Anderson, A.G. Chabaud and S. Willmott). Commonwealth Agricultural Bureaux, Farnham Royal, UK, iv + 15 pp.

HARTWICH, G. 1975. *Nemathelminthes, Nematoda I. Rhabditida and Ascaridida*. VEB Gustov Fischer Verlag, Jena, Germany, 256 pp.

HARTWICH, G. 1986. Zum *Strongylus tetracanthus* – problem und zur systemik der Cyathostominea (Nematoda: Strongyloidea). *Mitteilungen Zoologischen Museum in Berlin*, **62**, 61–102.

HARTWICH, G. 1994. *Schlauchwürmer, Nemathelminthes Rund-oder Fadenwürmer, Nematoda Parasitische Rundwürmer von Wirbeltieren II. Strongylida: Strongyloidea und Ancylostomatoidea. Die Tierwelt Deutschlands 68 Teil*. Gustav Fischer Verlag, Jena, Germany, 157 pp.

HASEGAWA, H. 1988. *Wakubitinema toyamai* gen. and n.sp. (Nematoda: Seuratoidea: Quimperiidae) from the intestine of *Rana (Limnonectes) namiyei* (Amphibia: Ranidae) on Okinawa Island, Japan. *Journal of Parasitology*, **74**, 332–335.

HASEGAWA, H. 1989. *Neoentomelas asatoi* gen. et sp.n. (Nematoda: Rhabdiasidae) and *Hedruris miyakoensis* sp.n. (Nematoda: Hedruridae) from skinks of the Ryukyu Archipelago, Japan. *Proceedings of the Helminthological Society of Washington*, **56**, 145–150.

HASEWGAWA, H. & MANGALI, A. 1996. Two new nematode species of *Bunomystrongylus* n. gen. (Trichostrongylina: Heligmonellidae) collected from *Bunomys* spp. (Rodentia: Muridae) of Sulawesi, Indonesia. *Journal of Parasitology*, **82** (**6**), 998–1004.

HASEGAWA, H. & SYAFRUDDIN 1994a. *Hasanuddinia maxomyos* n.gen., n.sp. and *Heligmonoides musseri* n.sp. (Nematoda: Heligmonellidae) collected from endemic murines of Sulawesi, Indonesia. *Journal of Parasitology*, **80**, 781–788.

HASEGAWA, H. & SYAFRUDDIN 1994b. *Cyclodontostomum purvisi* (syn. *Ancistronema coronatum)* (Nematoda: Strongyloidea: Chabertiidae) from rats of Kalimantan and Sulawesi, Indonesia. *Journal of Parasitology* **80** (**4**), 657–660.

HASEGAWA, H. & SYAFRUDDIN 1997. *Maxomystrongylus yasumai* gen. and sp. n. (Nematoda: Trichostrongylina: Heligmonellidae) collected from murid rodents in Kalimantan, Indonesia. *Journal of the Helminthological Society of Washington*, **64** (**2**), 263–268.

HASEGAWA, H., SHIRAISHI, S. & ROCHMAN 1992. *Tikusnema javaense* n.gen., n.sp. (Nematoda: Acuarioidea) and other nematodes from *Rattus argentiventer*, collected in West Java, Indonesia. *Journal of Parasitology*, **78**, 800–804.

HASEGAWA, H., DOI, T., ARAKI, J. & MIYATA, A. 2000. *Kamegainema cingulum* (Linstow, 1902) n. gen., n. comb. (Nematoda: Dracunculidae), a subcutaneous parasite of cryptobranchids (Amphibia: Caudata). *Journal of Parasitology* **86** (**3**), 583–587.

HEGT, J.N. 1910. *Chlamydonema felineum* nov. gen., nov spec. eine neue parasitische lebende nematode. *Nederlandsch Tijdschrift voor de Dierkunde Vereeniging*, ser. 2, **12**, 5–44.

HERNANDEZ-RODRIGUEZ, S. & GUTIÉRREZ-PALOMINO, P.N. 1992. *Gongylonema (Progongylonema) pacoi* n.subgen. n.sp. (Spiruroidea: Gongylonematidae), parasite d'oiseaux Corvidae. *Annales de Parasitologie Humaine et Comparée*, **67**, 188–193.

HOBERG, E.P. 1996. Emended description of *Mazamastrongylus peruvianus* (Nematoda: Trichostrongylidae), with comments on the relationships of genera *Mazamastrongylus* and *Spiculopteragia*. *Journal of Parasitology*, **82**, 470–477.

HOBERG, E.P. & ABRAMS, A. 2007. Emended description and redetermination of *Sarwaria caballeroi* n. comb. (Nematoda: Ostertagiinae) with details of the synlophe and esophageal characters. *Journal of Parasitology*, **93**, 1140–1150.

HOBERG, E.P. & ABRAMS, A. 2008. *Hamulonema* gen. nov. for *Teladorsagia hamata* and *Ostertagia kenyensis* in the Ostertagiinae fauna (Nematode: Trichostrongyloidea) from African ungulates. *Journal of Parasitology*, **94**, 866–879.

HOBERG, E.P. & KHRUSTALEV, A.V. 1996. Re-evaluation of *Mazamastrongylus dagestanica* (Trichostrongylidae) with descriptions of the synlophe, genital cone, and other structural characters. *Journal of Parasitology*, **82**, 778–787.

HOBERG, E.P. & LICHTENFELS, J.R. 1992. Morphology of the synlophe and genital cone of *Parostertagia heterospiculum* (Trichostrongylidae) with comments on the subfamilial placement of the genus. *Systematic Parasitology*, **22**, 1–16.

HOBERG, E.P. & LICHTENFELS, J.R. 1994. Phylogenetic systematic analysis of the Trichostrongylidae (Nematoda), with an initial assessment of coevolution and biogeography. *Journal of Parasitology*, **80**, 976–996.

HOBERG, E.P., LICHTENFELS, J.R & PILITT, P.A. 1993a. Synlophe of *Cooperia neitzi* Trichostrongylidae: Cooperiinae) with comments on vulval inflations and hypertrophy of cuticular ridges among the trichostrongylids. *Journal of the Helminthological Society of Washington*, **60**, 153–161.

HOBERG, E.P., LICHTENFELS, J.R & PILITT, P.A. 1993b. Affiliation of *Hyostrongylus rubidus* (Nematoda: Trichostrongylidae) with the Ostertagiinae, and evaluation of the synlophe and other structural characters. *Journal of the Helminthological Society of Washington*, **60**, 219–233.

HOBERG, E.P., POLLEY, L. GUNN, A. & NISHI, J.S. 1995. *Umingmakstrongylus pallikuukensis* gen. nov. et sp. nov. (Nematoda: Protostrongylidae) from Muskoxen, *Ovibos moschatus*, in the central Canadian Arctic, with comments on biology and biogeography. *Canadian Journal of Zoology*, **73**, 2266–2282.

HOBERG, E.P., ABRAMS, A., CARRENO, R. & LICHTENFELS, J.R. 2002. *Ashworthius patriciapilittae* n. sp. (Trichostrongyloidea: Haemonchinae), an abomasal nematode in *Odocoileus virginianus* from Costa Rica, and a first record for the genus in the Western Hemisphere. *Journal of Parasitology*, **88**, 1187–1199.

HOBERG, E.P., LICHTENFELS, J.R. & GIBBONS, L.M. 2004. Phylogeny for species of the genus *Haemonchus* (Nematoda: Trichostrongyloidea): Considerations of their evolutionary history and global biogeography among Camelidae and Pecora (Artiodactyla). *Journal of Parasitology*, **90**, 1085–1102.

HOBERG, E.P., LICHTENFELS, J.R. & RICKARD, L.G. 2005. Phylogeny for genera of Nematodirinae (Nematoda: Trichostrongylina). *Journal of Parasitology*, **91**, 382–389.

HOBERG, E.P., ABRAMS, A. & EZENWA, V.O. 2008. An exploration of diversity among Ostertagiinae (Nematoda: Trichostrongyloidea) in ungulates from sub-Saharan Africa with a proposal for a new genus. *Journal of Parasitology*, **94**, 230–251.

HOBERG, E.P., ABRAMS, A. & PILITT, P.A. 2009, *Robustostrongylus aferensis* gen. nov. et sp. nov. (Nematoda: Trichostrongyloidea) in kob (*Kobus kob*) and hartebeest (*Alcelaphus buselaphus jacksoni*) (Artiodactyla) from sub-Saharan Africa, with further ruminations on Ostertagiinae. *Journal of Parasitology*, **95**, 702–717.

HORNERO, M.J. & ROCA, V. 1992 Redescription of *Skrjabinodon medinae* (García-Calvente, 1948) (Nematoda: Pharyngodonidae) from the cloaca of *Podarcis pityusensis* (Boscam, 1883) (Sauria; Lacertidae) of the Balearic Islands (Spain). *Systematic Parasitology*, **23** (**1**), 3–35.

HUFFMAN, D.G. & MORAVEC, F. 1988. First description of adult *Huffmanela huffmani* Moravec, 1987 (Nematoda: Trichosomoididae) from the swimbladder of centrarchid fishes of the Upper San Marcos River, Central Texas. *Folia Parasitologica*, **35**, 227–234.

HUGOT, J.P. 1980. Morphologie et position systématique de *Rauschterina citelli* (Tiner et Rausch, 1950) n.gen., n.cb., et *R. eutamii* (Tiner, 1948) n.cb. (Nematoda, Oxyurinae), parasites de Rongeurs Sciurinae. *Bulletin Museum National d'Histoire Naturelle*, ser.4, section A, **2**, 99–107.

HUGOT, J.P. 1981. Les Syphaciinae (Oxyuridae) parasites de sciuridés évolution des genres *Syphatineria* et *Syphabulea*. *Mémoires du Muséum National d'Histoire Naturelle*, ser.A, **121**, 1–64.

HUGOT, J.P. 1982. Sur le genre *Wellcomia* (Oxyuridae, Nematoda), parasite de Rongeurs archaiques. *Bulletin du Muséum National d'Histoire Naturelle, Paris* 4, A, **4** (**1/2**), 25–48.

HUGOT, J.P. 1983a. Redescription *d'Helminthoxys tiflophila* et considérations sur la systématique de Oxyuridae parasites de Rongeurs. *Annales de Parasitologie Humaine et Comparée*, **58**, 255–265.

HUGOT, J.P. 1983b. Deux Oxyures parasites de *Petromus typicus*, un Rongeur sub-africain archaïque. *Bulletin Museum National d'Histoire Naturelle*, **5**, 187–199.

HUGOT, J.P. 1984. Sur le genre *Trypanoxyuris* (Oxyuridae, Nematoda). I. Parasites of Sciuridae: sous-genre *Rodentoxyuris*. *Bulletin du Muséum National d'Histoire Naturelle. Série A. Zoologie*, **6**, 711–720.

HUGOT, J.P. 1985. Sur le genre *Sypharista* (Oxyuridae, Nematoda). Étude morphologique – Systématique numérique – Paléobiogéographie. *Annales de Parasitologie Humaine et Comparée*, **60**, 165–193.

HUGOT, J.P. 1987. Sur le genre *Enterobius* (Oxyuridae, Nematoda): s.g. *Colobenterobius*. I. Parasites de Primates Colobinae en région éthiopienne. *Bulletin du Museum National d'Histoire Naturelle*, **9**, 341–352.

HUGOT, J.P. 1988. Les Nématodes Syphaciinae, parasites de rongeurs et de lagomorphs. Taxonomie. Zoogéographie. Evolution. *Mémoires du Muséum National d'Histoire Naturelle. Série A, Zoologie* **141**, 1–153, 47 figs.

HUGOT, J.P. 1991. Les Syphaciinae (Oxyuridae, Nematoda) parasites de rongeurs et de lagomorphs: un example de co-évolution hôte–parasite. *Bulletin de la Société Zoologique de France* **116** (**1**), 57–62.

HUGOT, J.P. 1995. Redescription of *Xeroxyuris parallela* (Linstow, 1908) n.gen., n.comb. parasite of *Xerus inauris*. *Parasite*, **2** (**3**), 323–329.

HUGOT, J.P., GARDNER, S.L. & MORAND, S. 1996. The Enterobiinae subfam. nov. (Nematoda, Oxyurida) pinworm parasites of primates and rodents. *International Journal for Parasitology* **26** (**2**), 147–159.

HUMPHREY-SMITH, I. & DURETTE-DESSET, M.-C. 1981. [*Patricialina*, a genus parasitic in Australian marsupials.] *Bulletin du Muséum National d'Histoire Naturelle 4e Serie, A*. **3** (**1**), 133–137.

INGLIS, W.G. 1967. The evolution, host relationships and classification of the nematode superfamily Heterakoidea. *Bulletin of the British Museum (Natural History) Zoology*, **15 (1)**, 1–28.

INGLIS, W.G. 1983. An outline classification on the Phylum Nematoda. *Australian Journal of Zoology*, **31**, 243–255.

INGLIS, W.G. 1991a. A revision of the nematode genus *Odontoterakis* Skrjabin & Schikhobalova, 1947 (Heterakoidea). *Systematic Parasitology*, **20 (1)**, 69–79.

INGLIS, W.G. 1991b. *Mammalakis* n.g. and Mammalakinae n.subfam. (Nematoda: Heterakoidea: Kiwinematidae): parasites of mole rats (Rodentia: Bathyergidae and Spalacidae). *Systematic Parasitology*, **20**, 89–95.

INGLIS, W.G. & COSGROVE, C.E. 1965. The pinworm parasites (Nematoda: Oxyuridae) of the Hapalidae (Mammalia: Primates). *Parasitology*, **55 (4)**, 731–737.

INGLIS, W.G. & HARRIS, E.A. 1990. Kiwinematidae n.fam. (Nematoda) for *Kiwinema* n.g. and *Hatterianema* Chabaud & Dollfus, 1966; heterakoids of native New Zealand vertebrates. *Systematic Parasitology*, **15**, 75–79.

INGLIS, W.G. & MAWSON, P.M. 1967. Miscellanea Nematodologica. VI. *Echinonema australis* sp. nov. from *Sminthopsis crassicaudata* in South Australia. *Journal of Natural History*, **1**, 173–175.

INTERNATIONAL COMMISSION ON ZOOLOGICAL NOMENCLATURE 2001. Opinion 1972. *Strongylus tetracanthus* Mehlis, 1831 (currently *Cyathostomum tetracanthum*) and *C. catinatum* Looss, 1900 (Nematoda): conserved by the designation of a neotype for *C. tetracanthum*. *Bulletin of Zoological Nomenclature*, **58**, 148–149.

ISLAM, A.-ul-, FAROOQ, M. & KHANUM, Z. 1979. Two new genera of nematode parasites (Oxyurida: Oxysomatiinae) from toad *Bufo andersoni* of Pakistan. *Pakistan Journal of Zoology*, **11**, 69–73.

IVASHKIN, V.M. & KHROMOVA, L.A. 1976. [Essentials of Nematodology, Vol. XXVII. Cucullanata and Gnathostomata of animals and man and the diseases caused by them (edited by K.I. Skrjabin)]. Academy of Sciences of the USSR, Moscow, 436 pp.

IVASHKIN, V.M., SOBOLEV, A.A. & KHROMOVA, L.A. 1971. [Essentials of Nematodology, Vol. XXII. Camallanata of animals and man and the diseases caused by them (edited by K.I. Skrjabin).] Academy of Sciences of the USSR, Moscow (Israel Program for Scientific Translation Ltd 1977) iv+381 pp.

JACKSON, J.A. & TINSLEY, R.C. 1995. Representatives of *Batrachocamallanus* n.g. (Nematoda: Procamallaninae) from *Xenopus* spp. (Anura: Pipidae): geographical distribution, host range and evolutionary relationships. *Systematic Parasitology*, **31**, 159–188.

JANSEN, J. 1985. *Passeristrongylus amandavae* n.g., n.sp. (Nematoda: Trichostrongyloidea) from the small intestine of the passeriform bird *Amandava subflava*. *Systematic Parasitology*, **7**, 291–294.

JANSEN, J. 1986. Redescription of *Mazamastrongylus trinitatis* Cameron, 1935 and a discussion on the systematic position and species composition of the genus *Mazamastrongylus* Cameron, 1935 (Nematoda: Trichostrongyloidea). *Systematic Parasitology*, **8 (4)**, 279–283.

JANSEN, J. 1989 (publ. 1990). A concise history of the Ostertagiinae Lopez-Neyra, 1947 (Nematoda: Trichostrongyloidea) and a discussion on its composition. *Acta Leidensia*, **58 (2)**, 151–159.

JIANG JIN-SHU, YIN PEI-YUN & KUNG FAN-YAO, 1986. [*Rhinocerotonema rhinocerotis* gen. et sp. nov. from the *Rhinoceros unicornis* in Beijing Zoo (Nematoda, Strongylidae: Globocephalinae)]. *Acta Zootaxonomica Sinica*, **11**, 365–369.

JOHNSTON, T.H. & MAWSON, P.M. 1939a. Strongyle nematodes from Queensland marsupials. *Transactions of the Royal Society of South Australia*, **63**, 121–148.

JOHNSTON, T.H. & MAWSON, P.M. 1939b. Strongylate nematodes from marsupials in New South Wales. *Proceedings of the Linnean Society of New South Wales*, **64**, 513–536.

JOHNSTON, T.H. & MAWSON, P.M. 1943. Some nematodes from Australian elasmobranchs. *Transactions of the Royal Society of South Australia*, **67**, 187–190.

JOHNSTON, T.H. & MAWSON, P.M. 1948. Some new records of nematodes from Australian snakes. *Records of the South Australian Museum*, **9**, 101–106.

JONES, H.I. 1985. Two new species of nematode (Spirurida: Physalopteridae) from Australian lizards (Reptilia: Scincidae: Gekkonidae). *Journal of Natural History*, **19**, 1231–1237.

JONES, H.I. 1987. *Wanaristrongylus* gen.n. (Nematoda: Trichostrongyloidea) from Australian lizards, with descriptions of three new species. *Proceedings of the Helminthological Society of Washington*, **54**, 40–47.

JOYEUX, C. & BAER, J.G. 1937. Quelques helminthes nouveux et peu connus de la musa raigne, *Crocidura russula* Herm. (Deuxième partie Nématodes et Acanthocéphales). *Revue Suisse de Zoologie, Genève*, **44**, 27–40.

JURÁŠEK, V. 1977. *Hovorkonema gastrofilaria* sp.n. and gen.nov. (Filariata, Skrjabin, 1915) in the Carpathian wild boar (*Sus scrofa atilla* Thomas, 1912) in south Slovakia. *Biológia Bratislava*, **32**, 105–109.

JUSTINE, J.-L, CASSONE, J. & PETTER, A. 2002. *Moravecnema segonzaci* gen. et sp. n. (Nematoda: Cystidicolidae) from *Pachycara thermophilum* (Zoarcidae), a deep-sea hydrothermal vent fish from the Mid-Atlantic Ridge. *Folia Parasitologica*, **49**, 299–303.

KALYANKAR, S.D. 1969. On *Mehdiascaris* n.g., and two new species of ascarids from a marine fish, *Muraenesox talabonoides. Marathwada University Journal of Science*, **8**, 89–95.

KALYANKAR, S.D. 1970. On a new nematode, *Alibagascaris pseudosciaenae* gen. et sp.nov. (Ascarididea: Heterochilidae) from marine fishes of Alibag and Bombay. *Japanese Journal of Parasitology*, **19**, 549–551.

KALYANKAR, S.D. & PALLADWAR, V.D. 1989. On two new nematode parasites of vertebrates from India. *Indian Journal of Helminthology,* **41 (Suppl.)** Fotedar Commemoration Volume, 180–186.

KARVE, J.N. & NAIK, G.G. 1951. Some parasitic nematodes of fishes II. *Journal of the University of Bombay, Section B Biological Science*, **19**, 1–37.

KAUR, J. & KHERA, S. 1991. *Goezia (Goezia) alii* sp.n. (Nematoda) from a freshwater fish, *Channa striatus* (Bloch), from Ludhiana (Punjab, India), and a revision of the genus. *Acta Parasitologica Polonica*, **36 (1)**, 51–54.

KHALIL, L.F. & GIBBONS, L.M. 1988. Two nematodes, *Paratractis hystrix* (Diesing, 1851) and *Buckleyatractis marinkelli* n.g., n.sp. (Atractidae: Cosmocercoidea) from *Podocnemis* spp. in Colombia. *Systematic Parasitology*, **12**, 187–198.

KHALIL, L.F. & OYETAYO, A.S. 1988. *Rhapidascaroides africanus* sp. nov. (Nematoda: Anisakidae) from the fish *Bostrychus africanus* in Nigeria and a key to the species of the genus *Rhaphidascaroides* Yamaguti, 1914. *Revue de Zoologie Africaine*, **102**, 85–91.

KHERA, S. 1954 (publ. 1956). Nematode parasites of some Indian vertebrates. *Indian Journal of Helminthology*, **6**, 27–113.

KHERA, S. 1955. On some species of *Procamallanus* Baylis, 1923 from India. *Anales de la Escuela Nacional de Ciencias Biológicas*, **8**, 243–252.

KO, R.C. 1985. *Ascarophis minuta* n.sp. (Nematoda: Cystidicolidae) from scorpaenid fish, *Sebasticus marmoratus,* in Hong Kong, Southern China. *Tropical Biomedicine*, **2 (2)**, 99–105.

KO, R.C. 1986. *A Preliminary Review of the Genus* Ascarophis *van Beneden, 1871 (Nematoda: Cystidicolidae) of the gastrointestinal tract of fishes*. Department of Zoology, University of Hong Kong, 54 pp.

KO, R.C., MARGOLIS, L. & MACHIDA, M. 1985. *Pseudascarophis kyphosi* n.gen., n.sp. (Nematoda: Cystidicolidae) from the stomach of fish *Kyphosus cinerascens* (Forskål) from Japan. *Canadian Journal of Zoology*, **63**, 2684–2688.

KONTRIMAVICHUS, V.L. 1979. [*Chabaudistrongylus* Kontrimavichus gen.nov. *Filaroididae of Domestic and Wild Animals* (Essentials of Nematology, Volume XXIX. Edited by V.L. Kontrimavichus & S.L. Delyamure).] Ryzhikar, Moscow, pp. 76–80. [In Russian.]

KONTRIMAVICHUS, V.L. & DELYAMURE, S.L. (eds) 1979. [Essentials of Nematology, Volume XXIX. Filaroididae of domestic and wild animals.] Ryzhikar, Moscow, pp. 131–132.

KONTRIMAVICHUS, V.L., DELYAMURE, S.L. & BOEV, S.N. 1976. [Essentials of Nematology, Volume XXVI. Metastrongyloidea of Domestic and Wild Animals (edited by K.M. Ryzhikov).] Moscow, USSR, Izdatel'stvo 'Nauka', 240 pp.

KREIS, H.A. 1932 A new pathogenic nematode of the family Oxyuroidea, *Oxyuronema atelophora* n.g., n.sp. in the red spider monkey, *Ateles geoffroyi. Journal of Parasitology*, **18**, 295–304.

KREIS, H.A. 1940. Beiträge zur Kenntnis parasitischer Nematoden. IX. Parasitische Nematoden aus dem Naturhistorischen Museum Basel. *Zentralblatt für Bakteriologie, Parasitenkunde und Infektionskrankheiten*, **145**, 163–208.

KRISHNASAMY, M., SINGH, M. & IYAMPERUMAL, R. 1981. *Makifilaria inderi* gen. et sp.nov. (Filarioidea: Onchocercidae) from the island flying fox, *Pteropus hypomelanus* Temminick in Malaysia. *Southeast Asian Journal of Tropical Medicine and Public Health*, **12**, 185–188.

KUMAR, V. & THIENPONT, D. 1974. *Neoheterakis chapini* n.g., n.sp. (Nematoda: Heterakoidea) from *Afropavo congensis* Chapin, 1936. *Journal of Helminthology*, **48**, 119–123.

KUNG, C.C. 1948. Some new nematodes from the Australian wallaby (*Macropus rufogrisea fruticus*) with a note on the synonymy of the genera *Zoniolaimus, Labiostrongylus* and *Buccostrongylus. Journal of Helminthology*, **22**, 93–108.

K'UNG, F.Y. 1959. The validity of *Oesophagostomum hsiungi* Ling, 1959, and a description of a new nematode genus. *Acta Zoologica Sinica*, **11**, 507–508. [In Chinese.]

LANKESTER, M.W. 1983. *Skrjabingylus* Petrov, 1927 (Nematoda: Metastrongyloidea) emended with redescriptions of *S. nasicola* (Leuckart, 1842) and *S. chitwoodorum* Hill, 1939 from North American Mustelids. *Canadian Journal of Zoology,* **61** (**9**), 2168–2178.

LANKESTER, M.W. & SNIDER, J.B. 1982. *Rumenfilaria andersoni* n.gen., n.sp. (Nematoda: Filarioidea) in moose, *Alces alces* (L.), from northwestern Ontario, Canada. *Canadian Journal of Zoology*, **60**, 2455–2458.

LEIPER, R.T. 1926. Discussion on the validity of certain generic names at present in use in medical helminthology. *Archiv für Schiffs – und Tropen-hygiene*, **30**, 484–491.

LE ROUX, P.L. 1940. On the division of the genus *Oesophagostomum* Molin, 1861 into subgenera and the creation of a new genus for the oesophagostomes of the warthog. *Journal of Helminthology*, **18**, 1–20.

LHERMITTE-VALLARINO, N., BAIN, O., DEHARO, E., BERTANI, S., VOZA, T., ATTOUT, T., & GAUCHER, P. 2005. A new rhabdiasid nematode, *Chabirenia cayennsis* n. g., n. sp., parasitic in the glands of the buccal mucosa of a South American saurian. *Systematic Parasitology*, **62**, 151–160.

LICHTENFELS, J.R. 1980a. Keys to genera of the superfamily Strongyloidea. No. 7 *CIH Keys to the Nematode Parasites of Vertebrates* (edited by R.C. Anderson, A.G. Chabaud and S. Willmott). Commonwealth Agricultural Bureaux, Farnham Royal, UK, 41 pp.

LICHTENFELS, J.R. 1980b. Keys to genera of the superfamilies Ancylostomatoidea and Diaphanocephaloidea. No. 8 In: *CIH Keys to the Nematode Parasites of Vertebrates* (edited by R.C. Anderson, A.G. Chabaud and S. Willmott). Commonwealth Agricultural Bureaux, Farnham Royal, UK, 26 pp.

LICHTENFELS, J.R. & PILITT, P.A. 2000. Synlophe patterns of the Haemonchinae of ruminants (Nematoda: Trichostrongyloidea). *Journal of Parasitology*, **86**, 1093–1098.

LICHTENFELS, J.R., PILITT, P.A. & HOBERG, E.P. 1994. New morphological characters for identifying individual specimens of *Haemonchus* spp. (Nematoda: Trichostrongyloidea) and a key to species in ruminants of North America. *Journal of Parasitology*, **80**, 107–119.

LICHTENFELS, J.R., PILITT, P.A. & CRAIG, T.M. 1996. The synlophe and other structural characteristics of *Sarwaria bubalis* (Nematoda: Trichostrongyloidea) from cattle in Guyana. *Journal of Parasitology*, **82**, 146–154.

LICHTENFELS, J.R., KHARCHENKO, V.A., KRECEK, R.C. & GIBBONS, L.M. 1998. An annotated checklist by genus and species of 93 species-level names for 51 recognized species of small strongyles (Nematoda: Strongyloidea: Cyathostominea) of horses, asses and zebras of the world. *Veterinary Parasitology*, **79**, 65–79.

LICHTENFELS, J.R., GIBBONS, L.M. & KRECEK, R.C. 2002. Recommended terminology and advances in the systematics of the Cyathostominea (Nematoda: Strongyloidea) of horses. *Veterinary Parasitology*, **107**, 337–342.

LICHTENFELS, J.R., KHARCHENKO, V.A. & DVOǏNOS, G.M. 2008. Ilustrated identification keys to strongylid parasites (Strongylidae: Nematoda) of horses, zebras and asses (Equidae). *Veterinary Parasitology*, **156**, 4–161.

LING, M.T. 1959. [One new species of *Oesophagostomum* (Nematoda: Trichonematidae) from swine in Chekiang.] *Acta Zoologica Sinica*, **11**, 24–28.

LITTLE, M.D. & ALI, M.M. 1980. *Chandleronema longigutturata* (Chandler, 1942) gen. et comb.n. (Nematoda: Acuariidae) from raccoons, *Procyon lotor*, in Louisiana. *Journal of Parasitology*, **66**, 555–558.

LOMAKIN, V.V. 1991. [Revision of the family Amidostomatidae (Nematoda, Strongylida).] *Trudy Gel'mintologicheskoi Laboratorii Akademiya, Nauk SSSR, Moscow and Leningrad* **38**, 70–85 [In Russian.].

LOMAKIN, V.V. 1993. [Revision of nematodes of the subfamily Amidostomatinae Travassos, 1919 (Amidostomatidae; Strongylida).] *Trudy Gel'mintologicheskoi Laboratorii*, **39**, 92–122. [In Russian.]

LOMAKIN, V.V. & CHERNOVA, T.N. 1980. [A new nematode genus, *Garkavillanus* n.g. (Camallanata; Skrjabillanidae.] *Trudi Gel'mintologicheskoĭ Laboratorii (Gel'minty vodnykh i nazemnykh biotsenozov)*, **30**, 45–51. [In Russian.]

LOMAKIN, V.V. & ROMASHOV, B.V. 1987. [Morphological–taxonomic analysis and phylogenetic relationships in nematodes of the family Capillariidae Railliet, 1915.] *Trudi Gel'mintologicheskoĭ Laboratorii (Morfologiya, Taksonomiya I Ekologiya Gel'mintov Zhivotnykh I Rastenii)*, **35**, 87–95. [In Russian.]

LOPEZ CABALLERO, E.J. & JIMENEZ MILLAN, F. 1979. Sobre la taxonomia del gen. *Splendidofilaria*, Skrjabin, 1923. *Revista Ibérica de Parasitologia*, **79**, 559–566.

LÓPEZ-NEYRA, C.R. 1945. Estudios y revisión de la familia Subuluridae, con descriptión de especies neuvas. *Revista Ibérica de Parasitologia*, **5**, 271–329.

LÓPEZ-NEYRA, C.R. 1947. Generos y especies neuvas o mal conocidas de Capillarinae. *Revista Ibérica de Parasitologia*, **7**, 192–238.

LUO, J., DUO, J. & CHENG, G. 1988. [Studies on a new genus and a new species of Protostrongylidae from China (Strongylata)]. *Acta Veterinaria et Zootechnica Sinica*, **19**, 270–273. [In Chinese, English summary.]

MACHIDA, M. 1975. Nematodes from the deep-sea fishes of Suruga Bay. I. A new rhabdochonid nematode, *Johnstonmawsonoides nemichthyos* from the nemichthyid fish. *Bulletin of the National Science Museum, Tokyo, Series A (Zoology)*, **1**, 1–14.

MACHIDA, M. 1976. Nematodes from the deep-sea fishes of Suruga Bay. II. Two new rhabdochonid nematodes from the macrourid fishes. *Bulletin of the National Science Museum, Tokyo, Series A (Zoology)*, **2**, 1–6.

MACHIDA, M., OGAWA, K. & OKIYAMA, M. 1982. A new nematode (Spirurida, Physalopteridae) from Frill Shark of Japan. *Bulletin of the National Science Museum, Tokyo, Series A (Zoology)*, **8**, 1–5.

MADSEN, H. 1945. The species of *Capillaria* (Nematoda: Trichinelloidea) parasitic in the digestive tract of Danish gallinaceous and anatine game birds with a revised list of species of *Capillaria* in birds. *Danish Review of Game Biology*, **1**, 1–112.

MAGALHAES PINTO, R. & CORREA GOMES, D. 1985. Nematodes of Amazonian birds, with a description of *Hoazinstrongylus amazonensis* n.gen. n.sp. (Trichostrongylidae, Libyostrongylinae). *Memórias do Instituto Oswaldo Cruz*, **80**, 213–217.

MAGGENTI, A.R. 1982. Nemata. *Synopsis and Classification of Living Organisms* (edited by S.P. Parker). McGraw-Hill Book Inc., New York, pp. 897–929.

MAGGENTI, A.R. 1983. Nematode higher classification as influenced by species and family concepts. *Concepts in Nematode Systematics* (edited by A.R. Stone, H.M. Platt & L.F. Khalil). Systematics Association Special Vol., **22**, pp. 25–40.

MAGZOUB, M., KASIM, A.A. & SHAWA, Y. 1980. A new cestode species of the genus *Oochoristica* Lühe, 1898 and a new nematode species of a new genus *Parathelandros* from Dabb Lizard, *Uromastyx aegyptia*. *Journal of the College of Science, University of Riyadh*, **11**, 111–118.

MALAKHOV, V.V., RYZHIKOV, K.M. & SONIN, M.D. 1982. [System of the major nematode taxa: subclasses, orders, suborders.] *Zoologicheskii Zhurnal*, **61** (**8**), 1125–1134.

MARGOLIS, L. 1977. *Caballeronema* gen.nov. for *Metabronema* wardlei Smedley, 1934 (Nematoda, Spiruroidea) from the marine fish *Scorpaenichthys marmoratus* from the Pacific Coast of Canada. *Insituto do Biologia Publicaciones Especiales*, **4**, 447–454.

MARKOV, G.S. 1962. *Tezisy Dokladov Nauchnoi Konfererentsii (Vsesoyuznogo Obshchestva Gel'mintologov)*. Moscow, **2**, 118–120. [In Russian.]

MAS-COMA, S. & ESTEBAN, J.G. 1982. *Kahmannia eliomydis* n.gen., n.sp. (Nematoda: Heteroxynematidae), parasite intestinal de *Eliomys quercinus* (Linnaeus, 1766) (Rodentia: Gliridae) a Minorque (Baléares). *Annales de Parasitologie Humaine et Comparée*, **57**, 487–495.

MAS-COMA, S. & ESTEBAN, J.G. 1985. *Tenoranema alcoveri* g.n., sp.n. (Trichuridae: Capillariinae), a nematode parasitizing *Eliomys quercinus* (Rodentia: Gliridae). *Folia Parasitologica*, **32**, 35–42.

MAWSON, P.M. 1977. The genus *Cyclostrongylus* Johnston and Mawson (Nematoda: Trichonematidae). *Transactions of the Royal Society of South Australia*, **101**, 121–124.

MAWSON, P.M. 1978. A new genus *Adelonema* (Nematoda: Oxyuridae) from Australian phalangerid marsupials. *Transactions of the Royal Society of South Australia*, **102**, 223–226.

MAWSON, P.M. 1979. *Alcostoma* new genus (Nematoda: Trichonematidae). *Transactions of the Royal Society of South Australia*, **103**, 123–126.

MAWSON, P.M. 1980. *Beveridgea* new genus (Nematoda: Strongylida) from the agile wallaby from northern Australia. *Transactions of the Royal Society of South Australia*, **104**, 81–82.

MAWSON, P.M. 1982. Some Acuariinae (Nematoda) from Australian birds. *Transactions of the Royal Society of South Australia*, **106**, 19–30.

MCVICAR, A.H. & GIBSON, D.I. 1975. *Pancreatonema torriensis* gen.nov., sp.nov. (Nematoda: Rhabdochonidae) from the pancreatic duct of *Raja naevus*. *International Journal of Parasitology*, **5**, 529–535.

MEJÍA-MADRID, H. H. & PERÉZ-PONCE DE LEÓN, G. 2007. A new rhabdochonid from the blue striped chubb *Sectator ocyurus* (Osteichthyes: Kyphosidae) in Chamela Bay, Mexico. *Journal of Parasitology*, **93**, 166–170.

MOLIN, R. 1860. Trenta specie di Nematoidi. *Sitzungsberichte der Akademie der Wissenschaften, Mathematisch-Naturwissenschaftliche Classe, Wien*, **40**, 331–358.

MÖNNIG, H.O. 1932. *Wild Antelopes as Carriers of Nematode Parasites of Domestic Ruminants*. Part II. 18th Report of the Director of Veterinary Services and Animal Industry, Union of South Africa, August 1932, pp. 153–172.

MOORHOUSE, D.-E., BAIN, O. & WOLF, G. 1979. *Josefilaria mackerrasae* gen. et sp.nov. (Nematoda: Filarioidea) parasite de la Chauve-souris *Macroderma gigas* Dobson. *Annales de Parasitologie Humaine et Comparée*, **54**, 645–651.

MOORTHY, V.N. 1937. *Camallanus sweeti* nov. sp., a new species of Camallanidae (Nematoda). *Journal of Parasitology*, **23**, 302–306.

MORAVEC, F. 1975. Reconstruction of the nematode genus *Rhabdochona* Raiiliet, 1916 with a review of the species parasitic in fishes of Europe and Asia. *Studie CSAV Number 8*. Academia, Prague, 104 pp.

MORAVEC, F. 1977. A new nematode parasite, *Esocinema bohemicum* gen. et sp.nov. (Skrjabillanidae) of the European pike. *Folia Parasitologica*, **24**, 86–90.

MORAVEC, F. 1982a. Proposal of a new systematic arrangement of nematodes of the family Capillariidae. *Folia Parasitologica*, **29**, 119–132.

MORAVEC, F. 1982b. Systematic status of *Sinoichthyonema itenopharyngodoni* Wu, 1973 (Nematoda). *Folia Parasitologia* **29** (**4**), 314.

MORAVEC, F. 1986a. Review of capillariid nematodes (Capillariinae) parasitic in amphibians and reptiles. Part 1. General introduction. Genera *Capillaria, Aonchotheca* and *Paratrichosoma. Vestnik Ceskoslovenské Spolecnosti Zoologické*, **50** (**2**), 120–131.

MORAVEC, F. 1986b. Review of capillariid nematodes (Capillariinae) parasitic in amphibians and reptiles. Part 3. Genus *Paracapillaria. Vestník Ceskoslovenské Spolecnosti Zoologické*, **50** (**4**), 273–283.

MORAVEC, F. 1987. Revision of capillariid nematodes (subfamily Capillariinae) parasitic in fishes. *Studie CSAV Number 3*. Prague, Czechoslovakia; Ceskoslovenská Akademie Ved, **7**, 141.

MORAVEC, F. 1994. *Parasitic Nematodes of Freshwater Fishes of Europe*. Kluwer Academic Publishers, Dordrecht, The Netherlands, 473 pp.

MORAVEC, F. 2001a. *Trichinelloid Nematodes Parasitic in Cold-blooded Vertebrates. Academia*, Prague, 430 pp.

MORAVEC, F. 2001b. Redescription and systematic status of *Capillaria philippinensis*, an intestinal parasite of human beings. *Journal of Parasitology*, **87**, 161–164.

MORAVEC, F. 2004. Some aspects of the taxonomy and biology of dracunculoid nematodes parasitic in fishes: a review. *Folia Parasitologica*, **51**, 1–13.

MORAVEC, F. 2006. Dracunculoid and Anguillicoloid nematodes parasitic in vertebrates. *Academia*, Prague, 634 pp.

MORAVEC, F. 2007. Some aspects of the taxonomy and biology of adult spirurine nematodes parasitic in fishes: a review. *Folia Parasitologica*, **54**, 239–257.

MORAVEC, F. & BARUŠ, V, 1991. Systematic status of *Thominx platyrrhinorum* Baruš, 1961 (Nematoda: Capillariidae). *Folia Parasitologica*, **38 (2)**, 155–162.

MORAVEC, F. & COSGROVE, G.E. 1982. *Pseudocapillaroides xenopi* gen. et sp.nov. from the skin of the South African clawed frog, *Xenopus laevis* Daud. (Nematoda: Capillariidae). *Revista de Zoologica Africa*, **96**, 129–137.

MORAVEC, F. & GONZÁLEZ-SOLÍS, D. 2007. Structure of the cephalic end of *Ascarophis Mexicana* (Nematoda: Cystidicolidae), as revealed by SEM. *Folia Parasitologica*, **54**, 155–156.

MORAVEC, F. & GUI TANG WANG, 2002. *Dentiphilometra monopteri* n. gen., n. sp. (Nematoda: Philometridae) from the abdominal cavity of the ricefield eel *Monopterus albus* in China. *Journal of Parasitology*, **88**, 961–966.

MORAVEC, F. & JUSTINE, J.-L., 2007. A new species of *Ascarophis* (Nematoda, Cystidicolidae) from the stomach of marine scorpaeniform fish *Hoplichtys citrinus* from a seamount off Chesterfield Islands, New Caledonia. *Acta Parasitologica*, **52**, 238–246.

MORAVEC, F. & KAISER, H. 1994. Description of *Cosmocerca longispicula* sp.nov. (Nematoda: Cosmocercidae), a parasite of dendrobatid frogs from Martinique, French Antilles. *Parasitology Research*, **80 (1)**, 29–32.

MORAVEC, F. & KØIE, M. 1987. *Daniconema anguillae* gen. et sp.n., a new nematode of a new family Daniconematidae fam.n. parasitic in European eels. *Folia Parasitologica*, **34**, 335–340.

MORAVEC, F. & LITTLE, M.D. 1988. *Granulinema* gen.n. a new dracunculoid genus with two new species (*G. carcharhini* sp.n. and *G. simile* sp.n.) from the bull shark, *Carcharhinus leucas* (Valenciennes), from Louisiana, USA. *Folia Parasitologica*, **35**, 113–120.

MORAVEC, F. & NAGASAWA, K. 1999. Morphology and taxonomy of *Salvelinema* species (Nematoda: Cystidicolidae), swimbladder parasites of Pacific area salmonids. *Folia Parasitologica*, **46**, 123–131.

MORAVEC, F. & PUYLAERT, F.A. 1970. On *Johnstonmawsonia africana* n. sp. (Nematoda: Rhabdochonidae) from the freshwater fish *Haplochromis schwetzi* of Angola. *Revue de Zoologie et de Botanique Africaines*, **82 (3/4)**, 306–314.

MORAVEC, F. & SCHOLZ, T. 1991. Observations on some nematodes parasitic in freshwater fishes in Laos. *Folia Parasitologica*, **38 (2)**, 163–178.

MORAVEC, F. & SEY, O. 1986. Three new nematode species from *Phrynomantis* spp. (Amphibia, Microhylidae) from Papua New Guinea. *Folia Parasitologica*, **33 (4)**, 343–351.

MORAVEC, F & SEY, O. 1988a. Nematodes of freshwater fishes from North Vietnam. Part 1. Camallanoidea and Habronematoidea. *Vestník Československé Spolecnosti Zoologické*, **52 (2)**, 128–148.

MORAVEC, F. & SEY, O. 1988b. Nematodes of freshwater fishes from North Vietnam. Part 2. Thelazioidea, Physalopteroidea and Gnathostomatoidea. *Véstník Československé Spolecnosti Zoologické*, **52 (3)**, 176–191.

MORAVEC, F. & SEY, O. 1988c. Nematodes of freshwater fishes from North Vietnam. Part 3. Cosmocercoidea, Seuratoidea, Atractoidea, Heterakoidea and Ascaridoidea. *Vestník Československé Spolecnosti Zoologické*, **52 (4)**, 250–265.

MORAVEC, F. & SEY, O. 1990. Some nematode parasites of frogs from Papua New Guinea and Australia. *Acta Societatis Zoologicae Bohemoslovacae*, **54**, 268–286.

MORAVEC, F. & SHAHAROM-HARRISON, F. 1989. *Paraphilometroides nemipteri* gen. et sp.n. (Nematoda: Philometridae) from the marine fish *Nemipterus peronii* (Valenciennes), from Malaysia. *Folia Parasitologica*, **36**, 345–350.

MORAVEC, F. & SPRATT, D.M. 1998. *Crocodylocapillaria longiovata* n. gen., n. sp. (Nematoda: Capillaridae) from the stomach of crocodiles in Australia and New Guinea. *Journal of Parasitology*, **84**, 426–430.

MORAVEC, F. & TARASCHEWSKI, H. 1988. Revision of the genus *Anguillicola* Yamaguti, 1935 (Nematoda: Anguillicolidae) of the swimbladder of eels, including descriptions of two new species, *A. novaezelandiae* sp.n. and *A. papernai* sp.n. *Folia Parasitologia*, **35 (2)**, 125–146.

MORAVEC, F. & THATCHER, V.E. 1996. *Myleusnema bicornis* gen et sp.n. (Nematoda: Kathlaniidae), an intestinal parasite of a freshwater serrasalmid fish, *Myleus ternetzi*, from French Guiana. *Folia Parasitologica*, **43**, 53–59.

MORAVEC, F. & THATCHER, V.E. 1997. *Procamallanus* (*Denticamallanus* subg. n.) *dentatus* n. sp. (Nematoda: Camallanidae) from the characid fish, *Bryconops alburnoides*, in the Brazilian Amazon. *Parasite*, **4**, 239–243.

MORAVEC, F. & VAN AS, J.G. 2004. Nematodes from the squeaker fishes *Synodontis nigromaculatus* and *S. vanderwaali* from the Okavango River, Botswana, including three new species. *Systematic Parasitology*, **59**, 169–187.

MORAVEC, F., KOHN, A. & FERNANDES, B.M.M. 1990. First record of *Raphidascaris* (*Sprentascaris*) *hypostomi* (Petter et Cassone, 1984) comb.n. and *R.* (*S.*) *mahnerti* (Petter et Cassone, 1984) comb.n. (Nematoda: Anisakidae) from Brazil with remarks on the taxonomic status of the genus *Sprentascaris* Petter et Cassone, 1984. *Folia Parasitologica*, **37 (2)**, 131–140.

MORAVEC, F., KOHN, A. & FERNANDES, B.M.M. 1992a. Three new species of oxyuroid nematodes, including two new genera, from freshwater catfishes in Brazil. *Systematic Parasitology*, **21**, 189–201.

MORAVEC, F., KOHN, A. & FERNANDES, B.M.M. 1992b. *Neoparaseuratum travassosi* n.g., n.sp. (Nematoda: Quimperiidae), a new parasite from thorny catfish *Pterododas granulosis* in Brazil. *Memórias do Instituto Oswaldo Cruz*, **87 (Suppl. I)**, 145–150.

MORAVEC, F., VIDAL, V. & SALGADO MALDONADO, G. 1992c. *Mexiconema cichlasomae* gen. et sp. n. (Nematoda: Daniconematidae) from *Cichlasoma* spp. (Pisces) from Mexico. *Folia Parasitologica*, **39**, 33–40.

MORAVEC, F., KOHN, A. & FERNANDES, B.M.M. 1993. Nematode parasites of fishes of the Paraná River, Brazil. Part 2. Seuratoidea, Ascaridoidea, Habronematoidea and Acuaroidea. *Folia Parasitologica,* **40 (2)**, 115–134.

MORAVEC, F., KOHN, A. & FERNANDES, B.M.M. 1994. Structure of the cephalic end of two little-known oxyurid genera, *Travnema* Pereira, 1938 and *Cosmoxynemoides* Travassos, 1949, parasites of fishes, as revealed by SEM. *Journal of Helminthology,* **68 (4)**, 319–322.

MORAVEC, F., MOLNÁR, K. & SZÉKELY, C. 1998. *Lucionema balatonese* gen. et sp. n., a new nematode of a new family Lucionematidae fam. n. (Dracunculoidea) from the swim-bladder of the European pikeperch, *Stizostedian incioperca* (Pisces). *Folia Parasitologica*, **45**, 57–61.

MORAVEC, F., SPANGENBERG, J.V. & FRASCA Jr, S. 2001. *Syngnathinema californiense* n. gen., n. sp. (Nematoda: Dracunculoidea) from the circulatory system in the Bay pipefish *Syngnathus leptorhynchus* in California. *Journal of Parasitology*, **87**, 1429–1432.

MORAVEC, F., SALGADO-MALDONADO,G. & AGUILAR-AGUILAR, R. 2002a. Two new nematodes, *Paraseuratoides ophisterni* gen. et sp. n. (Quimperiidae) and *Philometra ophisterni* sp. n. (Philometridae), from the swamp-eel *Ophisternon aenigmaticum* in Mexico. *Folia Parasitologica*, **49**, 109–117.

MORAVEC, F., SALGADO-MALDONADO,G. & AGUILAR-AGUILAR, R. 2002b. *Gibsonnema* nom. n., a new name for the nematode genus *Paraseuratoides* Moravec, Salgado-Maldonado et Aguilar-Aguilar, 2002. *Folia Parasitologia*, **49**, 217.

MORAVEC, F., SALGADO-MALDONADO, G. & AGUILAR-AGUILAR, R. 2002c. *Neophilometroides* n. gen. (Nematoda: Philometridae) for *Philometroides caudatus* Moravec, Scholz and Vivas-Rodríguez, 1995, with erection of Neophilometroidinae n. subfam. *Journal of Parasitology*, **88**, 774–777.

MORAVEC, F., CHARA, J. & SHINN, A.P. 2004. Two new nematodes, *Dentinema trichomycteri* n. g., n. sp. (Cosmocercidae) and *Procamallanus chimusensis* Freitas & Ibáñez, 1968 (Camallanidae), from catfishes *Trichomycterus* spp. (Pisces) in Colombia. *Systematic Parasitology*, **59**, 189–197.

MORAVEC, F., HANZELOVÁ, V. & GERDEAUX, D. 2007. New data on the morphology of *Comephoronema oschmarini* (Nematoda, Cystidicolidae), a little known gastro-intestinal parasite of *Lota lota* (Teleosti) in Palaearctic Eurasia. *Acta Parasitologica*, **52**, 135–141.

MORAVEC, F., MONTOYA-MENDOZA, J. & SALGADO-MALDONADO, G. 2008a. A new genus and species of philometrid (Nematoda) from the subcutaneous tissue of the crevalle jack, *Caranx hippos* (Osteichthyes), from the Southern Gulf of Mexico. *Journal of Parasitology*, **94**, 1342–1345.

MORAVEC, F., NIE, P. & WANG, G. 2008b. Redescription and systematic status of *Neocapillaria phoxini* Yu and Wang, 1994 (Nematoda: Capillariidae). *Folia Parasitologica*, **55**, 316–318.

MORAVEC, F., SANTOS, M.D. & BRASIL-SATO, M.C. 2008c. Redescription of *Cystidicoloides fischeri* based on specimens from Piranhas in Brazil, and erection of a new genus (Nematoda: Cystidicolidae). *Journal of Parasitology*, **94**, 889–897.

MORAVEC, F., CHARO-KARISA, H. & JIRKŮ, M. 2009. Philometrids (Nematoda: Philometridae) from Lake Turkana, Kenya, including two new species of *Philometra* and the erection of *Afrophilometra* gen. n. *Folia Parasitologica*, **56**, 41–54.

MOSGOVOI, A.A. 1951. *Opredelitel' Parasiticheskikh Nematod*, **2**, 405–566. [In Russian.]

MOSKVIN, S.N. 1939. [Diagnostics *in vivo* and the experimental therapy of Eucoleasis of silver foxes.] Thesis, 1939. (From Skrjabin, Shikhobalova & Orlov, 1957.)

MOZGOVOI, A.A. & SHAKHMATOVA, V.I. 1971. [Revision of nematodes of the suborder Ascaridata Skrjabin, 1915.] *Trudy Gel'mintologicheskoi Labratorii*, **22**, 129–145.

MULLER, R.L. 1971. *Dracunculus* and dracunculiasis. *Advances in Parasitology*, **9**, 73–151.

MUÑOZ, G., GONZÁLEZ, M.T., & GEORGE-NASCIMENTO, M. 2004. *Similascarophis* n. gen. n. spp. (Nematoda: Cystidicolidae) parasitizing marine fishes off the Chilean coast. *Journal of Parasitology*, **90**, 823–834.

NADTOCHIĬ, E. 1977. [The erection of a new genus, *Heligmophera* n.g. (Nematoda: Heligmosomatidae) from the Manchurian Zokor (*Myospalax psilurus*) from the Primorsk region.] *3rd National Conference of Parasitology*, Albena, Bulgaria, 12–14 October 1977, Abstract, Summaries, pp. 134–135.

NAIDU, T.S.V. & THAKARE, V.K. 1981. On two new nematodes from fishes and amphibians of Vidarbha Region, Maharashtra State, India. *Rivista di Parassitologia*, **42 (3)**, 415–424.

NAQVI, N.H. & GUPTA, S.P. 1987 (publ. 1989). Nematode parasites of vertebrates. III. On two new species of the *Neoichthyostrongylus* gen.nov (Ichthyostrongylidae Yamaguti, 1961) from marine fishes of Panjim, Goa, India. *Rivista di Parassitologia*, **4**, 191–195.

NAVONE, G.T., SURIANO, D.M. & PUJOL, C.A. 1990. *Neohilgertia venusti* gen.n. sp.n. (Nematoda: Oxyuridae) from *Thylamys venustus cinderellus* (Thomas) Reig, Kirsch & Marshall, 1985 (Marsupialia: Didelphidae) in Burruyacu, Tucuman, Argentina – systematic position and possible evolution. *Memórias do Instituto Oswaldo Cruz*, **85**, 185–189.

NODA, R. & NAGATA, S. 1976. *Struthiofilaria megalocephala* gen. et sp.n. (Nematoda: Filarioidea) from the body cavity of an Ostrich. *Bulletin of University of Osaka Prefecture, ser. B (Agriculture & Biology)*, **28**, 1–4.

OHBAYASHI, M., KAMIYA, M. & BHAIBULAYA, M. 1979. Studies on the parasite fauna of Thailand. I. Two new metastrongylid nematodes, *Angiostrongylus siamensis* sp.n. and *Thaistrongylus harinasutai* gen. et sp.n. (Metastrongyloidea: Angiostrongylidae) from wild rats. *Japanese Journal of Veterinary Research*, **27**, 5–10.

OKULEWICZ, A. & PEREC, A. 2004. Evolution and systematics of nematodes based on molecular investigation. *Wiadomosci Parazytologicnze*, **50**, 101–108. [from *Helminthological Abstracts*.]

ORIHEL, T.C. & EBERHARD M.L. 1982. *Mansonella ozzardi:* a redescription with comments on its taxonomic relationships. *American Journal of Tropical Medicine and Hygiene*, **31 (6)**, 1142–1147.

ORTLEPP, R.J. 1922. A new species of *Oesophagostomum* (*O. xeri* n.sp.) from a rodent (*Xerus setosus*). *Proceedings of the Zoological Society of London*, **June**, 461–469.

OSHMARIN, P.G & PARUKHIN, A.M. 1963. [Trematodes and nematodes of birds and mammals of the Sikhote-Alinsk preserve.] *Trudi Sikhote–Alinkskogo Gosudarstvennogo Zapovednika*, **3**,121–181.

OW YANG, C.K. 1967. Systematics and life history studies on nematodes of Malayan rodents. PhD Thesis, University of London.

OW YANG, C.K., DURETTE-DESSET, M.-C. & OHBAYASHI, M. 1983. Sur les nematodes parasites de rongeurs de malaisie. II. Les Trichostrongyloidea. *Annales de Parasitologie Humaine et Comparée*, **58**, 467–492.

PAGGI, L., NASCETTI, G., ORECCHIA, P., MATTIUCCI, S. & BULLINI, L. 1985. Biochemical taxonomy of ascaridoid nematodes. *Parassitologia*, **27** (**1–2**), 105–112.

PARUKHIN, A.M. 1978. [Systematic position of nematodes from the genus *Sobolevicephalus* Parukhin, 1964.] *Vestnik Zoologii*, **2**, 92–93.

PEARSE, A.S. 1933. Parasites of Siamese fishes and crustaceans. *Journal of the Siam Society, Natural History Supplement*, **9**, 179–191.

PETROVA, K. 1987. [Species composition and morphology of the genus *Amidostomum* Railliet et Henry, 1909 (Strongylata: Amidostomatidae) in Bulgaria.] *Khelmintologiya*, **24**, 53–72.

PETTER, A.J. 1976. Keys to genera of the superfamily Oxyuroidea No. 4: *CIH Keys to the Nematode Parasites of Vertebrates*. Edited by R.C. Anderson, A.G. Chabaud, S. Willmott. Commonwealth Agricultural Bureaux, Farnham Royal, UK, 30 pp.

PETTER, A.J. 1979a. Essai de classification de la sous-famille des Procamallaninae (Nematoda, Camallanidae). *Bulletin du Muséum National d'Histoire Naturelle, Sér. 4 Sect. A*, **1**, 219–239.

PETTER, A.J. 1979b. Trois Nématodes parasites de poissons en Malaisie. *Bulletin du Muséum National d'Histoire Naturelle*, **1**, 585–596.

PETTER, A.J. 1979c. Essai de classification de la sous-famille des Camallaninae (Nematoda, Camallanidae). *Bulletin du Muséum National d'Histoire Naturelle, 4th Série 1, Sect. A*, **4**, 991–1008.

PETTER, A.J. 1984. Nematodes de poissons du Paraguay II. Habronematoidea (Spirurida) description de 4 espéces nouvelles de la famille des Cystidicolidae. *Revue Suisse de Zoologie*, **9** (**4**), 935–952.

PETTER, A.J. 1987. Nématodes des poissons de l'Equateur. *Revue Suisse de Zoologie*, **94** (**1**), 61–76.

PETTER, A.J. 1989. Nématodes de poissons du Paraguay. V. Cucullanidae. Description de deux espéces nouvelles et redéfinition du genre *Neocucullanus* Travassos *et al. Revue Suisse de Zoologie*, **96** (**3**), 591–603.

PETTER, A.J. 1994. [Nematode parasites of Paraguayan Fishes VIII. Oxyuroidea. *Spinoxyuris oxydoras* gen. n., sp.n.] *Revue Suisse de Zoologie*, **101** (**3**), 761–769.

PETTER, A.J. & BROCHIER, B. 1989. *Cercogylus africanus* n.g., n.sp. (Angiostrongylidae, Metastrongyloidea, Nematoda), parasite du primate africain *Cercocebus galeritus chrysogaster. Systematic Parasitology*, **13**, 197–200.

PETTER, A.J. & CASSONE, J. 1984. Nématodes de poissons du Paraguay; I. Ascaridoidea: *Sprentascaris* n.gen. *Revue Suisse de Zoologie*, **91**, 617–634.

PETTER, A.J. & DOUGLASS, J.F. 1976. Études des populations d'Oxyures du côlon des *Gopherus* (Testudinidae). *Bulletin du Muséum National d'Histoire Naturelle, Série 3*, **271** (**389**), 731–768.

PETTER, A.J. & KØIE, M. 1993. *Fellicola longispiculus* gen.nov., sp.nov. (Nematoda, Rhabdochonidae) from the gall bladder of the marine fish *Coryphaenoides rupestris. Annales de Parasitologie Humaine et Comparée*, **68**, 226–228.

PETTER, A.J. & PLANELLES, G. 1986. Un nouveau genre de Dracunculidae (Nematoda) parasite d'Amphibien. *Bulletin du Muséum National d'Histoire Naturelle*, **8** (**1**), 123–132.

PETTER, A.J. & QUENTIN, J.-C. 1976. Keys to genera of the superfamily Oxyuroidea. No. 4. *CIH Keys to the Nematode Parasites of Vertebrates* (edited by R.C. Anderson, A.G. Chabaud and S. Willmott). Commonwealth Agricultural Bureaux, Farnham Royal, UK. 30 pp.

PETTER, A.J., CHERMETTE, R. & VASSART, M. 1988. *Cagourakis dorsalata* n.g., n. sp., Heterakidae (Nematoda) parasite du Cagou (*Rhynochtes jubatus*) (Ralliformes) en Nouvelle-Calédonie. *Bulletin du Muséum d'Histoire Naturelle, Paris 4e Sér. 10 Sect. A*, **4**, 675–683.

PETTER, A.J., PARADIZNIK, V., RADUJKOVIC, B.M. & CASSONE, J. 1991. Redescription d'*Acanthocheilus rotundatus* (Ascaridoidea, Nematoda) parasite de requins (Pleurotremata, Selachii). Considérations sur les affinités du genre *Acanthocheilus* et conclusions taxonomiques. *Annales de Parasitologie Humaine et Comparée*, **66** (5), 187–194.

PILAR, F.R. 2002. Estudio faunístico y sistemático de helmintos de aves Canarias. Departamento de Parasitología, Ecología y Genética, Universidad de la Laguna, Tenerife, Spain, [Doctoral thesis] 295 pp.

PINTO, C. 1935. Sobre o genero. *Bunostomum* Railliet, 1902 Nematoda: Ancylostomidae. *O Jornal* (Vida dos Campos) Rio de Janeiro, **17**, No. 5070, 11.

PINTO, R.M. & GOMEZ, D.R. 1985. Nematodes of Amazonian birds, with a description of *Hoazinstrongylus amazonensis* n. gen. n.sp. (Trichostrongylidae, Libyostrongylinae). *Memórias do Instituto Oswaldo Cruz*, **80** (**2**), 213–217.

POINAR, G.O. 1982. *The Natural History of Nematodes*. Prentice Hall, Upper Saddle River, New Jersey, 323 pp.

POLJANSKY, Y.I. 1952. Some new or little known parasitic nematodes of sea fish. *Trudy Zoologicheskogo Instituta., Akademiya Nauk SSSR*, **12**, 133–147. (Translated from Russian by Fisheries Research Board of Canada, Translation Series, No. 32.)

POPOVA, T.I. 1958. *Essentials of Nematology. Volume VII. Strongyloids of Animals and Man. Trichonematidae* (edited by K.I. Skrjabin). Academy of Sciences of the USSR, Moscow. Israel program for scientific translations, Jerusalem, 1965, x + 414 pp.

PORTER, A. 1948. Report of the honorary parasitologist for 1946. *Proceedings of the Zoological Society of London*, **17**, 673–674.

PUYLAERT, F.A. 1973. *Rhabdochonidae* parasites de poissons Africains d'eau douce et discussion sur la position systématique de ci group (Vermes – Nematoda). *Revue Zoologique et Botanique Africaines*, **87**, 647–665.

QUENTIN, J.C. & BEVERIDGE, I. 1986. Comparative morphogenesis of the cephalic structures of the acuariid nematodes *Stammerinema soricis* (Tiner, 1951), *Antechiniella suffodiax* (Beveridge & Barker, 1975) n.g., n.comb. and *Skrjabinoclava thapari* (Teixeira de Freitas, 1953.) *Systematic Parasitology*, **8**, 163–171.

QUENTIN, J.C., BETTERTON, C. & KRISHNASAMY, M. 1979. Oxyures nouveaux on peu connus, parasites de primates, de rongeur et de Dermopteres en Malaisie. Création du sous-genre *Colobenterobius* n. subgen. *Bulletin du Muséum National Histoire Naturelle, Paris, 4e série, 1, (Sect. A)*, **4**, 1031–1050.

RAILLIET, A. 1925. Les helminthes des animaux domestiques et de l'homme en Indochine. (2e partie). *Bulletin de la Société Zoologique de France*, **50**, 7–26.

RAINA, M.K. & RAINA, P. 1985 (publ. 1986). On two capillarid nematodes from shrew in Kashmir with a note on the synonymy of *Skrjabinocapillaria* Skarbilovitsch, 1946. *Indian Journal of Helminthology*, **37** (**2**), 109–115.

RAO, R. 1977. On a new genus *Stewartia* n.g. (Nematoda) from amphibian hosts in India with *Stewartia macintoshii* (Steward, 1914) as type and a second new species *Stewartia chabaudi. Abstracts of the 1st National Congress of Parasitology*, Baroda, India, 24–26 February 1977, Indian Society for Parasitology, p. 37.

RAO, R. 1978. On *Narsingiella narsingi*, a new genus and species of aspidoderid nematode from *Bufo viridis* found in Berhampur, India. *Proceedings of the Helminthological Society of Washington*, **45**, 246–248.

RAO, R. 1979. On a new genus and a new species *Amphibiogoezia spinosoma* n.g., n.sp. of the family Goeziidae Skrjabin et Karokhin, 1945 from frogs in Bangalore, India. *Rivista de Parassitologia*, **40**, 5–8.

RAO, R. 1989. New genus *Stewartia* and a new subfamily Stewartiinae with *Stewartia macintoshii* as type species and *Stewartia chabaudi* sp nov. as the second species. *Indian Journal of Helminthology* **41** (**1**), 60–68.

RATHORE, G.S. & NAMA, H.S. 1985. Nematode parasites of birds found in Rajasthan with the description of a new genus. *Journal of Research of the Assam Agricultural University*, **6**, 38–42.

RAUSCH, R.L. & RAUSCH, V.R. 1983. *Maseria vespertilionis* n.g., n.sp. (Dorylaimina: Muspiceidae), a nematode from nearactic bats (Vespertilionidae). *Annales de Parasitologie Humaine et Comparée*, **58**, 361–376.

RAUTELA, A.S. & MALHOTRA, S.K. 1985. *Kheraia himalayii* n.g., n.sp. from a hill bird, *Sturnus pagodarum* (Gmelin). *Indian Journal of Helminthology*, **2**, 89–93.

RIBU, D.L. AND LESTER, R.J.G. 2004. *Moravecia australiensis* n. g., n. sp. (Dracunculoidea: Guyanemidae) from the gills of the green porcupine fish *Tragulichthys jaculiferus* (Cuvier) in Australia. *Systematic Parasitology*, **57**, 59–65.

RICCI, M. 1939. Helminthes. *Missiore Biologica nel Paes dei Borana, Roma*, **3**, 342–449.

RICCI, M. 1987 (publ. 1988). Parassiti gel gen. *Speleomantes* (Amphibia, Urodela: Plethodontidae) in Italia. *Rivista di Parassitologia*, **4** (**48**), (**1**), 5–25.

RICHARDSON, J.P.M. & ADAMSON, M.L. 1988. *Megalobatrachonema (Chabaudgolvania) waldeni* n.sp. (Nematoda: Kathlaniidae) from the intestine of the north western salamander, *Ambystoma gracile* (Baird). *Canadian Journal of Zoology*, **66** (**7**), 1502–1505.

RICHARDSON, J.P.M. & ADAMSON, M.L. 1990. Phylogenetic analysis of *Megalobatrachonema* Yamaguti, 1941 (Ascaridida: Cosmocercoidea: Kathlaniidae): reinstatement of *Chabaudgolvania* Freitas, 1958. *Canadian Journal of Zoology*, **68**, 531–538.

RICKARD, L.G. & HOBERG, E.P. 2000. Reassignment of *Lamanema* from Nematodirinae to Molineinae (Nematoda: Trichostrongyloidea). *Journal of Parasitology*, **86**, 647–650.

RIVOLTA, S. 1884. La natura parassitica elle piaghe estive ogli effeti morbosi di una specie di filaria che si puodenominaire *Dermofilaria irritans. Geornale di Anatomoa Fisiologia e Patologia di Animali, Pisa*, **16**, 128–134.

R'KHA, S. & DURETTE-DESSET, M.-C. 1991. Trois espèces (dont deux nouvelles) de Nématodes Trichostrongyloïdes coparasites de *Proechimys semispinosus* en Colombie: description de *Justinema* n. gen. *Bulletin du Muséum d'Histoire Naturelle, 4e Série (Zoologie) (1990)*, **12**, 555–562.

ROCA, V. & GARCIA-ADELL, G. 1988. [Description of *Skrjabinelazia pyrenaica* n.sp. (Nematoda: Seuratidae) and a new diagnosis proposed for the genus *Skrjabinelazia*.] *Annales de Parasitologie Humaine et Comparée*, **63** (**6**), 414–419.

RODRIGUES, H. DE O., PINTO, R.M. & NORONHA, D. 1991. Key to the species of Brazilian *Procamallanus* with general considerations (Nematoda, Camallanoidea). *Mémorias do Instituto Oswaldo Cruz*, **86** (**1**), 107–113.

ROMANOVITCH, M. 1915. Quelques helminthes de renne (*Tarandus rangifer*). *Compte Rendu des Séances de la Societé de Biologie*, **78**, 451–453.

ROUND, M.C. 1962. A new species of the genus *Strongylus* Müller, 1780 from the black rhinoceros, *Diceros bicornus* L. and a note on other species occurring in Kenya. *Journal of Helminthology*, **36**, 189–200.

RUDOLPHI, C.A. 1819. Genus VI. *Spiroptera. Entozoorum Synopsis cui Accedunt Mantissa Duplex et Indices Locupletissimi.* Sumtibus Augusti Rücker, Berlin, pp. 235–255.

RYZHIKOV, K.M. and SONIN, M.D. 1981. [The classification of nematode parasites of vertebrates.] *Parazitologiya* **15** (**6**), 510–518.

RYZHIKOV, K.M., SHARPILO, V.P. & SHEVCHENKO, N.N. 1980. [*Helminths of Amphibia in the USSR*.] Izdat 'Nauka', Moscow, 279 pp.

SAHAY, U. & PRASAD, D. 1993. Some comments on the genus *Pseudoproleptus* Khera, 1953 along with a revised key to the species. *Bioved*, **4**, 105–110.

SAKAMOTO, T. & MALGOR, R. 1995. *Cardiostrongylus sikae* n.gen., n.sp. (Molineoidea; Dictyocaulidae) from the heart muscle of sika deer in Japan. *Japanese Journal of Parasitology*, **44**, (**6**), 462–469.

SANDOSHAM, A.A. 1950. On *Enterobius vermicularis* (Linnaeus, 1758) and some related species from primates and rodents. *Journal of Helminthology*, **24**, 171–204.

SARWAR, M.M. 1957. Some observations on the taxonomy of dictyocaulids. *Acta Tropica*, **14**, 228–230.

SAUNDERS, D.C. 1955. The classification of microfilariae in birds. *Avifilaris tyrannidarum* and *A. fringillidarum*, two new species. *Transactions of the American Microscopical Society*, **74**, 37–45.

SCHACHER, J.F. 1975. *Waltonella* nom. n. for subgenus *Waltonia* (Nematoda: Filarioidea) Schacher and Crans, 1973 preoccupied by *Waltonia* Davidson, 1850 (Brachiopoda). *Journal of Parasitology*, **61**, 58.

SCHACHER, J.F. & CRANS, W.J. 1973. *Foleyella flexicauda* sp.n. (Nematoda: Filarioidea) from *Rana catesbiana* in New Jersey. *Journal of Parasitology*, **59 (4)**, 685–691.

SCHMIDT, G.D. & KUNTZ, R.E. 1971. Nematode parasites of Oceanica. XI. *Madelinema angelae* gen. et sp. n. and *Inglisonema mawsonae* sp. n. (Heterakoidea: Inglisonematidae) from birds. *Journal of Parasitology*, **57**, 479–484.

SCHULZ, R.E. 1927. [Die familie Physalopteridae Leiper, 1908 (Nematodes) und die Prinzipien ihrer Klassifikation. [Papers on helminthology presented to Professor Konstantin Ivanovich Skryabin, by his students. 10-year jubilee Vol.] Izdanie Avtorov, Moscow, pp. 287–312.

SHAKHMATOVA, V.I. 1971. [*Cloeoascaris skrjabini* n.sp. in a new subfamily Cloeoascarinae n.subfam (Ascaridata, Anisakidae)] *Trudy Gel'mintologicheskoi Laboratorii* (*Teoreticheskie Voprosy Obshcheĭ Gel'mintologii*), **22**, 224–227.

SHAMIM, S. & DESHMUKH, P.G. 1980 (publ. 1981). *Gallioxynema thapari* n.gen., n.sp. from *Gallus gallus domesticus*. *Indian Journal of Helminthology*, **32**, 112–116.

SHEENKO, P.S. 1985. [The systematics of parasitic nematodes of the genus *Porrocaecum* from fishes with description of a new genus *Mariporrocaecum*.] *Biologiya Morya, Vladivostok*, **6**, 18–23. [In Russian.]

SINGH, S.N. 1934. *Gastronodus strasseni* g. et sp.n. A new nematode parasite of *Crocidura coerulea* – the common musk-shrew or, as it is usually called, must-rat (Vernacular: Chachundar). *Current Science*, **11**, 287.

SKRJABIN, K.I. 1942. The ways of the phylogenetic evolution of nematodes of the family Pseudaliidae, parasitic of the auditory apparatus, circulatory system and respiratory organs of marine mammals. *Doklady Akad. Nauk SSSR*, **37**, 35–40.

SKRJABIN, K.I. (ed.). 1949. *Key to Parasitic Nematodes. Vol. 1. Spirurata and Filariata*. Academy of Sciences of the USSR, Moscow. (Israel Program for Scientific Translation Ltd., 1969.) xi + 497 pp.

SKRJABIN, K.I. & IVASHKIN, V.M. 1973. [Composition of the superfamily Gnathostomatoidea and its raising to the rank of suborder.] *Trudi Gel'mintologicheskoi Laboratorii*, **23**, 144–148.

SKRJABIN, K.I. & MOZGOVOI, A.A. 1973. [Erection of a new nematode superfamily Ascaridoidea n.superf.] *Parazitologiya*, **7 (6)**, 518–520.

SKRJABIN, K.I. & SHIKHOBALOVA, N.P. 1948. [*Filariae of animals and man*.] Ogiz-Sel'khozgiz, Moscow, 608pp. [In Russian.]

SKRJABIN, K.I. & SHIKHOBALOVA, N.P. 1951. [Essentials of nematodology. Vol. VIII. Oxyuroidea of animals and man. Part I (edited by K.I. Skrjabin, N.P. Shikhobalova & E.A. Lagodovskaya).]. Academy of Sciences of the USSR, Moscow, 1960, 557 pp. (also translation 1974).

SKRJABIN, K.I. & SOBOLEV, A.A. 1963. [Essentials of nematodology. Vol. XI. Spirurata of animals and man and the diseases caused by them. Part I. Spiruroidea (edited by K.I. Skrjabin).] Academy of Sciences of the USSR, Moscow, 511 pp.

SKRJABIN, K.I. & SOBOLEV, A.A. 1964. [Essentials of nematodology. Volume XII. Spirurata of animals and man and the diseases caused by them. Part II.Physalopteroidea (edited by K.I. Skrjabin).] Academy of Sciences of the USSR, Moscow, 334 pp.

SKRJABIN, K.I., SHIKHOBALOVA, N.P. & MOZGOVOI, A.A. 1951. [Descriptive catalogue of parasitic nematodes. Vol. 2. Oxyurata and Ascaridata.] *Opredelitel' Paraziticheskikh* Nematod, 7–403 [In Russian.]

SKRJABIN, K.I., SHIKHOBALOVA, N.P. & SCHULZ, R.S. 1954a. [Essentials of nematodology. Vol. IV. Dictyocaulidae, Heligmosomatidae and Ollulanidae of animals (edited by K.I. Skrjabin).] Academy of Sciences of the USSR, Moscow, 323 pp.

SKRJABIN, K.I., SHIKHOBALOVA, N.P. & SCHULZ, R.S. 1954b. [Essentials of nematodology. Vol. III: Trichostrongylids of animals and man (edited by K.I. Skrjabin).] Academy of Sciences of the USSR, Moscow, 704 pp.

SKRJABIN, K.I. SHIKHOBALOVA, N.P. & ORLOV, I.V. 1957. [Essentials of nematodology. Vol. VI. Trichocephalidae and Capillariidae of animals and man and the diseases caused by them (edited by K.I. Skrjabin).] Academy of Sciences of the USSR, Moscow [Transl. 1971], 587 pp.

SKRJABIN, K.I., SHIKHOBALOVA, N.P. & LAGODOVSKAYA, E.A. 1960. [Essentials of Nematodology. Volume VIII. Oxyuroidea of animals and man. Part I. Edited by K.I. Skrjabin.] Academy of Sciences of the USSR, Moscow. 557 pp. (also translation 1974).

SKRJABIN, K.I., SHIKHOBALOVA, N.P. & LAGODOVSKAYA, E.A. 1961. [Essentials of Nematodology. Vol. X. Oxyuroidea of animals and man. Part 2 (edited by K.I. Skrjabin).] Academy of Sciences of the USSR, Moscow (Israel Program for Scientific Translation Ltd., 1974.), v + 460 pp.

SKRJABIN, K.I., SHIKHOBALOVA, N.P. & LAGODOVSKAYA, E.A. 1964. [Essentials of Nematodology. Vol. XIII. Oxyuroidea of animals and man. Part 3 (edited by K.I. Skrjabin).] Academy of Sciences of the USSR, Moscow (Israel Program for Scientific Translation Ltd., 1976.), v + 486 pp.

SKRJABIN, K.I., SOBOLEV, A.A. & IVASHKIN, V.M. 1967. [Essentials of Nematodology. Vol. XVI. Spirurata of animals and man and the diseases caused by them. Part 4. Thelazioidea (edited by K.I. Skrjabin).] Academy of Sciences of the USSR, Moscow, 624 pp. (also translation 1971).

SMALES, L.R. 1982a. A new genus and three new species of nematode parasites (Strongylidea: Cloacininae) from Macropodid marsupials from Papua New Guinea. *Systematic Parasitology*, **4**, 361–371.

SMALES, L.R. 1982b. *Dorcopsistrongylus* new genus (Nematoda: Strongyloidea) from the grey scrub wallaby *Dorcopsis veterum* Lesson, 1827 from Papua New Guinea. *Transactions of the Royal Society of South Australia*, **106**, 31–34.

SMALES, L.R. 1983. Helminth type specimens in the South Australian Museum 1. Nematoda. *Records of the South Australian Museum*, **18**, 385–413.

SMALES, L.R. 1992. A survey of the helminths of *Rattus sordidus*, the canefield rat, together with a description of *Ancistronema coronatum* n.g., n.sp. (Nematoda: Chabertiidae). *Systematic Parasitology*, **22**, 73–80.

SMALES, L.R. 1994. A taxonomic revision of *Labiostrongylus (Labiostrongylus)* (Yorke & Maplestone, 1926) and *Labiostrongylus (Labiomultiplex)* n.subg. (Nematoda: Cloacinidae) from macropodid and potoroid marsupials. *Systematic Parasitology*, **29**, 181–216.

SMALES, L.R. 1995a. A revision of the subgenus *Labiostrongylus* (*Labiosimplex*) (Nematoda: Cloacinidae) from macropodid marsupials, with descriptions of twelve new species and a key to the species of the subgenus. *Invertebrate Taxonomy*, **9**, 181–242.

SMALES, L.R. 1995b. A revision of the genus *Tikusinema* (Nematoda: Acuarioidea) with the description of a new species from the false water rat *Xeromys myoides* from Queensland. *Transactions of the Royal Society of South Australia*, **119 (2)**, 89–94.

SMALES, L.R. 1997a. A revision of the genus *Potorostrongylus* Johnston & Mawson, 1939 (Nematoda: Cloacinidae) from potoroos and bettongs (Marsupialia: Potoroidae), with the description of two new species. *Systematic Parasitology*, **38**, 185–196.

SMALES, L.R. 1997b. A revision of the Echinonematinae (Nematoda: Seuratidae) from bandicoots (Marsupialia: Paramelidae). *Transactions of the Royal Society of South Australia*, **121 (1)**, 1–27.

SMALES, L.R. 1999a. *Chabaudechina* n.g. (Nematoda: Seuratidae), with the description of two new species from dasyurid marsupials from Australia. *Systematic Parasitology*, **44**, 145–151.

SMALES, L.R. 1999b. *Bainechina rossiae* gen. et sp. nov. (Nematoda: Seuratidae) from Australian dasyurid marsupials. *Transactions of the Royal Society of Southern Australia*, **123 (1)**, 37–41.

SMALES, L.R. 2000. *Durettechina beveridgei* n.g., n.sp. (Nematoda: Seuratidae) from *Antechinus* spp. (Dasyuridae: Marsupialia) from Australia. *Systematic Parasitology*, **45 (1)**, 25–28.

SMALES, L.R. 2002. A cladistic analysis of the tribe Labiostrongylinea Beveridge, 1983 (Nematoda: Cloacinidae) parasitic in macropodoid marsupials (Marsupialia: Macropodoidea), with a redescription of *Parazoniolaimus collaris* Johnston and Mawson, 1939. *Systematic Parasitology*, **51**, 179–197.

SMALES, L.R. 2004. *Spirosprattus scyphiformis* n.g., n.sp. (Nematoda: Spirurida), from the Cape York Rat, *Rattus leucopus* (Gray, 1867) (Rodentia: Muridae), in Cape York, Australia. *Comparative Parasitology*, **71**, 184–189.

SMITH, J.D. 1984. Taxonomy of *Raphidascaris* spp. (Nematoda, Anisakidae) of fishes, with a redescription of *R. acus* (Bloch, 1772). *Canadian Journal of Zoology*, **62**, 685–694.

SOKOLOV, S.G. 2006. [Dracunculoid nematodes (Spirurida: Dracunculoidea) of fishes in the Volga River delta.] [In Russian.] *Parazitologiya*, **40**, 355–362.

SOLEIM, Ø. 1984. A synopsis of the genera *Thynnascaris* and *Contracaecum* (Nematoda, Ascaridoidea) with an emendation of the generic definitions. *Acta Parasitica Polonica*, **29 (9–19)**, 85–96.

SOLOV'EVA, G.F. 1987. [*Crenatobronema* gen.nov. (Spinitectidae), a new genus of nematodes from fishes of the Pacific.] *Parazitologiya*, **21**, 591–594.

SONIN, M.D. 1971. [Phylogenetic interrelationships of different groups of Filariata and a modification of the classification of nematodes of this suborder.] *Trudi Gel'mintologicheskoi Laboratorii*, **22**, 162–181.

SOOD, M.L. & PARSHAD, V.R. 1972. *Isospirura meltadi* gen. et sp. nov. from rodents in Punjab. Annual Number 1972. *National Academy of Sciences, India*, pp.123–124.

SOOTA, T.D. & CHATURVEDI, Y. 1971. The nematoda fauna of Goa. *Indian Journal of Helminthology*, **23**, 11–20.

SPECIAN, R.D., UBELAKER, J.E. & DAILEY, M.D. 1975. *Neoleptus* gen.n. and a revision of the genus *Proleptus* Dujardin, 1845. *Proceedings of the Helminthological Society of Washington*, **42**, 14–21.

SPENCER JONES, M.E. & GIBSON, D.I. 1987. A list of old and recently erected genus-group names not included in the *CIH Keys to Nematode Parasites of Vertebrates and Invertebrates. Sytematic Parasitology*, **9**, 125–136.

SPRATT, D.M. 1981. *Antechinostrongylus disgubernaculus* gen. et sp.n. (Nematoda: Angiostrongylidae) from the marsupial mouse, *Antechinus swainsonii* (Marsupialia: Dasyuridae). *Journal of Parasitology*, **67**, 90–93.

SPRATT, D.M. 1984. Further studies of lung parasites (Nematoda) from Australian marsupials. *Australian Journal of Zoology*, **32**, 283–310.

SPRATT, D.M., BEVERIDGE, I. & WALTER, E. 1991. A catalogue of Australian monotremes and marsupials and their recorded helminth parasites. *Records of the South Australian Museum Monograph Series*, **1**, 105 pp.

SPRENT, J.F.A. 1978a. Ascaridoid nematodes of amphibia and reptiles: *Paraheterotyphlum. Journal of Helminthology*, **52**, 163–170.

SPRENT, J.F.A. 1978b. Ascaridoid nematodes of amphibians and reptiles: *Gedoelstascaris* n.g. and *Ortleppascaris* n.g. *Journal of Helminthology*, **52**, 261–282.

SPRENT, J.F.A. 1978c. Ascaridoid nematodes of amphibians and reptiles: *Polydelphis, Travassosascaris* n.g. and *Hexametra. Journal of Helminthology*, **52**, 355–384.

SPRENT, J.F.A. 1979a. Ascaridoid nematodes of amphibians and reptiles: *Multicaecum* and *Brevicaecum. Journal of Helminthology*, **53**, 91–116.

SPRENT, J.F.A. 1979b. Ascaridoid nematodes of amphibians and reptiles: *Terranova. Journal of Helminthology*, **53**, 265–282.

SPRENT, J.F.A. 1980a. Ascaridoid nematodes of Sirenians – the Heterocheilinae redefined. *Journal of Helminthology*, **54**, 309–327.

SPRENT, J.F.A. 1980b. Ascaridoid nematodes of amphibians and reptiles: *Angusticaecum* and *Krefftascaris* n.g. *Journal of Helminthology*, **54**, 55–73.

SPRENT, J.F.A. 1982. Ascaridoid nematodes of South American mammals, with a definition of a new genus. *Journal of Helminthology*, **56**, 275–295.

SPRENT, J.F.A. 1983a. Ascaridoid nematodes of amphibians and reptiles: *Typhlophorus, Hartwichia* and *Trispiculascaris. Journal of Helminthology*, **57**, 179–189.

SPRENT, J.F.A. 1983b. Observations on the systematics of Ascaridoid nematodes *Systematics Association Special Vol.*, **22** (edited by A.R. Stone, H.M. Platt & L.F. Khalil), 303–319.

SPRENT, J.F.A. 1983c. Ascaridoid nematodes of amphibians and reptiles: *Freitasascaris* n.g. *Journal of Helminthology*, **57**, 283–290.

SPRENT, J.F.A. 1985a. Ascaridoid nematodes of amphibians and reptiles: *Orneoascaris. Annales de Parasitologie Humaine et Comparée*, **60 (1)**, 33–55.

SPRENT, J.F.A. 1985b. Ascaridoid nematodes of amphibians and reptiles: *Seuratascaris* n.g. *Annales de Parasitologie Humaine et Comparée*, **60**, 231–246.

SPRENT, J.F.A. 1985c. Ascaridoid nematodes of amphibians and reptiles: *Raillietascaris* n.g. *Annales de Parasitologie Humaine et Comparée*, **60**, 601–611.

SPRENT, J.F.A. 1990a. Some ascaridoid nematodes of fishes: *Paranisakis* and *Mawsonascaris* n.g. *Systematic Parasitology*, **15**, 41–63.

SPRENT, J.F.A. 1990b. Some ascaridoid nematodes of fishes: Heterocheilinae. *Systematic Parasitology*, **16** (2), 149–161.

SRIVASTAVA, A.B. & GUPTA, S.P. 1976 (publ.1978). Nematode parasites of marine fishes. On some nematode parasites of marine fishes from Pentkota, Puri, Orissa. *Indian Journal of Helminthology*, **28**, 148–155.

STILES, C.W. 1907. *Agamofilaria georgiana* n.sp., an apparently new roundworm parasite from the ankle of a negress. *Hygienic Laboratory Bulletin*, **34**, 9–30.

STILES, C.W. & HASSALL, A. 1920. Index-catalogue of medical and veterinary zoology subjects: roundworms (Nematoda, Gordiacea and Acanthocephala) and the diseases they cause. United States Public Health Service Laboratory. Bulletin No. 114. Government printing office, Washington, DC. 886 pp.

STRAND, E. 1942. Miscellanea nomenclatorica zoologica et palaeontologica. *Folia Zoologica et Hydrobiologica*, **11**, 387–388.

STROM, J. 1937. [A new roundworm of birds, *Cardiofilaria pavlovskyi* n.gen., n.sp.] *Trudy Soveta Polzucheniyu Proizvoditelnuikh Sil, Seriya Turkmenskaya*, **9**, 217–221. [In Russian.]

SUTTON, C.A. & DURETTE-DESSET, M.-C. 1987. Contribution à la connaissance de la faune parasitologique argentine. XVIII. *Lagostonema ecasiense* n.gen., n.sp. (Trichostrongyloidea, Nematoda), parasite de *Lagostomus maximus* (Chinchillidae, Caviomorpha). *Bulletin du Muséum National d'Histoire Naturelle*, **9**, 127–131.

SUTTON, C.A. & DURETTE-DESSET, M.-C. 1991. Nippostrongylinae (Nematoda: Trichostrongyloidea). parasites d'*Oryzomys flavescens* en Argentine et en Uruguay. *Revue Suisse de Zoologie*, **98** (**3**), 535–553.

SUTTON, C.A. & DURETTE-DESSET, M.-C. 1995. A description of *Graphidioides kravetzi* n.sp. and the revision of *Graphidioides* Cameron, 1923 (Nematoda: Trichostrongyloidea), parasites of neotropical rodents. *Systematic Parasitology*, **31**, 133–145.

SUTTON, C.A., CHABAUD, A.G. & DURETTE-DESSET, M.-C. 1980. Contribucion al conocimiento de la fauna parasitológica Argentina. VI. Sobre un nuevo Lauroiinae (Nematoda, Ascaridida). *Bulletin de Muséum National d'Histoire Naturelle*, **1**, 81–85.

TADROS, G. 1964. On *Chabertiella pesteri* gen. et sp.nov. (Cyathostomidae) from a Giraffe and on the validity of *Chabertia rishati* Akhtar, 1937. *Journal of Helminthology*, **38**, 109–116.

TEIXEIRA DE FREITAS, J.F. & NEGRÃO DE SOUZA FRANCO, S.R. 1967. Novo parasito de roeder; *Mariostrongylus pessoai* gen. n., sp.n. (Nematoda, Strongyloidea). *Atas da Sociedade de Biologia do Rio de Janeiro*, **11**, 15–17.

TEIXEIRA DE FREITAS, J.F. & NICANOR IBAÑEZ, H. 1968. Fauna helmintologica del Peru – nuovo genero y nueva especie de la subfamilia Splendofilariinae (Nematoda, Filarioidea). *Archivos Peruanos de Patologia y Clinica*, **22**, 1–8.

THIEME, H. 1961. Über die verwurmung der leber vom kaulbarsch (*Acerina cernua*) mit *Hepatospina acerinae* in abhängigkeit vom alter und geschlecht. *Wiadomosci Parazytologiczne*, **7**, 856–864.

THWAITE, J.W. 1926. Notes on some nematodes in the Museum of the Liverpool School of Tropical Medicine. *Annals of Tropical Medicine and Parasitology*, **20**, 273–278.

TIKHOMIROVA, V.A. 1975. [The erection of a new superfamily of nematodes Skrjabillanoidea n.superfam. (Camallanata Chitwood, 1936).] *Vosprosy Ekologii Zhivotnykh Vypusk*, **2**, 114–117. [In Russian.]

TRAVASSOS, L. 1920. Contributions à l'étude de la faune helminthologique de Brésil. *Memorias do Instituto Oswaldo Cruz*, **12**, 66–77.

TRAVASSOS, L. 1937. Revisão da familia Trichostrongylidae Leiper, 1912. Monograph of the Institute of Oswaldo Cruz, Rio de Janeiro, Brazil, No. 1, 512 pp.

TRAVASSOS, L., TEIXEIRA DE FREITAS, J.F & MACHADO DE MENDONÇA, J. 1964. Relatório da excursão do Instituto Oswaldo Cruz ao Parque de Reserva e Refúgio Sóoretama, no Estado do Espírito Santo, em outubro de 1963. *Boletim do Museu de Biologia Prof. Mello-Leitão, Zoologia*, **23**, 1–26.

TRIVEDI, K.K. 1987 (publ. 1989). Nematode parasites of vertebrates. IX. On a new nematode *Pulmostrongylus stunkardi* sp.nov. (Strongylidea: Protostrongylidae) from mongoose, *Herpestes ferrugineous* from Udaipur, Rajasthan, India. *Rivista di Parassitologia*, **4** (**48**) (**2**), 207–214.

UBELAKER, J.E. 1986. Systematics of species referred to the genus *Angiostrongylus*. *Journal of Parasitology*, **72**, 237–244.

UCHIDA, A. & ITAGAKI, H. 1979. Studies on the amphibian helminths in Japan. VI. *Pseudoxyascaris japonicus* n.g. and n.sp. (Oxyascarididae; Nematoda) and *Pseudopolystoma dendriticum* (Monogenea; Trematoda) from a salamander. *Japanese Journal of Parasitology*, **28**, 43–50.

UNI, S., SUZUKI, Y., BABA, M., MITANI, N., TAKAOKA, H., KATSUMI, A. & BAIN, O. 2001. Coexistence of five *Cercopithifilaria* species in the Japanese rupicaprine bovid, *Capricornis crispus*. *Parasite*, **8**, 197–213.

UNI, S., BAIN, O. TAKAOKA, H. 2004. Affinities between *Cutifilaria* (Nematoda: Filarioidea), parasites of deer, and *Mansonella* as seen in a new onchocercid, *M.* (*C.*) *perforata* n. sp., from Japan. *Parasite*, **11**, 131–140.

UNI, S., BAIN, O., AGATSUMA, T., KATSUMI, A., BABA, M., YANAI, T. & TAKAOKA, H. 2006. New filarial nematode from Japanese serows (*Naemorhedus crispus*: Bovidae) close to parasites from elephants. *Parasite*, **13**, 193–200.

VAUCHER, C. & DURETTE-DESSET, M.-C. 1986. Trichostrongyloidea (Nematoda) parasites de Chiroptéres néotropicaux. I. *Websternema parnelli* (Webster, 1971) n.gen. n.comb. et *Linustrongylus pteronoti* n.gen. n.sp., parasites de *Pteronotus* au Nicaragua. *Revue Suisse de Zoologie*, **93**, 237–246.

VICENTE, J.J. & DOS SANTOS, E. 1972. Sobre um novo gênero da subfamilia Filocapsulariinae Yamaguti, 1961 (Nematoda, Ascaridoidea). *Atas da Sociedade de Biologia do Rio de Janeiro*, **16**, 17–19.

VICENTE, J.J. MAGALHÃES PINTO, R & NORONHA, D. 1980. Sobre um novo gênero e duas novas espécies de nematódeos parasitas de aves Brasileiras. *Atas da Sociedade de Biologia do Rio de Janeiro*, **21**, 75–79.

VICENTE, J.J., RODRIGUES, H.DE O., GOMES, D.C. & PINTO, R.M. 1997. Nematóides do Brasil. Parte V: Nematóides de Mamíferos. *Revista Brasileira de Zoologia*, **14** (**Suppl. 1**), 1–452.

VON LINSTOW, O. 1876. Helminthologische Beobachtungen. *Archiv für Naturgeschichte*, **42**, 1–18.

WALTON, A.C. 1928. A revision of the nematodes of the Leidy collection. *Proceedings of the Academy of Natural Sciences of Philadelphia*, **79**, 49–163.

WANG, P.Q. 1980. [Studies on some new nematodes of the suborder Oxyurata from Fujian, China.] *Acta Zootaxonomica Sinica*, **5**, 242–252. [In Chinese.]

WANG, P.Q. 1984. [Some nematodes of fishes from Fujian Province, China.] *Acta Zootaxonomica Sinica*, **9**, 228–237. [In Chinese.]

WANG, P.Q., ZHAO, Y.R. & CHEN, C.C. 1978. [On some nematodes from vertebrates in South China.] *Fujian Shida Xuebao*, **2**, 75–90.

WANG, P.Q., ZHAO, Y.R., WANG, X.Y., ZHANG, J.Y. & CHANG, C.Y. 1979. Report on some nematodes from vertebrates in Central and South China. *Fujian Shida Xuebao*, **2**, 78–92.

WARD, H.B. & MAGATH, T.B. 1916. Notes on some nematodes from fresh-water fishes. *Journal of Parasitology*, **3**, 57–63.

WASON, A. & JOHNSON, S. 1977. A new physalopterid nematode from the small bat *Rhinopoma microphyllum*. *Revista Brazil Biologica*, **37**, 563–565.

WEHR, E.E. 1936. A monograph of the Filarioidea and Dracunculoidea (Nematoda) of birds. *The George Washington University Bulletin, Summaries of Doctoral Theses, (1934–1936)*, pp. 59–61.

WERTHEIM, G. & CHABAUD, A.G. 1977. Scanning electron microscopy of the cephalic structures of Pneumospiruridae (Thelazioidea: Nematoda), Revision of the family. *Annales de Parasitologie Humaine et Comparée*, **52, (b)**, 647–657.

WONG, P.L. & ANDERSON, R.C. 1983. Revision of the genera *Cordonema* Schmidt and Kuntz, 1972 and *Skrjabinoclava* Sobolev, 1943 (Nematoda: Acuaroidea). *Canadian Journal of Zoology*, **61** (**2**), 339–348.

WONG, P.L. & ANDERSON, R.C. 1993. New and described species of nematodes from shorebirds (Charadriiformes) collected in spring in Iceland. *Systematic Parasitology*, **25**, 187–202.

WONG, P.L. & LANKESTER, M.W. 1984. Revision of the genus *Viktocara* Guschanskaya, 1950 (Nematoda: Acuaroidea). *Canadian Journal of Zoology*, **62**, 2541–2549.

WONG, P.L. & LANKESTER, M.W. 1985a. Revision of the genus *Ancyracanthopsis* Diesing, 1861 and description of a new genus *Molinacuaria* n.gen. (Nematoda: Acuarioidea). *Canadian Journal of Zoology*, **63**, 1556–1564.

WONG, P.L. & LANKESTER, M.W. 1985b. Revision of the genus *Sobolevicephalus* Parukhin, 1964 (Nematoda: Acuarioidea). *Canadian Journal of Zoology*, **63**, 1576–1581.

WONG, P.L., WATSON, T. & ANDERSON, R.C. 1980. *Vigisopirura potekhina* (Petrow and Potekhina, 1953) (Nematoda: Spiruroidea) from the bobcat, *Lynx rufus* (Shriber), in the South-eastern U.S.A. *Canadian Journal of Zoology*, **58**, 1612–1616.

WONG, P.L., ANDERSON, R.C. & BARTLETT, C.M. 1986. Revision of the genus *Syncuaria* Gilbert, 1927 (Nematoda: Acuarioidea). *Canadian Journal of Zoology*, **64** (**5**), 1186–1196.

WU, S.-Q., YUN, L., JIA, X.-G., XU, Z.-X. & XIAO, N. 1979. A new genus and species of Dipetalonematidae (Nematoda: Filariata). *Acta Zootaxonomica Sinica*, **4**, 113–117. [In Chinese.]

YAMAGUTI, S. 1961. *Systema Helminthum. 3. The Nematodes of Vertebrates*. Interscience, London & New York, pp. 1261 (p. 14).

YATES, J.A. 1980. Filariae of the capybara *Hydrochoerus hydrochoeris* from Colombia, South America. Dissertation, 26 June 1980. Tulane University Dr. Philosophy, 81 p polycopiées.

YATES, J.A. & JORGENSON, J.P. 1983. *Dipetalonema (Alafilaria) hydrochoerus* subgen. et sp.n. (Nematoda: Filarioidea) from Colombian capybaras. *Journal of Parasitology*, **69**, 606–609.

YIN, P.Y., JIANG, J.S. & KUNG, F.Y. 1986. *Parastrongylus paradoxus* gen. et sp.nov., a new nematode from the large intestine of the rhinoceros (Nematoda: Strongylidae: Strongylinae). *Acta Zootaxonomica Sinica*, **11**, 139–143. [In Chinese.]

YIN, W.-Z.. & ZHANG, N.-X. 1984. A new genus and species of quimperiid nematodes (Nematoda: Cucullanta). *Acta Zootaxonomica Sinica*, **9**, 347–350.

YORKE, W. & MAPLESTONE, P.A. 1926. *The Nematode Parasites of Vertebrates*. J. & A. Churchill, London, pp. xi + 536 pp.

YU, Y. & WANG, G.T. 1992. [Studies on fish parasitic nematodes of *Wuinema* gen.nov. and *W. mysti* sp.nov.] *Acta Hydrobiologica Sinica*, **16**, 299–303. [In Chinese.]

YU, Y. & WANG, G.T. 1994. [Study of a new capillariid genus *Neocapillaria* gen.nov. with a new species (*N. phoxin*) from freshwater fish.] *Acta Hydrobiologica Sinica*, **18**, 71–75. [In Chinese.]

ZAPTERO, R.L.M., CASTAÑO FERNANDEZ, C., LOPEZ MARTINEZ, P. & MARTIN RUEDA, J. 1990 (publ. 1991). *Blanusia pseudorhabditis n.g., n.sp.* (Nematoda, Cosmocercidae) parasite *Blanus cinereus* (Vandelli, 1797) (Reptilia, Amphisbaenidae). *Revista Iberica de Parasitologia*, **50** (**3–4**), 281–288.

INDEX

Note: The index includes all orders and superfamilies not listed in the Contents, and all families, subfamilies, genera and subgenera. Species names are not included. Page numbers in *italics* refer to figures.